AF884884

Legal Notice

For information on bulk purchases and licensing agreements, please email

support@SATPrepGet800.com

This is the Solution Guide to the book
"Pure Mathematics for Beginners – Accelerated and Expanded Edition."

GET 800
PURE
MATHEMATICS
FOR BEGINNERS
Accelerated and Expanded Edition
Logic
Set Theory
Abstract Algebra
Number Theory
Real Analysis
Topology
Complex Analysis
Linear Algebra
By Dr. Steve Warner

Also Available from Dr. Steve Warner

CONNECT WITH DR. STEVE WARNER

www.facebook.com/SATPrepGet800

www.youtube.com/TheSATMathPrep

www.twitter.com/SATPrepGet800

www.linkedin.com/in/DrSteveWarner

www.pinterest.com/SATPrepGet800

Also Available from Dr. Steve Warner

CONNECT WITH DR. STEVE WARNER

www.facebook.com/SATPrepGet800

www.youtube.com/TheSATMathPrep

www.twitter.com/SATPrepGet800

www.linkedin.com/in/DrSteveWarner

www.pinterest.com/SATPrepGet800

Pure Mathematics for Beginners

Accelerated and Expanded Edition

Solution Guide

Dr. Steve Warner

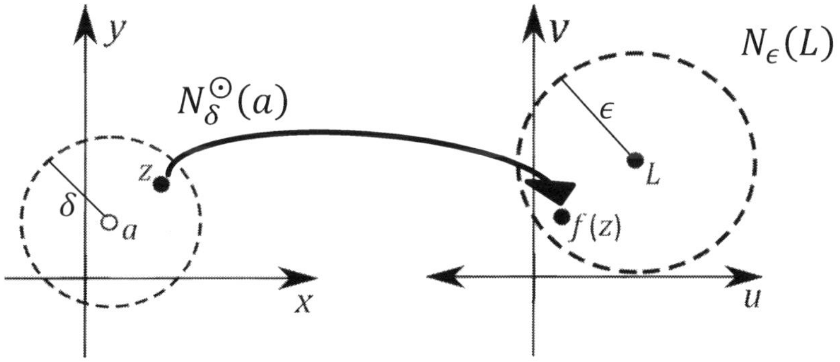

Table of Contents

Problem Set 1

LEVEL 1

Determine whether each of the following sentences is an atomic statement, a compound statement, or not a statement at all:

1. Henry does not know where he is going.

This is a **compound statement**. It has the form $\neg p$, where p is the statement "Henry knows where he is going."

2. What are you doing?

This is **not a statement**. It is a question.

3. Stop doing that!

This is **not a statement**. It is a command.

4. $x \neq 53$.

This is **not a statement**. It has an unknown variable.

5. I like the song *Crimson and Clover*.

This is an **atomic statement**. Even though the word "and" appears in the statement, here it is part of the name of the song. It is not being used as a connective.

6. If lobsters are birds, then we have a problem.

This is a **compound statement**. It has the form $p \to q$, where p is the statement "Lobsters are birds," and q is the statement "We have a problem."

7. $5 = 6$ or $5 \neq 6$.

This is a **compound statement**. It has the form $p \vee \neg p$, where p is the statement "$5 = 6$."

8. This sentence has five words.

This is **not a statement** because it is self-referential. Self-referential sentences can cause problems. For example, observe that the negation of this sentence would be "This sentence does not have five words." The sentence and its negation both appear to be true. That would be a problem. It's a good thing they're not statements!

9. A quadrilateral is a rhombus if and only if the diagonals of the quadrilateral are perpendicular.

This is a **compound statement**. It has the form $p \leftrightarrow q$, where p is the statement "A quadrilateral is a rhombus," and q is the statement "The diagonals of the quadrilateral are perpendicular."

10. I cannot speak Japanese, but I can speak Italian.

This is a **compound statement**. It has the form $\neg p \land q$, where p is the statement "I can speak Japanese," and q is the statement "I can speak Italian." Note that in sentential logic, the word "but" has the same meaning as the word "and." In English, the word "but" is used to introduce contrast with the part of the sentence that has already been mentioned. However, logically it is no different from the word "and."

What is the negation of each of the following statements?

11. A tomato is a fruit.

A tomato is not a fruit.

12. Cats have nine lives.

Cats do not have nine lives.

13. $76 \geq 72$.

$\mathbf{76 < 72}$

Note: In words, $76 \geq 72$ can be read, "Seventy six is greater than or equal to seventy two."

The negation of this statement is "Seventy six is not greater than or equal to seventy two" (symbolically, we can write $76 \ngeq 72$). Using a basic property of the number systems we are most familiar with, $76 \ngeq 72$ is equivalent to $76 < 72$.

It follows that the negation of $76 \geq 72$ is equivalent to $76 < 72$.

14. You will not believe what I just heard.

You will believe what I just heard.

Note: Technically, the negation is "It is not the case that you will not believe what I just heard." However, by the law of double negation, this is logically equivalent to "You will believe what I just heard."

15. The function f has a discontinuity at $x = 3$.

The function f does not have a discontinuity at $x = 3$.

16. The fundamental group of the circle is isomorphic to the set of integers.

The fundamental group of the circle is not isomorphic to the set of integers.

Let p, q, and r represent true statements. Compute the truth value of each of the following compound statements:

17. $(p \vee q) \vee r$

Detailed solution: $(p \vee q) \vee r \equiv (\mathrm{T} \vee \mathrm{T}) \vee \mathrm{T} \equiv \mathrm{T} \vee \mathrm{T} \equiv \mathbf{T}$.

Quicker solution: $(p \vee q) \vee r \equiv (p \vee q) \vee \mathrm{T} \equiv \mathbf{T}$.

18. $(p \vee q) \wedge \neg r$

Detailed solution: $(p \vee q) \wedge \neg r \equiv (\mathrm{T} \vee \mathrm{T}) \wedge \neg \mathrm{T} \equiv \mathrm{T} \wedge \mathrm{F} \equiv \mathbf{F}$.

Quicker solution: $(p \vee q) \wedge \neg r \equiv (p \vee q) \wedge \mathrm{F} \equiv \mathbf{F}$.

19. $\neg p \to (q \vee r)$

Detailed solution: $\neg p \to (q \vee r) \equiv \neg \mathrm{T} \to (\mathrm{T} \vee \mathrm{T}) \equiv \mathrm{F} \to \mathrm{T} \equiv \mathbf{T}$.

Quicker solution: $\neg p \to (q \vee r) \equiv \mathrm{F} \to (q \vee r) \equiv \mathbf{T}$.

20. $\neg(p \leftrightarrow \neg q) \wedge r$

Detailed solution: $\neg(p \leftrightarrow \neg q) \wedge r \equiv \neg(\mathrm{T} \leftrightarrow \neg \mathrm{T}) \wedge \mathrm{T} \equiv \neg(\mathrm{T} \leftrightarrow \mathrm{F}) \wedge \mathrm{T} \equiv \neg \mathrm{F} \wedge \mathrm{T} \equiv \mathrm{T} \wedge \mathrm{T} \equiv \mathbf{T}$.

Quicker solution: $\neg(p \leftrightarrow \neg q) \wedge r \equiv \neg(\mathrm{T} \leftrightarrow \mathrm{F}) \wedge \mathrm{T} \equiv \neg \mathrm{F} \wedge \mathrm{T} \equiv \mathrm{T} \wedge \mathrm{T} \equiv \mathbf{T}$.

21. $\neg[p \wedge (\neg q \to r)]$

Detailed solution: $\neg[p \wedge (\neg q \to r)] \equiv \neg[\mathrm{T} \wedge (\neg \mathrm{T} \to \mathrm{T})] \equiv \neg[\mathrm{T} \wedge (\mathrm{F} \to \mathrm{T})] \equiv \neg[\mathrm{T} \wedge \mathrm{T}] \equiv \neg \mathrm{T} \equiv \mathbf{F}$.

Quicker solution: $\neg[p \wedge (\neg q \to r)] \equiv \neg[p \wedge (\mathrm{F} \to r)] \equiv \neg[p \wedge \mathrm{T}] \equiv \neg[\mathrm{T} \wedge \mathrm{T}] \equiv \neg \mathrm{T} \equiv \mathbf{F}$.

22. $\neg[(\neg p \vee \neg q) \leftrightarrow \neg r]$

Detailed solution:

$$\neg[(\neg p \vee \neg q) \leftrightarrow \neg r] \equiv \neg[(\neg \mathrm{T} \vee \neg \mathrm{T}) \leftrightarrow \neg \mathrm{T}] \equiv \neg[(\mathrm{F} \vee \mathrm{F}) \leftrightarrow \mathrm{F}] \equiv \neg[\mathrm{F} \leftrightarrow \mathrm{F}] \equiv \neg \mathrm{T} \equiv \mathbf{F}.$$

Quicker solution: $\neg[(\neg p \vee \neg q) \leftrightarrow \neg r] \equiv \neg[\mathrm{F} \leftrightarrow \mathrm{F}] \equiv \neg \mathrm{T} \equiv \mathbf{F}$.

23. $p \to (q \to \neg r)$

Detailed solution: $p \to (q \to \neg r) \equiv \mathrm{T} \to (\mathrm{T} \to \neg \mathrm{T}) \equiv \mathrm{T} \to (\mathrm{T} \to \mathrm{F}) \equiv \mathrm{T} \to \mathrm{F} \equiv \mathbf{F}$.

Quicker solution: $p \to (q \to \neg r) \equiv \mathrm{T} \to (\mathrm{T} \to \mathrm{F}) \equiv \mathrm{T} \to \mathrm{F} \equiv \mathbf{F}$.

24. $\neg[\neg p \to (q \to \neg r)]$

Detailed solution:

$$\neg[\neg p \to (q \to \neg r)] \equiv \neg[\neg \mathrm{T} \to (\mathrm{T} \to \neg \mathrm{T})] \equiv \neg[\mathrm{F} \to (\mathrm{T} \to \mathrm{F})] \equiv \neg[\mathrm{F} \to \mathrm{F}] \equiv \neg \mathrm{T} \equiv \mathbf{F}.$$

Quicker solution: $\neg[\neg p \to (q \to \neg r)] \equiv \neg[\mathrm{F} \to (q \to \neg r)] \equiv \neg \mathrm{T} \equiv \mathbf{F}$.

Let ϕ be the following statement: $(p \wedge \neg q) \leftrightarrow \neg[p \vee (\neg r \to q)]$.

25. The statement ϕ is abbreviated. Write ϕ in its unabbreviated form.

$\big((p \wedge (\neg q)) \leftrightarrow (\neg[p \vee ((\neg r) \to q)])\big)$

26. Write down all the substatements of ϕ in both abbreviated and unabbreviated form.

Abbreviated forms: $p, q, r, \neg q, \neg r, p \wedge \neg q, \neg r \to q, p \vee (\neg r \to q), \neg[p \vee (\neg r \to q)]$

Unabbreviated forms: p, q, r, $(\neg q)$, $(\neg r)$, $(p \wedge (\neg q))$, $((\neg r) \to q)$, $\big(p \vee ((\neg r) \to q)\big)$, $(\neg[p \vee ((\neg r) \to q)])$

LEVEL 2

Let p represent the statement "5 is an odd integer," let q represent the statement "Brazil is in Europe," and let r represent the statement "A lobster is an insect." Rewrite each of the following symbolic statements in words, and state the truth value of each statement:

Note: p has truth value T, while q and r both have truth value F (a lobster is a crustacean, not an insect).

27. $p \vee q$

$p \vee q$ represents "**5 is an odd integer or Brazil is in Europe.**" Since p has truth value T, it follows that $p \vee q$ has truth value **T**.

28. $\neg r$

$\neg r$ represents "**A lobster is not an insect.**" Since r has truth value F, it follows that $\neg r$ has truth value **T**.

29. $p \to q$

$p \to q$ represents "**If 5 is an odd integer , then Brazil is in Europe.**" Since p has truth value T and q has truth value F, it follows that $p \to q$ has truth value **F**.

30. $p \leftrightarrow r$

$p \leftrightarrow r$ represents "**5 is an odd integer if and only if a lobster is an insect.**" Since p and r have opposite truth values, $p \leftrightarrow r$ has truth value **F**.

31. $\neg q \wedge r$

$\neg q \wedge r$ represents "**Brazil is not in Europe and a lobster is an insect.**" Since r has truth value F, it follows that $\neg q \wedge r$ has truth value **F**.

32. $\neg(p \wedge q)$

$\neg(p \wedge q)$ represents "**It is not the case that 5 is an odd integer and Brazil is in Europe.**" Since $p \wedge q$ has truth value F (because q has truth value F), it follows that $\neg(p \wedge q)$ has truth value **T**.

33. $\neg p \vee \neg q$

$\neg p \vee \neg q$ represents "**5 is not an odd integer or Brazil is not in Europe.**" Since $\neg q$ has truth value T (do you see why?), it follows that $\neg p \vee \neg q$ has truth value **T**.

34. $(p \wedge q) \rightarrow r$

$(p \wedge q) \rightarrow r$ represents "**If 5 is an odd integer and Brazil is in Europe, then a lobster is an insect.**" Since $p \wedge q$ has truth value F (because q has truth value F), it follows that $(p \wedge q) \rightarrow r$ has truth value **T**.

Consider the compound sentence "You can have a cookie or ice cream." In English this would most likely mean that you can have one or the other but not both. The word "or" used here is generally called an "exclusive or" because it excludes the possibility of both. The disjunction is an "inclusive or."

35. Using the symbol $\oplus$ for exclusive or, draw the truth table for this connective.

p	q	$p \oplus q$
T	T	F
T	F	T
F	T	T
F	F	F

36. Using only the logical connectives $\neg$, $\wedge$, and $\vee$, produce a statement using the propositional variables p and q that has the same truth values as $p \oplus q$.

We want to express that p is true or q is true, but p and q are not both true. Expressed in symbols, this is $\boldsymbol{(p \vee q) \wedge \neg(p \wedge q)}$.

Note: (1) Let's check that $(p \vee q) \wedge \neg(p \wedge q)$ behaves as desired.

If p and q are both true, then $\neg(p \wedge q) \equiv \mathrm{F}$, and so, $(p \vee q) \wedge \neg(p \wedge q) \equiv (p \vee q) \wedge \mathrm{F} \equiv \mathrm{F}$.

If p and q are both false, then $p \vee q \equiv \mathrm{F}$, and so, $(p \vee q) \wedge \neg(p \wedge q) \equiv \mathrm{F} \wedge \neg(p \wedge q) \equiv \mathrm{F}$.

Finally, if p and q have opposite truth values, then $p \vee q \equiv \mathrm{T}$ and $\neg(p \wedge q) \equiv \mathrm{T}$ (because $p \wedge q \equiv \mathrm{F}$). Therefore, $(p \vee q) \wedge \neg(p \wedge q) \equiv \mathrm{T} \wedge \mathrm{T} \equiv \mathrm{T}$.

(2) Recall that the word "but" is logically the same as the word "and" (see Problem 10 above).

(3) Another way to see that $p \oplus q$ has the same truth values as $(p \vee q) \wedge \neg(p \wedge q)$ is to draw the truth tables for each and observe that row by row they have the same truth values. We do this below.

p	q	$p \oplus q$	$p \vee q$	$p \wedge q$	$\neg(p \wedge q)$	$(p \vee q) \wedge \neg(p \wedge q)$
T	T	F	T	T	F	F
T	F	T	T	F	T	T
F	T	T	T	F	T	T
F	F	F	F	F	T	F

Observe that the third column of the truth table corresponds to $p \oplus q$, the last (seventh) column corresponds to $(p \vee q) \wedge \neg(p \wedge q)$, and both these columns have the same truth values.

(4) In this problem, we showed that $p \oplus q \equiv (p \vee q) \wedge \neg(p \wedge q)$.

Verify the following:

37. Each logical equivalence in List 1.20.

Solutions:

1. **Law of double negation:** $p \equiv \neg(\neg p)$: This was done in Example 1.16.
2. **De Morgan's laws:** $\neg(p \wedge q) \equiv \neg p \vee \neg q$: This was done in Example 1.19.

   $\neg(p \vee q) \equiv \neg p \wedge \neg q$: Let $\phi = \neg(p \vee q)$ and let $\psi = \neg p \wedge \neg q$. If $p \equiv \mathrm{T}$ or $q \equiv \mathrm{T}$, then $\phi \equiv \neg\mathrm{T} \equiv \mathrm{F}$ and $\psi \equiv \mathrm{F}$ (because $\neg p \equiv \mathrm{F}$ or $\neg q \equiv \mathrm{F}$). If $p \equiv \mathrm{F}$ and $q \equiv \mathrm{F}$, then $\phi \equiv \neg\mathrm{F} \equiv \mathrm{T}$ and $\psi \equiv \mathrm{T} \wedge \mathrm{T} \equiv \mathrm{T}$. So, all four possible truth assignments of p and q lead to the same truth value for ϕ and ψ. It follows that $\phi \equiv \psi$.
3. **Commutative laws:** $p \wedge q \equiv q \wedge p$, $p \vee q \equiv q \vee p$: We can see this from the following truth tables:

p	q	$p \wedge q$	$q \wedge p$
T	T	T	T
T	F	F	F
F	T	F	F
F	F	F	F

p	q	$p \vee q$	$q \vee p$
T	T	T	T
T	F	T	T
F	T	T	T
F	F	F	F

4. **Associative laws:** $(p \wedge q) \wedge r \equiv p \wedge (q \wedge r)$, $(p \vee q) \vee r \equiv p \vee (q \vee r)$: Draw truth tables.
5. **Distributive laws:** $p \wedge (q \vee r) \equiv (p \wedge q) \vee (p \wedge r)$, $p \vee (q \wedge r) \equiv (p \vee q) \wedge (p \vee r)$: Draw truth tables.

6. **Identity laws:** $p \wedge \mathrm{T} \equiv p$: Let $\phi = p \wedge \mathrm{T}$ and let $\psi = p$. If $p \equiv \mathrm{T}$, then $\phi \equiv \mathrm{T} \wedge \mathrm{T} \equiv \mathrm{T}$ and $\psi \equiv \mathrm{T}$. If $p \equiv \mathrm{F}$, then $\phi \equiv \mathrm{F} \wedge \mathrm{T} \equiv \mathrm{F}$ and $\psi \equiv \mathrm{F}$. So, both possible truth assignments of p lead to the same truth value for ϕ and ψ. It follows that $\phi \equiv \psi$.

   $p \wedge \mathrm{F} \equiv \mathrm{F}$: Let $\phi = p \wedge \mathrm{F}$ and let $\psi = \mathrm{F}$. If $p \equiv \mathrm{T}$, then $\phi \equiv \mathrm{T} \wedge \mathrm{F} \equiv \mathrm{F}$ and $\psi \equiv \mathrm{F}$. If $p \equiv \mathrm{F}$, then $\phi \equiv \mathrm{F} \wedge \mathrm{F} \equiv \mathrm{F}$ and $\psi \equiv \mathrm{F}$. So, both possible truth assignments of p lead to the same truth value for ϕ and ψ. It follows that $\phi \equiv \psi$.

   $p \vee \mathrm{T} \equiv \mathrm{T}$: Let $\phi = p \vee \mathrm{T}$ and let $\psi = \mathrm{T}$. If $p \equiv \mathrm{T}$, then $\phi \equiv \mathrm{T} \vee \mathrm{T} \equiv \mathrm{T}$ and $\psi \equiv \mathrm{T}$. If $p \equiv \mathrm{F}$, then $\phi \equiv \mathrm{F} \vee \mathrm{T} \equiv \mathrm{T}$ and $\psi \equiv \mathrm{T}$. So, both possible truth assignments of p lead to the same truth value for ϕ and ψ. It follows that $\phi \equiv \psi$.

   $p \vee \mathrm{F} \equiv p$: Let $\phi = p \vee \mathrm{F}$ and let $\psi = p$. If $p \equiv \mathrm{T}$, then $\phi \equiv \mathrm{T} \vee \mathrm{F} \equiv \mathrm{T}$ and $\psi \equiv \mathrm{T}$. If $p \equiv \mathrm{F}$, then $\phi \equiv \mathrm{F} \vee \mathrm{F} \equiv \mathrm{F}$ and $\psi \equiv \mathrm{F}$. So, both possible truth assignments of p lead to the same truth value for ϕ and ψ. It follows that $\phi \equiv \psi$.

7. **Negation laws:** $p \wedge \neg p \equiv \mathrm{F}$: Let $\phi = p \wedge \neg p$ and let $\psi = \mathrm{F}$. If $p \equiv \mathrm{T}$, then $\phi \equiv \mathrm{T} \wedge \mathrm{F} \equiv \mathrm{F}$ and $\psi \equiv \mathrm{F}$. If $p \equiv \mathrm{F}$, then $\phi \equiv \mathrm{F} \wedge \mathrm{T} \equiv \mathrm{F}$ and $\psi \equiv \mathrm{F}$. So, both possible truth assignments of p lead to the same truth value for ϕ and ψ. It follows that $\phi \equiv \psi$.

   $p \vee \neg p \equiv \mathrm{T}$: Let $\phi = p \vee \neg p$ and let $\psi = \mathrm{T}$. If $p \equiv \mathrm{T}$, then $\phi \equiv \mathrm{T} \vee \mathrm{F} \equiv \mathrm{T}$ and $\psi \equiv \mathrm{T}$. If $p \equiv \mathrm{F}$, then $\phi \equiv \mathrm{F} \vee \mathrm{T} \equiv \mathrm{T}$ and $\psi \equiv \mathrm{T}$. So, both possible truth assignments of p lead to the same truth value for ϕ and ψ. It follows that $\phi \equiv \psi$.

8. **Redundancy laws:** $p \wedge p \equiv p$: Let $\phi = p \wedge p$ and let $\psi = p$. If $p \equiv \mathrm{T}$, then $\phi \equiv \mathrm{T} \wedge \mathrm{T} \equiv \mathrm{T}$ and $\psi \equiv \mathrm{T}$. If $p \equiv \mathrm{F}$, then $\phi \equiv \mathrm{F} \wedge \mathrm{F} \equiv \mathrm{F}$ and $\psi \equiv \mathrm{F}$. So, both possible truth assignments of p lead to the same truth value for ϕ and ψ. It follows that $\phi \equiv \psi$.

   $p \vee p \equiv p$: Let $\phi = p \vee p$ and let $\psi = p$. If $p \equiv \mathrm{T}$, then $\phi \equiv \mathrm{T} \vee \mathrm{T} \equiv \mathrm{T}$ and $\psi \equiv \mathrm{T}$. If $p \equiv \mathrm{F}$, then $\phi \equiv \mathrm{F} \vee \mathrm{F} \equiv \mathrm{F}$ and $\psi \equiv \mathrm{F}$. So, both possible truth assignments of p lead to the same truth value for ϕ and ψ. It follows that $\phi \equiv \psi$.

9. **Absorption laws:** $(p \vee q) \wedge p \equiv p$: Let $\phi = (p \vee q) \wedge p$ and let $\psi = p$. If $p \equiv \mathrm{T}$, then $p \vee q \equiv \mathrm{T} \vee q \equiv \mathrm{T}$. So, $\phi \equiv \mathrm{T} \wedge \mathrm{T} \equiv \mathrm{T}$. Also, $\psi \equiv \mathrm{T}$. If $p \equiv \mathrm{F}$, then $\phi \equiv (p \vee q) \wedge \mathrm{F} \equiv \mathrm{F}$ and $\psi \equiv \mathrm{F}$. So, all four possible truth assignments of p and q lead to the same truth value for ϕ and ψ. It follows that $\phi \equiv \psi$.

   $(p \wedge q) \vee p \equiv p$: Let $\phi = (p \wedge q) \vee p$ and let $\psi = p$. If $p \equiv \mathrm{T}$, then $\phi \equiv (p \wedge q) \vee \mathrm{T} \equiv \mathrm{T}$ and $\psi \equiv \mathrm{T}$. If $p \equiv \mathrm{F}$, then $p \wedge q \equiv \mathrm{F} \wedge q \equiv \mathrm{F}$. So, $\phi \equiv \mathrm{F} \vee \mathrm{F} \equiv \mathrm{F}$. Also, $\psi \equiv \mathrm{F}$. So, all four possible truth assignments of p and q lead to the same truth value for ϕ and ψ. It follows that $\phi \equiv \psi$.

10. **Law of the conditional:** $p \to q \equiv \neg p \vee q$: Let $\phi = p \to q$ and let $\psi = \neg p \vee q$. If $p \equiv \mathrm{F}$, then $\phi \equiv \mathrm{F} \to q \equiv \mathrm{T}$ and $\psi \equiv \mathrm{T} \vee q \equiv \mathrm{T}$. If $q \equiv \mathrm{T}$, then $\phi \equiv p \to \mathrm{T} \equiv \mathrm{T}$ and $\psi \equiv \neg p \vee \mathrm{T} \equiv \mathrm{T}$. Finally, if $p \equiv \mathrm{T}$ and $q \equiv \mathrm{F}$, then $\phi \equiv \mathrm{T} \to \mathrm{F} \equiv \mathrm{F}$ and $\psi \equiv \mathrm{F} \vee \mathrm{F} \equiv \mathrm{F}$. So, all four possible truth assignments of p and q lead to the same truth value for ϕ and ψ. It follows that $\phi \equiv \psi$.

11. **Law of the contrapositive:** $p \to q \equiv \neg q \to \neg p$: Let $\phi = p \to q$ and let $\psi = \neg q \to \neg p$. If $p \equiv \mathrm{F}$, then $\phi \equiv \mathrm{F} \to q \equiv \mathrm{T}$ and $\psi \equiv \neg q \to \mathrm{T} \equiv \mathrm{T}$. If $q \equiv \mathrm{T}$, then $\phi \equiv p \to \mathrm{T} \equiv \mathrm{T}$ and $\psi \equiv \mathrm{F} \to \neg p \equiv \mathrm{T}$. Finally, if $p \equiv \mathrm{T}$ and $q \equiv \mathrm{F}$, then $\phi \equiv \mathrm{T} \to \mathrm{F} \equiv \mathrm{F}$ and $\psi \equiv \mathrm{T} \to \mathrm{F} \equiv \mathrm{F}$. So, all four possible truth assignments of p and q lead to the same truth value for ϕ and ψ. It follows that $\phi \equiv \psi$.

12. **Law of the biconditional:** $p \leftrightarrow q \equiv (p \to q) \wedge (q \to p)$: Let $\phi = p \leftrightarrow q$ and let $\psi = (p \to q) \wedge (q \to p)$. If $p \equiv \mathrm{T}$, then $\phi \equiv \mathrm{T} \leftrightarrow q \equiv q$, $\psi \equiv (\mathrm{T} \to q) \wedge (q \to \mathrm{T}) \equiv q \wedge \mathrm{T} \equiv q$. If $p \equiv \mathrm{F}$, then $\phi \equiv \mathrm{F} \leftrightarrow q \equiv \neg q$, $\psi \equiv (\mathrm{F} \to q) \wedge (q \to \mathrm{F}) \equiv \mathrm{T} \wedge \neg q \equiv \neg q$. So, all four possible truth assignments of p and q lead to the same truth value for ϕ and ψ. It follows that $\phi \equiv \psi$.

38. Each rule of inference in List 1.27.

Modus Ponens

$$\begin{array}{l} p \to q \\ p \\ \hline q \end{array}$$

Modus Tollens

$$\begin{array}{l} p \to q \\ \neg q \\ \hline \neg p \end{array}$$

Disjunctive Syllogism

$$\begin{array}{l} p \vee q \\ \neg p \\ \hline q \end{array}$$

Hypothetical Syllogism

$$\begin{array}{l} p \to q \\ q \to r \\ \hline p \to r \end{array}$$

Modus Ponens: This was done in Example 1.24.

Modus Tollens: Suppose that $p \to q \equiv \mathrm{T}$ and $\neg q \equiv \mathrm{T}$. Then $q \equiv \mathrm{F}$, and therefore, we must have $p \equiv \mathrm{F}$. So, $\neg p \equiv \mathrm{T}$.

Disjunctive Syllogism: Suppose that $p \vee q \equiv \mathrm{T}$ and $\neg p \equiv \mathrm{T}$. Then $p \equiv \mathrm{F}$, and therefore, we must have $q \equiv \mathrm{T}$.

Hypothetical Syllogism: Suppose that $p \to q \equiv \mathrm{T}$ and $q \to r \equiv \mathrm{T}$. If $p \equiv \mathrm{T}$, then since $p \to q \equiv \mathrm{T}$, we must have $q \equiv \mathrm{T}$. Since $q \to r \equiv \mathrm{T}$, we must have $r \equiv \mathrm{T}$. Since $p \equiv \mathrm{T}$ and $r \equiv \mathrm{T}$, we have $p \to r \equiv \mathrm{T}$. If $p \equiv \mathrm{F}$, then $p \to r \equiv \mathrm{T}$.

Conjunctive Introduction

$$\begin{array}{l} p \\ q \\ \hline p \wedge q \end{array}$$

Disjunctive Introduction

$$\begin{array}{l} p \\ \hline p \vee q \end{array}$$

Biconditional Introduction

$$\begin{array}{l} p \to q \\ q \to p \\ \hline p \leftrightarrow q \end{array}$$

Constructive Dilemma

$$\begin{array}{l} p \to q \\ r \to s \\ p \vee r \\ \hline q \vee s \end{array}$$

Conjunctive Introduction: Suppose that $p \equiv \mathrm{T}$ and $q \equiv \mathrm{T}$. Then $p \wedge q \equiv \mathrm{T}$.

Disjunctive Introduction: Suppose that $p \equiv \mathrm{T}$. Then $p \vee q \equiv \mathrm{T}$.

Biconditional Introduction: Suppose that $p \to q \equiv \mathrm{T}$ and $q \to p \equiv \mathrm{T}$. If $p \equiv \mathrm{T}$, then since $p \to q \equiv \mathrm{T}$, we must have $q \equiv \mathrm{T}$. Since $p \equiv \mathrm{T}$ and $q \equiv \mathrm{T}$, we have $p \leftrightarrow q \equiv \mathrm{T}$. If $p \equiv \mathrm{F}$, then since $q \to p \equiv \mathrm{T}$, we must have $q \equiv \mathrm{F}$. Since $p \equiv \mathrm{F}$ and $q \equiv \mathrm{F}$, we have $p \leftrightarrow q \equiv \mathrm{T}$.

Constructive Dilemma: Suppose that $p \to q \equiv \mathrm{T}$, $r \to s \equiv \mathrm{T}$, and $p \vee r \equiv \mathrm{T}$. If $q \equiv \mathrm{F}$ and $s \equiv \mathrm{F}$, then since $p \to q \equiv \mathrm{T}$, we must have $p \equiv \mathrm{F}$. Since $r \to s \equiv \mathrm{T}$, we must have $r \equiv \mathrm{F}$. But then $p \vee r \equiv \mathrm{F}$. So, there is no truth assignment satisfying $p \to q$, $r \to s$, and $p \vee r$ that will make $q \vee s \equiv \mathrm{F}$.

Conjunctive Elimination

$$\begin{array}{l} p \wedge q \\ \hline p \end{array}$$

Disjunctive Resolution

$$\begin{array}{l} p \vee q \\ \neg p \vee r \\ \hline q \vee r \end{array}$$

Biconditional Elimination

$$\begin{array}{l} p \leftrightarrow q \\ \hline p \to q \end{array}$$

Destructive Dilemma

$$\begin{array}{l} p \to q \\ r \to s \\ \neg q \vee \neg s \\ \hline \neg p \vee \neg r \end{array}$$

Conjunctive Elimination: Suppose that $p \wedge q \equiv \mathrm{T}$. Then $p \equiv \mathrm{T}$ and $q \equiv \mathrm{T}$. In particular, $p \equiv \mathrm{T}$.

Disjunctive Resolution: Suppose that $p \vee q \equiv \text{T}$ and $\neg p \vee r \equiv \text{T}$. If $q \equiv \text{F}$ and $r \equiv \text{F}$, then since $p \vee q \equiv \text{T}$, we must have $p \equiv \text{T}$. Since $\neg p \vee r \equiv \text{T}$, we must have $\neg p \equiv \text{T}$. But then $p \wedge \neg p \equiv \text{T}$, which is impossible. So, there is no truth assignment satisfying $p \vee q$ and $\neg p \vee r$ that will make $q \vee r \equiv \text{F}$.

Biconditional Elimination: Suppose that $p \leftrightarrow q \equiv \text{T}$. If $p \rightarrow q \equiv \text{F}$, then $p \equiv \text{T}$ and $q \equiv \text{F}$. But then $p \leftrightarrow q \equiv \text{F}$. So, there is no truth assignment satisfying $p \leftrightarrow q$ that will make $p \rightarrow q \equiv \text{F}$.

Destructive Dilemma: Suppose that $p \rightarrow q \equiv \text{T}$, $r \rightarrow s \equiv \text{T}$, and $\neg q \vee \neg s \equiv \text{T}$. If $\neg p \equiv \text{F}$ and $\neg r \equiv \text{F}$, then $p \equiv \text{T}$ and $r \equiv \text{T}$. Since $p \rightarrow q \equiv \text{T}$, we must have $q \equiv \text{T}$. Since $r \rightarrow s \equiv \text{T}$, we must have $s \equiv \text{T}$. Then $\neg q \equiv \text{F}$ and $\neg s \equiv \text{F}$. Thus, $\neg q \vee \neg s \equiv \text{F}$. So, there is no truth assignment satisfying $p \rightarrow q$, $r \rightarrow s$, and $\neg q \vee \neg s$ that will make $\neg p \vee \neg r \equiv \text{F}$.

LEVEL 3

Consider the five distinct propositional variables p, q, r, s, and t.

39. How many different truth assignments are there for this list of propositional variables?

$2^5 = 2 \cdot 2 \cdot 2 \cdot 2 \cdot 2 = \mathbf{32}$

40. How many different truth assignments are there for this list of propositional variables such that p is true and q is false?

$2^3 = 2 \cdot 2 \cdot 2 = \mathbf{8}$

41. How many different truth assignments are there for this list of propositional variables such that q, r, and s are all true?

$2^2 = 2 \cdot 2 = \mathbf{4}$

42. How many different truth assignments are there for a list of 6 propositional variables?

$2^6 = 2 \cdot 2 \cdot 2 \cdot 2 \cdot 2 \cdot 2 = \mathbf{64}$

Determine if each of the following statements is a tautology, a contradiction, or neither.

43. $p \wedge p$

If $p \equiv \text{T}$, then $p \wedge p \equiv \text{T} \wedge \text{T} \equiv \text{T}$. If $p \equiv \text{F}$, then $p \wedge p \equiv \text{F} \wedge \text{F} \equiv \text{F}$. **Neither**

44. $p \wedge \neg p$

$p \wedge \neg p \equiv \text{F}$. **Contradiction**

45. $(p \vee \neg p) \rightarrow (p \wedge \neg p)$

$(p \vee \neg p) \rightarrow (p \wedge \neg p) \equiv \text{T} \rightarrow \text{F} \equiv \text{F}$. **Contradiction**

46. $\neg(p \vee q) \leftrightarrow (\neg p \wedge \neg q)$

Since $\neg(p \vee q) \equiv \neg p \wedge \neg q$ (De Morgan's law), $\neg(p \vee q) \leftrightarrow (\neg p \wedge \neg q)$ is a **Tautology**.

47. $p \rightarrow (\neg q \wedge r)$

If $p \equiv \mathrm{F}$, then we have $p \rightarrow (\neg q \wedge r) \equiv \mathrm{F} \rightarrow (\neg q \wedge r) \equiv \mathrm{T}$. If $p \equiv \mathrm{T}$ and $r \equiv \mathrm{F}$, then we have $p \rightarrow (\neg q \wedge r) \equiv \mathrm{T} \rightarrow (\neg q \wedge \mathrm{F}) \equiv \mathrm{T} \rightarrow \mathrm{F} \equiv \mathrm{F}$. **Neither**

48. $(p \leftrightarrow q) \rightarrow (p \rightarrow q)$

If p and q have the same truth value, then we have $p \leftrightarrow q \equiv \mathrm{T}$ and $p \rightarrow q \equiv \mathrm{T}$, and therefore, $(p \leftrightarrow q) \rightarrow (p \rightarrow q) \equiv \mathrm{T} \rightarrow \mathrm{T} \equiv \mathrm{T}$. If p and q have opposite truth values, then $p \leftrightarrow q \equiv \mathrm{F}$, and therefore, we have $(p \leftrightarrow q) \rightarrow (p \rightarrow q) \equiv \mathrm{F} \rightarrow (p \rightarrow q) \equiv \mathrm{T}$. Since all possible truth assignments of the propositional variables lead to a truth value of T, $(p \leftrightarrow q) \rightarrow (p \rightarrow q)$ is a **Tautology**.

Let p represent a true statement. Decide if this is enough information to determine the truth value of each of the following statements. If so, state that truth value.

49. $p \vee q$

$(p \vee q) \equiv \mathrm{T} \vee q \equiv \mathbf{T}$.

50. $p \rightarrow q$

$p \rightarrow q \equiv \mathrm{T} \rightarrow q$. If $q \equiv \mathrm{T}$, we get $\mathrm{T} \rightarrow \mathrm{T} \equiv \mathrm{T}$. If $q \equiv \mathrm{F}$, we get $\mathrm{T} \rightarrow \mathrm{F} \equiv \mathrm{F}$. **There is not enough information**.

51. $\neg p \rightarrow \neg(q \vee \neg r)$

$\neg p \rightarrow \neg(q \vee \neg r) \equiv \mathrm{F} \rightarrow \neg(q \vee \neg r) \equiv \mathbf{T}$.

52. $\neg(\neg p \wedge q) \leftrightarrow p$

$\neg(\neg p \wedge q) \leftrightarrow p \equiv \neg(\mathrm{F} \wedge q) \leftrightarrow \mathrm{T} \equiv \neg\mathrm{F} \leftrightarrow \mathrm{T} \equiv \mathrm{T} \leftrightarrow \mathrm{T} \equiv \mathbf{T}$.

53. $(p \leftrightarrow q) \leftrightarrow \neg p$

$(p \leftrightarrow q) \leftrightarrow \neg p \equiv (\mathrm{T} \leftrightarrow q) \leftrightarrow \mathrm{F}$. If $q \equiv \mathrm{T}$, we get $(\mathrm{T} \leftrightarrow \mathrm{T}) \leftrightarrow \mathrm{F} \equiv \mathrm{T} \leftrightarrow \mathrm{F} \equiv \mathrm{F}$. If $q \equiv \mathrm{F}$, we get $(\mathrm{T} \leftrightarrow \mathrm{F}) \leftrightarrow \mathrm{F} \equiv \mathrm{F} \leftrightarrow \mathrm{F} \equiv \mathrm{T}$. **There is not enough information**.

54. $\neg[(\neg p \wedge \neg q) \leftrightarrow \neg r]$

$\neg[(\neg p \wedge \neg q) \leftrightarrow \neg r] \equiv \neg[(\mathrm{F} \wedge \neg q) \leftrightarrow \neg r] \equiv \neg(\mathrm{F} \leftrightarrow \neg r)$. If $r \equiv \mathrm{T}$, we get $\neg\mathrm{T} \equiv \mathrm{F}$. If $r \equiv \mathrm{F}$, we get $\neg\mathrm{F} \equiv \mathrm{T}$. **There is not enough information**.

55. $[(p \wedge \neg p) \rightarrow p] \wedge (p \vee \neg p)$

$[(p \wedge \neg p) \rightarrow p] \wedge (p \vee \neg p) \equiv [(\mathrm{T} \wedge \mathrm{F}) \rightarrow \mathrm{T}] \wedge (\mathrm{T} \vee \mathrm{F}) \equiv [\mathrm{F} \rightarrow \mathrm{T}] \wedge \mathrm{T} \equiv \mathrm{T} \wedge \mathrm{T} \equiv \mathbf{T}$.

56. $r \rightarrow [\neg q \rightarrow (\neg p \rightarrow \neg r)]$

$r \rightarrow [\neg q \rightarrow (\neg p \rightarrow \neg r)] \equiv r \rightarrow [\neg q \rightarrow (\text{F} \rightarrow \neg r)] \equiv r \rightarrow [\neg q \rightarrow \text{T}] \equiv r \rightarrow \text{T} \equiv \mathbf{T}$.

LEVEL 4

Assume that the given compound statement is true. Determine the truth value of each propositional variable.

57. $p \wedge q$

If $p \equiv \text{F}$ or $q \equiv \text{F}$, then $p \wedge q \equiv \text{F}$. Therefore, $\boldsymbol{p} \equiv \mathbf{T}$ and $\boldsymbol{q} \equiv \mathbf{T}$.

58. $\neg(p \rightarrow q)$

Since $\neg(p \rightarrow q)$ is true, $p \rightarrow q$ is false. Therefore, $\boldsymbol{p} \equiv \mathbf{T}$ and $\boldsymbol{q} \equiv \mathbf{F}$.

59. $p \leftrightarrow [\neg(p \wedge q)]$

If $p \equiv \text{F}$, then $p \wedge q \equiv \text{F}$, and so, $p \leftrightarrow [\neg(p \wedge q)] \equiv \text{F} \leftrightarrow \text{T} \equiv \text{F}$. So, $\boldsymbol{p} \equiv \mathbf{T}$. It follows that $\neg(p \wedge q) \equiv \text{T}$, and so $p \wedge q \equiv \text{F}$. Since $p \equiv \text{T}$, we must have $\boldsymbol{q} \equiv \mathbf{F}$.

60. $[p \wedge (q \vee r)] \wedge \neg r$

As in Problem 57, we must have $p \wedge (q \vee r) \equiv \text{T}$ and $\neg r \equiv \text{T}$. So, $\boldsymbol{p} \equiv \mathbf{T}$, $q \vee r \equiv \text{T}$, and $\boldsymbol{r} \equiv \mathbf{F}$. Since $q \vee r \equiv \text{T}$ and $r \equiv \text{F}$, we must have $\boldsymbol{q} \equiv \mathbf{T}$.

Simplify each statement.

61. $p \vee (p \wedge \neg p)$

$p \vee (p \wedge \neg p) \equiv p \vee \text{F} \equiv \boldsymbol{p}$.

62. $(p \wedge q) \vee \neg p$

$(p \wedge q) \vee \neg p \equiv (p \vee \neg p) \wedge (q \vee \neg p) \equiv \text{T} \wedge (q \vee \neg p) \equiv q \vee \neg p \equiv \neg\boldsymbol{p} \vee \boldsymbol{q}$.

63. $\neg p \rightarrow (\neg q \rightarrow p)$

$\neg p \rightarrow (\neg q \rightarrow p) \equiv p \vee (\neg q \rightarrow p) \equiv p \vee (q \vee p) \equiv p \vee (p \vee q) \equiv (p \vee p) \vee q \equiv \boldsymbol{p} \vee \boldsymbol{q}$.

64. $(p \wedge \neg q) \vee p$

$(p \wedge \neg q) \vee p \equiv \boldsymbol{p}$ (Absorption).

65. $[(q \wedge p) \vee q] \wedge [(q \vee p) \wedge p]$

$[(q \wedge p) \vee q] \wedge [(q \vee p) \wedge p] \equiv [(q \wedge p) \vee q] \wedge [(p \vee q) \wedge p] \equiv q \wedge p$ (Absorption twice) $\equiv \boldsymbol{p} \wedge \boldsymbol{q}$.

Let ϕ, ψ, and τ be statements. Show each of the following:

66. $\phi \vdash \psi$ and $\psi \vdash \tau$ implies $\phi \vdash \tau$.

Let ϕ, ψ, and τ be statements with $\phi \vdash \psi$ and $\psi \vdash \tau$. Let a be a truth assignment that satisfies ϕ. Since $\phi \vdash \psi$, a satisfies ψ. Since $\psi \vdash \tau$, a satisfies τ. Since a was an arbitrary truth assignment that satisfies ϕ, we see that $\phi \vdash \tau$.

Notes: (1) Recall that the symbol $\vdash$ is pronounced "tautologically implies."

(2) Recall that if a truth assignment a makes a statement ϕ true, we say that a **satisfies** ϕ. If a makes ϕ false, we say that a **does not satisfy** ϕ.

67. $\phi \vdash \psi$ if and only if $\phi \to \psi$ is a tautology.

Let ϕ and ψ be statements and assume that $\phi \vdash \psi$. Let a be a truth assignment of the propositional variables appearing in ϕ or ψ or both. If a satisfies ϕ, then a satisfies ψ (because $\phi \vdash \psi$). It follows that $\phi \to \psi \equiv \mathrm{T} \to \mathrm{T} \equiv \mathrm{T}$. If a does not satisfy ϕ, then $\phi \to \psi \equiv \mathrm{F} \to \psi \equiv \mathrm{T}$. So, we have shown that every truth assignment satisfies $\phi \to \psi$. Therefore, $\phi \to \psi$ is a tautology.

Conversely, assume that $\phi \to \psi$ is a tautology, and let a be a truth assignment that satisfies ϕ. If a does not satisfy ψ, then we would have $\phi \to \psi \equiv \mathrm{T} \to \mathrm{F} \equiv \mathrm{F}$. So, a must satisfy ψ. Since a was an arbitrary truth assignment that satisfies ϕ, $\phi \vdash \psi$.

LEVEL 5

Without drawing a truth table or using List 1.20, show that each of the following is a tautology.

68. $[p \wedge (q \vee r)] \leftrightarrow [(p \wedge q) \vee (p \wedge r)]$

If $p \equiv \mathrm{F}$, then $p \wedge (q \vee r) \equiv \mathrm{F}$, $p \wedge q \equiv \mathrm{F}$, and $p \wedge r \equiv \mathrm{F}$. So, $(p \wedge q) \vee (p \wedge r) \equiv \mathrm{F}$. It follows that $[p \wedge (q \vee r)] \leftrightarrow [(p \wedge q) \vee (p \wedge r)] \equiv \mathrm{F} \leftrightarrow \mathrm{F} \equiv \mathbf{T}$.

If $p \equiv \mathrm{T}$ and $q \equiv \mathrm{T}$, then $p \wedge (q \vee r) \equiv \mathrm{T} \wedge \mathrm{T} \equiv \mathrm{T}$ and $(p \wedge q) \vee (p \wedge r) \equiv \mathrm{T} \vee (p \wedge r) \equiv \mathrm{T}$. It follows that $[p \wedge (q \vee r)] \leftrightarrow [(p \wedge q) \vee (p \wedge r)] \equiv \mathrm{T} \leftrightarrow \mathrm{T} \equiv \mathbf{T}$.

If $p \equiv \mathrm{T}$ and $q \equiv \mathrm{F}$, then $p \wedge (q \vee r) \equiv \mathrm{T} \wedge r \equiv r$ and $(p \wedge q) \vee (p \wedge r) \equiv \mathrm{F} \vee r \equiv r$. It follows that $[p \wedge (q \vee r)] \leftrightarrow [(p \wedge q) \vee (p \wedge r)] \equiv r \leftrightarrow r \equiv \mathbf{T}$.

Notes: (1) We can display this reasoning visually as follows:

$[p \wedge (q \vee r)]$	$\leftrightarrow$	$[(p \wedge q) \vee (p \wedge r)]$
F F	**T**	F F F F F
T T T T	**T**	T T T T
T r F r	**T**	T F F r T r

Each row of truth values is placed in the order suggested by the solution above. For example, for the first row, we start by writing F under each p because we are assuming that $p \equiv \text{F}$. Next, since the conjunction of F with anything else is F, we write F under each $\wedge$ (there are three that appear). Next, since $\text{F} \vee \text{F} \equiv \text{F}$, we write F under the rightmost $\vee$. Finally, since $\text{F} \leftrightarrow \text{F} \equiv \text{T}$, we write T under $\leftrightarrow$. This is the truth value of the entire statement, and therefore, we are done with the case $p \equiv \text{F}$. The other two rows work the same way.

(2) We used two of the identity laws in the third part of the solution: $\text{T} \wedge r \equiv r$ and $\text{F} \vee r \equiv r$.

69. $\big[[(p \wedge q) \to r] \to s\big] \to [(p \to r) \to s]$

If $s \equiv \text{T}$, then $(p \to r) \to s \equiv \text{T}$, and therefore, $\big[[(p \wedge q) \to r] \to s\big] \to [(p \to r) \to s] \equiv \mathbf{T}$.

Now, assume $s \equiv \text{F}$, and either $p \equiv \text{F}$ or $q \equiv \text{F}$. Then $p \wedge q \equiv \text{F}$, and so $(p \wedge q) \to r \equiv \text{T}$. Therefore, $[(p \wedge q) \to r] \to s \equiv \text{F}$, and so, $\big[[(p \wedge q) \to r] \to s\big] \to [(p \to r) \to s] \equiv \mathbf{T}$.

Finally, assume $s \equiv \text{F}$, $p \equiv \text{T}$, and $q \equiv \text{T}$. Then $p \wedge q \equiv \text{T}$, and so, $(p \wedge q) \to r \equiv r$. Therefore, $[(p \wedge q) \to r] \to s \equiv \neg r$. Also, $p \to r \equiv r$, and so $(p \to r) \to s \equiv \neg r$. So, we get $\neg r \to \neg r \equiv \mathbf{T}$.

Note: The dedicated reader should display this reasoning visually, as was done in Note 1 following the solution to Problem 68 above.

Determine whether each of the following logical arguments is valid or invalid. If the argument is valid, provide a deduction. If the argument is invalid, provide a counterexample.

70.

$$\begin{array}{l} p \vee q \\ q \\ \hline p \end{array}$$

If we let $p \equiv \text{F}$ and $q \equiv \text{T}$, then $p \vee q \equiv \text{F} \vee \text{T} \equiv \text{T}$. So, we have found a truth assignment that makes the premises true and the conclusion false. Therefore, the argument is **invalid**.

71.

$$\begin{array}{l} \neg(p \wedge q) \\ q \\ \hline \neg p \end{array}$$

Here is a derivation:

1	$\neg(p \wedge q)$	Premise
2	q	Premise
3	$\neg p \vee \neg q$	De Morgan's law (1)
4	$\neg(\neg q)$	Law of double negation (2)
5	$\neg p$	Disjunctive syllogism (3, 4)

Therefore, the argument is **valid**.

72.

$$\begin{array}{l} \neg p \\ p \vee r \\ \underline{q \to \neg r} \\ \neg q \end{array}$$

Here is a derivation.

1	$\neg p$	Premise
2	$p \vee r$	Premise
3	$q \to \neg r$	Premise
4	r	Disjunctive syllogism (2, 1)
5	$\neg(\neg r)$	Law of double negation (4)
6	$\neg q$	Modus tollens (3, 5)

Therefore, the argument is **valid**.

73.

$$\begin{array}{l} p \to q \\ \underline{r \to \neg q} \\ p \to r \end{array}$$

If we let $p \equiv \text{T}$, $q \equiv \text{T}$ and $r \equiv \text{F}$, then $p \to q \equiv \text{T} \to \text{T} \equiv \text{T}$, $r \to \neg q \equiv \text{F} \to \text{F} \equiv \text{T}$, and $p \to r \equiv \text{T} \to \text{F} \equiv \text{F}$. So, we have found a truth assignment that makes the premises true and the conclusion false. Therefore, the argument is **invalid**.

74. If a piano has 88 keys, then the box is empty.
If a piano does not have 88 keys, then paintings are white.
If we are in immediate danger, then the box is not empty.
Therefore, paintings are white or we are not in immediate danger.

Let p represent "A piano has 88 keys," let b represent "The box is empty," let w represent "Paintings are white," and let d represent "We are in immediate danger." To the right, we give a deduction showing that the argument is **valid**.

1	$p \to b$	Premise
2	$\neg p \to w$	Premise
3	$d \to \neg b$	Premise
4	$\neg w \to \neg(\neg p)$	Law of the contrapositive (2)
5	$\neg w \to p$	Law of double negation (4)
6	$\neg w \to b$	Hypothetical syllogism (5, 1)
7	$\neg(\neg b) \to \neg d$	Law of the contrapositive (3)
8	$b \to \neg d$	Law of double negation (7)
9	$\neg w \to \neg d$	Hypothetical syllogism (6, 8)
10	$\neg(\neg w) \vee \neg d$	Law of the conditional (9)
11	$w \vee \neg d$	Law of double negation (10)

75. Tangs have fangs or tings have wings.
It is not the case that tangs have fangs and tings do not have wings.
It is not the case that tangs do not have fangs and tings have wings.
Therefore, tangs have fangs and either tings have wings or tangs do not have fangs.

Let f represent "Tangs have fangs," let w represent "Tings have wings." We now give a deduction showing that the argument is **valid**.

1	$f \vee w$	Premise
2	$\neg(f \wedge \neg w)$	Premise
3	$\neg(\neg f \wedge w)$	Premise
4	$\neg f \vee \neg(\neg w)$	De Morgan's law (2)
5	$\neg f \vee w$	Law of double negation (4)
6	$w \vee w$	Disjunctive resolution (1, 5)
7	w	Redundancy law (6)
8	$w \vee \neg f$	Disjunctive introduction (7)
9	$w \vee f$	Commutative law (1)
10	$\neg(\neg f) \vee \neg w$	De Morgan's law (3)
11	$f \vee \neg w$	Law of double negation (10)
12	$\neg w \vee f$	Commutative law (11)
13	$f \vee f$	Disjunctive resolution (9, 12)
14	f	Redundancy law (13)
15	$f \wedge (w \vee \neg f)$	Conjunctive introduction (14, 8)

Problem Set 2

LEVEL 1

Determine whether each of the following statements is true or false:

1. $t \in \{t\}$

$\{t\}$ has exactly 1 element, namely t. So, $t \in \{t\}$ is **true**.

2. $4 \in \{0, 2, 4, 6\}$

$\{0, 2, 4, 6\}$ has exactly 4 elements, namely 0, 2, 4, and 6. In particular, $4 \in \{0, 2, 4, 6\}$ is **true**.

3. $-7 \in \{7\}$

$\{7\}$ has exactly 1 element, namely 7. So, $-7 \notin \{7\}$. Therefore, $-7 \in \{7\}$ is **false**.

4. $0 \in \mathbb{Z}$

$\mathbb{Z} = \{\dots, -4, -3, -2, -1, 0, 1, 2, 3, 4, \dots\}$. In particular, $0 \in \mathbb{Z}$ is **true**.

5. $-26 \in \mathbb{N}$

$\mathbb{N} = \{0, 1, 2, 3, \dots\}$. Therefore, $-26 \in \mathbb{N}$ is **false**.

6. $\frac{7}{2} \in \mathbb{Q}$

Since $7, 2 \in \mathbb{Z}$ and $2 \neq 0$, $\frac{7}{2} \in \mathbb{Q}$ is **true**.

7. $\emptyset \subseteq \{1, 2, 3\}$

The empty set is a subset of every set. So, $\emptyset \subseteq \{1, 2, 3\}$ is **true**.

8. $\{\square\} \subseteq \{\square, \Delta\}$

The only element of $\{\square\}$ is $\square$. Since $\square$ is also an element of $\{\square, \Delta\}$, $\{\square\} \subseteq \{\square, \Delta\}$ is **true**.

9. $\{x, y, z, w\} \subset \{x, y, z, w\}$

No set is a **proper** subset of itself. So, $\{x, y, z, w\} \subset \{x, y, z, w\}$ is **false**.

10. $\{0, 1, \{2, 3\}\} \subseteq \{0, 1, 2, 3\}$

$\{2, 3\} \in \{0, 1, \{2,3\}\}$, but $\{2, 3\} \notin \{0, 1, 2, 3\}$. So, $\{0, 1, \{2,3\}\} \subseteq \{0, 1, 2, 3\}$ is **false**.

Determine the cardinality of each of the following sets:

11. $\{\text{hammer}, \text{screwdriver}, \text{saw}\}$

$|\{\text{hammer}, \text{screwdriver}, \text{saw}\}| = \mathbf{3}$.

12. $\{4, 16, 27, 46, 59, 201\}$

$|\{4, 16, 27, 46, 59, 201\}| = \mathbf{6}$.

13. $\{1, 2, \ldots, 236\}$

$|\{1, 2, \ldots, 236\}| = \mathbf{236}$.

14. $\left\{\frac{1}{2}, \frac{1}{3}, \ldots, \frac{1}{15}\right\}$

$\left|\left\{\frac{1}{2}, \frac{1}{3}, \ldots, \frac{1}{15}\right\}\right| = \mathbf{14}$.

15. $\emptyset$

$|\emptyset| = \mathbf{0}$.

For each pair of sets A and B below, determine if $A \subseteq B$, $B \subseteq A$, both, or neither.

16. $A = 2\mathbb{N}$, $B = \mathbb{Z}$

$\boldsymbol{A \subseteq B}$

17. $A = \{0, 1\}$, $B = \{1, 0\}$

$\boldsymbol{A = B}$ **(Both)**

18. $A = \emptyset$, $B = \{\,\}$

$\boldsymbol{A = B}$ **(Both)**

19. $A = \{0, 1, 2, 3, 4\}$, $B = \{1, 3\}$

$\boldsymbol{B \subseteq A}$

20. $A = \{a\}$, $B = \{b\}$

Neither (unless $a = b$, in which case $A = B$ and the answer is **Both**)

Let $A = \{x, y, z, w\}$ and $B = \{s, t, y, w\}$. Determine each of the following:

21. $A \cup B$

$A \cup B = \{\boldsymbol{x}, \boldsymbol{y}, \boldsymbol{z}, \boldsymbol{w}, \boldsymbol{s}, \boldsymbol{t}\}$

22. $A \cap B$

$A \cap B = \{\boldsymbol{y}, \boldsymbol{w}\}$

23. $A \setminus B$

$A \setminus B = \{\boldsymbol{x}, \boldsymbol{z}\}$

24. $B \setminus A$

$B \setminus A = \{\boldsymbol{s}, \boldsymbol{t}\}$

25. $A \Delta B$

$A \Delta B = (A \setminus B) \cup (B \setminus A) = \{x, z\} \cup \{s, t\} = \{\boldsymbol{x}, \boldsymbol{z}, \boldsymbol{s}, \boldsymbol{t}\}$

Use the roster method to describe each of the following sets.

26. $\{u\} \times \{v\} \times \{w\} \times \{z\} \times \{t\}$

$\{(\boldsymbol{u}, \boldsymbol{v}, \boldsymbol{w}, \boldsymbol{z}, \boldsymbol{t})\}$

27. $\{u, v, w\} \times \{z, t\}$

$\{(\boldsymbol{u}, \boldsymbol{z}), (\boldsymbol{u}, \boldsymbol{t}), (\boldsymbol{v}, \boldsymbol{z}), (\boldsymbol{v}, \boldsymbol{t}), (\boldsymbol{w}, \boldsymbol{z}), (\boldsymbol{w}, \boldsymbol{t})\}$

28. $\{z, t\}^2$

$\{z, t\}^2 = \{z, t\} \times \{z, t\} = \{(\boldsymbol{z}, \boldsymbol{z}), (\boldsymbol{z}, \boldsymbol{t}), (\boldsymbol{t}, \boldsymbol{z}), (\boldsymbol{t}, \boldsymbol{t})\}$

29. $\{z, t\}^3$

$\{z, t\}^3 = \{z, t\} \times \{z, t\} \times \{z, t\} = \{(\boldsymbol{z}, \boldsymbol{z}, \boldsymbol{z}), (\boldsymbol{z}, \boldsymbol{z}, \boldsymbol{t}), (\boldsymbol{z}, \boldsymbol{t}, \boldsymbol{z}), (\boldsymbol{z}, \boldsymbol{t}, \boldsymbol{t}), (\boldsymbol{t}, \boldsymbol{z}, \boldsymbol{z}), (\boldsymbol{t}, \boldsymbol{z}, \boldsymbol{t}), (\boldsymbol{t}, \boldsymbol{t}, \boldsymbol{z}), (\boldsymbol{t}, \boldsymbol{t}, \boldsymbol{t})\}$

30. $\{z, t\}^4$

$$\begin{aligned}\{z, t\}^4 &= \{z, t\} \times \{z, t\} \times \{z, t\} \times \{z, t\} \\ &= \{(\boldsymbol{z}, \boldsymbol{z}, \boldsymbol{z}, \boldsymbol{z}), (\boldsymbol{z}, \boldsymbol{z}, \boldsymbol{z}, \boldsymbol{t}), (\boldsymbol{z}, \boldsymbol{z}, \boldsymbol{t}, \boldsymbol{z}), (\boldsymbol{z}, \boldsymbol{z}, \boldsymbol{t}, \boldsymbol{t}), (\boldsymbol{z}, \boldsymbol{t}, \boldsymbol{z}, \boldsymbol{z}), (\boldsymbol{z}, \boldsymbol{t}, \boldsymbol{z}, \boldsymbol{t}), (\boldsymbol{z}, \boldsymbol{t}, \boldsymbol{t}, \boldsymbol{z}), (\boldsymbol{z}, \boldsymbol{t}, \boldsymbol{t}, \boldsymbol{t}), \\ &\quad (\boldsymbol{t}, \boldsymbol{z}, \boldsymbol{z}, \boldsymbol{z}), (\boldsymbol{t}, \boldsymbol{z}, \boldsymbol{z}, \boldsymbol{t}), (\boldsymbol{t}, \boldsymbol{z}, \boldsymbol{t}, \boldsymbol{z}), (\boldsymbol{t}, \boldsymbol{z}, \boldsymbol{t}, \boldsymbol{t}), (\boldsymbol{t}, \boldsymbol{t}, \boldsymbol{z}, \boldsymbol{z}), (\boldsymbol{t}, \boldsymbol{t}, \boldsymbol{z}, \boldsymbol{t}), (\boldsymbol{t}, \boldsymbol{t}, \boldsymbol{t}, \boldsymbol{z}), (\boldsymbol{t}, \boldsymbol{t}, \boldsymbol{t}, \boldsymbol{t})\},\end{aligned}$$

31. $\{u, v, w\}^2$

$\{u, v, w\}^2 = \{u, v, w\} \times \{u, v, w\} = \{(\boldsymbol{u}, \boldsymbol{u}), (\boldsymbol{u}, \boldsymbol{v}), (\boldsymbol{u}, \boldsymbol{w}), (\boldsymbol{v}, \boldsymbol{u}), (\boldsymbol{v}, \boldsymbol{v}), (\boldsymbol{v}, \boldsymbol{w}), (\boldsymbol{w}, \boldsymbol{u}), (\boldsymbol{w}, \boldsymbol{v}), (\boldsymbol{w}, \boldsymbol{w})\}$

32. $\{u, v, w\}^3$

$$\{u, v, w\}^3 = \{u, v, w\} \times \{u, v, w\} \times \{u, v, w\}$$

$$= \{(\boldsymbol{u},\boldsymbol{u},\boldsymbol{u}), (\boldsymbol{u},\boldsymbol{u},\boldsymbol{v}), (\boldsymbol{u},\boldsymbol{u},\boldsymbol{w}), (\boldsymbol{u},\boldsymbol{v},\boldsymbol{u}), (\boldsymbol{u},\boldsymbol{v},\boldsymbol{v}), (\boldsymbol{u},\boldsymbol{v},\boldsymbol{w}), (\boldsymbol{u},\boldsymbol{w},\boldsymbol{u}), (\boldsymbol{u},\boldsymbol{w},\boldsymbol{v}), (\boldsymbol{u},\boldsymbol{w},\boldsymbol{w}),$$
$$(\boldsymbol{v},\boldsymbol{u},\boldsymbol{u}), (\boldsymbol{v},\boldsymbol{u},\boldsymbol{v}), (\boldsymbol{v},\boldsymbol{u},\boldsymbol{w}), (\boldsymbol{v},\boldsymbol{v},\boldsymbol{u}), (\boldsymbol{v},\boldsymbol{v},\boldsymbol{v}), (\boldsymbol{v},\boldsymbol{v},\boldsymbol{w}), (\boldsymbol{v},\boldsymbol{w},\boldsymbol{u}), (\boldsymbol{v},\boldsymbol{w},\boldsymbol{v}), (\boldsymbol{v},\boldsymbol{w},\boldsymbol{w}),$$
$$(\boldsymbol{w},\boldsymbol{u},\boldsymbol{u}), (\boldsymbol{w},\boldsymbol{u},\boldsymbol{v}), (\boldsymbol{w},\boldsymbol{u},\boldsymbol{w}), (\boldsymbol{w},\boldsymbol{v},\boldsymbol{u}), (\boldsymbol{w},\boldsymbol{v},\boldsymbol{v}), (\boldsymbol{w},\boldsymbol{v},\boldsymbol{w}), (\boldsymbol{w},\boldsymbol{w},\boldsymbol{u}), (\boldsymbol{w},\boldsymbol{w},\boldsymbol{v}), (\boldsymbol{w},\boldsymbol{w},\boldsymbol{w})\}$$

33. $\emptyset^2$

$\emptyset^2 = \emptyset \times \emptyset = \emptyset$

34. $\emptyset^3$

$\emptyset^3 = \emptyset \times \emptyset \times \emptyset = \emptyset$

35. $\left(\mathcal{P}(\{\emptyset\})\right)^2$

Since $\mathcal{P}(\{\emptyset\}) = \{\emptyset, \{\emptyset\}\}$, we have

$$\mathcal{P}(\{\emptyset\})^2 = \mathcal{P}(\{\emptyset\}) \times \mathcal{P}(\{\emptyset\}) = \{(\emptyset,\emptyset), (\emptyset,\{\emptyset\}), (\{\emptyset\},\emptyset), (\{\emptyset\},\{\emptyset\})\}.$$

Let A, B, and C be sets, let $X = (A \setminus B) \setminus C$, and let $Y = A \setminus (B \setminus C)$.

36. Draw Venn Diagrams for X and Y.

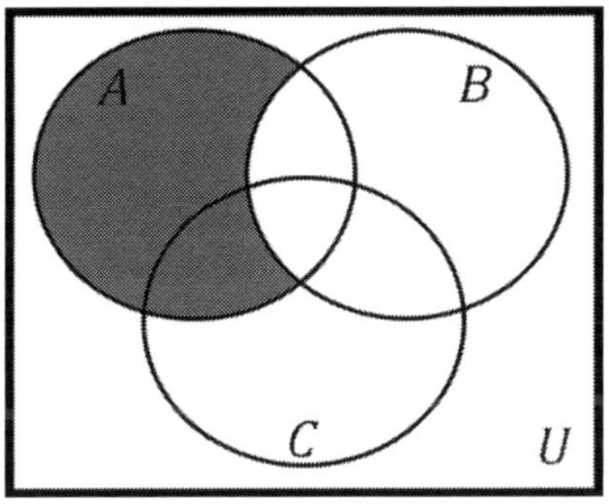

$\boldsymbol{A \setminus B}$

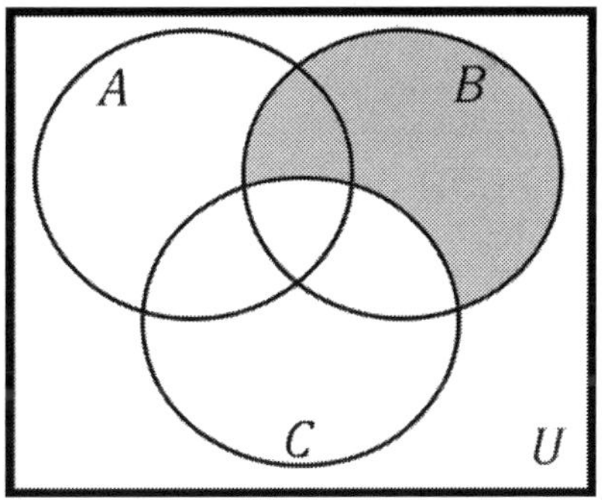

$\boldsymbol{B \setminus C}$

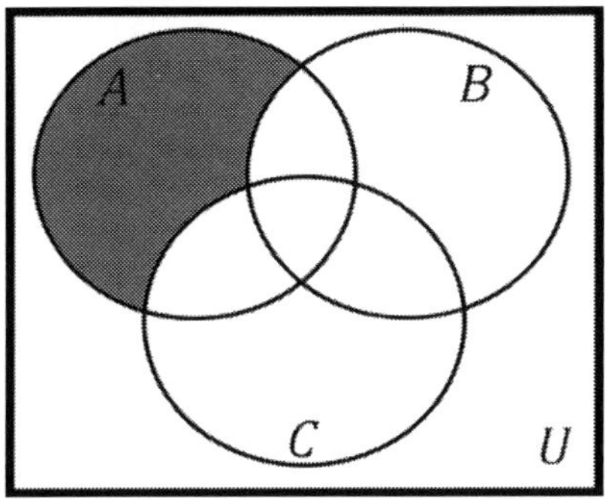

$\boldsymbol{X = (A \setminus B) \setminus C}$

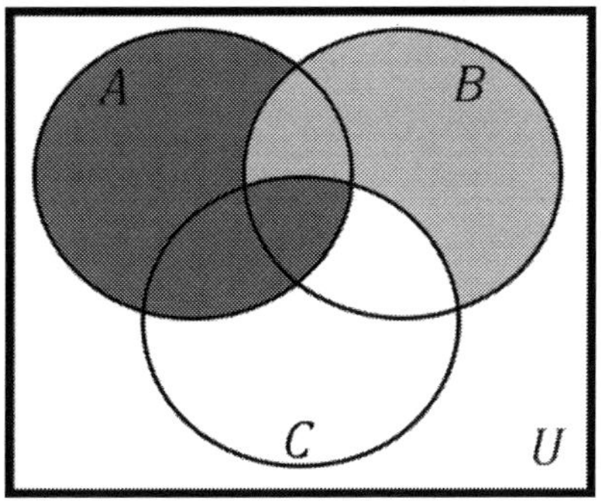

$\boldsymbol{Y = A \setminus (B \setminus C)}$

37. Is $X \subseteq Y$?

Yes

Note: Although it was not asked in the question, let's prove that $X \subseteq Y$. Let $x \in (A \setminus B) \setminus C$. Then $x \in A \setminus B$ and $x \notin C$. Since $x \in A \setminus B$, $x \in A$ and $x \notin B$. In particular, $x \in A$. Since $x \notin B$, $x \notin B \setminus C$ (because if $x \in B \setminus C$, then $x \in B$). So, we have $x \in A$ and $x \notin B \setminus C$. Therefore, $x \in A \setminus (B \setminus C)$. Since $x \in (A \setminus B) \setminus C$ was arbitrary, $(A \setminus B) \setminus C \subseteq A \setminus (B \setminus C)$. That is, $X \subseteq Y$. □

38. Is $Y \subseteq X$?

No

Counterexample: Let $A = \{0, 1\}$, $B = \{0, 2\}$, and $C = \{0, 3\}$. Then we have

$$X = (A \setminus B) \setminus C = (\{0, 1\} \setminus \{0, 2\}) \setminus \{0, 3\} = \{1\} \setminus \{0, 3\} = \{1\}$$
$$Y = A \setminus (B \setminus C) = \{0, 1\} \setminus (\{0, 2\} \setminus \{0, 3\}) = \{0, 1\} \setminus \{2\} = \{0, 1\}$$

Since $0 \in Y$, but $0 \notin X$, $Y \nsubseteq X$

39. Is $X = Y$?

No

Let S be the set of people on a checkout line at a supermarket and define a relation $\prec$ on X by $x \prec y$ if x will check out before y.

40. Is $(S, \prec)$ a poset?

No because $\prec$ is not reflexive.

41. Is $(S, \prec)$ a strict poset?

Yes (because $\prec$ is antireflexive, antisymmetric, and transitive on S).

42. Is $(S, \prec)$ a strict linearly ordered set?

Yes (because $(S, \prec)$ is a poset that satisfies trichotomy).

Let $X = \{a, b, c\}$. Determine if each of the following is an equivalence relation on X.

43. $\{(a, a), (b, b), (c, c)\}$

Yes. This is the equality relation on X.

44. $\{(a, a), (a, b), (a, c)\}$

No. This relation is not reflexive (for example, it does not include (b, b)) or symmetric (for example, it includes (a, b), but not (b, a)).

45. $\{(a,a),(a,b),(a,c),(b,a),(b,b),(b,c),(c,a),(c,b),(c,c)\}$

Yes. This is the trivial equivalence relation on X.

46. $\{(a,a),(a,b),(a,c),(b,b),(b,c),(c,c)\}$

No. This relation is not symmetric (for example, it includes (a,b), but not (b,a)).

47. $\{(a,a),(a,b),(b,b),(b,a),(c,c)\}$

Yes. This relation is reflexive, symmetric, and transitive.

48. $\{(a,a),(a,b),(b,a),(b,b),(b,c),(c,b),(c,c)\}$

No. This relation is not transitive (for example, it includes (a,b) and (b,c), but not (a,c)).

49. $\{(a,a),(b,b),(b,c),(c,c),(c,b)\}$

Yes. This relation is reflexive, symmetric, and transitive.

For each of the following equivalence relations on $A = \{a,b,c,d\}$, find $[a]$, $[b]$, $[c]$, and $[d]$.

50. $\{(a,a),(b,b),(c,c),(d,d)\}$

$[\boldsymbol{a}] = \{\boldsymbol{a}\}$ $[\boldsymbol{b}] = \{\boldsymbol{b}\}$ $[\boldsymbol{c}] = \{\boldsymbol{c}\}$ $[\boldsymbol{d}] = \{\boldsymbol{d}\}$

51. $\{(a,a),(a,b),(b,a),(b,b),(c,c),(d,d)\}$

$[\boldsymbol{a}] = [\boldsymbol{b}] = \{\boldsymbol{a},\boldsymbol{b}\}$ $[\boldsymbol{c}] = \{\boldsymbol{c}\}$ $[\boldsymbol{d}] = \{\boldsymbol{d}\}$

52. $\{(a,a),(a,b),(a,c),(b,a),(b,b),(b,c),(c,a),(c,b),(c,c),(d,d)\}$

$[\boldsymbol{a}] = [\boldsymbol{b}] = [\boldsymbol{c}] = \{\boldsymbol{a},\boldsymbol{b},\boldsymbol{c}\}$ $[\boldsymbol{d}] = \{\boldsymbol{d}\}$

53. $\{(a,a),(a,c),(b,b),(b,d),(c,a),(c,c),(d,b),(d,d)\}$

$[\boldsymbol{a}] = [\boldsymbol{c}] = \{\boldsymbol{a},\boldsymbol{c}\}$ $[\boldsymbol{b}] = [\boldsymbol{d}] = \{\boldsymbol{b},\boldsymbol{d}\}$

Find all partitions of each of the following sets:

54. $\{x\}$

There is only one partition of $\{x\}$, namely $\{\{\boldsymbol{x}\}\}$.

55. $\{x,y\}$

The partitions of $\{x,y\}$ are $\{\{\boldsymbol{x}\},\{\boldsymbol{y}\}\}$ and $\{\{\boldsymbol{x},\boldsymbol{y}\}\}$.

56. $\{x, y, z\}$

The partitions of $\{x, y, z\}$ are $\{\{x\},\{y\},\{z\}\}$, $\{\{x\},\{y, z\}\}$, $\{\{y\},\{x, z\}\}$, $\{\{z\},\{x, y\}\}$, and $\{\{x, y, z\}\}$.

57. $\{x, y, z, w\}$

The partitions of $\{x, y, z, w\}$ are $\{\{x\},\{y\},\{z\},\{w\}\}$, $\{\{x\},\{y\},\{z, w\}\}$, $\{\{x\},\{z\},\{y, w\}\}$, $\{\{x\},\{w\},\{y, z\}\}$, $\{\{y\},\{z\},\{x, w\}\}$, $\{\{y\},\{w\},\{x, z\}\}$, $\{\{z\},\{w\},\{x, y\}\}$, $\{\{x, y\},\{z, w\}\}$, $\{\{x, z\},\{y, w\}\}$, $\{\{x, w\},\{y, z\}\}$, $\{\{x, y, z\},\{w\}\}$, $\{\{x, y, w\},\{z\}\}$, $\{\{x, z, w\},\{y\}\}$, $\{\{y, z, w\},\{x\}\}$, and $\{\{x, y, z, w\}\}$.

Determine if each of the following relations is a function. State the domain and range of each such function.

58. $\{(0,0),(1,0)\}$

Function; $\text{domain} = \{0,1\}, \text{range} = \{0\}$

59. $\{(0,0),(0,1)\}$

Not a function

60. $\{(a, a),(b, b),(c, c)\}$

Function; $\text{domain} = \{a, b, c\}, \text{range} = \{a, b, c\}$

61. $\{(a, a),(a, b),(a, c)\}$

Not a function

62. $\{(a, a),(b, a),(c, a)\}$

Function; $\text{domain} = \{a, b, c\}, \text{range} = \{a\}$

63. $\{(a, b),(b, b),(c, d),(e, a)\}$

Function; $\text{domain} = \{a, b, c, e\}, \text{range} = \{a, b, d\}$

64. $\{(a, a),(a, b),(b, a)\}$

Not a function

65. $\{(0,0),(1,1),(2,2),(3,3),(4,4)\}$

Function; $\text{domain} = \{0,1,2,3,4\}, \text{range} = \{0,1,2,3,4\}$

66. $\{(0, a),(a, 0),(b, 1),(1, c),(c, 0)\}$

Function; $\text{domain} = \{0,1, a, b, c\}, \text{range} = \{0,1, a, c\}$

67. $\{(0,a),(a,0),(0,1),(b,2)\}$

Not a function

Determine if each of the following functions is injective.

68. $\{(s,0),(t,0)\}$

Not injective

69. $\{(s,0),(t,1)\}$

Injective

70. $\{(a,a),(b,b),(c,c)\}$

Injective

71. $\{(a,a),(b,a),(c,b)\}$

Not injective

72. $\{(s,s),(t,s),(u,s)\}$

Not injective

73. $\{(a,b),(b,a),(c,d),(d,c)\}$

Injective

For each of the following, determine if A and B are equinumerous. If so, provide a bijection from one set to the other. If not, explain why.

74. $A = \emptyset, B = \{\emptyset\}$

$\boldsymbol{A \nsim B}$ because $|A| = 0$ and $|B| = 1$.

75. $A = \mathcal{P}(\emptyset), B = \{\emptyset\}$

$\boldsymbol{A \sim B}$ because $A = B$.

76. $A = \{0,1,2\}, B = \{\Delta, \square, \perp\}$

$\boldsymbol{A \sim B}$ via the bijection $\{(0,\Delta),(1,\square),(2,\perp)\}$.

77. $A = \{n \in \mathbb{N} \mid 5 \leq n \leq 250\}, B = \{6,7,8,\ldots,250,251\}$

$\boldsymbol{A \sim B}$ via the bijection $f: A \to B$ defined by $f(n) = n + 1$.

78. $A = \{2n \mid n \in \mathbb{Z}\}$, $B = \{2n \in \mathbb{Z} \mid -100 < n < 9000\}$

$\boldsymbol{A} \nsim \boldsymbol{B}$ because A is infinite and B is finite.

State whether each of the following sets is finite, denumerable, or uncountable.

79. $\{3, 4, 5, \dots\}$

Denumerable

80. $\{3, 4, 5, \dots, 888, 889\}$

Finite

81. $\mathcal{P}(\{3, 4, 5, \dots\})$

Uncountable (The power set of an infinite set is uncountable.)

82. $\mathcal{P}(\{3, 4, 5, \dots, 888, 889\})$

Finite (The power set of a finite set is finite.)

83. $\mathcal{P}\big(\mathcal{P}(\{3, 4, 5, \dots, 888, 889\})\big)$

Finite (The power set of a finite set is finite.)

84. $\bigcup\big\{\{n\} \bigm| n \in \mathbb{Z}\big\}$

Denumerable (This set is equal to $\mathbb{Z}$.)

85. $\mathbb{Z} \setminus \mathbb{N}$

Denumerable

86. $\mathcal{P}(\mathbb{Q} \setminus \mathbb{Z})$

Uncountable ($\mathbb{Q} \setminus \mathbb{Z}$ is infinite and the power set of an infinite set is uncountable.)

LEVEL 2

Compute the power set of each of the following sets:

87. $\emptyset$

$\mathcal{P}(\emptyset) = \{\emptyset\}$

88. $\{\emptyset\}$

$\mathcal{P}(\{0\}) = \big\{\emptyset, \{\emptyset\}\big\}$

89. {lion, tiger, jaguar}

$\mathcal{P}(\{\text{lion, tiger, jaguar}\}) =$

$\{\emptyset, \{\text{lion}\}, \{\text{tiger}\}, \{\text{jaguar}\}, \{\text{lion, tiger}\}, \{\text{lion, jaguar}\}, \{\text{tiger, jaguar}\}, \{\text{lion, tiger, jaguar}\}\}$

90. $\{\emptyset, \{\emptyset\}\}$

$\mathcal{P}(\{\emptyset, \{\emptyset\}\}) = \{\emptyset, \{\emptyset\}, \{\{\emptyset\}\}, \{\emptyset, \{\emptyset\}\}\}$

91. $\{\{\emptyset\}\}$

$\mathcal{P}(\{\{\emptyset\}\}) = \{\emptyset, \{\{\emptyset\}\}\}$

92. $\{\emptyset, \{\emptyset\}, \{\emptyset, \{\emptyset\}\}\}$

$\mathcal{P}(\{\emptyset, \{\emptyset\}, \{\emptyset, \{\emptyset\}\}\}) =$

$\{\emptyset, \{\emptyset\}, \{\{\emptyset\}\}, \{\{\emptyset, \{\emptyset\}\}\}, \{\emptyset, \{\emptyset\}\}, \{\emptyset, \{\emptyset, \{\emptyset\}\}\}, \{\{\emptyset\}, \{\emptyset, \{\emptyset\}\}\}, \{\emptyset, \{\emptyset\}, \{\emptyset, \{\emptyset\}\}\}\}$

Determine whether each of the following statements is true or false:

93. $a \in \emptyset$

The empty set has no elements. So, $x \in \emptyset$ is false for any x. In particular, $a \in \emptyset$ is **false**.

94. $\emptyset \in \emptyset$

The empty set has no elements. So, $x \in \emptyset$ is false for any x. In particular, $\emptyset \in \emptyset$ is **false**.

95. $\emptyset \in \{\emptyset, \{\emptyset\}\}$

The set $\{\emptyset, \{\emptyset\}\}$ has exactly 2 elements, namely $\emptyset$ and $\{\emptyset\}$. In particular, $\emptyset \in \{\emptyset, \{\emptyset\}\}$ is **true**.

96. $\{\emptyset\} \in \emptyset$

The empty set has no elements. So, $x \in \emptyset$ is false for any x. In particular, $\{\emptyset\} \in \emptyset$ is **false**.

97. $\{\emptyset\} \in \{\emptyset\}$

The set $\{\emptyset\}$ has 1 element, namely $\emptyset$. Since $\{\emptyset\} \neq \emptyset$, $\{\emptyset\} \in \{\emptyset\}$ is **false**.

98. $7 \in 2\mathbb{N}$

$2\mathbb{N} = \{0, 2, 4, 6, 8, 10, 12, 14, \dots\}$. In particular, $7 \in 2\mathbb{N}$ is **false**.

99. $\emptyset \subseteq \emptyset$

The empty set is a subset of every set. So, $\emptyset \subseteq X$ is true for any X. In particular, $\emptyset \subseteq \emptyset$ is **true**. (This can also be done by using the fact that every set is a subset of itself.)

100. $\emptyset \subseteq \{\emptyset\}$

Again, (as in Problem 99), $\emptyset \subseteq X$ is true for any X. In particular, $\emptyset \subseteq \{\emptyset\}$ is **true**.

101. $\{\emptyset\} \subseteq \emptyset$

The only subset of $\emptyset$ is $\emptyset$. So, $\{\emptyset\} \subseteq \emptyset$ is **false**.

102. $\{\emptyset\} \subseteq \{\emptyset\}$

Every set is a subset of itself. So, $\{\emptyset\} \subseteq \{\emptyset\}$ is **true**.

Determine the cardinality of each of the following sets:

103. $\{a, a, b, c, c, c, d, d\}$

$\{a, a, b, c, c, c, d, d\} = \{a, b, c, d\}$. Therefore, $|\{a, a, b, c, c, c, d, d\}| = |\{a, b, c, d\}| = \mathbf{4}$.

104. $\{\{x, y\}, \{z, u, v\}, \{w\}\}$

$\{\{x, y\}, \{z, u, v\}, \{w\}\}$ consists of the 3 elements $\{x, y\}$, $\{z, u, v\}$, and $\{w\}$. Therefore, we have $|\{\{x, y\}, \{z, u, v\}, \{w\}\}| = \mathbf{3}$.

105. $\{5, 6, 7, \dots, 4322, 4323\}$

$|\{5, 6, 7, \dots, 4322, 4323\}| = 4323 - 5 + 1 = \mathbf{4319}$ (notice that we used the fence-post formula here).

Determine if each of the following real numbers is rational or irrational:

106. $5.\overline{7}$

rational (repeating decimal)

107. $-72.8101214161820222426 28\dots$

irrational

108. $463.6543215432154321 54321\dots$

rational ($463.654321543215432154321\dots = 463.6\overline{54321}$, repeating decimal)

109. 0

rational (terminating decimal)

Let $A = \{\emptyset, \{\emptyset, \{\emptyset\}\}\}$ and $B = \{\emptyset, \{\emptyset\}\}$. Compute each of the following:

110. $A \cup B$

$A \cup B = \{\emptyset, \{\emptyset\}, \{\emptyset, \{\emptyset\}\}\}$

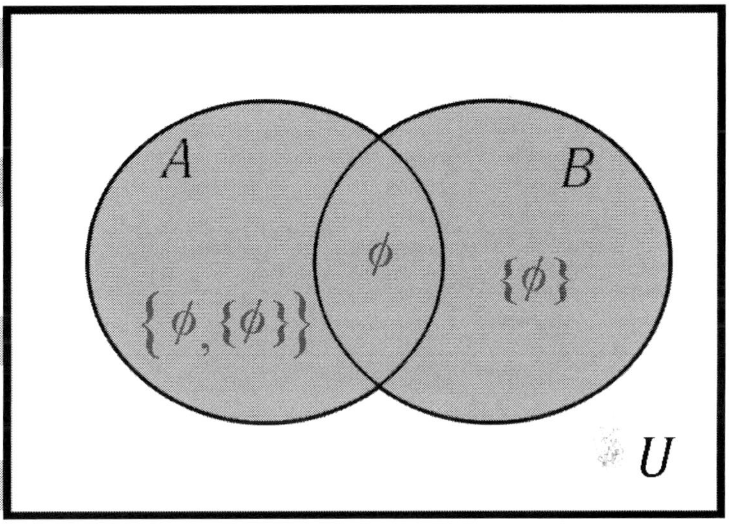

111. $A \cap B$

$A \cap B = \{\emptyset\}$

112. $A \setminus B$

$A \setminus B = \{\{\emptyset, \{\emptyset\}\}\}$

113. $B \setminus A$

$B \setminus A = \{\{\emptyset\}\}$

114. $A \,\Delta\, B$

$A \,\Delta\, B = \{\{\emptyset, \{\emptyset\}\}\} \cup \{\{\emptyset\}\} = \{\{\emptyset\}, \{\emptyset, \{\emptyset\}\}\}$

Let $A = (-17, 6)$ and $B = (-1, 17)$. Compute each of the following:

115. $A \cup B$

$(-\mathbf{17}, \mathbf{17})$

116. $A \cap B$

$(-\mathbf{1}, \mathbf{6})$

117. $A \setminus B$

$(-\mathbf{17}, -\mathbf{1}]$

118. $B \setminus A$

$[\mathbf{6}, \mathbf{17})$

119. $A \,\Delta\, B$

$(-\mathbf{17}, -\mathbf{1}] \cup [\mathbf{6}, \mathbf{17})$

Let $A = [14, \infty)$ and $B = (-\infty, 15)$. Compute each of the following:

120. $A \cup B$

$(-\infty, \infty)$

121. $A \cap B$

$[\mathbf{14}, \mathbf{15})$

122. $A \Delta B$

$(-\infty, \mathbf{14}) \cup [\mathbf{15}, \infty)$

Find the domain, range, and field of each of the following relations:

123. $R = \{(a, b), (c, d), (e, f), (f, a)\}$

$\mathbf{dom}\ \boldsymbol{R} = \{\boldsymbol{a}, \boldsymbol{c}, \boldsymbol{e}, \boldsymbol{f}\}$; $\mathbf{ran}\ \boldsymbol{R} = \{\boldsymbol{a}, \boldsymbol{b}, \boldsymbol{d}, \boldsymbol{f}\}$; $\mathbf{field}\ \boldsymbol{R} = \{\boldsymbol{a}, \boldsymbol{b}, \boldsymbol{c}, \boldsymbol{d}, \boldsymbol{e}, \boldsymbol{f}\}$

124. $S = \{(2k, 2t + 1) \mid k, t \in \mathbb{Z}\}$

$\mathbf{dom}\ \boldsymbol{S} = \mathbf{2}\mathbb{Z} = \mathbb{E}$; $\mathbf{ran}\ \boldsymbol{S} = \{\mathbf{2}\boldsymbol{t} + \mathbf{1} \mid \boldsymbol{t} \in \mathbb{Z}\} = \mathbf{2}\mathbb{Z} + \mathbf{1} = \mathbb{O}$; $\mathbf{field}\ \boldsymbol{S} = \mathbb{Z}$

125. $T = \left\{(a, b) \,\middle|\, \frac{a}{b} \in \mathbb{Q}\right\}$

$\mathbf{dom}\ \boldsymbol{T} = \mathbb{Z}$; $\mathbf{ran}\ \boldsymbol{T} = \mathbb{Z}^*$; $\mathbf{field}\ \boldsymbol{T} = \mathbb{Z}$

126. $U = \mathbb{R} \times \mathbb{Z}^+$

$\mathbf{dom}\ \boldsymbol{T} = \mathbb{R}$; $\mathbf{ran}\ \boldsymbol{T} = \mathbb{Z}^+$; $\mathbf{field}\ \boldsymbol{T} = \mathbb{R}$

Determine a and b so that each of the following equations is true, or state that the equation has no solution.

127. $\{a, b\} = \{8\}$

$\boldsymbol{a} = \mathbf{8}, \boldsymbol{b} = \mathbf{8}$

128. $\{a, b\} = \{7, 9\}$

$\boldsymbol{a} = \mathbf{7}, \boldsymbol{b} = \mathbf{9}$ or $\boldsymbol{a} = \mathbf{9}, \boldsymbol{b} = \mathbf{7}$

129. $(a, b) = (7, 9)$

$\boldsymbol{a} = \mathbf{7}, \boldsymbol{b} = \mathbf{9}$

130. $(a, b) = \{\{7\}, \{9, 7\}\}$

$\boldsymbol{a} = \mathbf{7}, \boldsymbol{b} = \mathbf{9}$ (note that $\{\{7\}, \{9, 7\}\} = \{\{7\}, \{7, 9\}\}$)

131. $(a, b) = \{5\}$

No solution.

132. $(a, b) = \{\{5\}\}$

$\boldsymbol{a} = \mathbf{5}, \boldsymbol{b} = \mathbf{5}$

133. $(a, a, b) = (1, b, a)$

$\boldsymbol{a} = \mathbf{1}, \boldsymbol{b} = \mathbf{1}$

134. $(a, b, c, d) = (1, c, 2, a)$

$\boldsymbol{a} = \mathbf{1}, \boldsymbol{b} = \mathbf{2}, \boldsymbol{c} = \mathbf{2}, \boldsymbol{d} = \mathbf{1}$

Suppose that $|A| = 5$, $|B| = 10$, $|C| = 50$, and $|D| = 100$. Compute each of the following:

135. $|A \times B|$

$|A \times B| = |A| \cdot |B| = 5 \cdot 10 = \mathbf{50}$

136. $|A \times B \times C \times D|$

$|A \times B \times C \times D| = |A| \cdot |B| \cdot |C| \cdot |D| = 5 \cdot 10 \cdot 50 \cdot 100 = \mathbf{250{,}000}$

137. $|C^2|$

$|C^2| = 50^2 = \mathbf{2500}$

138. $|\mathcal{P}(C \times D)|$

$|\mathcal{P}(C \times D)| = 2^{|C \times D|} = 2^{|C| \cdot |D|} = 2^{50 \cdot 100} = \mathbf{2^{5000}}$

139. $|\{R \mid R \text{ is a relation on } B\}|$

$|\mathcal{P}(B \times B)| = 2^{|B \times B|} = 2^{|B| \cdot |B|} = 2^{10 \cdot 10} = \mathbf{2^{100}}$

140. $|\{R \mid R \text{ is a relation on } B \times C\}|$

$\left|\mathcal{P}\big((B \times C) \times (B \times C)\big)\right| = 2^{|(B \times C) \times (B \times C)|} = 2^{|B \times C| \cdot |B \times C|} = 2^{500 \cdot 500} = \mathbf{2^{250{,}000}}$

Find an equivalence relation R on $A = \{0, 1, 2\}$ with the given property.

141. If S is any equivalence relation on A, then $R \subseteq S$.

$\boldsymbol{R} = \{(\mathbf{0}, \mathbf{0}), (\mathbf{1}, \mathbf{1}), (\mathbf{2}, \mathbf{2})\}$

142. If S is any equivalence relation on A, then $S \subseteq R$.

$\boldsymbol{R} = \{(\mathbf{0}, \mathbf{0}), (\mathbf{0}, \mathbf{1}), (\mathbf{0}, \mathbf{2}), (\mathbf{1}, \mathbf{0}), (\mathbf{1}, \mathbf{1}), (\mathbf{1}, \mathbf{2}), (\mathbf{2}, \mathbf{0}), (\mathbf{2}, \mathbf{1}), (\mathbf{2}, \mathbf{2})\}$

143. $|R| = 5$ and $0R1$.

$\boldsymbol{R} = \{(\mathbf{0}, \mathbf{0}), (\mathbf{0}, \mathbf{1}), (\mathbf{1}, \mathbf{0}), (\mathbf{1}, \mathbf{1}), (\mathbf{2}, \mathbf{2})\}$

Determine if each of the following is a partition of $\{0, 1, 2, 3, 4, 5, 6, 7, 8\}$.

144. $\{\{0\}, \{1\}, \{2\}, \{3\}, \{4\}, \{5\}, \{6\}, \{7\}, \{8\}\}$

Yes

145. $\{\{0, 1\}, \{1, 2, 3\}, \{3, 4, 5, 6\}, \{6, 7, 8\}\}$

No because for example, $\{0, 1\} \cap \{1, 2, 3\} = \{1\} \neq \emptyset$.

146. $\{\{0, 1, 2, 3, 4, 5, 6, 7, 8\}\}$

Yes

147. $\{\{0, 1, 2, 3\}, \{5, 6, 7, 8\}\}$

No because $\{0, 1, 2, 3\} \cup \{5, 6, 7, 8\} = \{1, 2, 3, 5, 6, 7, 8\} \neq \{1, 2, 3, 4, 5, 6, 7, 8\}$.

148. $\{\{0, 2, 4, 6, 8\}, \{1, 3, 5, 7\}\}$

Yes

Determine if each of the following relations is a function. State the domain and range of each such function.

149. $\{(a, b) \in \mathbb{R}^2 \mid b \leq 0 \wedge a^2 + b^2 = 9\}$

Function; domain $= [-3, 3]$, range $= [-3, 0]$

150. $\{(a, b) \in \mathbb{R}^2 \mid a < 0 \wedge a^2 + b^2 = 9\}$

Not a function

151. $\left\{\left(\frac{a}{b}, c\right) \in \mathbb{Q} \times \mathbb{Z} \,\middle|\, c = a - b\right\}$

Not a function (for example, $\left(\frac{1}{2}, -1\right), \left(\frac{2}{4}, -2\right)$ are both in the set and $\frac{1}{2} = \frac{2}{4}$).

152. $\{(a + bi, c) \in \mathbb{C} \times \mathbb{R} \mid c = a\}$

Function; $\text{domain} = \mathbb{C}, \text{range} = \mathbb{R}$

153. $\{(c, a + bi) \in \mathbb{R} \times \mathbb{C} \mid c = \max\{a, b\}\}$, where $\max\{a, b\}$ is the larger of a or b.

Not a function (for example, $(2, 1 + 2i), (2, 2 + i)$ are both in the set).

Determine if each of the following functions is injective, surjective, both, or neither.

154. $\{(a, b) \in \mathbb{R}^2 \mid b > 0 \wedge a^2 + b^2 = 1\}$

Neither

155. $\{(a, b) \in \mathbb{R} \times (0, 1] \mid b > 0 \wedge a^2 + b^2 = 1\}$

Surjective

156. $\{(a + bi, c) \in \mathbb{C} \times \mathbb{R} \mid c = a\}$

Surjective

157. $\{(n, m) \in \mathbb{N}^2 \mid m = 5n\}$

Injective

158. $\{(n, z) \in \mathbb{Z} \times \mathbb{C} \mid z = n + ni\}$

Injective

Let $f: \{a, b, c, d\} \to \{0, 1, 2\}$ be defined by $f = \{(a, 0), (b, 0), (c, 1), (d, 2)\}$. Compute $f[A]$ for each of the following sets A:

159. $A = \{a\}$

$f[A] = f[\{a\}] = \{\mathbf{0}\}$

160. $A = \{a, b\}$

$f[A] = f[\{a, b\}] = \{\mathbf{0}\}$

161. $A = \{a, b, c\}$

$f[A] = f[\{a, b, c\}] = \{\mathbf{0}, \mathbf{1}\}$

162. $A = \{a, c\}$

$f[A] = f[\{a, c\}] = \{\mathbf{0}, \mathbf{1}\}$

163. $A = \{b, c, d\}$

$f[A] = f[\{b, c, d\}] = \{\mathbf{0}, \mathbf{1}, \mathbf{2}\}$

Let $f: \{a, b, c, d\} \to \{0, 1, 2\}$ be defined by $f = \{(a, 0), (b, 0), (c, 1), (d, 2)\}$. Compute $f^{-1}[B]$ for each of the following sets B:

164. $B = \{0\}$

$f^{-1}[B] = f^{-1}[\{0\}] = \{\boldsymbol{a}, \boldsymbol{b}\}$

165. $B = \{1\}$

$f^{-1}[B] = f^{-1}[\{1\}] = \{\boldsymbol{c}\}$

166. $B = \{0, 1\}$

$f^{-1}[B] = f^{-1}[\{0, 1\}] = \{\boldsymbol{a}, \boldsymbol{b}, \boldsymbol{c}\}$

167. $B = \{0, 2\}$

$f^{-1}[B] = f^{-1}[\{0, 2\}] = \{\boldsymbol{a}, \boldsymbol{b}, \boldsymbol{d}\}$

168. $B = \{0, 1, 2\}$

$f^{-1}[B] = f^{-1}[\{0, 1, 2\}] = \{\boldsymbol{a}, \boldsymbol{b}, \boldsymbol{c}, \boldsymbol{d}\}$

Define $\mathcal{P}_k(\mathbb{N})$ for each $k \in \mathbb{N}$ by $\mathcal{P}_0(\mathbb{N}) = \mathbb{N}$ and $\mathcal{P}_{k+1}(\mathbb{N}) = \mathcal{P}(\mathcal{P}_k(\mathbb{N}))$ for $k > 0$.

169. Explain why $\mathcal{P}_k(\mathbb{N}) \prec \mathcal{P}_{k+1}(\mathbb{N})$ for each $k \in \mathbb{N}$.

This follows from Cantor's Theorem (Theorem 2.68).

170. Find a set B such that for all $k \in \mathbb{N}$, $\mathcal{P}_k(\mathbb{N}) \prec B$. You may use Problem 247 below.

$\boldsymbol{B} = \bigcup\{\boldsymbol{\mathcal{P}_n}(\mathbb{N}) \mid \boldsymbol{n} \in \mathbb{N}\}$

To see that $\mathcal{P}_k(\mathbb{N}) \prec B$ for all $k \in \mathbb{N}$, begin by letting $k \in \mathbb{N}$. Since $\mathcal{P}_k(\mathbb{N}) \subseteq B$, $\mathcal{P}_k(\mathbb{N}) \preccurlyeq B$. Since k was arbitrary, we have $\mathcal{P}_k(\mathbb{N}) \preccurlyeq B$ for all $k \in \mathbb{N}$.

Again, let $k \in \mathbb{N}$. We have $\mathcal{P}_k(\mathbb{N}) \prec \mathcal{P}_{k+1}(\mathbb{N})$ and $\mathcal{P}_{k+1}(\mathbb{N}) \preccurlyeq B$. By Problem 247 below, $\mathcal{P}_k(\mathbb{N}) \prec B$. Since $k \in \mathbb{N}$ was arbitrary, we have shown that for all $k \in \mathbb{N}$, $\mathcal{P}_k(\mathbb{N}) \prec B$.

171. Find a set C such that $B \prec C$.

$\boldsymbol{C = \mathcal{P}(B)}$

172. Is there a set X such that $A \preccurlyeq X$ for all sets A?

No. Given any set X, $D = \mathcal{P}(X)$ is a set such that $D \not\preccurlyeq X$.

LEVEL 3

Use set-builder notation to describe each of the following sets.

173. $\{2, 4, 6, 8, 10, 12, 14, 16, 18, 20, 22\}$

$\{2, 4, 6, 8, 10, 12, 14, 16, 18, 20, 22\} = \{\boldsymbol{n \in \mathbb{N} \mid n} \textbf{ is even and } \boldsymbol{2 \leq n \leq 22}\}$

174. $2\mathbb{N}$

$2\mathbb{N} = \{\boldsymbol{n \in \mathbb{N} \mid n} \textbf{ is even}\}$

175. $\mathbb{Z}^-$

$\mathbb{Z}^- = \{\boldsymbol{n \in \mathbb{Z} \mid n} \textbf{ is negative}\}$

Determine the cardinality of each of the following sets:

176. $\{\{\{x, y, z\}\}\}$

The only element of $\{\{\{x, y, z\}\}\}$ is $\{\{x, y, z\}\}$. So, $|\{\{\{x, y, z\}\}\}| = \mathbf{1}$.

177. $\{\{0, 1\}, 0, \{0\}, \{0, \{0, 1, 2\}\}\}$

The elements of $\{\{0, 1\}, 0, \{0\}, \{0, \{0, 1, 2\}\}\}$ are $\{0, 1\}$, 0, $\{0\}$, and $\{0, \{0, 1, 2\}\}$. So, we see that $|\{\{0, 1\}, 0, \{0\}, \{0, \{0, 1, 2\}\}\}| = \mathbf{4}.$

178. $\{a, \{a\}, \{a, a\}, \{a, a, a, a\}, \{a, a, \{a\}\}, \{a, \{a\}, \{a\}\}\}$

$$\begin{aligned}&\{a, \{a\}, \{a, a\}, \{a, a, a, a\}, \{a, a, \{a\}\}, \{a, \{a\}, \{a\}\}\}\\ &= \{a, \{a\}, \{a\}, \{a\}, \{a, \{a\}\}, \{a, \{a\}\}\}\\ &= \{a, \{a\}, \{a, \{a\}\}\}.\end{aligned}$$

So, $|\{a, \{a\}, \{a, a\}, \{a, a, a, a\}, \{a, a, \{a\}\}, \{a, \{a\}, \{a\}\}\}| = |\{a, \{a\}, \{a, \{a\}\}\}| = \mathbf{3}.$

For each set X, determine $|\mathcal{P}(X)|$

179. $X = \{0, 1, 2, 3, 4, 5, 6\}$

Since $|X| = 7$, $|\mathcal{P}(X)| = 2^7 = \mathbf{128}$.

180. $X = \{\emptyset, \{\emptyset\}, \{\emptyset, \{\emptyset\}\}\}$

Since $|X| = 3$, $|\mathcal{P}(X)| = 2^3 = \mathbf{8}$.

181. $X = \{26, 27, 28, \dots, 203, 204\}$

Since $|X| = 204 - 26 + 1 = 179$, $|\mathcal{P}(X)| = \mathbf{2^{179}}$.

For each of the following, compute $\bigcup \boldsymbol{X}$ and $\bigcap \boldsymbol{X}$.

182. $\boldsymbol{X} = \{\{0, 1, 2\}, \{1, 2, 3\}, \{2, 3, 4\}\}$

$\bigcup \boldsymbol{X} = \{\mathbf{0, 1, 2, 3, 4}\}$ $\bigcap \boldsymbol{X} = \{\mathbf{2}\}$

183. $\boldsymbol{X} = \{\mathbb{N}, \mathbb{Z}, \mathbb{Q}, \mathbb{R}\}$

$\bigcup \boldsymbol{X} = \mathbb{R}$ $\bigcap \boldsymbol{X} = \mathbb{N}$

184. $\boldsymbol{X} = \{(0, 10), (1, 11), (2, 12), (3, 13), (4, 14)\}$ (Note that $\boldsymbol{X}$ consists of open intervals and **not** ordered pairs.)

$\bigcup \boldsymbol{X} = (\mathbf{0, 14})$ $\bigcap \boldsymbol{X} = (\mathbf{4, 10})$

185. $\boldsymbol{X} = \{(-\infty, 20), [-5, 17), (4, 100]\}$

$\bigcup \boldsymbol{X} = (-\infty, \mathbf{100}]$ $\bigcap \boldsymbol{X} = (\mathbf{4, 17})$

186. $\boldsymbol{X} = \{(0, 1], (1, 2], (2, 3], (3, 4]\}$

$\bigcup \boldsymbol{X} = (\mathbf{0, 4}]$ $\bigcap \boldsymbol{X} = \emptyset$

Find the natural number that each of the following expressions is equal to.

187. 7^+

$7^+ = 7 \cup \{7\} = \{0, 1, 2, 3, 4, 5, 6\} \cup \{7\} = \{0, 1, 2, 3, 4, 5, 6, 7\} = \mathbf{8}$

188. $12 \cup \{12\}$

$12 \cup \{12\} = \{0, 1, 2, 3, 4, 5, 6, 7, 8, 9, 10, 11\} \cup \{12\} = \{0, 1, 2, 3, 4, 5, 6, 7, 8, 9, 10, 11, 12\} = \mathbf{13}$

189. $8 \cap 3$

$8 \cap 3 = \{0,1,2,3,4,5,6,7\} \cap \{0,1,2\} = \{0,1,2\} = \mathbf{3}$

Write each of the following in its unabbreviated form:

190. (x,x,z)

$(x,x,z) = \big((x,x),z\big) = \{\{(x,x)\},\{(x,x),z\}\} = \Big\{\big\{\{\{\boldsymbol{x}\}\}\big\},\big\{\{\{\boldsymbol{x}\}\},\boldsymbol{z}\big\}\Big\}$

191. (x,z,x)

$(x,z,x) = \big((x,z),x\big) = \{\{(x,z)\},\{(x,z),x\}\} = \Big\{\big\{\{\{\boldsymbol{x}\},\{\boldsymbol{x},\boldsymbol{z}\}\}\big\},\big\{\{\{\boldsymbol{x}\},\{\boldsymbol{x},\boldsymbol{z}\}\},\boldsymbol{x}\big\}\Big\}$

192. (x,x,x,x)

$(x,x,x,x) = \big((x,x,x),x\big) = \Bigg\{\bigg\{\Big\{\big\{\{\{\boldsymbol{x}\}\}\big\},\big\{\{\{\boldsymbol{x}\}\},\boldsymbol{x}\big\}\Big\}\bigg\},\bigg\{\Big\{\big\{\{\{\boldsymbol{x}\}\}\big\},\big\{\{\{\boldsymbol{x}\}\},\boldsymbol{x}\big\}\Big\},\boldsymbol{x}\bigg\}\Bigg\}$

193. (x,y,z,w)

$(x,y,z,w) = \big((x,y,z),w\big) = \{\{(x,y,z)\},\{(x,y,z),w\}\}$

$$= \Bigg\{\bigg\{\Big\{\big\{\{\{\boldsymbol{x}\},\{\boldsymbol{x},\boldsymbol{y}\}\}\big\},\big\{\{\{\boldsymbol{x}\},\{\boldsymbol{x},\boldsymbol{y}\}\},\boldsymbol{z}\big\}\Big\}\bigg\},\bigg\{\Big\{\big\{\{\{\boldsymbol{x}\},\{\boldsymbol{x},\boldsymbol{y}\}\}\big\},\big\{\{\{\boldsymbol{x}\},\{\boldsymbol{x},\boldsymbol{y}\}\},\boldsymbol{z}\big\}\Big\},w\bigg\}\Bigg\}$$

Let $A = \{a,b\}$ with $a \neq b$, consider the strict posets $(\mathbb{Z},<)$ and $(\mathcal{P}(A),\subset)$, and let $<_D$ be the dictionary order on $\mathbb{Z} \times \mathcal{P}(A)$.

194. Is $(\mathbb{Z} \times \mathcal{P}(A),<_D)$ a poset?

No because $<_D$ is not reflexive.

195. Is $(\mathbb{Z} \times \mathcal{P}(A),<_D)$ a strict poset?

Yes (because $<_D$ is antireflexive, antisymmetric, and transitive on $\mathbb{Z} \times \mathcal{P}(A)$).

196. Is $(\mathbb{Z} \times \mathcal{P}(A),<_D)$ a strict linearly ordered set?

No (For example, $(0,\{a\})$ and $(0,\{b\})$ are incomparable.)

Determine if each of the following is true or false.

197. $\{\emptyset,\{\emptyset\}\} \in 3$

True because $\{\emptyset,\{\emptyset\}\} = \{0,1\} = 2$ and $2 \in 3$ (or equivalently, $\{\emptyset,\{\emptyset\}\} \in \big\{\emptyset,\{\emptyset\},\{\emptyset,\{\emptyset\}\}\big\}$).

198. $3 \in \{\emptyset, \{\emptyset\}, \{\emptyset, \{\emptyset\}\}\}$

False because $\{\emptyset, \{\emptyset\}, \{\emptyset, \{\emptyset\}\}\} = \{0, 1, 2\} = 3$ and $3 \notin 3$.

199. $5 \in 8$

True because $(\mathbb{N}, <)$ is an ordered set and $5 \in 8$ is equivalent to $5 < 8$.

200. $5 <_{\mathbb{N}} \{0, 1, 2, 3, 4, 5, 6, 7\}$

True because $\{0, 1, 2, 3, 4, 5, 6, 7\} = 8$ and $5 \in 8$.

Let $X = \{0, 1, 2, 3, 4\}$. For each of the following, find the equivalence relation R on X of least cardinality containing the given elements.

201. $(0, 1) \in R$.

$\boldsymbol{R = \{(0,0), (0,1), (1,0), (1,1), (2,2), (3,3), (4,4)\}}$

202. $(0, 1), (2, 3) \in R$.

$\boldsymbol{R = \{(0,0), (0,1), (1,0), (1,1), (2,2), (2,3), (3,2), (3,3), (4,4)\}}$

203. $(0, 1), (1, 2), (2, 3) \in R$.

$\boldsymbol{R = \{(0,0), (0,1), (0,2), (0,3), (1,0), (1,1), (1,2), (1,3), (2,0), (2,1), (2,2), (2,3), (3,0),}$
$\boldsymbol{(3,1), (3,2), (3,3), (4,4)\}}$

204. $(0, 1), (2, 1), (3, 4), (4, 2) \in R$.

$\boldsymbol{R = \{(0,0), (0,1), (0,2), (0,3), (0,4), (1,0), (1,1), (1,2), (1,3), (1,4), (2,0), (2,1), (2,2),}$
$\boldsymbol{(2,3), (2,4), (3,0), (3,1), (3,2), (3,3), (3,4), (4,0), (4,1), (4,2), (4,3), (4,4)\}}$

Determine if each of the following is a partition of $\mathbb{Z}$.

205. $\{\mathbb{Z}\}$

Yes

206. $\{3\mathbb{Z}, 3\mathbb{Z} + 1\}$

No because $2 \notin 3\mathbb{Z} \cup (3\mathbb{Z} + 1)$.

207. $\{\mathbb{Z}^-, \mathbb{Z}^+\}$

No because $0 \notin \mathbb{Z}^- \cup \mathbb{Z}^+$.

208. $\{\mathbb{N}, \{-n \mid n \in \mathbb{N}\}\}$

No because $\mathbb{N} \cap \{-n \mid n \in \mathbb{N}\} = \{0\} \neq \emptyset$.

209. $\{5\mathbb{Z}, 5\mathbb{Z}+1, 5\mathbb{Z}+2, 5\mathbb{Z}+3, 5\mathbb{Z}+4\}$

Yes

Let $R = \{((a,b),(c,d)) \in (\mathbb{N} \times \mathbb{N})^2 \mid a+d = b+c\}$, so that $\mathbb{Z} = \{[(a,b)] \mid (a,b) \in \mathbb{N} \times \mathbb{N}\}$. Determine if each of the following is true or false.

210. $[(11,12)] = [(12,11)]$

False: $11 + 11 = 22$ and $12 + 12 = 24$.

211. $[(9,9)] = 0$

True

212. $[(0,1)] = 1$

False: $[(0,1)] = -1$

213. $[(9,7)] < [(7,9)]$

False: $9 + 9 = 18$, $7 + 7 = 14$, and $18 > 14$.

214. $[(0,2)] < [(3,5)]$

False: $0 + 5 = 2 + 3$

215. $[(4,4)] > [(3,4)]$

True: $4 + 4 > 4 + 3$

Let $f: \mathbb{R} \to \mathbb{R}$ be defined by $f(x) = x^4$. Compute $f[A]$ for each of the following sets A:

216. $A = \mathbb{R}$

$f[A] = f[\mathbb{R}] = [\mathbf{0}, \infty)$

217. $A = \{-2, 0, 3\}$

$f[A] = f[\{-2,0,3\}] = \{\mathbf{0}, \mathbf{16}, \mathbf{81}\}$

218. $A = (-3, 2]$

$f[A] = f[(-3,2]] = [\mathbf{0}, \mathbf{81})$

219. $A = (4, \infty)$

$f[A] = f[(4, \infty)] = (\mathbf{256}, \infty)$

220. $A = (-\infty, 1]$

$f[A] = f\big[(-\infty, 1]\big] = [\mathbf{0}, \infty)$

Let $f: \mathbb{R} \to \mathbb{R}$ be defined by $f(x) = x^4$. Compute $f^{-1}[B]$ for each of the following sets B:

221. $B = \{16\}$

$f^{-1}[B] = f^{-1}[\{16\}] = \{-\mathbf{2}, \mathbf{2}\}$

222. $B = \{-1\}$

$f^{-1}[B] = f^{-1}[\{-1\}] = \emptyset$

223. $B = [0, \infty)$

$f^{-1}[B] = f^{-1}\big[[0, \infty)\big] = \mathbb{R}$

224. $B = (-\infty, 0)$

$f^{-1}[B] = f^{-1}[(-\infty, 0)] = \emptyset$

225. $B = (0, \infty)$

$f^{-1}[B] = f^{-1}[(0, \infty)] = (-\infty, 0) \cup (0, \infty)$

For each of the following sets X and Y, compute ${}^XY$:

226. $X = \emptyset$, $Y = \emptyset$

${}^XY = \emptyset$

227. $X = \{a\}$, $Y = \{b\}$

${}^XY = \big\{\{(\boldsymbol{a}, \boldsymbol{b})\}\big\}$

228. $X = \{a, b\}$, $Y = \{c\}$

${}^XY = \big\{\{(\boldsymbol{a}, \boldsymbol{c}), (\boldsymbol{b}, \boldsymbol{c})\}\big\}$

229. $X = \{a, b\}$, $Y = \{a\}$

${}^XY = \big\{\{(\boldsymbol{a}, \boldsymbol{a}), (\boldsymbol{b}, \boldsymbol{a})\}\big\}$

230. $X = \{a\}, Y = \{0, 1\}$

$^{X}Y = \{\{(\boldsymbol{a}, \mathbf{0})\}, \{(\boldsymbol{a}, \mathbf{1})\}\}$

231. $X = \{a, b\}, Y = \{a, b, c\}$

$^{X}Y = \{\{(\boldsymbol{a}, \boldsymbol{a}), (\boldsymbol{b}, \boldsymbol{a})\}, \{(\boldsymbol{a}, \boldsymbol{a}), (\boldsymbol{b}, \boldsymbol{b})\}, \{(\boldsymbol{a}, \boldsymbol{a}), (\boldsymbol{b}, \boldsymbol{c})\}, \{(\boldsymbol{a}, \boldsymbol{b}), (\boldsymbol{b}, \boldsymbol{a})\}, \{(\boldsymbol{a}, \boldsymbol{b}), (\boldsymbol{b}, \boldsymbol{b})\}, \{(\boldsymbol{a}, \boldsymbol{b}), (\boldsymbol{b}, \boldsymbol{c})\}, \{(\boldsymbol{a}, \boldsymbol{c}), (\boldsymbol{b}, \boldsymbol{a})\}, \{(\boldsymbol{a}, \boldsymbol{c}), (\boldsymbol{b}, \boldsymbol{b})\}, \{(\boldsymbol{a}, \boldsymbol{c}), (\boldsymbol{b}, \boldsymbol{c})\}\}$

232. $X = \{a, b, c\}, Y = \{0, 1\}$

$^{X}Y = \{\{(\boldsymbol{a}, \mathbf{0}), (\boldsymbol{b}, \mathbf{0}), (\boldsymbol{c}, \mathbf{0})\}, \{(\boldsymbol{a}, \mathbf{0}), (\boldsymbol{b}, \mathbf{0}), (\boldsymbol{c}, \mathbf{1})\}, \{(\boldsymbol{a}, \mathbf{0}), (\boldsymbol{b}, \mathbf{1}), (\boldsymbol{c}, \mathbf{0})\}, \{(\boldsymbol{a}, \mathbf{0}), (\boldsymbol{b}, \mathbf{1}), (\boldsymbol{c}, \mathbf{1})\}$
$\{(\boldsymbol{a}, \mathbf{1}), (\boldsymbol{b}, \mathbf{0}), (\boldsymbol{c}, \mathbf{0})\}, \{(\boldsymbol{a}, \mathbf{1}), (\boldsymbol{b}, \mathbf{0}), (\boldsymbol{c}, \mathbf{1})\}, \{(\boldsymbol{a}, \mathbf{1}), (\boldsymbol{b}, \mathbf{1}), (\boldsymbol{c}, \mathbf{0})\}, \{(\boldsymbol{a}, \mathbf{1}), (\boldsymbol{b}, \mathbf{1}), (\boldsymbol{c}, \mathbf{1})\}\}$

233. $X = \{a, b, c\}, Y = \{0, 1, 2\}$

$^{X}Y = \{\{(\boldsymbol{a}, \mathbf{0}), (\boldsymbol{b}, \mathbf{0}), (\boldsymbol{c}, \mathbf{0})\}, \{(\boldsymbol{a}, \mathbf{0}), (\boldsymbol{b}, \mathbf{0}), (\boldsymbol{c}, \mathbf{1})\}, \{(\boldsymbol{a}, \mathbf{0}), (\boldsymbol{b}, \mathbf{0}), (\boldsymbol{c}, \mathbf{2})\}, \{(\boldsymbol{a}, \mathbf{0}), (\boldsymbol{b}, \mathbf{1}), (\boldsymbol{c}, \mathbf{0})\},$
$\{(\boldsymbol{a}, \mathbf{0}), (\boldsymbol{b}, \mathbf{1}), (\boldsymbol{c}, \mathbf{1})\}, \{(\boldsymbol{a}, \mathbf{0}), (\boldsymbol{b}, \mathbf{1}), (\boldsymbol{c}, \mathbf{2})\}, \{(\boldsymbol{a}, \mathbf{0}), (\boldsymbol{b}, \mathbf{2}), (\boldsymbol{c}, \mathbf{0})\}, \{(\boldsymbol{a}, \mathbf{0}), (\boldsymbol{b}, \mathbf{2}), (\boldsymbol{c}, \mathbf{1})\},$
$\{(\boldsymbol{a}, \mathbf{0}), (\boldsymbol{b}, \mathbf{2}), (\boldsymbol{c}, \mathbf{2})\}, \{(\boldsymbol{a}, \mathbf{1}), (\boldsymbol{b}, \mathbf{0}), (\boldsymbol{c}, \mathbf{0})\}, \{(\boldsymbol{a}, \mathbf{1}), (\boldsymbol{b}, \mathbf{0}), (\boldsymbol{c}, \mathbf{1})\}, \{(\boldsymbol{a}, \mathbf{1}), (\boldsymbol{b}, \mathbf{0}), (\boldsymbol{c}, \mathbf{2})\},$
$\{(\boldsymbol{a}, \mathbf{1}), (\boldsymbol{b}, \mathbf{1}), (\boldsymbol{c}, \mathbf{0})\}, \{(\boldsymbol{a}, \mathbf{1}), (\boldsymbol{b}, \mathbf{1}), (\boldsymbol{c}, \mathbf{1})\}, \{(\boldsymbol{a}, \mathbf{1}), (\boldsymbol{b}, \mathbf{1}), (\boldsymbol{c}, \mathbf{2})\}, \{(\boldsymbol{a}, \mathbf{1}), (\boldsymbol{b}, \mathbf{2}), (\boldsymbol{c}, \mathbf{0})\},$
$\{(\boldsymbol{a}, \mathbf{1}), (\boldsymbol{b}, \mathbf{2}), (\boldsymbol{c}, \mathbf{1})\}, \{(\boldsymbol{a}, \mathbf{1}), (\boldsymbol{b}, \mathbf{2}), (\boldsymbol{c}, \mathbf{2})\}, \{(\boldsymbol{a}, \mathbf{2}), (\boldsymbol{b}, \mathbf{0}), (\boldsymbol{c}, \mathbf{0})\}, \{(\boldsymbol{a}, \mathbf{2}), (\boldsymbol{b}, \mathbf{0}), (\boldsymbol{c}, \mathbf{1})\},$
$\{(\boldsymbol{a}, \mathbf{2}), (\boldsymbol{b}, \mathbf{0}), (\boldsymbol{c}, \mathbf{2})\}, \{(\boldsymbol{a}, \mathbf{2}), (\boldsymbol{b}, \mathbf{1}), (\boldsymbol{c}, \mathbf{0})\}, \{(\boldsymbol{a}, \mathbf{2}), (\boldsymbol{b}, \mathbf{1}), (\boldsymbol{c}, \mathbf{1})\}, \{(\boldsymbol{a}, \mathbf{2}), (\boldsymbol{b}, \mathbf{1}), (\boldsymbol{c}, \mathbf{2})\},$
$\{(\boldsymbol{a}, \mathbf{2}), (\boldsymbol{b}, \mathbf{2}), (\boldsymbol{c}, \mathbf{0})\}, \{(\boldsymbol{a}, \mathbf{2}), (\boldsymbol{b}, \mathbf{2}), (\boldsymbol{c}, \mathbf{1})\}, \{(\boldsymbol{a}, \mathbf{2}), (\boldsymbol{b}, \mathbf{2}), (\boldsymbol{c}, \mathbf{2})\}\}$

Determine if each of the following sequences is a Cauchy sequence. Are any of the Cauchy sequences equivalent?

234. $(x_n) = \left(1 + \frac{1}{n+1}\right)$

Cauchy sequence

235. $(y_n) = (2^n)$

Not a Cauchy sequence

236. $(z_n) = \left(1 - \frac{1}{2n+1}\right)$

Cauchy sequence

The Cauchy sequences (x_n) in Problem 234 and (z_n) in Problem 236 are equivalent.

Let $f\colon \mathbb{N} \to \mathbb{Z}$ be defined by $f(n) = \begin{cases} \frac{n}{2} & \text{if } n \text{ is even.} \\ -\frac{n+1}{2} & \text{if } n \text{ is odd.} \end{cases}$

237. Prove that f is injective.

Proof: Let $n, m \in \mathbb{N}$ with $f(n) = f(m)$. If n and m are both even, we have $\frac{n}{2} = \frac{m}{2}$, and so, $2 \cdot \frac{n}{2} = 2 \cdot \frac{m}{2}$. Thus, $n = m$. If n and m are both odd, we have $-\frac{n+1}{2} = -\frac{m+1}{2}$, and so, $\frac{n+1}{2} = \frac{m+1}{2}$. Therefore, $2 \cdot \frac{n+1}{2} = 2 \cdot \frac{m+1}{2}$. So, $n + 1 = m + 1$, and therefore, $n = m$. If n is even and m is odd, then we have $\frac{n}{2} = -\frac{m+1}{2}$. So, $2 \cdot \frac{n}{2} = 2\left(-\frac{m+1}{2}\right)$. Therefore, $n = -(m + 1)$. Since $m \in \mathbb{N}$, $m \geq 0$. So, $m + 1 \geq 1$. Therefore, $n = -(m + 1) \leq -1$, which is impossible because natural numbers cannot be negative. So, it is not possible for n to be even, m to be odd, and $f(n) = f(m)$. Similarly, we cannot have n odd and m even. So, f is injective. □

238. Prove that f is surjective.

Proof: Let $k \in \mathbb{Z}$. If $k \geq 0$, then $2k \in \mathbb{N}$ and $f(2k) = \frac{2k}{2} = k$. If $k < 0$, then $-2k > 0$, and so, $-2k - 1 \in \mathbb{N}$. Then $f(-2k - 1) = -\frac{(-2k-1)+1}{2} = -\frac{-2k}{2} = k$. So, f is surjective. □

239. Prove that f is bijective.

Proof: By Problem 237, f is injective. By Problem 238, f is surjective. So, f is bijective. □

Note: It should really be checked that whenever $n \in \mathbb{N}$, $f(n) \in \mathbb{Z}$. To see this, first observe that if n is even, then there is $k \in \mathbb{N}$ with $n = 2k$, and so, $\frac{n}{2} = \frac{2k}{2} = k \in \mathbb{Z}$. Now, if n is odd, there is $k \in \mathbb{N}$ with $n = 2k + 1$, and so, $-\frac{n+1}{2} = -\frac{(2k+1)+1}{2} = -\frac{2k+2}{2} = -\frac{2(k+1)}{2} = -(k + 1) \in \mathbb{Z}$. So, f does in fact take each natural number to an integer.

For $f, g \in {}^{\mathbb{R}}\mathbb{R}$, define $f \preccurlyeq g$ if and only if for all $x \in \mathbb{R}$, $f(x) \leq g(x)$.

240. Is $\preccurlyeq$ a partial ordering on ${}^{\mathbb{R}}\mathbb{R}$?

Yes

If $f \in {}^{\mathbb{R}}\mathbb{R}$, then for all $x \in \mathbb{R}$, $f(x) = f(x)$. So, $f \preccurlyeq f$, and therefore, $\preccurlyeq$ is reflexive.

Let $f, g \in {}^{\mathbb{R}}\mathbb{R}$ with $f \preccurlyeq g$ and $g \preccurlyeq f$. Then for all $x \in \mathbb{R}$, $f(x) \leq g(x)$ and $g(x) \leq f(x)$. So, $f = g$, and therefore, $\preccurlyeq$ is antisymmetric.

Let $f, g, h \in {}^{\mathbb{R}}\mathbb{R}$ with $f \preccurlyeq g$ and $g \preccurlyeq h$. Then for all $x \in \mathbb{R}$, $f(x) \leq g(x)$ and $g(x) \leq h(x)$. So, by the transitivity of $\leq$, for all $x \in \mathbb{R}$, $f(x) \leq h(x)$. Thus, $f \preccurlyeq h$, and therefore, $\preccurlyeq$ is transitive.

Since $\preccurlyeq$ is reflexive, antisymmetric, and transitive on ${}^{\mathbb{R}}\mathbb{R}$, $({}^{\mathbb{R}}\mathbb{R}, \preccurlyeq)$ is a poset.

241. Is $\preccurlyeq$ a linear ordering on ${}^{\mathbb{R}}\mathbb{R}$?

No

We will show that $\preccurlyeq$ does not satisfy the comparability condition on ${}^{\mathbb{R}}\mathbb{R}$.

Let $f(x)=x$ and $g(x)=x^2$. Then $f(2)=2$ and $g(2)=4$. So, $f(2)<g(2)$. Therefore, $g \not\preccurlyeq f$. We also have $f\left(\frac{1}{2}\right)=\frac{1}{2}$ and $g\left(\frac{1}{2}\right)=\frac{1}{4}$. So, $g\left(\frac{1}{2}\right)<f\left(\frac{1}{2}\right)$. Therefore, $f \not\preccurlyeq g$. So, f and g are incomparible with respect to $\preccurlyeq$. Therefore, $\preccurlyeq$ is **not** a linear ordering on ${}^{\mathbb{R}}\mathbb{R}$.

For $f,g \in {}^{\mathbb{R}}\mathbb{R}$, define $f \preccurlyeq^* g$ if and only if there is an $x \in \mathbb{R}$ such that $f(x) \leq g(x)$.

242. Is $\preccurlyeq^*$ a partial ordering on ${}^{\mathbb{R}}\mathbb{R}$?

No because $\preccurlyeq^*$ is not antisymmetric or transitive on ${}^{\mathbb{R}}\mathbb{R}$. To see that $\preccurlyeq^*$ is not antisymmetric on ${}^{\mathbb{R}}\mathbb{R}$, observe that $f(x)=x$ and $g(x)=x^2$ satisfy $f \preccurlyeq^* g$ and $g \preccurlyeq^* f$, but $f \neq g$. To see that $\preccurlyeq^*$ is not transitive on ${}^{\mathbb{R}}\mathbb{R}$, let $f(x)=\begin{cases}0 & \text{if } x \neq 1\\ 5 & \text{if } x=1\end{cases}$, $g(x)=1$, and $h(x)=\begin{cases}-1 & \text{if } x \neq 1\\ 2 & \text{if } x=1.\end{cases}$ Then $f \preccurlyeq^* g$ and $g \preccurlyeq^* h$, but $f \not\preccurlyeq^* h$.

Note: We only needed to show that $\preccurlyeq^*$ is not antisymmetric on ${}^{\mathbb{R}}\mathbb{R}$ **or** $\preccurlyeq^*$ is not transitive on ${}^{\mathbb{R}}\mathbb{R}$ to verify that $\preccurlyeq^*$ is not a partial ordering on ${}^{\mathbb{R}}\mathbb{R}$.

243. Is $\preccurlyeq^*$ a linear ordering on ${}^{\mathbb{R}}\mathbb{R}$?

No (because $\preccurlyeq^*$ is not a partial ordering on ${}^{\mathbb{R}}\mathbb{R}$).

Let A, B, C, and D be sets. Prove each of the following.

244. $\preccurlyeq$ is transitive.

Proof: Suppose that $A \preccurlyeq B$ and $B \preccurlyeq C$. Then there are functions $f\colon A \hookrightarrow B$ and $g\colon B \hookrightarrow C$. By Theorem 2.57, $g \circ f\colon A \hookrightarrow C$. So, $A \preccurlyeq C$. Therefore, $\preccurlyeq$ is transitive. □

245. $\prec$ is transitive.

Proof: Suppose that $A \prec B$ and $B \prec C$. Then $A \preccurlyeq B$ and $B \preccurlyeq C$. By Problem 244, $A \preccurlyeq C$. Assume toward contradiction that $A \sim C$. Since $\sim$ is symmetric, $C \sim A$. In particular, $C \preccurlyeq A$. Since $C \preccurlyeq A$ and $A \preccurlyeq B$, by Problem 244, $C \preccurlyeq B$. Since $B \preccurlyeq C$ and $C \preccurlyeq B$, by the Cantor-Schroeder-Bernstein Theorem, $B \sim C$, contradicting $B \prec C$. It follows that $A \nsim C$, and thus, $A \prec C$. □

246. If $A \preccurlyeq B$ and $B \prec C$, then $A \prec C$.

Proof: Suppose that $A \preccurlyeq B$ and $B \prec C$. Then $B \preccurlyeq C$. By Problem 244, $A \preccurlyeq C$. Assume toward contradiction that $A \sim C$. The rest of the argument is the same as in the solution to Problem 245. □

247. If $A \prec B$ and $B \preccurlyeq C$, then $A \prec C$.

Proof: Suppose that $A \prec B$ and $B \preccurlyeq C$. Then $A \preccurlyeq B$. By Problem 244, $A \preccurlyeq C$. Assume toward contradiction that $A \sim C$. Since $\sim$ is symmetric, $C \sim A$. In particular, $C \preccurlyeq A$. Since $B \preccurlyeq C$ and $C \preccurlyeq A$, by Problem 244, $B \preccurlyeq A$. Since $A \preccurlyeq B$ and $B \preccurlyeq A$, by the Cantor-Schroeder-Bernstein Theorem, $A \sim B$, contradicting $A \prec B$. It follows that $A \nsim C$, and thus, $A \prec C$. □

248. If $A \sim B$, then $\mathcal{P}(A) \sim \mathcal{P}(B)$.

Proof: Suppose that $A \sim B$. Then there exists a bijection $h: A \to B$. Define $F: \mathcal{P}(A) \to \mathcal{P}(B)$ by $F(X) = \{h(a) \mid a \in X\}$ for each $X \in \mathcal{P}(A)$.

Suppose $X, Y \in \mathcal{P}(A)$ with $F(X) = F(Y)$. Let $a \in X$. Then $h(a) \in F(X)$. Since $F(X) = F(Y)$, $h(a) \in F(Y)$. So, there is $b \in Y$ such that $h(a) = h(b)$. Since h is injective, $a = b$. So, $a \in Y$. Since $a \in X$ was arbitrary, $X \subseteq Y$. By a symmetrical argument, $Y \subseteq X$. Therefore, $X = Y$. Since $X, Y \in \mathcal{P}(A)$ were arbitrary, F is injective.

Let $Y \in \mathcal{P}(B)$ and let $X = \{a \in A \mid h(a) \in Y\}$. Then $b \in F(X)$ if and only if $b = h(a)$ for some $a \in X$ if and only if $b \in Y$ (because h is surjective). So, $F(X) = Y$. Since $Y \in \mathcal{P}(B)$ was arbitrary, F is surjective.

Since F is injective and surjective, $\mathcal{P}(A) \sim \mathcal{P}(B)$. □

249. If $A \subseteq B$, then $\mathcal{P}(A) \preccurlyeq \mathcal{P}(B)$.

Proof: Suppose that $A \subseteq B$. We show that $\mathcal{P}(A) \subseteq \mathcal{P}(B)$. Let $X \in \mathcal{P}(A)$. Then $X \subseteq A$. Since $X \subseteq A$ and $A \subseteq B$, and $\subseteq$ is transitive (Theorem 2.15), we have $X \subseteq B$. Therefore, $X \in \mathcal{P}(B)$. Since X was an arbitrary element of $\mathcal{P}(A)$, we have shown that every element of $\mathcal{P}(A)$ is an element of $\mathcal{P}(B)$. Therefore, $\mathcal{P}(A) \subseteq \mathcal{P}(B)$. By Note 1 following Example 2.72, $\mathcal{P}(A) \preccurlyeq \mathcal{P}(B)$. □

250. If $A \sim B$ and $C \sim D$, then $A \times C \sim B \times D$.

Proof: Suppose that $A \sim B$ and $C \sim D$. Then there exist bijections $h: A \to B$ and $k: C \to D$. Define $f: A \times C \to B \times D$ by $f(a, c) = \big(h(a), k(c)\big)$.

Suppose $(a, c), (a', c') \in A \times C$ with $f\big((a, c)\big) = f\big((a', c')\big)$. Then $\big(h(a), k(c)\big) = \big(h(a'), k(c')\big)$. So, $h(a) = h(a')$ and $k(c) = k(c')$. Since h is an injection, $a = a'$. Since k is an injection, $c = c'$. Since $a = a'$ and $c = c'$, $(a, c) = (a', c')$. Since $(a, c), (a', c') \in A \times C$ were arbitrary, f is an injection.

Now, let $(b, d) \in B \times D$. Since h and k are bijections, h^{-1} and k^{-1} exist. Let $a = h^{-1}(b)$, $c = k^{-1}(d)$. Then $f(a, c) = \big(h(a), k(c)\big) = \Big(h\big(h^{-1}(b)\big), k\big(k^{-1}(d)\big)\Big) = (b, d)$. Since $(b, d) \in B \times D$ was arbitrary, f is a surjection.

Since f is both an injection and a surjection, $A \times C \sim B \times D$. □

251. There is a partition $\boldsymbol{P}$ of $\mathbb{N}$ such that $\boldsymbol{P} \sim \mathbb{N}$ and for each $X \in \boldsymbol{P}$, $X \sim \mathbb{N}$.

Proof: For each $n \in \mathbb{N}$, let P_n be the set of natural numbers ending with exactly n zeros and let $\boldsymbol{P} = \{P_n \mid n \in \mathbb{N}\}$. For example, $5231 \in P_0$, $0 \in P_1$, and $26{,}200 \in P_2$. Let's define $\widetilde{m, n}$ to be the natural number consisting of m 1's followed by n 0's. For example, $\widetilde{3, 0} = 111$ and $\widetilde{2, 5} = 1{,}100{,}000$. For each $n \in \mathbb{N}$, $\{\widetilde{m, n} \mid m \in \mathbb{N}\} \subseteq P_n$ showing that each P_n is equinumerous to $\mathbb{N}$. Also, if $k \in P_n \cap P_m$, then k ends with exactly n zeros and exactly m zeros, and so, $n = m$. Therefore, $\boldsymbol{P}$ is pairwise disjoint. This also shows that the function $f: \mathbb{N} \to \boldsymbol{P}$ defined by $f(n) = P_n$ is a bijection. So, $\boldsymbol{P} \sim \mathbb{N}$. Finally, if $k \in \mathbb{N}$, then there is $n \in \mathbb{N}$ such that k ends with exactly n zeros. So, $\bigcup \boldsymbol{P} = \mathbb{N}$. □

252. A countable union of countable sets is countable.

Proof: For each $n \in \mathbb{N}$, let A_n be a countable set. By replacing each A_n by $A_n \times \{n\}$, we can assume that $\{A_n \mid n \in \mathbb{N}\}$ is a pairwise disjoint collection of sets ($A_n \sim A_n \times \{n\}$ via the bijection f sending x to (x, n)). By Problem 251, there is a partition $\boldsymbol{P}$ of $\mathbb{N}$ such that $\boldsymbol{P} \sim \mathbb{N}$ and for each $X \in \boldsymbol{P}$, $X \sim \mathbb{N}$. Let's say $\boldsymbol{P} = \{P_n \mid n \in \mathbb{N}\}$. Since each A_n is countable, for each $n \in \mathbb{N}$ there are injective functions $f_n: A_n \to P_n$. Define $f: \bigcup\{A_n \mid n \in \mathbb{N}\} \to \mathbb{N}$ by $f(x) = f_n(x)$ if $x \in A_n$.

Since $\{A_n \mid n \in \mathbb{N}\}$ is pairwise disjoint, f is well-defined. Now, suppose that $x, y \in \bigcup\{A_n \mid n \in \mathbb{N}\}$ with $f(x) = f(y)$. There exist $n, m \in \mathbb{N}$ such that $x \in A_n$ and $y \in A_m$. So, $f(x) = f_n(x) \in P_n$ and $f(y) = f_m(y) \in P_m$. Since $f(x) = f(y)$, we have $f_n(x) = f_m(y)$. Since for $n \neq m$, $P_n \cap P_m = \emptyset$, we must have $n = m$. So, we have $f_n(x) = f_n(y)$. Since f_n is injective, $x = y$. Since $x, y \in \bigcup\{A_n \mid n \in \mathbb{N}\}$ were arbitrary, f is an injective function. Therefore, $\bigcup\{A_n \mid n \in \mathbb{N}\}$ is countable. □

253. Every pairwise disjoint set of sets consisting of at least two sets is disjoint.

Proof: Let $\boldsymbol{X}$ be a pairwise disjoint set of sets with at least two sets, say $A, B \in \boldsymbol{X}$. Suppose towards contradiction that $x \in \bigcap \boldsymbol{X}$. Then $x \in A$ and $x \in B$. So, $x \in A \cap B$. But $A \cap B = \emptyset$ because $\boldsymbol{X}$ is pairwise disjoint. This contradiction shows that the statement $x \in \bigcap \boldsymbol{X}$ is false. Therefore, $\boldsymbol{X}$ is disjoint. □

LEVEL 4

Explicitly write down each of the following natural numbers using only set brackets and the empty set.

254. 5

$$5 = \{0, 1, 2, 3, 4\} = \Big\{\emptyset, \{\emptyset\}, \{\emptyset, \{\emptyset\}\}, \big\{\emptyset, \{\emptyset\}, \{\emptyset, \{\emptyset\}\}\big\}, \big\{\emptyset, \{\emptyset\}, \{\emptyset, \{\emptyset\}\}, \big\{\emptyset, \{\emptyset\}, \{\emptyset, \{\emptyset\}\}\big\}\big\}\Big\}$$

255. 6

$$6 = \{0, 1, 2, 3, 4, 5\} =$$

$$\Big\{\emptyset, \{\emptyset\}, \{\emptyset, \{\emptyset\}\}, \big\{\emptyset, \{\emptyset\}, \{\emptyset, \{\emptyset\}\}\big\}, \big\{\emptyset, \{\emptyset\}, \{\emptyset, \{\emptyset\}\}, \big\{\emptyset, \{\emptyset\}, \{\emptyset, \{\emptyset\}\}\big\}\big\}, \Big\{\emptyset, \{\emptyset\}, \{\emptyset, \{\emptyset\}\}, \big\{\emptyset, \{\emptyset\}, \{\emptyset, \{\emptyset\}\}\big\}, \big\{\emptyset, \{\emptyset\}, \{\emptyset, \{\emptyset\}\}, \big\{\emptyset, \{\emptyset\}, \{\emptyset, \{\emptyset\}\}\big\}\big\}\Big\}\Big\}$$

For each of the following, determine if $\boldsymbol{X}$ is disjoint, pairwise disjoint, both, or neither.

256. $\{(n, n+1] \mid n \in \mathbb{Z}\}$

Both

257. $\left\{\left(\frac{1}{n+1}, \frac{1}{n}\right) \mid n \in \mathbb{Z}^+\right\}$

Both

258. $\{\mathbb{Q}\}$

Pairwise disjoint, but not disjoint

259. $\{\mathbb{Q}^-, 2\mathbb{N}, \{n \in \mathbb{N} \mid n \text{ is a prime number greater than } 2\}\}$

Both

Let A, B, C, and D be sets. Determine if each of the following statements is true or false. If true, provide a proof. If false, provide a counterexample.

260. $B \subseteq A$ if and only if $A \cap B = B$.

This is **true**.

Proof: Suppose that $B \subseteq A$. First let $x \in A \cap B$. Then $x \in A$ and $x \in B$. In particular, $x \in B$. Since x was an arbitrary element of $A \cap B$, we have shown that every element of $A \cap B$ is an element of B. That is, $\forall x(x \in A \cap B \rightarrow x \in B)$ is true. Therefore, $A \cap B \subseteq B$. Next, let $x \in B$. Since $B \subseteq A$, $x \in A$. Therefore, $x \in A$ and $x \in B$. So, $x \in A \cap B$. Since x was an arbitrary element of B, we have shown that every element of B is an element of $A \cap B$. That is, $\forall x(x \in B \rightarrow x \in A \cap B)$. Therefore, $B \subseteq A \cap B$. Since $A \cap B \subseteq B$ and $B \subseteq A \cap B$, it follows that $A \cap B = B$.

Now, suppose that $A \cap B = B$ and let $x \in B$. Then $x \in A \cap B$. So, $x \in A$ and $x \in B$. In particular, $x \in A$. Since x was an arbitrary element of B, we have shown that every element of B is an element of A. That is, $\forall x(x \in B \rightarrow x \in A)$. Therefore, $B \subseteq A$. □

261. $A \cap (B \cup C) = (A \cap B) \cup (A \cap C)$.

This is **true**.

Proof: $x \in A \cap (B \cup C) \Leftrightarrow x \in A$ and $x \in B \cup C \Leftrightarrow x \in A$ and either $x \in B$ or $x \in C \Leftrightarrow x \in A$ and $x \in B$ or $x \in A$ and $x \in C \Leftrightarrow x \in A \cap B$ or $x \in A \cap C \Leftrightarrow x \in (A \cap B) \cup (A \cap C)$. □

262. $A \cup (B \cap C) = (A \cup B) \cap (A \cup C)$.

This is **true**.

Proof: $x \in A \cup (B \cap C) \Leftrightarrow x \in A$ or $x \in B \cap C \Leftrightarrow$ either $x \in A$ or we have both $x \in B$ and $x \in C \Leftrightarrow$ we have both $x \in A$ or $x \in B$ and $x \in A$ or $x \in C \Leftrightarrow x \in A \cup B$ and $x \in A \cup C$ $\Leftrightarrow x \in (A \cup B) \cap (A \cup C)$. □

263. $C \setminus (A \cup B) = (C \setminus A) \cap (C \setminus B)$.

This is **true**.

Proof: $x \in C \setminus (A \cup B) \Leftrightarrow x \in C$ and $x \notin A \cup B \Leftrightarrow x \in C$ and $x \notin A$ and $x \notin B \Leftrightarrow x \in C$ and $x \notin A$ and $x \in C$ and $x \notin B \Leftrightarrow x \in C \setminus A$ and $x \in C \setminus B \Leftrightarrow x \in (C \setminus A) \cap (C \setminus B)$. □

264. $C \setminus (A \cap B) = (C \setminus A) \cup (C \setminus B)$.

This is **true**.

Proof: $x \in C \setminus (A \cap B) \Leftrightarrow x \in C$ and $x \notin A \cap B \Leftrightarrow x \in C$ and $x \notin A$ or $x \notin B \Leftrightarrow x \in C$ and $x \notin A$ or $x \in C$ and $x \notin B \Leftrightarrow x \in C \setminus A$ or $x \in C \setminus B \Leftrightarrow x \in (C \setminus A) \cup (C \setminus B)$. □

265. $(A \times B) \cap (C \times D) = (A \cap C) \times (B \cap D)$.

This is **true**.

Proof: $(x, y) \in (A \times B) \cap (C \times D)$ if and only if $(x, y) \in A \times B$ and $(x, y) \in C \times D$ if and only if $x \in A$, $y \in B$, $x \in C$, and $y \in D$ if and only if $x \in A \cap C$ and $y \in B \cap D$ if and only if $(x, y) \in (A \cap C) \times (B \cap D)$. Therefore, $(A \times B) \cap (C \times D) = (A \cap C) \times (B \cap D)$. □

266. $(A \times B) \cup (C \times D) = (A \cup C) \times (B \cup D)$.

This is **false**. If $A = \{0\}, B = \{1\}, C = \{2\}, D = \{3\}$, then $A \times B = \{(0,1)\}$, $C \times D = \{(2,3)\}$, and so, $(A \times B) \cup (C \times D) = \{(0,1),(2,3)\}$. Also, $A \cup C = \{0,2\}$, $B \cup D = \{1,3\}$, and therefore, $(A \cup C) \times (B \cup D) = \{(0,1),(0,3),(2,1),(2,3)\}$. Since $(2,1) \in (A \cup C) \times (B \cup D)$, but $(2,1) \notin (A \times B) \cup (C \times D)$, we see that $(A \times B) \cup (C \times D) \neq (A \cup C) \times (B \cup D)$.

267. If $A \subseteq B$ and $C \subseteq D$, then $A \times C \subseteq B \times D$.

This is **true**.

Proof: Let A, B, C, and D be sets with $A \subseteq B$ and $C \subseteq D$ and let $(x, y) \in A \times C$. Then $x \in A$ and $y \in C$. Since $x \in A$ and $A \subseteq B$, $x \in B$. Since $y \in C$ and $C \subseteq D$, $y \in D$. Therefore, $(x, y) \in B \times D$. Since $(x, y) \in A \times C$ was arbitrary, $A \times C \subseteq B \times D$. □

Verify that each of the following is true.

268. If $x = u$, $y = v$, and $z = w$, then $(x, y, z) = (u, v, w)$.

If $x = u$, $y = v$, and $z = w$, then $(x, y) = (u, v)$ and $z = w$. So, $\big((x, y), z\big) = \big((u, v), w\big)$. Therefore, $(x, y, z) = \big((x, y), z\big) = \big((u, v), w\big) = (u, v, w)$.

269. If $(x, y, z) = (u, v, w)$, then $x = u$, $y = v$, and $z = w$.

If $(x, y, z) = (u, v, w)$, then $\big((x, y), z\big) = \big((u, v), w\big)$. By Theorem 2.27, $(x, y) = (u, v)$ and $z = w$. Again by Theorem 2.27, $x = u$ and $y = v$.

270. If $x_1 = y_1$, $x_2 = y_2$, $x_3 = y_3$, and $x_4 = y_4$, then $(x_1, x_2, x_3, x_4) = (y_1, y_2, y_3, y_4)$.

If $x_1 = y_1, x_2 = y_2, x_3 = y_3$, and $x_4 = y_4$, then by Problem 268, $(x_1, x_2, x_3) = (y_1, y_2, y_3)$. So, we have $(x_1, x_2, x_3, x_4) = \big((x_1, x_2, x_3), x_4\big) = \big((y_1, y_2, y_3), y_4\big) = (y_1, y_2, y_3, y_4)$.

271. If $(x_1, x_2, x_3, x_4) = (y_1, y_2, y_3, y_4)$, then $x_1 = y_1$, $x_2 = y_2$, $x_3 = y_3$, and $x_4 = y_4$.

If $(x_1, x_2, x_3, x_4) = (y_1, y_2, y_3, y_4)$, then $\big((x_1, x_2, x_3), x_4\big) = \big((y_1, y_2, y_3), y_4\big)$. By Theorem 2.27, we have $(x_1, x_2, x_3) = (y_1, y_2, y_3)$ and $x_4 = y_4$. By Problem 269, $x_1 = y_1$, $x_2 = y_2$, and $x_3 = y_3$.

For $n, m \in \mathbb{Z}$, we say that n divides m, written $n|m$, if there is an integer k such that $m = nk$. For $n \in \mathbb{Z}^+$, define $\equiv_n$ on $\mathbb{Z}$ by $\equiv_n = \{(a, b) \in \mathbb{Z}^2 \mid n|b - a\}$.

272. Prove that $\equiv_n$ is reflexive on $\mathbb{Z}$.

Proof: Let $a \in \mathbb{Z}$. Then $a - a = 0 = n \cdot 0$. So, $n|a - a$. Therefore, $a \equiv_n a$, and so, $\equiv_n$ is reflexive. □

273. Prove that $\equiv_n$ is symmetric on $\mathbb{Z}$.

Proof: Let $a, b \in \mathbb{Z}$ and suppose that $a \equiv_n b$. Then $n|b - a$. So, there is $k \in \mathbb{Z}$ such that $b - a = nk$. Thus $-(b - a) = -nk = n(-k)$. Since $k \in \mathbb{Z}$, $-k \in \mathbb{Z}$. So, $n|a - b$, and therefore, $b \equiv_n a$. So, $\equiv_n$ is symmetric. □

274. Prove that $\equiv_n$ is transitive on $\mathbb{Z}$.

Proof: Let $a, b, c \in \mathbb{Z}$ with $a \equiv_n b$ and $b \equiv_n c$. Then $n|b - a$ and $n|c - b$. So, there are $j, k \in \mathbb{Z}$ such that $b - a = nj$ and $c - b = nk$. So, $c - a = (c - b) + (b - a) = nk + nj = n(k + j)$. Since $j, k \in \mathbb{Z}$, $k + j \in \mathbb{Z}$. Therefore, $n|c - a$. So, $a \equiv_n c$. Thus, $\equiv_n$ is transitive. □

275. Is $\equiv_n$ an equivalence relation on $\mathbb{Z}$?

Yes: Since $\equiv_n$ is reflexive, symmetric, and transitive (by Problems 272 – 272), $\equiv_n$ is an equivalence relation on $\mathbb{Z}$.

Let $\boldsymbol{X}$ be a nonempty set of equivalence relations on a set A. For each of the following statements, either verify that the statement is true or provide a counterexample showing that the statement is false.

276. $\bigcap \boldsymbol{X}$ is reflexive on A.

True: Let $x \in A$ and let $R \in \boldsymbol{X}$. Since R is reflexive, $(x, x) \in R$. Since $R \in \boldsymbol{X}$ was arbitrary, $\forall R \in \boldsymbol{X}\big((x, x) \in R\big)$. So, $(x, x) \in \bigcap \boldsymbol{X}$. Since $x \in A$ was arbitrary, $\bigcap \boldsymbol{X}$ is reflexive.

277. $\bigcup \boldsymbol{X}$ is reflexive on A.

True: Let $x \in A$ and let $R \in \boldsymbol{X}$ be arbitrary. Since R is reflexive, $(x, x) \in R$. So, $\exists R \in \boldsymbol{X}\big((x, x) \in R\big)$. Therefore, $(x, x) \in \bigcup \boldsymbol{X}$. Since $x \in A$ was arbitrary, $\bigcup \boldsymbol{X}$ is reflexive.

278. $\bigcap \boldsymbol{X}$ is symmetric on A.

True: Let $(x, y) \in \bigcap \boldsymbol{X}$ and let $R \in \boldsymbol{X}$. Then $(x, y) \in R$. Since R is an equivalence relation, R is symmetric. Therefore, $(y, x) \in R$. Since $R \in \boldsymbol{X}$ was arbitrary, $\forall R \in \boldsymbol{X}\big((y, x) \in R\big)$. So, $(y, x) \in \bigcap \boldsymbol{X}$. Since $(x, y) \in \bigcap \boldsymbol{X}$ was arbitrary, $\bigcap \boldsymbol{X}$ is symmetric.

279. $\bigcup \boldsymbol{X}$ is symmetric on A.

True: Let $(x,y) \in \bigcup \boldsymbol{X}$. Then there is $R \in \boldsymbol{X}$ with $(x,y) \in R$. Since R is an equivalence relation, R is symmetric. Therefore, $(y,x) \in R$. So, $\exists R \in \boldsymbol{X}\big((y,x) \in R\big)$. Therefore, $(y,x) \in \bigcup \boldsymbol{X}$. Since $(x,y) \in \bigcup \boldsymbol{X}$ was arbitrary, $\bigcup \boldsymbol{X}$ is symmetric.

280. $\bigcap \boldsymbol{X}$ is transitive on A.

True: Let $(x,y),(y,z) \in \bigcap \boldsymbol{X}$ and let $R \in \boldsymbol{X}$. Then $(x,y),(y,z) \in R$. Since R is an equivalence relation, R is transitive. Therefore, $(x,z) \in R$. Since $R \in \boldsymbol{X}$ was arbitrary, $\forall R \in \boldsymbol{X}\big((x,z) \in R\big)$. So, $(x,z) \in \bigcap \boldsymbol{X}$. Since $(x,y),(y,z) \in \bigcap \boldsymbol{X}$ was arbitrary, $\bigcap \boldsymbol{X}$ is transitive.

281. $\bigcup \boldsymbol{X}$ is transitive on A.

False: Let $A = \{0, 1, 2\}$.

Let $R = \{(0,0),(0,1),(1,0),(1,1),(2,2)\}$ and $S = \{(0,0),(1,1),(1,2),(2,1),(2,2)\}$. Then R and S are both equivalence relations.

$R \cup S = \{(0,0),(0,1),(1,0),(1,1),(1,2),(2,1),(2,2)\}$.

We have $(0,1),(1,2) \in R \cup S$, but $(0,2) \notin R \cup S$. So, $R \cup S$ is **not** transitive.

282. $\bigcap \boldsymbol{X}$ is an equivalence relation on A.

True: Since $\bigcap \boldsymbol{X}$ is reflexive, symmetric, and transitive (by Problems 276, 278, and 280), $\bigcap \boldsymbol{X}$ is an equivalence relation.

283. $\bigcup \boldsymbol{X}$ is an equivalence relation on A.

False: By Problem 281, $\bigcup \boldsymbol{X}$ is not necessarily transitive on A.

Let $R = \{(x,y) \in \mathbb{R} \times \mathbb{R} \mid x - y \in \mathbb{Z}\}$.

284. Prove that R is reflexive on $\mathbb{R}$.

Proof: If $x \in \mathbb{R}$, then $x - x = 0 \in \mathbb{Z}$. So, xRx, and therefore, R is reflexive on $\mathbb{R}$. □

285. Prove that R is symmetric on $\mathbb{R}$.

Proof: If xRy, then $x - y \in \mathbb{Z}$. It follows that $y - x = -(x - y) \in \mathbb{Z}$. So, yRx, and therefore, R is symmetric on $\mathbb{R}$. □

286. Prove that R is transitive on $\mathbb{R}$.

Proof: If xRy and yRz, then $x - y \in \mathbb{Z}$ and $y - z \in \mathbb{Z}$. It follows that $x - z = (x - y) + (y - z) \in \mathbb{Z}$ (because the sum of two integers is an integer). So, xRy, and therefore, R is transitive. □

287. Is R an equivalence relation on $\mathbb{R}$?

Yes. Since R is reflexive, symmetric, and transitive (by Problems 284 – 286), R is an equivalence relation.

Let $A = \{(x_n) \mid (x_n) \text{ is a Cauchy sequence of rational numbers}\}$ and define the relation R on A by $(x_n)R(y_n)$ if and only if for every $k \in \mathbb{N}^+$, there is $K \in \mathbb{N}$ such that $n > K$ implies $|x_n - y_n| < \frac{1}{k}$..

288. Prove that R is reflexive on $\mathbb{R}$.

Proof: Let $(x_n) \in A$, let $k \in \mathbb{N}^+$ and let $K = 0$. Then $n > K$ implies $|x_n - x_n| = 0 < \frac{1}{k}$. So, $(x_n)R(x_n)$, and therefore, R is reflexive. □

289. Prove that R is symmetric on $\mathbb{R}$.

Proof: Since $|x_n - y_n| = |y_n - x_n|$, it is clear that R is symmetric. □

290. Prove that R is transitive on $\mathbb{R}$.

Proof: Let $(x_n), (y_n), (z_n) \in A$ with $(x_n)R(y_n)$ and $(y_n)R(z_n)$ and let $k \in \mathbb{N}^+$. Since $(x_n)R(y_n)$, there is $K_1 \in \mathbb{N}$ such that $n > K_1$ implies $|x_n - y_n| < \frac{1}{2k}$. Since $(y_n)R(z_n)$, there is $K_2 \in \mathbb{N}$ such that $n > K_2$ implies $|y_n - z_n| < \frac{1}{2k}$. Let $K = \max\{K_1, K_2\}$. Let $n > K$. Since $K \geq K_1$, $n > K_1$, and therefore, we have $|x_n - y_n| < \frac{1}{2k}$. Since $K \geq K_2, n > K_2$, and therefore, we have $|y_n - z_n| < \frac{1}{2k}$. It follows from the Triangle Inequality that

$$|x_n - z_n| = |x_n - y_n + y_n - z_n| \leq |x_n - y_n| + |y_n - z_n| < \frac{1}{2k} + \frac{1}{2k} = 2 \cdot \frac{1}{2k} = \frac{1}{k}.$$

So, $(x_n)R(z_n)$, and therefore, R is transitive. □

Notes: (1) For the equation $x_n - z_n = x_n - y_n + y_n - z_n$, we used the "**Standard Advanced Calculus Trick**" (SACT – see the analysis after Theorem 5.10). In this case, we want $x_n - y_n$ to appear (as well as $y_n - z_n$). So, we subtract y_n from x_n and then add back y_n to "undo the damage we just did."

(2) Observe that after applying SACT to get $|x_n - z_n| = |x_n - y_n + y_n - z_n|$, we then applied the Triangle Inequality to get $|x_n - y_n + y_n - z_n| \leq |x_n - y_n|$.

291. Is R an equivalence relation on $\mathbb{R}$?

Since R is reflexive, symmetric, and transitive (by Problems 288 – 290), it follows that R is an equivalence relation.

Prove each of the following:

292. $\mathbb{N} \times \mathbb{N} \sim \mathbb{N}$.

Proof: $\mathbb{N} \times \mathbb{N} = \bigcup\{\mathbb{N} \times \{n\} \mid n \in \mathbb{N}\}$. This is a countable union of countable sets. By Problem 252, $\mathbb{N} \times \mathbb{N}$ is countable. □

293. $\mathbb{Q} \sim \mathbb{N}$.

Proof: $\mathbb{Q}^+ = \left\{\frac{a}{b} \,\middle|\, a \in \mathbb{N} \wedge b \in \mathbb{N}^+\right\} = \bigcup\left\{\left\{\frac{a}{b} \,\middle|\, a \in \mathbb{N}\right\} \,\middle|\, b \in \mathbb{N}^+\right\}$. This is a countable union of countable sets. By Problem 252, $\mathbb{Q}^+$ is countable. Now, $\mathbb{Q} = \mathbb{Q}^+ \cup \{0\} \cup \mathbb{Q}^-$, where $\mathbb{Q}^- = \{q \in \mathbb{Q} \mid -q \in \mathbb{Q}^+\}$. This is again a countable union of countable sets, thus countable. So, $\mathbb{Q} \sim \mathbb{N}$. □

294. Any two intervals of real numbers are equinumerous (including $\mathbb{R}$ itself).

Proof: The function $f: \mathbb{R} \to (0, \infty)$ defined by $f(x) = 2^x$ is a bijection. So, $\mathbb{R} \sim (0, \infty)$. The function $g: (0, \infty) \to (0, 1)$ defined by $g(x) = \frac{1}{x^2+1}$ is a bijection. So, $(0, \infty) \sim (0, 1)$. If $a, b \in \mathbb{R}$ with $a < b$, the function $h: (0, 1) \to (a, b)$ defined by $h(x) = (b - a)x + a$ is a bijection. So, $(0, 1) \sim (a, b)$. It follows that all bounded open intervals are equinumerous with each other and $\mathbb{R}$.

We have, $[a, b] \subseteq (a - 1, b + 1) \sim (a, b) \subseteq [a, b) \subseteq [a, b]$ and $(a, b) \subseteq (a, b] \subseteq [a, b]$. It follows that all bounded intervals are equinumerous with each other and $\mathbb{R}$.

We also have the following:

$$(a, \infty) \subseteq [a, \infty) \subseteq \mathbb{R} \sim (a, a + 1) \subseteq (a, \infty)$$

$$(-\infty, b) \subseteq (-\infty, b] \subseteq \mathbb{R} \sim (b - 1, b) \subseteq (-\infty, b)$$

Therefore, all unbounded intervals are equinumerous with $\mathbb{R}$. It follows that any two intervals of real numbers are equinumerous. □

Notes: (1) In the proof above, we used the fact that equinumerosity is an equivalence relation, the Cantor-Schroeder-Bernstein Theorem, and Problem 244 many times without mention. For example, we have $\mathbb{R} \sim (0, \infty)$ and $(0, \infty) \sim (0, 1)$. So, by the transitivity of $\sim$, we have $\mathbb{R} \sim (0, 1)$. As another example, the sequence $(a, \infty) \subseteq [a, \infty) \subseteq \mathbb{R} \sim (a, a + 1) \subseteq (a, \infty)$ together with Note 1 following Example 2.72 gives us that $(a, \infty) \preccurlyeq \mathbb{R}$ and $\mathbb{R} \preccurlyeq (a, \infty)$. By the Cantor-Schroeder-Bernstein Theorem, $(a, \infty) \sim \mathbb{R}$.

(2) Once we showed that for all $a, b \in \mathbb{R}$, $(0, 1) \sim (a, b)$, it follows from the fact that $\sim$ is an equivalence relation that any two bounded open intervals are equinumerous. Indeed, if (a, b) and (c, d) are bounded open intervals, then $(0, 1) \sim (a, b)$ and $(0, 1) \sim (c, d)$. By the symmetry of $\sim$, we have $(a, b) \sim (0, 1)$, and finally, by the transitivity of $\sim$, we have $(a, b) \sim (c, d)$.

(3) It's easy to prove that two specific intervals of real numbers are equinumerous using just the fact that any two bounded open intervals are equinumerous with each other, together with the fact that $\mathbb{R} \sim (0, 1)$. For example, to show that $[3, \infty)$ is equinumerous with $(-2, 5]$, simply consider the following sequence: $[3, \infty) \subseteq \mathbb{R} \sim (0, 1) \sim (-2, 5) \subseteq (-2, 5] \subseteq (-2, 6) \sim (3, 4) \subseteq [3, \infty)$.

(4) In the proof above, we claimed that the function $f: \mathbb{R} \to (0, \infty)$ defined by $f(x) = 2^x$ is a bijection. See Lesson 13 for a formal definition of this function and an explanation as to why it is a bijection.

295. ${}^{\mathbb{N}}\mathbb{N} \sim \mathcal{P}(\mathbb{N})$.

Proof: ${}^{\mathbb{N}}\mathbb{N} \subseteq \mathcal{P}(\mathbb{N} \times \mathbb{N})$ by the definition of ${}^{\mathbb{N}}\mathbb{N}$. So, ${}^{\mathbb{N}}\mathbb{N} \preccurlyeq \mathcal{P}(\mathbb{N} \times \mathbb{N})$ by Note 1 following Example 2.72. By Problem 292, $\mathbb{N} \times \mathbb{N} \sim \mathbb{N}$. So, by Problem 248, $\mathcal{P}(\mathbb{N} \times \mathbb{N}) \sim \mathcal{P}(\mathbb{N})$. Thus, $\mathcal{P}(\mathbb{N} \times \mathbb{N}) \preccurlyeq \mathcal{P}(\mathbb{N})$. Since $\preccurlyeq$ is transitive, ${}^{\mathbb{N}}\mathbb{N} \preccurlyeq \mathcal{P}(\mathbb{N})$.

Now, $\mathcal{P}(\mathbb{N}) \sim {}^{\mathbb{N}}\{0,1\}$ (see Example 2.66 (part 5)). So, $\mathcal{P}(\mathbb{N}) \preccurlyeq {}^{\mathbb{N}}\{0,1\}$. Also, ${}^{\mathbb{N}}\{0,1\} \subseteq {}^{\mathbb{N}}\mathbb{N}$, and so, by Note 1 following Example 2.72, ${}^{\mathbb{N}}\{0,1\} \preccurlyeq {}^{\mathbb{N}}\mathbb{N}$. Since $\preccurlyeq$ is transitive, $\mathcal{P}(\mathbb{N}) \preccurlyeq {}^{\mathbb{N}}\mathbb{N}$.

By the Cantor-Schroeder-Bernstein Theorem, ${}^{\mathbb{N}}\mathbb{N} \sim \mathcal{P}(\mathbb{N})$. □

296. For any sets A, B, C, and D, if $A \sim B$ and $C \sim D$, then ${}^{A}C \sim {}^{B}D$.

Proof: Suppose that $A \sim B$ and $C \sim D$. Then there exist bijections $h: A \to B$ and $k: C \to D$. Define $F: {}^{A}C \to {}^{B}D$ by $F(f)(b) = k\left(f\left(h^{-1}(b)\right)\right)$.

Suppose $f, g \in {}^{A}C$ with $F(f) = F(g)$. Let $a \in A$, and let $b = h(a)$. We have $F(f)(b) = F(g)(b)$, or equivalently, $k\left(f\left(h^{-1}(b)\right)\right) = k\left(g\left(h^{-1}(b)\right)\right)$. Since k is injective, $f\left(h^{-1}(b)\right) = g\left(h^{-1}(b)\right)$. Since $b = h(a)$, $a = h^{-1}(b)$. So, $f(a) = g(a)$. Since $a \in A$ was arbitrary, $f = g$. Since $f, g \in {}^{A}C$ were arbitrary, F is injective.

Now, let $g \in {}^{B}D$ and let's define $f \in {}^{A}C$ by $f(a) = k^{-1}\left(g\left(h(a)\right)\right)$. Let $b \in B$. Then we have $F(f)(b) = k\left(f\left(h^{-1}(b)\right)\right) = k\left(k^{-1}\left(g\left(h\left(h^{-1}(b)\right)\right)\right)\right) = g(b)$. Since $b \in B$ was arbitrary, we have $F(f) = g$. Since $g \in {}^{B}D$ was arbitrary, F is surjective.

Since F is injective and surjective, ${}^{A}C \sim {}^{B}D$. □

LEVEL 5

We say that a set A is **transitive** if every element of A is a subset of A. Determine if each of the following sets is transitive:

297. $\emptyset$

Since $\emptyset$ has no elements, $\emptyset$ **is transitive**. (The statement "$x \in \emptyset \to x \subseteq \emptyset$" is true simply because "$x \in \emptyset$" is always false. In other words, the statement is **vacuously true**.)

298. $\{\emptyset\}$

The only element of $\{\emptyset\}$ is $\emptyset$, and $\emptyset \subseteq \{\emptyset\}$ is true. So, $\{\emptyset\}$ **is transitive**.

299. $\{\{\emptyset\}\}$

$\{\emptyset\} \in \{\{\emptyset\}\}$ and $\emptyset \in \{\emptyset\}$, but $\emptyset \notin \{\{\emptyset\}\}$. So, $\{\{\emptyset\}\}$ **is not transitive**.

300. $\{\emptyset, \{\emptyset\}\}$

$\{\emptyset, \{\emptyset\}\}$ has 2 elements, namely $\emptyset$ and $\{\emptyset\}$. Both sets are subsets of $\{\emptyset, \{\emptyset\}\}$. It follows that $\{\emptyset, \{\emptyset\}\}$ **is transitive**.

301. $\{\emptyset, \{\emptyset\}, \{\{\emptyset\}\}\}$

$\{\emptyset, \{\emptyset\}, \{\{\emptyset\}\}\}$ has 3 elements, namely $\emptyset$, $\{\emptyset\}$, and $\{\{\emptyset\}\}$. All three of these sets are subsets of $\{\emptyset, \{\emptyset\}, \{\{\emptyset\}\}\}$. It follows that $\{\emptyset, \{\emptyset\}, \{\{\emptyset\}\}\}$ **is transitive**.

302. $\{\{\emptyset\}, \{\emptyset, \{\emptyset\}\}\}$

$\{\emptyset\} \in \{\{\emptyset\}, \{\emptyset, \{\emptyset\}\}\}$ and $\emptyset \in \{\emptyset\}$, but $\emptyset \notin \{\{\emptyset\}, \{\emptyset, \{\emptyset\}\}\}$. So, $\{\{\emptyset\}, \{\emptyset, \{\emptyset\}\}\}$ **is not transitive**.

Let $\boldsymbol{X}$ be a nonempty set of sets. Prove each of the following:

303. For all $A \in \boldsymbol{X}$, $A \subseteq \bigcup \boldsymbol{X}$.

Proof: Let $\boldsymbol{X}$ be a nonempty set of sets, let $A \in \boldsymbol{X}$, and let $x \in A$. Then there is $B \in \boldsymbol{X}$ such that $x \in B$ (namely A). So, $x \in \bigcup \boldsymbol{X}$. Since x was an arbitrary element of A, we have shown that $A \subseteq \bigcup \boldsymbol{X}$. Since A was an arbitrary element of $\boldsymbol{X}$, we have shown that for all $A \in \boldsymbol{X}$, we have $A \subseteq \bigcup \boldsymbol{X}$. □

304. For all $A \in \boldsymbol{X}$, $\bigcap \boldsymbol{X} \subseteq A$.

Proof: Let $\boldsymbol{X}$ be a nonempty set of sets, let $A \in \boldsymbol{X}$, and let $x \in \bigcap \boldsymbol{X}$. Then for every $B \in \boldsymbol{X}$, we have $x \in B$. In particular, $x \in A$ (because $A \in \boldsymbol{X}$). Since x was an arbitrary element of $\bigcap \boldsymbol{X}$, we have shown that $\bigcap \boldsymbol{X} \subseteq A$. Since A was an arbitrary element of $\boldsymbol{X}$, we have shown that for all $A \in \boldsymbol{X}$, we have $\bigcap \boldsymbol{X} \subseteq A$. □

Let A be a set and let $\boldsymbol{X}$ be a nonempty set of sets. Prove each of the following:

305. $A \cap \bigcup \boldsymbol{X} = \bigcup\{A \cap B \mid B \in \boldsymbol{X}\}$.

Proof: $x \in A \cap \bigcup \boldsymbol{X} \Leftrightarrow x \in A$ and $x \in \bigcup \boldsymbol{X} \Leftrightarrow x \in A$ and there is a $B \in \boldsymbol{X}$ with $x \in B \Leftrightarrow x \in A \cap B$ for some $B \in \boldsymbol{X} \Leftrightarrow x \in \bigcup\{A \cap B \mid B \in \boldsymbol{X}\}$. □

306. $A \cup \bigcap \boldsymbol{X} = \bigcap\{A \cup B \mid B \in \boldsymbol{X}\}$.

Proof: $x \in A \cup \bigcap \boldsymbol{X} \Leftrightarrow x \in A$ or $x \in \bigcap \boldsymbol{X} \Leftrightarrow x \in A$ or $x \in B$ for every $B \in \boldsymbol{X} \Leftrightarrow x \in A \cup B$ for every $B \in \boldsymbol{X} \Leftrightarrow x \in \bigcap\{A \cup B \mid B \in \boldsymbol{X}\}$. □

307. $A \setminus \bigcup \boldsymbol{X} = \bigcap\{A \setminus B \mid B \in \boldsymbol{X}\}$.

Proof: $x \in A \setminus \bigcup \boldsymbol{X} \Leftrightarrow x \in A$ and $x \notin \bigcup \boldsymbol{X} \Leftrightarrow x \in A$ and $x \notin B$ for every $B \in \boldsymbol{X} \Leftrightarrow x \in A \setminus B$ for every $B \in \boldsymbol{X} \Leftrightarrow x \in \bigcap\{A \setminus B \mid B \in \boldsymbol{X}\}$. □

308. $A \setminus \bigcap \boldsymbol{X} = \bigcup\{A \setminus B \mid B \in \boldsymbol{X}\}$.

Proof: $x \in A \setminus \bigcap \boldsymbol{X} \Leftrightarrow x \in A$ and $x \notin \bigcap \boldsymbol{X} \Leftrightarrow x \in A$ and $x \notin B$ for some $B \in \boldsymbol{X} \Leftrightarrow x \in A \setminus B$ for some $B \in \boldsymbol{X} \Leftrightarrow x \in \bigcup\{A \setminus B \mid B \in \boldsymbol{X}\}$. □

Find the natural number that each of the following expressions is equal to or explain why the given expression is not a natural number.

309. $\bigcup\{2k \mid 0 \leq k \leq 100\}$

$\bigcup\{2k \mid 0 \leq k \leq 100\} = 0 \cup 2 \cup 4 \cup \cdots \cup 200 = 0 \cup \{0,1\} \cup \{0,1,2,3\} \cup \cdots \cup \{0,1,2,\dots,199\}$
$= \{0,1,2,\dots,199\} = \mathbf{200}$.

310. $\bigcup\{\{k\} \mid 0 \leq k \leq 100\}$

$\bigcup\{\{k\} \mid 0 \leq k \leq 100\} = \{0\} \cup \{1\} \cup \{2\} \cup \cdots \cup \{100\} = \{0,1,2,\dots,100\} = \mathbf{101}$.

311. $\bigcup\{\{2k\} \mid 0 \leq k \leq 100\}$

$\bigcup\{\{2k\} \mid 0 \leq k \leq 100\} = \{0\} \cup \{2\} \cup \{4\} \cup \cdots \cup \{200\} = \{0,2,4,\dots,200\}$. This is **not** a natural number.

312. $202 \setminus \bigcup\{2k \mid 0 \leq k \leq 100\}$

By Problem 309, $\bigcup\{2k \mid 0 \leq k \leq 100\} = 200$. Therefore,

$$202 \setminus \bigcup\{2k \mid 0 \leq k \leq 100\} = 202 \setminus 200 = \{0,1,2,\dots,201\} \setminus \{0,1,2,\dots,199\} = \{200,201\}.$$

This is **not** a natural number.

313. $(202 \setminus \bigcup\{2k \mid 0 \leq k \leq 100\}) \cup 201$

By Problem 312, we have $202 \setminus \bigcup\{2k \mid 0 \leq k \leq 100\} = \{200,201\}$. Therefore, we see that $(202 \setminus \bigcup\{2k \mid 0 \leq k \leq 100\}) \cup 201 = \{200,201\} \cup \{0,1,2,\dots,200\} = \{0,1,2,\dots,201\} = \mathbf{202}$.

For $a, b \in \mathbb{N}$, we will say that a divides b, written $a|b$, if there is a natural number k such that $b = ak$. Notice that $|$ is a binary relation on $\mathbb{N}$.

314. Is $(\mathbb{N}, |)$ a poset?

Yes

Proof: If $a \in \mathbb{N}$ then $a = a \cdot 1$, so that $a|a$. Therefore, $|$ is reflexive. If $a|b$ and $b|a$, then there are natural numbers j and k such that $b = ja$ and $a = kb$. If $a = 0$, then $b = j \cdot 0 = 0$, and so, $a = b$. Suppose $a \neq 0$. We have $a = k(ja) = (kj)a$. Thus, $(kj - 1)a = (kj)a - 1a = 0$. So, $kj - 1 = 0$, and therefore, $kj = 1$. So, $k = j = 1$. Thus, $b = ja = 1a = a$. Therefore, $|$ is antisymmetric. If $a|b$ and $b|c$, then there are natural numbers j and k such that $b = ja$ and $c = kb$. Then $c = kb = k(ja) = (kj)a$. Since the product of two natural numbers is a natural number, $kj \in \mathbb{N}$. So, $a|c$. Therefore, $|$ is transitive. Since $|$ is reflexive, antisymmetric, and transitive on $\mathbb{N}$, $(\mathbb{N}, |)$ is a partially ordered set. □

315. Is $(\mathbb{N}, |)$ a strict poset?

No

Since $2|2$, $|$ is **not** antireflexive. Therefore, $(\mathbb{N}, |\,)$ is **not** a strict poset.

316. Is $(\mathbb{N}, |)$ a linearly ordered set?

No

Since 2 and 3 do not divide each other, $(\mathbb{N}, |\,)$ is **not** linearly ordered.

Recall that we define the ordering $<_{\mathbb{Z}}$ on $\mathbb{Z}$ by $[(a,b)] <_{\mathbb{Z}} [(c,d)]$ if and only if $a + d <_{\mathbb{N}} b + c$, where $<_{\mathbb{N}}$ is the usual ordering on $\mathbb{N}$.

317. Prove that $<_{\mathbb{Z}}$ is a well-defined relation on $\mathbb{Z}$.

Proof: Suppose that $(a,b) \sim (a',b')$ and $(c,d) \sim (c',d')$. Since $(a,b) \sim (a',b')$, $a + b' = b + a'$. Since $(c,d) \sim (c',d')$, $c + d' = d + c'$.

We need to check that $[(a,b)] <_{\mathbb{Z}} [(c,d)]$ if and only if $[(a',b')] <_{\mathbb{Z}} [(c',d')]$. We have

$[(a,b)] <_{\mathbb{Z}} [(c,d)]$ if and only if $a + d <_{\mathbb{N}} b + c$ if and only if $a + d + b' + c' <_{\mathbb{N}} b + c + b' + c'$ if and only if $a + b' + d + c' <_{\mathbb{N}} b + c + b' + c'$ if and only if $b + a' + c + d' <_{\mathbb{N}} b + c + b' + c'$ if and only if $b + c + a' + d' <_{\mathbb{N}} b + c + b' + c'$ if and only if $a' + d' < b' + c'$ if and only if $[(a',b')] <_{\mathbb{Z}} [(c',d')]$, as desired. □

318. Prove that $<_{\mathbb{Z}}$ is transitive on $\mathbb{Z}$.

Proof: Suppose that $[(a,b)] <_{\mathbb{Z}} [(c,d)]$ and $[(c,d)] <_{\mathbb{Z}} [(e,f)]$. Then $a + d <_{\mathbb{N}} b + c$ and $c + f <_{\mathbb{N}} d + e$. By adding each side of these two inequalities we get the inequality $a + d + c + f <_{\mathbb{N}} b + c + d + e$. Cancelling c and d from each side of this last inequality yields $a + f <_{\mathbb{N}} b + e$. Therefore, $[(a,b)] <_{\mathbb{Z}} [(e,f)]$. □

319. Prove that $(\mathbb{Z}, <_{\mathbb{Z}})$ satisfies trichotomy.

Proof: Suppose $[(a,b)] \not<_{\mathbb{Z}} [(c,d)]$ and $[(a,b)] \neq [(c,d)]$. Then $a + d \not<_{\mathbb{N}} b + c$ and $a + d \neq b + c$. Since trichotomy holds for $<_{\mathbb{N}}$, we have $b + c <_{\mathbb{N}} a + d$, or equivalently, $c + b <_{\mathbb{N}} d + a$. Therefore, $[(c,d)] <_{\mathbb{Z}} [(a,b)]$. □

320. Prove that $(\mathbb{Z}, <_{\mathbb{Z}})$ is a strict linearly ordered set.

Proof: By Problems 317 – 319, $<_{\mathbb{Z}}$ is a well-defined binary relation on $\mathbb{Z}$ that is transitive and such that $(\mathbb{Z}, <_{\mathbb{Z}})$ satisfies trichotomy. In order to show that $<_{\mathbb{Z}}$ is a strict linear ordering of $\mathbb{Z}$, we still need to show that $<_{\mathbb{Z}}$ is antireflexive and antisymmetric.

To see that $<_{\mathbb{Z}}$ is antireflexive, note that $a + b \not<_{\mathbb{N}} a + b$ because $\not<_{\mathbb{N}}$ is antireflexive. Therefore, $a + b \not<_{\mathbb{N}} b + a$. So, $[(a,b)] \not<_{\mathbb{Z}} [(a,b)]$.

To see that $<_{\mathbb{Z}}$ is antisymmetric, suppose that $[(a,b)] <_{\mathbb{Z}} [(c,d)]$ and $[(c,d)] <_{\mathbb{Z}} [(a,b)]$. Then we have $a + d <_{\mathbb{N}} b + c$ and $c + b <_{\mathbb{N}} d + a$, or equivalently, $b + c <_{\mathbb{N}} a + d$. This is impossible, and so, it is vacuously true that $<_{\mathbb{Z}}$ is antisymmetric. □

Let $\sim$ be an equivalence relation on a set S.

321. Prove that $\bigcup\{[x] \mid x \in S\} \subseteq S$.

Proof: Let $y \in \bigcup\{[x] \mid x \in S\}$. Then there is $x \in S$ with $y \in [x]$. By definition of $[x]$, $y \in S$. Therefore, $\bigcup\{[x] \mid x \in S\} \subseteq S$. □

322. Prove that $S \subseteq \bigcup\{[x] \mid x \in S\}$.

Proof: Let $y \in S$. Since $\sim$ is an equivalence relation, $y \sim y$. So, $y \in [y]$. Thus, $y \in \bigcup\{[x] \mid x \in S\}$. So, we have $S \subseteq \bigcup\{[x] \mid x \in S\}$. □

323. Suppose that $x, y \in S$ and $[x] \cap [y] \neq \emptyset$. Prove that $[x] \subseteq [y]$.

Proof: Since $[x] \cap [y] \neq \emptyset$, we can find $z \in [x] \cap [y]$. Then $x \sim z$ and $y \sim z$. Since $\sim$ is symmetric, $z \sim y$. Since $\sim$ is transitive, $x \sim y$. Since $\sim$ is symmetric, $y \sim x$.

Let $w \in [x]$. Then $x \sim w$. By transitivity, $y \sim w$. So, $w \in [y]$. Since $w \in [x]$ was arbitrary, $[x] \subseteq [y]$. □

324. Suppose that $x, y \in S$ and $[x] \cap [y] \neq \emptyset$. Prove that $[x] = [y]$.

Proof: By Problem 323, $[x] \subseteq [y]$. By a symmetric argument to that given in the solution to Problem 323, $[y] \subseteq [x]$.

Since $[x] \subseteq [y]$ and $[y] \subseteq [x]$, we have $[x] = [y]$. □

325. Prove that the equivalence classes of $\sim$ form a partition of S.

Proof: By Problems 321 and 322, $\bigcup\{[x] \mid x \in S\} = S$. By Problems 323 and 324, each pair of equivalence classes is either disjoint or equal. Therefore, the set of equivalence classes partitions S. □

Let $\boldsymbol{P}$ be a partition of S, and define the relation $\sim$ by $x \sim y$ if and only if there is $X \in \boldsymbol{P}$ with $x, y \in X$.

326. Prove that $\sim$ is reflexive on S.

Proof: Let $x \in S$. Since $\boldsymbol{P}$ is a partition of S, $S = \bigcup \boldsymbol{P}$. So, there is $X \in \boldsymbol{P}$ with $x \in X$. It follows that $x \sim x$. Therefore, $\sim$ is reflexive. □

327. Prove that $\sim$ is symmetric on S.

Proof: If $x \sim y$, then there is $X \in \boldsymbol{P}$ with $x, y \in X$. So, $y, x \in X$ (obviously!). Thus, $y \sim x$, and therefore, $\sim$ is symmetric. □

328. Prove that $\sim$ is transitive on S.

Proof: If $x \sim y$ and $y \sim z$, then there are $X, Y \in \boldsymbol{P}$ with $x, y \in X$ and $y, z \in Y$. Since $y \in X$ and $y \in Y$, we have $y \in X \cap Y$. Since $\boldsymbol{P}$ is a partition and $X \cap Y \neq \emptyset$, we must have $X = Y$. So, $z \in X$. Thus, $x, z \in X$, and therefore, $x \sim z$. So, $\sim$ is transitive. □

329. Prove that $\sim$ is an equivalence relation on S.

Proof: By Problems 326 – 328, $\sim$ is reflexive, symmetric, and transitive on S. Therefore, $\sim$ is an equivalence relation on S. □

330. Prove that $\boldsymbol{P} \subseteq \{[x] \mid x \in S\}$.

Proof: Let $X \in \boldsymbol{P}$ and let $x \in X$. We will show that $X = [x]$.

Let $y \in X$. Since $x, y \in X$, $x \sim y$. So $y \in [x]$. Thus, $X \subseteq [x]$.

Now, let $y \in [x]$. Then $x \sim y$. So, there is $Y \in \boldsymbol{P}$ such that $x, y \in Y$. Since $x \in X$ and $x \in Y$, $x \in X \cap Y$. Since $\boldsymbol{P}$ is a partition and $X \cap Y \neq \emptyset$, we must have $X = Y$. So, $y \in X$. Thus, $[x] \subseteq X$.

Since $X \subseteq [x]$ and $[x] \subseteq X$, we have $X = [x]$.

Since $X \in \boldsymbol{P}$ was arbitrary, we have shown $\boldsymbol{P} \subseteq \{[x] \mid x \in S\}$. □

331. Prove that $\{[x] \mid x \in S\} \subseteq \boldsymbol{P}$.

Proof: Let $X \in \{[x] \mid x \in S\}$. Then there is $x \in S$ such that $X = [x]$. Since $\boldsymbol{P}$ is a partition of S, $S = \bigcup \boldsymbol{P}$. So, there is $Y \in \boldsymbol{P}$ with $x \in Y$. We will show that $X = Y$.

Let $y \in X$. Then $x \sim y$. So, there is $Z \in \boldsymbol{P}$ with $x, y \in Z$. Since $x \in Y$ and $x \in Z$, $x \in Y \cap Z$. Since $\boldsymbol{P}$ is a partition and $Y \cap Z \neq \emptyset$, we must have $Y = Z$. So, $y \in Y$. Since $y \in X$ was arbitrary, $X \subseteq Y$.

Now, let $y \in Y$. Then $x \sim y$. So, $y \in [x] = X$. Since $y \in Y$ was arbitrary, $Y \subseteq X$.

Since $X \subseteq Y$ and $Y \subseteq X$, we have $X = Y$. Therefore, $X \in \boldsymbol{P}$.

Since $X \in \{[x] \mid x \in S\}$ was arbitrary, we have $\{[x] \mid x \in S\} \subseteq \boldsymbol{P}$. □

332. Prove that $\boldsymbol{P} = \{[x] \mid x \in S\}$.

Proof: By Problem 330, $\boldsymbol{P} \subseteq \{[x] \mid x \in S\}$. By Problem 331, $\{[x] \mid x \in S\} \subseteq \boldsymbol{P}$. So, $\boldsymbol{P} = \{[x] \mid x \in S\}$, as desired. □

Let $\boldsymbol{X}$ be a nonempty set of sets and let f be a function such that $\bigcup\boldsymbol{X} \subseteq \text{dom } f$. Prove each of the following:

333. $f[\bigcup\boldsymbol{X}] = \bigcup\{f[A] \mid A \in \boldsymbol{X}\}$

Proof: Let $y \in f[\bigcup\boldsymbol{X}]$. Then there is $x \in \bigcup\boldsymbol{X}$ such that $f(x) = y$. Since $x \in \bigcup\boldsymbol{X}$, there is $B \in \boldsymbol{X}$ such that $x \in B$. So, $y = f(x) \in f[B]$. Therefore, $y \in \bigcup\{f[A] \mid A \in \boldsymbol{X}\}$. Since $y \in f[\bigcup\boldsymbol{X}]$ was arbitrary, we see that $f[\bigcup\boldsymbol{X}] \subseteq \bigcup\{f[A] \mid A \in \boldsymbol{X}\}$.

Now, let $y \in \bigcup\{f[A] \mid A \in X\}$. Then there is $B \in X$ such that $y \in f[B]$. So, there is $x \in B$ such that $y = f(x)$. By Problem 303, $B \subseteq \bigcup\boldsymbol{X}$. Since $x \in B$ and $B \subseteq \bigcup\boldsymbol{X}$, $x \in \bigcup\boldsymbol{X}$. Thus, $y = f(x) \in f[\bigcup\boldsymbol{X}]$. Since $y \in \bigcup\{f[A] \mid A \in \boldsymbol{X}\}$ was arbitrary, we see that $\bigcup\{f[A] \mid A \in \boldsymbol{X}\} \subseteq f[\bigcup\boldsymbol{X}]$.

Since $f[\bigcup\boldsymbol{X}] \subseteq \bigcup\{f[A] \mid A \in \boldsymbol{X}\}$ and $\bigcup\{f[A] \mid A \in X\} \subseteq f[\bigcup\boldsymbol{X}]$, $f[\bigcup\boldsymbol{X}] = \bigcup\{f[A] \mid A \in \boldsymbol{X}\}$. □

334. $f[\bigcap\boldsymbol{X}] \subseteq \bigcap\{f[A] \mid A \in \boldsymbol{X}\}$

Proof: Let $y \in f[\bigcap\boldsymbol{X}]$. Then there is $x \in \bigcap\boldsymbol{X}$ such that $f(x) = y$. Let $B \in \boldsymbol{X}$. Since $x \in \bigcap\boldsymbol{X}$, $x \in B$. So, $y = f(x) \in f[B]$. Since $B \in \boldsymbol{X}$ was arbitrary, $y \in \bigcap\{f[A] \mid A \in \boldsymbol{X}\}$. Since $y \in f[\bigcap\boldsymbol{X}]$ was arbitrary, we see that $f[\bigcap\boldsymbol{X}] \subseteq \bigcap\{f[A] \mid A \in \boldsymbol{X}\}$. □

335. $f^{-1}[\bigcup\boldsymbol{X}] = \bigcup\{f^{-1}[A] \mid A \in \boldsymbol{X}\}$

Proof: $x \in f^{-1}[\bigcup\boldsymbol{X}]$ if and only if $f(x) \in \bigcup\boldsymbol{X}$ if and only if there is $A \in \boldsymbol{X}$ such that $f(x) \in A$ if and only if there is $A \in \boldsymbol{X}$ such that $x \in f^{-1}[A]$ if and only if $x \in \bigcup\{f^{-1}[A] \mid A \in \boldsymbol{X}\}$. Therefore, $f^{-1}[\bigcup\boldsymbol{X}] = \bigcup\{f^{-1}[A] \mid A \in \boldsymbol{X}\}$. □

336. $f^{-1}[\bigcap\boldsymbol{X}] = \bigcap\{f^{-1}[A] \mid A \in \boldsymbol{X}\}$

Proof: $x \in f^{-1}[\bigcap\boldsymbol{X}]$ if and only if $f(x) \in \bigcap\boldsymbol{X}$ if and only for all $A \in \boldsymbol{X}$, $f(x) \in A$ if and only if for all $A \in \boldsymbol{X}$, $x \in f^{-1}[A]$ if and only if $x \in \bigcap\{f^{-1}[A] \mid A \in \boldsymbol{X}\}$. Therefore, we see that $f^{-1}[\bigcap\boldsymbol{X}] = \bigcap\{f^{-1}[A] \mid A \in \boldsymbol{X}\}$. □

Prove each of the following:

337. For any sets A, B, and C, ${}^{B\times C}A \sim {}^{C}({}^{B}A)$.

Proof: Let A, B, and C be sets, and define $F: {}^{B\times C}A \to {}^{C}({}^{B}A)$ by $F(f)(c)(b) = f(b,c)$.

Suppose $f, g \in {}^{B\times C}A$ with $F(f) = F(g)$. Let $c \in C$. Since $F(f) = F(g)$, $F(f)(c) = F(g)(c)$. So, for all $b \in B$, $F(f)(c)(b) = F(g)(c)(b)$. So, for all $b \in B$, $f(b,c) = g(b,c)$. Since $c \in C$ was arbitrary, for all $b \in B$ and $c \in C$, $f(b,c) = g(b,c)$. Therefore, $f = g$. Since $f, g \in {}^{B\times C}A$ were arbitrary, F is injective.

Let $k \in {}^{C}({}^{B}A)$ and define $f \in {}^{B\times C}A$ by $f(b,c) = k(c)(b)$. Then $F(f)(c)(b) = f(b,c) = k(c)(b)$. So, $F(f) = k$. Since $k \in {}^{C}({}^{B}A)$ was arbitrary, F is surjective.

Since F is injective and surjective, ${}^{B\times C}A \sim {}^{C}({}^{B}A)$. □

338. $\mathcal{P}(\mathbb{N}) \sim \{f \in {}^{\mathbb{N}}\mathbb{N} \mid f \text{ is a bijection}\}$.

Proof: Let $S = \{f \in {}^{\mathbb{N}}\mathbb{N} \mid f \text{ is a bijection}\}$. Then $S \subseteq {}^{\mathbb{N}}\mathbb{N}$. So $S \preccurlyeq {}^{\mathbb{N}}\mathbb{N}$ by Note 1 following Example 2.72. By Problem 295, ${}^{\mathbb{N}}\mathbb{N} \sim \mathcal{P}(\mathbb{N})$. So, ${}^{\mathbb{N}}\mathbb{N} \preccurlyeq \mathcal{P}(\mathbb{N})$. By the transitivity of $\preccurlyeq$, $S \preccurlyeq \mathcal{P}(\mathbb{N})$.

Now, define $F: \mathcal{P}(\mathbb{N}) \to S$ by $F(A) = f_A$, where f_A is defined as follows: if $n \notin A$, then $f_A(2n) = 2n$ and $f_A(2n+1) = 2n+1$; if $n \in A$, then $f_A(2n) = 2n+1$ and $f_A(2n+1) = 2n$.

To see that F is injective, suppose that $A, B \in \mathcal{P}(\mathbb{N})$ and $A \neq B$. Without loss of generality, suppose that there is $n \in A \setminus B$. Then $f_A(2n) = 2n+1$ and $f_B(2n) = 2n$. So, $f_A \neq f_B$. Thus, $F(A) \neq F(B)$, and therefore, F is injective.

Since $S \preccurlyeq \mathcal{P}(\mathbb{N})$ and $\mathcal{P}(\mathbb{N}) \preccurlyeq S$, by the Cantor-Schroeder-Bernstein Theorem, $\mathcal{P}(\mathbb{N}) \sim S$. □

Note: To help us understand the function F defined above, let's draw a visual representation of $F(\mathbb{E})$, where $\mathbb{E}$ is the set of even natural numbers.

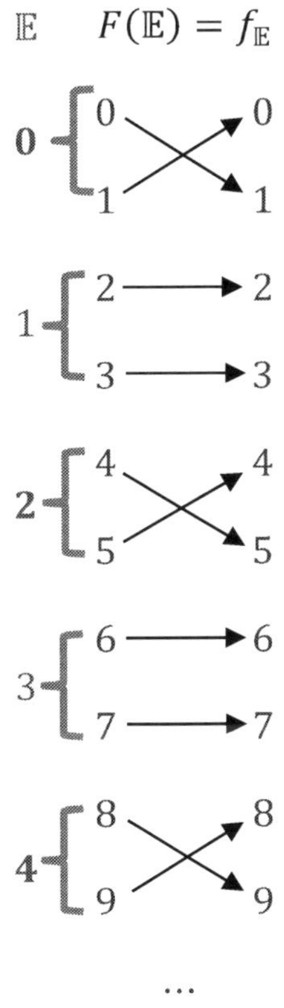

Along the left of the image we have listed the natural numbers $0, 1, 2, 3, 4, \ldots$ (we stopped at 4, but our intention is that they keep going). The elements of $\mathbb{E}$ are $0, 2, 4, \ldots$ We highlighted these in bold. We associate each natural number n with the pair $\{2n, 2n+1\}$. For example, $2 \cdot 4 = 8$ and $2 \cdot 4 + 1 = 9$. So, we associate 4 with the pair of natural numbers $\{8, 9\}$. We used left braces to indicate that association. The arrows give a visual representation of $f_{\mathbb{E}}$. Since $0 \in \mathbb{E}$, $f_{\mathbb{E}}$ swaps the corresponding pair 0 and 1. Since $1 \notin \mathbb{E}$, $f_{\mathbb{E}}$ leaves the corresponding pair 2 and 3 fixed. And so on, down the line...

The configuration of $f_{\mathbb{O}}$, where $\mathbb{O}$ is the set of odd natural numbers would be the opposite of the configuration for the evens. For example, 0 and 1 would remain fixed, while 2 and 3 would be swapped.

339. $\mathbb{R} \sim \mathcal{P}(\mathbb{N})$.

Proof: We first show that $\mathbb{R} \preccurlyeq {}^{\mathbb{N}}\mathbb{Q}$. Define $f: \mathbb{R} \to {}^{\mathbb{N}}\mathbb{Q}$ by $f(x) = (x_n)$, where (x_n) is some representative of x (remember that x is an equivalence class of Cauchy sequences of rational numbers). Clearly, f is injective. It follows that

$$\mathbb{R} \preccurlyeq {}^{\mathbb{N}}\mathbb{Q} \subseteq \mathcal{P}(\mathbb{N} \times \mathbb{Q}) \sim \mathcal{P}(\mathbb{N} \times \mathbb{N}) \sim \mathcal{P}(\mathbb{N}).$$

Thus, $\mathbb{R} \preccurlyeq \mathcal{P}(\mathbb{N})$.

Next, we show that $\mathcal{P}(\mathbb{N}) \preccurlyeq \mathbb{R}$. Let $A \in \mathcal{P}(\mathbb{N})$. For each $n \in \mathbb{N}$, let $a_n = 0$ if $n \notin A$ and let $a_n = 1$ if $n \in A$. Let $a_n^* = \frac{a_0 a_1 \cdots a_n}{10^n}$. Then (a_n^*) is a Cauchy sequence (Check this!). Define $g: \mathcal{P}(\mathbb{N}) \to \mathbb{R}$ by $g(A) = [(a_n^*)]$. Then g is injective (Check this!).

Since $\mathbb{R} \preccurlyeq \mathcal{P}(\mathbb{N})$ and $\mathcal{P}(\mathbb{N}) \preccurlyeq \mathbb{R}$, by the Cantor-Schroeder-Bernstein Theorem, we have $\mathbb{R} \sim \mathcal{P}(\mathbb{N})$. □

340. ${}^{\mathbb{N}}\mathbb{R} \nsim {}^{\mathbb{R}}\mathbb{N}$.

Proof: By Problem 339, $\mathbb{R} \sim \mathcal{P}(\mathbb{N})$. By Problem 296, ${}^{\mathbb{N}}\mathbb{R} \sim {}^{\mathbb{N}}\mathcal{P}(\mathbb{N})$.

Using previous equinumerosity results, we get the following:

$${}^{\mathbb{N}}\mathbb{R} \sim {}^{\mathbb{N}}\mathcal{P}(\mathbb{N}) \sim {}^{\mathbb{N}}({}^{\mathbb{N}}2) \sim {}^{\mathbb{N}\times\mathbb{N}}2 \sim {}^{\mathbb{N}}2 \sim \mathcal{P}(\mathbb{N}) \sim \mathbb{R} \prec \mathcal{P}(\mathbb{R}) \sim {}^{\mathbb{R}}2 \subseteq {}^{\mathbb{R}}\mathbb{N}.$$

It follows that ${}^{\mathbb{N}}\mathbb{R} \prec {}^{\mathbb{R}}\mathbb{N}$. □

Problem Set 3

LEVEL 1

I

$\star$	a	b
a	a	a
b	a	a

II

$\star$	a	b
a	a	b
b	c	a

III

$\star$	a	b
a	a	b
b	b	a

IV

$\star$	a	b
a	a	a
b	b	b

For each of the multiplication tables defined on the set $S = \{a, b\}$ above, determine if each of the following is true or false:

1. $\star$ defines a binary operation on S.

For tables I, III, and IV, $\star$ **does** define a binary operation because only a and b appear inside each of these tables. For table II, $\star$ does **not** define a binary operation because an element different from a and b appears in the table (assuming that $c \neq a$ and $c \neq b$).

2. $\star$ is commutative in S.

Since there are just two elements a and b, we need only check if a and b commute ($a \star b = b \star a$). This is very easy to see just by looking at the tables. We simply check if the entries on opposite sides of the main diagonal are the same.

$\star$	a	b
a	a	a
b	a	a

Yes

$\star$	a	b
a	a	b
b	c	a

No

$\star$	a	b
a	a	b
b	b	a

Yes

$\star$	a	b
a	a	a
b	b	b

No

We see that for tables I and III, $\star$ **is** commutative in S, whereas for tables II and IV, $\star$ **is not** commutative in S.

3. a is an identity with respect to $\star$.

To see if a is an identity with respect to $\star$, we need to check if $a \star a = a$, $a \star b = b$, and $b \star a = b$. This is also very easy to see just by looking at the tables. We simply check if the row corresponding to a is the same as the "input row," and if the column corresponding to a is the same as the "input column."

$\star$	a	b
a	a	a
b	a	a

No

$\star$	a	b
a	a	b
b	c	a

Maybe

$\star$	a	b
a	a	b
b	b	a

Maybe

$\star$	a	b
a	a	a
b	b	b

No

We see that for tables I and IV, the row corresponding to a is **not** the same as the "input row." So, for I and IV, a is **not** an identity with respect to $\star$.

We still need to check the columns for tables II and III.

$\star$	a	b
a	a	b
b	c	a

No

$\star$	a	b
a	a	b
b	b	a

Yes

We see that for table II, the column corresponding to a is **not** the same as the "input column." So, for II, a is **not** an identity with respect to $\star$.

For table III, a **is** an identity with respect to $\star$.

4. b is an identity with respect to $\star$.

To see if b is an identity with respect to $\star$, we need to check if $a \star b = a$, $b \star a = a$, and $b \star b = b$. Again, this is very easy to see just by looking at the tables. In this case, we see that for each table, the row corresponding to b is **not** the same as the "input row."

$\star$	a	b
a	a	a
b	a	a

No

$\star$	a	b
a	a	b
b	c	a

No

$\star$	a	b
a	a	b
b	b	a

No

$\star$	a	b
a	a	a
b	b	b

No

So, b is **not** an identity with respect to $\star$ in all four cases.

Notes on Problems 1 through 4: (1) Table I defines a semigroup $(S,\star)$. To see that $\star$ is associative in S, just observe that all the outputs are the same. Therefore, there cannot be a counterexample to associativity. For example, $(a \star b) \star b = a \star b = a$ and $a \star (b \star b) = a \star a = a$.

(2) Table I does **not** define a monoid. Problems 3 and 4 showed us that there is no identity with respect to $\star$.

(3) Table III defines a commutative group $(S,\star)$ with identity a. a and b are each their own inverses because $a \star a = a$ and $b \star b = a$ (remember that a is the identity). Associativity can be checked by brute force. There are eight instances that need to be verified. For example, $(a \star a) \star b = a \star b = b$ and $a \star (a \star b) = a \star b = b$. So, $(a \star a) \star b = a \star (a \star b)$. If you do it this way, make sure you check the other seven instances. Alternatively, we can observe that the multiplication table is identical to the multiplication table for $(C_2, +)$ (or $(\mathbb{Z}_2, +)$) if we replace a with 0 and b with 1.

$\star$	0	1
0	0	1
1	1	0

(4) Table IV defines a semigroup $(S,\star)$ known as the **left zero semigroup**. The name of this semigroup comes from the fact that $a \star a = a$ and $a \star b = a$, so that a is behaving just like 0 behaves when multiplying on the left (0 times anything equals 0). Notice that $b \star a = b \neq a$, so that a does **not** behave like 0 when multiplying on the right. Similar computations show that b also behaves like 0 from the left. The dedicated reader may want to check associativity by brute force, as described in Note 3.

(5) Table IV does **not** define a monoid. Problems 3 and 4 showed us that there is no identity with respect to $\star$.

$\circ$	a	b	c	d
a	a	a	a	b
b	d	d	b	b
c	a	b	c	d
d	c	c	d	d

The multiplication table above is defined on the set $S = \{a, b, c, d\}$.

5. Is S closed under $\circ$?

Yes. Only a,b, c, and d appear inside the table.

6. Compute $a \circ b$.

$a \circ b = \boldsymbol{a}$.

$\circ$	a	b	c	d
a	a	a	a	b
b	d	d	b	b
c	a	b	c	d
d	c	c	d	d

7. Is $\circ$ commutative in S? Why or why not?

No. For example, $a \circ b = a$, whereas $b \circ a = d$. So, $a \circ b \neq b \circ a$.

$\circ$	a	b	c	d
a	a	a	a	b
b	d	d	b	b
c	a	b	c	d
d	c	c	d	d

8. Is $\circ$ associative in S? Why or why not?

No. For example, $(b \circ d) \circ b = b \circ b = d$, whereas $b \circ (d \circ b) = b \circ c = b$. Therefore, we see that $(b \circ d) \circ b \neq b \circ (d \circ b)$.

9. Does S have an identity with respect to $\circ$? If so, what is it?

Yes. The identity is c

$\circ$	a	b	c	d
a	a	a	a	b
b	d	d	b	b
c	a	b	c	d
d	c	c	d	d

$\circ$	a	b	c	d
a	a	a	a	b
b	d	d	b	b
c	a	b	c	d
d	c	c	d	d

Let $S = \{e, a\}$.

10. Show that there are exactly two monoids on S with identity e.

Since $e \star x = x \star e = x$ for all x in the monoid, we can easily fill out the first row and the first column of the table.

$\star$	e	a
e	e	a
a	a	$\boxdot$

The entry labeled with $\boxdot$ must be either e or a because we need $\star$ to be a binary operation on S.

Case 1: If we let $\boxdot$ be a, we get the following table.

$\star$	e	a
e	e	a
a	a	a

Associativity holds because any computation of the form $(x \star y) \star z$ or $x \star (y \star z)$ will result in a if any of x, y, or z is a. So, all that is left to check is that $(e \star e) \star e = e \star (e \star e)$. But each side of that equation is equal to e.

So, with this multiplication table, $(S,\star)$ **is** a monoid.

Case 2: If we let $\boxdot$ be e, we get the following table.

$\star$	e	a
e	e	a
a	a	e

Let's check that associativity holds. There are eight instances to check.

$$\begin{array}{ll}
(e \star e) \star e = e \star e = e & e \star (e \star e) = e \star e = e \\
(e \star e) \star a = e \star a = a & e \star (e \star a) = e \star a = a \\
(e \star a) \star e = a \star e = a & e \star (a \star e) = e \star a = a \\
(a \star e) \star e = a \star e = a & a \star (e \star e) = a \star e = a \\
(e \star a) \star a = a \star a = e & e \star (a \star a) = e \star e = e \\
(a \star e) \star a = a \star a = e & a \star (e \star a) = a \star a = e \\
(a \star a) \star e = e \star e = e & a \star (a \star e) = a \star a = e \\
(a \star a) \star a = e \star a = a & a \star (a \star a) = e \star a = a
\end{array}$$

So, with this multiplication table, $(S,\star)$ **is** a monoid.

11. Determine if either of the two monoids on S is a group.

This first monoid is **not** a group because a has no inverse. Indeed, $a \star e = a \neq e$ and $a \star a = a \neq e$.

The second monoid **is** a group. Since $e \star e = e$, e is its own inverse. Since $a \star a = e$, a is also its own inverse.

12. Are either of the monoids on S commutative?

This first monoid is commutative because $a \star e = a$ and $e \star a = a$.

This second monoid is also commutative for the same reason.

Define $\star$ on $\mathbb{N}$ by $a \star b = b$.

13. Is $\star$ a binary operation on $\mathbb{N}$?

Yes. The output is always b.

14. Is $\star$ commutative in $\mathbb{N}$?

No. For example, $2 \star 3 = 3$, whereas $3 \star 2 = 2$.

15. Is $\star$ associative in $\mathbb{N}$?

Yes. $(a \star b) \star c = c = b \star c = a \star (b \star c)$.

16. Does $\mathbb{N}$ have an identity with respect to $\star$? If so, what is it?

No. Let $a \in \mathbb{N}$. Then $(a+1) \star a = a \neq a + 1$. So, a is not an identity. Since a was an arbitrary element of $\mathbb{N}$, we showed that there is no identity.

17. Is $(\mathbb{N},\star)$ a semigroup?

Yes by Problem 13 and Problem 15.

18. Is $(\mathbb{N},\star)$ a monoid?

No by Problem 16.

+	0	1
0	0	1
1	1	0

$\cdot$	0	1
0	1	0
1	0	1

The addition and multiplication tables above are defined on the set $S = \{0, 1\}$.

19. Does $+$ define a binary operation on S?

Yes because only 0 and 1 appears inside the addition table.

20. Does $\cdot$ define a binary operation on S?

Yes because only 0 and 1 appears inside the multiplication table.

21. Does S have an identity with respect to $+$? If so, what is it?

Yes. The identity is 0.

22. Does S have an identity with respect to $\cdot$? If so, what is it?

Yes. The identity is 1.

23. Is $\cdot$ distributive over $+$ in S?

No. For example, $0(1+1) = 0 \cdot 0 = 1$ and $0 \cdot 1 + 0 \cdot 1 = 0 + 0 = 0$. So, $0(1+1) \neq 0 \cdot 1 + 0 \cdot 1$.

24. Explain why $(S, +, \cdot)$ does **not** define a ring.

Explanation 1: By Problem 23, multiplication is **not** distributive over addition in S, and so, $(S, +, \cdot)$ does not define a ring.

Explanation 2: By Theorem 3.18, if $(S, +, \cdot)$ were a ring, we would have $0 \cdot 0 = 0$. So, $(S, +, \cdot)$ does not define a ring.

$+$	0	1	2
0	0	1	2
1	1	2	0
2	2	0	1

$\cdot$	0	1	2
0	0	0	0
1	0	1	2
2	0	2	2

The addition and multiplication tables above are defined on the set $S = \{0, 1, 2\}$.

25. Show that $(S, +, \cdot)$ does **not** define a field.

We have $2 \cdot 0 = 0$, $2 \cdot 1 = 2$, and $2 \cdot 2 = 2$. So, 2 has no multiplicative inverse, and therefore, $(S, +, \cdot)$ does **not** define a field.

26. Does $(S, +, \cdot)$ define a ring?

No because distributivity fails. For example, $2(1+1) = 2 \cdot 2 = 2$, whereas $2 \cdot 1 + 2 \cdot 1 = 2 + 2 = 1$.

Let $S = \{0, 1\}$ and suppose that $(S, +, \cdot)$ is a ring with additive identity 0 and multiplicative identity 1.

27. Draw the tables for addition and multiplication.

Since $(S, +)$ is a commutative group, by the solutions to Problems 10 and 11, the addition table must be the following.

$+$	0	1
0	0	1
1	1	0

Since $(S, \cdot)$ is a monoid and 1 is the multiplicative identity, by the solution to Problem 10, the multiplication table must be one of the following.

$\cdot$	0	1
0	1	0
1	0	1

$\cdot$	0	1
0	0	0
1	0	1

However, we showed in Problem 24 that if we use the table on the left, then $(S, +, \cdot)$ will **not** define a ring.

So, the addition and multiplication tables must be as follows:

$+$	0	1
0	0	1
1	1	0

$\cdot$	0	1
0	0	0
1	0	1

28. Verify that with the tables you drew in Problem 27 that $(S, +, \cdot)$ is a ring.

Since we already know that $(S, +)$ is a commutative group and $(S, \cdot)$ is a monoid, all we need to verify is that distributivity holds. Since $\cdot$ is commutative for S (by the solution to Problem 12), it suffices to verify left distributivity. We will do this by brute force. There are eight instances to check.

$$\begin{array}{ll}
0(0+0) = 0 \cdot 0 = 0 & 0 \cdot 0 + 0 \cdot 0 = 0 + 0 = 0 \\
0(0+1) = 0 \cdot 1 = 0 & 0 \cdot 0 + 0 \cdot 1 = 0 + 0 = 0 \\
0(1+0) = 0 \cdot 1 = 0 & 0 \cdot 1 + 0 \cdot 0 = 0 + 0 = 0 \\
0(1+1) = 0 \cdot 0 = 0 & 0 \cdot 1 + 0 \cdot 1 = 0 + 0 = 0 \\
1(0+0) = 1 \cdot 0 = 0 & 1 \cdot 0 + 1 \cdot 0 = 0 + 0 = 0 \\
1(0+1) = 1 \cdot 1 = 1 & 1 \cdot 0 + 1 \cdot 1 = 0 + 1 = 1 \\
1(1+0) = 1 \cdot 1 = 1 & 1 \cdot 1 + 1 \cdot 0 = 1 + 0 = 1 \\
1(1+1) = 1 \cdot 0 = 0 & 1 \cdot 1 + 1 \cdot 1 = 1 + 1 = 0
\end{array}$$

So, we see that left distributivity holds, and therefore $(S, +, \cdot)$ is a ring.

29. Is $(S, +, \cdot)$ a field?

Yes. It is easy to see from the multiplication table that multiplication is commutative in S. Also, 1 is invertible with $1^{-1} = 1$. It follows that $(S, +, \cdot)$ is a field.

LEVEL 2

Define $\star$ on $\mathbb{Z}$ by $a \star b = \max\{a, b\}$, where $\max\{a, b\}$ is the largest of a or b.

30. Is $\star$ a binary operation on $\mathbb{Z}$?

Yes. The output is always equal to one of the inputs.

31. Is $\star$ commutative in $\mathbb{Z}$?

Yes. There are two cases to consider:

Case 1 ($a \leq b$): $a \star b = \max\{a, b\} = b = \max\{b, a\} = b \star a$
Case 2 ($b \leq a$): $a \star b = \max\{a, b\} = a = \max\{b, a\} = b \star a$

32. Is $\star$ associative in $\mathbb{Z}$?

Yes. There are 6 cases to consider:

Case 1 ($a \leq b \leq c$):

$$(a \star b) \star c = \max\{a, b\} \star c = b \star c = \max\{b, c\} = c$$
$$a \star (b \star c) = a \star \max\{b, c\} = a \star c = \max\{a, c\} = c$$

Case 2 ($a \leq c \leq b$):

$$(a \star b) \star c = \max\{a, b\} \star c = b \star c = \max\{b, c\} = b$$
$$a \star (b \star c) = a \star \max\{b, c\} = a \star b = \max\{a, b\} = b$$

Case 3 ($b \leq a \leq c$):

$$(a \star b) \star c = \max\{a, b\} \star c = a \star c = \max\{a, c\} = c$$
$$a \star (b \star c) = a \star \max\{b, c\} = a \star c = \max\{a, c\} = c$$

Case 4 ($b \leq c \leq a$):

$$(a \star b) \star c = \max\{a, b\} \star c = a \star c = \max\{a, c\} = a$$
$$a \star (b \star c) = a \star \max\{b, c\} = a \star c = \max\{a, c\} = a$$

Case 5 ($c \leq a \leq b$):

$$(a \star b) \star c = \max\{a, b\} \star c = b \star c = \max\{b, c\} = b$$
$$a \star (b \star c) = a \star \max\{b, c\} = a \star b = \max\{a, b\} = b$$

Case 6 ($c \leq b \leq a$):

$$(a \star b) \star c = \max\{a, b\} \star c = a \star c = \max\{a, c\} = a$$
$$a \star (b \star c) = a \star \max\{b, c\} = a \star b = \max\{a, b\} = a$$

33. Does $\mathbb{Z}$ have an identity with respect to $\star$? If so, what is it?

No. Let $a \in \mathbb{Z}$. Then $(a - 1) \star a = a \neq a - 1$. So, a is not an identity. Since a was an arbitrary element of $\mathbb{Z}$, we showed that there is no identity.

34. Is $(\mathbb{Z},\star)$ a commutative semigroup?

Yes by Problems 30, 31, and 32.

35. Is $(\mathbb{Z},\star)$ a commutative monoid?

No by Problem 33.

Let $S = \{-1, 1, -i, i\}$, where i is the complex number such that $i^2 = -1$.

36. Is $+$ a binary operation on S? If so, draw the multiplication table.

No. For example, $1 + 1 = 2 \notin S$.

37. Is $\cdot$ a binary operation on S? If so, draw the multiplication table.

Yes.

$\cdot$	1	-1	i	$-i$
1	1	-1	i	$-i$
-1	-1	1	$-i$	i
i	i	$-i$	-1	1
$-i$	$-i$	i	1	-1

38. Explain why $\cdot$ is associative in S.

$S \subseteq \mathbb{C}$ and $\cdot$ is associative in $\mathbb{C}$.

39. Is $(S, \cdot)$ a group? If so, what is the identity, and what is the inverse of each element in S?

Yes. Identity is 1, $1^{-1} = 1$, $(-1)^{-1} = -1$, $i^{-1} = -i$, $(-i)^{-1} = i$

Find all zero divisors in each of the following rings.

40. $(C_5, +, \cdot)$

No zero divisors.

41. $(C_6, +, \cdot)$

$\mathbf{2}, \mathbf{3}$, and $\mathbf{4}$ because $2 \cdot 3 = 0$ and $4 \cdot 3 = 0$.

42. $(\mathbb{Z}, +, \cdot)$

No zero divisors.

Let F be a field and let $F^n = \{(a_1, a_2, \dots, a_n) \mid a_i \in F \text{ for each } i = 1, 2, \dots, n\}$. Define addition and scalar multiplication on F^n as follows:

$$(a_1, a_2, \dots, a_n) + (b_1, b_2, \dots, b_n) = (a_1 + b_1, a_2 + b_2, \dots, a_n + b_n).$$
$$k(a_1, a_2, \dots, a_n) = (ka_1, ka_2, \dots, ka_n).$$

Prove that each of the following is true

43. F^n is closed under addition.

Let $(a_1, a_2, \dots, a_n), (b_1, b_2, \dots, b_n) \in F^n$. Then $a_1, a_2, \dots, a_n, b_1, b_2, \dots, b_n \in F$. By definition, $(a_1, a_2, \dots, a_n) + (b_1, b_2, \dots, b_n) = (a_1 + b_1, a_2 + b_2, \dots, a_n + b_n)$. Since F is closed under addition, $a_1 + b_1, a_2 + b_2, \dots, a_n + b_n \in F$. Therefore, $(a_1, a_2, \dots, a_n) + (b_1, b_2, \dots, b_n) \in F$.

44. Addition is associative in F^n.

Let $(a_1, a_2, \ldots, a_n), (b_1, b_2, \ldots, b_n), (c_1, c_2, \ldots, c_n) \in F^n$. Since addition is associative in F, we have

$$\begin{aligned}[(a_1, a_2, \ldots, a_n) + (b_1, b_2, \ldots, b_n)] + (c_1, c_2, \ldots, c_n) &= (a_1 + b_1, a_2 + b_2, \ldots, a_n + b_n) + (c_1, c_2, \ldots, c_n) \\ &= \big((a_1 + b_1) + c_1, (a_2 + b_2) + c_2, \ldots, (a_n + b_n) + c_n\big) \\ &= (a_1 + (b_1 + c_1), a_2 + (b_2 + c_2), \ldots, a_n + (b_n + c_n)) \\ &= (a_1, a_2, \ldots, a_n) + (b_1 + c_1, b_2 + c_2, \ldots, b_n + c_n) \\ &= (a_1, a_2, \ldots, a_n) + [(b_1, b_2, \ldots, b_n) + (c_1, c_2, \ldots, c_n)].\end{aligned}$$

45. Addition is commutative in F^n.

Let $(a_1, a_2, \ldots, a_n), (b_1, b_2, \ldots, b_n) \in F$. Since addition is commutative in F, we have

$$\begin{aligned}(a_1, a_2, \ldots, a_n) + (b_1, b_2, \ldots, b_n) &= (a_1 + b_1, a_2 + b_2, \ldots, a_n + b_n) = (b_1 + a_1, b_2 + a_2, \ldots, b_n + a_n) \\ &= (b_1, b_2, \ldots, b_n) + (a_1, a_2, \ldots, a_n).\end{aligned}$$

46. There is an identity $\mathbf{0}$ in F^n.

We show that $(0, 0, \ldots, 0)$ is an additive identity for F^n. Let $(a_1, a_2, \ldots, a_n) \in F^n$. Since 0 is an additive identity for F, we have

$$(0, 0, \ldots, 0) + (a_1, a_2, \ldots, a_n) = (0 + a_1, 0 + a_2, \ldots, 0 + a_n) = (a_1, a_2, \ldots, a_n).$$

$$(a_1, a_2, \ldots, a_n) + (0, 0, \ldots, 0) = (a_1 + 0, a_2 + 0, \ldots, a_n + 0) = (a_1, a_2, \ldots, a_n).$$

47. Each element in F^n has an additive inverse in F^n.

Let $(a_1, a_2, \ldots, a_n) \in F^n$. Then $a_1, a_2, \ldots, a_n \in F$. Since F has the additive inverse property, $-a_1, -a_2, \ldots, -a_n \in F$. So, $(-a_1, -a_2, \ldots, -a_n) \in F^n$ and

$$(a_1, a_2, \ldots, a_n) + (-a_1, -a_2, \ldots, -a_n) = (a_1 - a_1, a_2 - a_2, \ldots, a_n - a_n) = (0, 0, \ldots, 0).$$

$$(-a_1, -a_2, \ldots, -a_n) + (a_1, a_2, \ldots, a_n) = (-a_1 + a_1, -a_2 + a_2, \ldots, -a_n + a_n) = (0, 0, \ldots, 0).$$

48. $(F^n, +)$ is a commutative group.

This follows from Problems 43 through 47.

49. F^n is closed under scalar multiplication.

Let $k \in F$ and let $(a_1, a_2, \ldots, a_n) \in F^n$. Then $a_1, a_2, \ldots, a_n \in F$. By definition, we have $k(a_1, a_2, \ldots, a_n) = (ka_1, ka_2, \ldots, ka_n)$. Since F is closed under multiplication, $ka_1, ka_2, \ldots, ka_n \in F$. Therefore, $k(a_1, a_2, \ldots, a_n) \in F^n$.

50. For each $x \in F^n$, $1x = x$.

Let $(a_1, a_2, \ldots, a_n) \in F^n$. Then $1(a_1, a_2, \ldots, a_n) = (1a_1, 1a_2, \ldots, 1a_n) = (a_1, a_2, \ldots, a_n)$.

51. For all $j,k \in F$ and each $x \in F^n$, $(jk)x = j(kx)$.

Let $j,k \in F$ and $(a_1, a_2, \dots, a_n) \in F$. Then since multiplication is associative in F, we have

$$(jk)(a_1, a_2, \dots, a_n) = \big((jk)a_1, (jk)a_2, \dots, (jk)a_n\big) = (j(ka_1), j(ka_2), \dots, j(ka_n))$$
$$= j(ka_1, ka_2, \dots, ka_n) = j\big(k(a_1, a_2, \dots, a_n)\big).$$

52. For each $k \in F$ and all $x, y \in F^n$, $k(x + y) = kx + ky$.

Let $k \in F$ and $(a_1, a_2, \dots, a_n), (b_1, b_2, \dots, b_n) \in F$. Since multiplication is distributive over addition in F, we have

$$k\big((a_1, a_2, \dots, a_n) + (b_1, b_2, \dots, b_n)\big) = k\big((a_1 + b_1, a_2 + b_2, \dots, a_n + b_n)\big)$$
$$= \big(k(a_1 + b_1), k(a_2 + b_2), \dots, k(a_n + b_n)\big) = \big((ka_1 + kb_1), (ka_2 + kb_2), \dots, (ka_n + kb_n)\big)$$
$$= (ka_1, ka_2, \dots, ka_n) + (kb_1, kb_2, \dots, kb_n) = k(a_1, a_2, \dots, a_n) + k(b_1, b_2, \dots, b_n).$$

53. For all $j,k \in F$ and each $x \in F$, $(j + k)x = jx + kx$.

Let $j,k \in F$ and $(a_1, a_2, \dots, a_n) \in F^n$. Since multiplication is distributive over addition in F, we have

$$(j + k)(a_1, a_2, \dots, a_n) = \big((j + k)a_1, (j + k)a_2, \dots, (j + k)a_n\big)$$
$$= (ja_1 + ka_1, ja_2 + ka_2, \dots, ja_n + ka_n) = (ja_1, ja_2, \dots, ja_n) + (ka_1, ka_2, \dots, ka_n)$$
$$= j(a_1, a_2, \dots, a_n) + k(a_1, a_2, \dots, a_n).$$

54. F^n is a vector space over F.

This follows from Problems 48 through 53.

LEVEL 3

Let $(G,\star)$ be a group with $G = \{e, a, b, c\}$, where e is the identity. Construct the multiplication table for G under the following conditions:

55. $a^2 = e$ and $b^2 = e$

Since $e \star x = x \star e = x$ for all x in the group, we can easily fill out the first row, the first column, and two more entries of the table.

$\star$	e	a	b	c
e	e	a	b	c
a	a	e	$\boxdot$	
b	b	$\boxdot$	e	
c	c			

Each of the entries labeled with $\boxdot$ cannot be a, e, or b (Why?), and so they must be c. So, we get the following:

$\star$	e	a	b	c
e	e	a	b	c
a	a	e	c	b
b	b	c	e	
c	c	b		

The rest of the table is now determined:

$\star$	e	a	b	c
e	e	a	b	c
a	a	e	c	b
b	b	c	e	a
c	c	b	a	e

Note: This table gives a group $(G,\star)$ called the **Klein four group**.

56. $a^2 = b$ and $b^2 = e$

Since $e \star x = x \star e = x$ for all x in the group, we can easily fill out the first row and the first column of the table.

$\star$	e	a	b	c
e	e	a	b	c
a	a			
b	b			
c	c			

We now add in $a \star a = a^2 = b$ and $b \star b = b^2 = e$.

$\star$	e	a	b	c
e	e	a	b	c
a	a	b	$\boxdot$	
b	b		e	
c	c			

Now, the entry labeled with $\boxdot$ cannot be a or b because a and b appear in that row. It also cannot be e because e appears in that column. Therefore, the entry labeled with $\boxdot$ must be c. It follows that the entry to the right of $\boxdot$ must be e, and the entry at the bottom of the column must be a.

$\star$	e	a	b	c
e	e	a	b	c
a	a	b	c	e
b	b	$\odot$	e	
c	c		a	

Now, the entry labeled with $\odot$ cannot be b or e because b and e appear in that row. It also cannot be a because a appears in that column. Therefore, the entry labeled with $\odot$ must be c. The rest of the table is then determined.

$\star$	e	a	b	c
e	e	a	b	c
a	a	b	c	e
b	b	c	e	a
c	c	e	a	b

Note: Observe that in the table we produced, $b = a \star a = a^2$ and $c = b \star a = a^2 \star a = a^3$. So, another way to draw the table is as follows:

$\star$	e	a	a^2	a^3
e	e	a	a^2	a^3
a	a	a^2	a^3	e
a^2	a^2	a^3	e	a
a^3	a^3	e	a	a^2

This group is the **cyclic group of order 4**.

Let A be a nonempty set.

57. Prove that $(\mathcal{P}(A), \ \cap)$ is a commutative monoid.

Proof: Let $X, Y \in \mathcal{P}(A)$, then $X \subseteq A$ nd $Y \subseteq A$. If $x \in X \cap Y$, then $x \in X$ and $x \in Y$. In particular, $x \in X$. Since $X \subseteq A$, $x \in A$. So, $\forall x(x \in X \cap Y \to x \in A)$. It follows that $X \cap Y \subseteq A$, and therefore, $X \cap Y \in \mathcal{P}(A)$. This shows that $\cap$ is a binary operation on $\mathcal{P}(A)$.

Let $X, Y \in \mathcal{P}(A)$. Then $x \in X \cap Y$ if and only if $x \in X$ and $x \in Y$ if and only if $x \in Y$ and $x \in X$ if and only if $x \in Y \cap X$. Since x was arbitrary, we have $\forall x(x \in X \cap Y \leftrightarrow x \in Y \cap X)$. Therefore, $X \cap Y = Y \cap X$. So, the operation of forming intersections is commutative.

Let $X, Y, Z \in \mathcal{P}(A)$. Then $x \in (X \cap Y) \cap Z$ if and only if $x \in X \cap Y$ and $x \in Z$ if and only if $x \in X$, $x \in Y$, and $x \in Z$ if and only if $x \in X$ and $x \in Y \cap Z$ if and only if $x \in X \cap (Y \cap Z)$. Since x was arbitrary, we have shown $\forall x\big(x \in (X \cap Y) \cap Z \leftrightarrow x \in X \cap (Y \cap Z)\big)$. Therefore, we have shown that $(X \cap Y) \cap Z = X \cap (Y \cap Z)$. So, the operation of forming intersections is associative.

A is an identity for $(\mathcal{P}(A), \ \cap)$ because if $X \in \mathcal{P}(A)$, then $X \cap A = X$ and $A \cap X = X$. It follows that $(\mathcal{P}(A), \ \cap)$ is a commutative monoid. □

58. Prove that $(\mathcal{P}(A), \ \cap)$ is not a group.

To see that $(\mathcal{P}(A), \ \cap)$ is **not** a group, note that $\emptyset$ has no inverse in $\mathcal{P}(A)$. Indeed, if $B \in \mathcal{P}(A)$, then by Theorem 2.11, $\emptyset \subseteq B$. So, by Problem 260 in Problem Set 2, $B \cap \emptyset = \emptyset \neq A$. □

Let $(R, +, \ \cdot)$ be a ring and define addition and multiplication on $R \times R$ componentwise. That is, for $a, b, c, d \in R$, we define addition and multiplication by $(a, b) + (c, d) = (a + c, b + d)$ and $(a, b)(c, d) = (ac, bd)$.

59. Prove that $(R \times R, +, \ \cdot)$ is a ring.

Proof: Let $(a,b),(c,d) \in R \times R$. Then $a,b,c,d \in R$. Since R is closed under addition and multiplication, $a+c, b+d, ac, bd \in R$. So, $(a+c, b+d), (ac, bd) \in R \times R$. So, $R \times R$ is closed under addition and multiplication.

Let $(a,b),(c,d),(e,f) \in R \times R$. Since addition and multiplication are associative in R, we have

$$(a,b) + \big((c,d)+(e,f)\big) = (a,b) + (c+e, d+f) = \big(a+(c+e), b+(d+f)\big)$$
$$= \big((a+c)+e, (b+d)+f\big) = (a+c, b+d) + (e,f) = \big((a,b)+(c,d)\big) + (e,f).$$
$$(a,b)\cdot\big((c,d)\cdot(e,f)\big) = (a,b)\cdot(ce, df) = \big(a(ce), b(df)\big)$$
$$= \big((ac)e, (bd)f\big) = (ac, bd)\cdot(e,f) = \big((a,b)\cdot(c,d)\big)\cdot(e,f).$$

So, addition and multiplication are associative in $R \times R$.

For all $(a,b) \in R \times R$, we have

$$(0,0) + (a,b) = (0+a, 0+b) = (a,b) \text{ and } (a,b) + (0,0) = (a+0, b+0) = (a,b).$$
$$(1,1)\cdot(a,b) = (1a, 1b) = (a,b) \text{ and } (a,b)\cdot(1,1) = (a\cdot 1, b\cdot 1) = (a,b).$$

So, $(0,0)$ is an additive identity and $(1,1)$ is a multiplicative identity.

For all $(a,b) \in R \times R$, we have

$$(a,b) + (-a,-b) = (a-a, b-b) = (0,0) \text{ and } (-a,-b) + (a,b) = (-a+a, -b+b) = (0,0).$$

So, $(-a,-b)$ is an additive inverse of (a,b).

Let $(a,b),(c,d) \in R \times R$. Since addition is commutative in R, we have

$$(a,b) + (c,d) = (a+c, b+d) = (c+a, d+b) = (c,d) + (a,b).$$

So, addition is commutative in $R \times R$.

Let $(a,b),(c,d),(e,f) \in R \times R$. Since multiplication is left distributive over addition in R, we have

$$(a,b)\big((c,d)+(e,f)\big) = (a,b)(c+e, d+f) = \big(a(c+e), b(d+f)\big)$$
$$= (ac+ae, bd+bf) = (ac, bd) + (ae, bf) = (a,b)(c,d) + (a,b)(e,f).$$

So, multiplication is left distributive over addition in $R \times R$.

Right distributivity is proved similarly.

Therefore, $(R \times R, +, \cdot)$ is a ring. □

60. If $(R, +, \cdot)$ is a domain, does it follow that $(R \times R, +, \cdot)$ must be a domain? If so, prove it. If not, provide a counterexample.

If $R \neq \{0\}$, then $(R \times R, +, \cdot)$ will **not** be a domain. To see this, let $a \in R$ with $a \neq 0$. Then we have $(a,0),(0,a) \in R \times R$ and $(a,0)\cdot(0,a) = (a\cdot 0, 0\cdot a) = (0,0)$. Since $(a,0),(0,a) \neq (0,0)$, we see that $(a,0)$ and $(0,a)$ are zero divisors.

Let V be a vector space over a field F. Prove each of the following:

61. For every $v \in V$, $-(-v) = v$.

Proof: Since $-v$ is the additive inverse of v, we have $v + (-v) = -v + v = 0$. But this equation also says that v is the additive inverse of $-v$. So, $-(-v) = v$. □

62. For every $v \in V$, $0v = 0$.

Proof: Let $v \in V$. Then $0v = (0 + 0)v = 0v + 0v$. So, we have

$$0 = -0v + 0v = -0v + (0v + 0v) = (-0v + 0v) + 0v = 0 + 0v = 0v.$$ □

63. For every $k \in F$, $k \cdot 0 = 0$.

Proof: Let $k \in F$. Then $k \cdot 0 = k(0 + 0) = k \cdot 0 + k \cdot 0$. So, we have

$$0 = -k \cdot 0 + k \cdot 0 = -k \cdot 0 + (k \cdot 0 + k \cdot 0) = (-k \cdot 0 + k \cdot 0) + k \cdot 0 = 0 + k \cdot 0 = k \cdot 0.$$ □

64. For every $v \in V$, $-1v = -v$.

Proof: Let $v \in V$. Then we have $v + (-1v) = 1v + (-1v) = \big(1 + (-1)\big)v = 0v = 0$ by Problem 62 and we have $-1v + v = -1v + 1v = (-1 + 1)v = 0v = 0$ again by Problem 62. So, $-1v = -v$. □

LEVEL 4

Let $(G,\star)$ be a group with $a, b \in G$, and let a^{-1} and b^{-1} be the inverses of a and b, respectively. Prove each of the following:

65. $(a \star b)^{-1} = b^{-1} \star a^{-1}$.

Proof: Let $a, b \in G$. Then we have

$$(a \star b) \star (b^{-1} \star a^{-1}) = a \star \big(b \star (b^{-1} \star a^{-1})\big) = a \star \big((b \star b^{-1}) \star a^{-1}\big) = a \star (e \star a^{-1}) = a \star a^{-1} = e.$$

So, $(a \star b)^{-1} = (b^{-1} \star a^{-1})$. □

Notes: (1) For the first and second equalities we used the associativity of $\star$ in G.

(2) For the third equality, we used the inverse property of $\star$ in G.

(3) For the fourth equality, we used the identity property of $\star$ in G.

(4) For the last equality, we again used the inverse property of $\star$ in G.

(5) Since multiplying $a \star b$ by $b^{-1} \star a^{-1}$ results in the identity element e, it follows that $b^{-1} \star a^{-1}$ is the inverse of $a \star b$.

(6) By Note 4 following Theorem 3.9, we get $(b^{-1} \star a^{-1}) \star (a \star b) = e$ for free.

66. The inverse of a^{-1} is a.

Proof: Let $a \in G$. Since a^{-1} is the inverse of a, we have $a \star a^{-1} = a^{-1} \star a = e$. But this sequence of equations also says that a is the inverse of a^{-1}. □

Let $(F, +, \cdot)$ be a field with $\mathbb{N} \subseteq F$.

67. Prove that $\mathbb{Z} \subseteq F$.

Proof: Let $n \in \mathbb{Z}$. If $n \in \mathbb{N}$, then $n \in F$ because $\mathbb{N} \subseteq F$. If $n \notin \mathbb{N}$, then $-n \in \mathbb{N}$. So, $-n \in F$. Since F is a field, we have $n = -(-n) \in F$. (Note that $-(-n) = n$ by Problem 66 above. □

68. Prove that $\left\{\frac{1}{n} \middle| n \in \mathbb{Z}^*\right\} \subseteq F$.

Proof: Let $n \in \mathbb{Z}^*$. By Problem 67, $n \in F$. Then $\frac{1}{n} = n^{-1} \in F$ because $n \in F$ and the multiplicative inverse property holds in F. □

69. Prove that $\mathbb{Q} \subseteq F$.

Proof: Let $\frac{m}{n} \in \mathbb{Q}$. Then $m \in \mathbb{Z}$ and $n \in \mathbb{Z}^*$. Since $\mathbb{Z} \subseteq F$, $m \in F$. Since $n \in \mathbb{Z}^*$ by Problem 68, we have $\frac{1}{n} \in F$. Therefore, $\frac{m}{n} = \frac{m \cdot 1}{1 \cdot n} = \frac{m}{1} \cdot \frac{1}{n} = m\left(\frac{1}{n}\right) \in F$ because F is closed under multiplication. Since $\frac{m}{n}$ was an arbitrary element of $\mathbb{Q}$, we see that $\mathbb{Q} \subseteq F$. □

Let $(F, +, \cdot)$ be a field.

70. Prove that $\cdot$ is commutative in F.

Proof: Let $x, y \in F$. If $x, y \in F^*$, then $xy = yx$. If $x = 0$, then $xy = 0y = 0$ and $yx = y \cdot 0 = 0$ by Theorem 3.18. If $y = 0$, then $xy = x \cdot 0 = 0$ and $yx = 0x = 0$ by Theorem 3.18. In all cases, we have $xy = yx$. □

71. Prove that $\cdot$ is associative in F.

Proof: Let $x, y, z \in F$. If $x, y, z \in F^*$, then $(xy)z = x(yz)$. If $x = 0$, then $(xy)z = (0y)z = 0z = 0$ and $x(yz) = 0(yz) = 0$. If $y = 0$, then $(xy)z = (x \cdot 0)z = 0z = 0$ and $x(yz) = x(0z) = x \cdot 0 = 0$. If $z = 0$, we have $(xy)z = (xy) \cdot 0 = 0$ and $x(yz) = x(y \cdot 0) = x \cdot 0 = 0$. In all cases, we have $(xy)z = x(yz)$. □

Note: We used Theorem 3.18 several times in the proof above.

72. Prove that 1 is a multiplicative identity in F.

Proof: Let $(F, +, \cdot)$ be a field. Then $\cdot$ is a binary operation on F and $(F^*, \cdot)$ is a commutative group.

Let $x \in F$. If $x \in F^*$, then $1x = x \cdot 1 = x$. If $x = 0$, then by Theorem 3.18, $1x = 1 \cdot 0 = 0$ and $x \cdot 1 = 0 \cdot 1 = 0$. In all cases, we have $1x = x \cdot 1 = x$. □

73. Prove that $(F, \cdot)$ is a commutative monoid.

Proof: We first show that $\cdot$ is a binary operation on F.

Let $x, y \in F$. If $x, y \in F^*$, then $xy \in F$ because $\cdot$ is a binary operation on F^*. If $x = 0$, then we have $xy = 0y = 0 \in F$ by Theorem 3.18. If $y = 0$, then $xy = x \cdot 0 = 0 \in F$ by Theorem 3.18. In all cases, we have $xy \in F$. So, $\cdot$ is a binary operation on F.

By Problems 70 – 72, $(F, \cdot)$ is a commutative monoid. □

Let $\mathbb{Q}$ be the set of rational numbers.

74. Prove that multiplication is commutative in $\mathbb{Q}$.

Proof: Let $\frac{a}{b}, \frac{c}{d} \in \mathbb{Q}$. Then $a, b, c, d \in \mathbb{Z}$ and $b, d \neq 0$. Since multiplication is commutative in $\mathbb{Z}$, we have

$$\frac{a}{b} \cdot \frac{c}{d} = \frac{ac}{bd} = \frac{ca}{db} = \frac{c}{d} \cdot \frac{a}{b}$$

Since $\frac{a}{b}, \frac{c}{d} \in \mathbb{Q}$ were arbitrary, multiplication is commutative in $\mathbb{Q}$. □

75. Prove that addition is commutative in $\mathbb{Q}$.

Let $\frac{a}{b}, \frac{c}{d} \in \mathbb{Q}$. Then $a, b, c, d \in \mathbb{Z}$ and $b, d \neq 0$. Since addition and multiplication are commutative in $\mathbb{Z}$,

$$\frac{a}{b} + \frac{c}{d} = \frac{ad + bc}{bd} = \frac{cb + da}{db} = \frac{c}{d} + \frac{a}{b}$$

Since $\frac{a}{b}, \frac{c}{d} \in \mathbb{Q}$ were arbitrary, addition is commutative in $\mathbb{Q}$. □

Let $(R, +, \cdot)$ be a ring. Prove each of the following:

76. If $a, b \in R$ with $a + b = b$, then $a = 0$.

Proof: Let $a, b \in R$ with $a + b = b$. Then we have

$$a = a + 0 = a + \big(b + (-b)\big) = (a + b) + (-b) = b + (-b) = 0.$$ □

77. If $a, b \in R$, b^{-1} exists, and $ab = b$, then $a = 1$.

Proof: Let $a, b \in R$ and $ab = b$. Then we have

$$a = a \cdot 1 = a(bb^{-1}) = (ab)b^{-1} = bb^{-1} = 1.$$ □

78. If $a, b \in R$, a^{-1} exists, and $ab = 1$, then $b = \frac{1}{a}$.

Proof: Let $a, b \in R$ and $ab = 1$. Then $b = 1b = (a^{-1}a)b = a^{-1}(ab) = a^{-1} \cdot 1 = a^{-1} = \frac{1}{a}$. □

79. If $a \in R$, then $-a = -1a = a(-1)$.

Proof: Let $a \in R$. Then $-1a + a = -1a + 1a = (-1+1)a = 0 \cdot a = 0$ (by Theorem 3.18). So, $-1a$ is the additive inverse of a. So, $-1a = -a$. Also, $a(-1) + a = a(-1) + a \cdot 1 = a(-1+1) = a \cdot 0 = 0$ (by Theorem 3.18). So, $a(-1)$ is also the additive inverse of a. Therefore, $a(-1) = -a$. □

80. $(-1)(-1) = 1$.

Proof: $(-1)(-1) + (-1) = (-1)(-1) + (-1) \cdot 1 = (-1)(-1+1) = (-1)(0) = 0$ (by Theorem 3.18). So, $(-1)(-1)$ is the additive inverse of -1. Therefore, $(-1)(-1) = -(-1)$. □

Let $(R, +, \cdot)$ be a ring.

81. Prove that if $(R, +, \cdot)$ is an integral domain, then $(R, +, \cdot)$ is **not** necessarily a field.

$(\mathbb{Z}, +, \cdot)$ is an integral domain that is not a field.

82. Prove that if $(R, +, \cdot)$ is a field, then $(R, +, \cdot)$ is an integral domain. Are all rings integral domains?

Proof: Let F be a field, $a, b \in F$ and $ab = 0$. Assume that $a \neq 0$. Then

$$b = 1b = (a^{-1}a)b = a^{-1}(ab) = a^{-1} \cdot 0.$$

By Theorem 3.18, $a^{-1} \cdot 0 = 0$. So, $b = 0$. □

Not all rings are integral domains. $(C_4, +, \cdot)$ is a counterexample (clock arithmetic with a 4-hour clock). Here are the addition and multiplication tables:

+	0	1	2	3
0	0	1	2	3
1	1	2	3	0
2	2	3	0	1
3	3	0	1	2

·	0	1	2	3
0	0	0	0	0
1	0	1	2	3
2	0	2	0	2
3	0	3	2	1

Observe that in this ring, $2 \cdot 2 = 0$.

Prove each of the following:

83. Let $(G, \star)$ be a group such that $a^2 = e$ for all $a \in G$. Then $(G, \star)$ is commutative.

Proof: Let $a, b \in G$. Then $(a \star a) \star (b \star b) = a^2 \star b^2 = e \star e = e = (ab)^2 = (a \star b) \star (a \star b)$. So, we have $(a \star a) \star (b \star b) = (a \star b) \star (a \star b)$.

We multiply on the left by a^{-1} and on the right by b^{-1} to get

$$a^{-1} \star (a \star a \star b \star b) \star b^{-1} = a^{-1} \star (a \star b \star a \star b) \star b^{-1}$$
$$(a^{-1} \star a) \star a \star b \star (b \star b^{-1}) = (a^{-1} \star a) \star b \star a \star (b \star b^{-1})$$
$$(e \star a) \star (b \star e) = (e \star b) \star (a \star e)$$
$$a \star b = b \star a$$ □

Note: To make the proof less tedious, we have omitted some of the parentheses starting in the fourth line. The associativity of $\star$ allows us to do this. In general, there are two possible meanings for the expression $x \star y \star z$. It could mean $(x \star y) \star z$ or it could mean $x \star (y \star z)$. Since both meanings produce the same result (by associativity), we can simply write $x \star y \star z$ without worrying about the notation being unclear.

We have done this several times in the proof above. For example, we wrote $a \star a \star b \star b$ in the fourth line. This could have multiple meanings, but all those meanings lead to the same result.

84. Let $\mathbb{Z}[x]_n = \{a_n x^n + a_{n-1}x^{n-1} + \cdots + a_1 x + a_0 \mid a_0, a_1, \ldots, a_n \in \mathbb{Z}\}$. In other words, $\mathbb{Z}[x]_n$ consists of all polynomials of degree at most n with integer coefficients. Then $(\mathbb{Z}[x]_n, +)$ is a commutative group for $n = 0$, 1, and 2, where addition is defined in the "usual way." What if we replace "polynomials of degree at most n" by polynomials of degree exactly n?" Do we still get a commutative group?

Proof: $\mathbb{Z}[x]_0 = \{a_0 \mid a_0 \in \mathbb{Z}\} = \mathbb{Z}$, and we already know that $(\mathbb{Z}, +)$ is a group.

$\mathbb{Z}[x]_1 = \{a_1 x + a_0 \mid a_0, a_1 \in \mathbb{Z}\}$. Let $a_1 x + a_0, b_1 x + b_0 \in \mathbb{Z}[x]_1$. We have

$$(a_1 x + a_0) + (b_1 x + b_0) = (a_1 + b_1)x + (a_0 + b_0).$$

Since $\mathbb{Z}$ is closed under addition, $a_1 + b_1, a_0 + b_0 \in \mathbb{Z}$. Therefore, $\mathbb{Z}[x]_1$ is closed under addition.

Simple computations show that addition is associative in $\mathbb{Z}[x]_1$, $0 = 0x + 0$ is an additive identity in $\mathbb{Z}[x]_1$, and the additive inverse of $a_1 x + a_0$ is $-a_1 x - a_0$. One more simple computation can be used to verify that addition is commutative in $\mathbb{Z}[x]_1$. So, $(\mathbb{Z}[x]_1, +)$ is a commutative group.

$\mathbb{Z}[x]_2 = \{a_2 x^2 + a_1 x + a_0 \mid a_0, a_1, a_2 \in \mathbb{Z}\}$. Let $a_2 x^2 + a_1 x + a_0, b_2 x^2 + b_1 x + b_0 \in \mathbb{Z}[x]_2$. We have

$$(a_2 x^2 + a_1 x + a_0) + (b_2 x^2 + b_1 x + b_0) = (a_2 + b_2)x^2 + (a_1 + b_1)x + (a_0 + b_0).$$

Since $\mathbb{Z}$ is closed under addition, $a_2 + b_2, a_1 + b_1, a_0 + b_0 \in \mathbb{Z}$. Therefore, $\mathbb{Z}[x]_2$ is closed under addition.

Simple computations show that addition is associative in $\mathbb{Z}[x]_2$, $0 = 0x^2 + 0x + 0$ is an additive identity in $\mathbb{Z}[x]_2$, and the additive inverse of $a_2 x^2 + a_1 x + a_0$ is $-a_2 x^2 - a_1 x - a_0$. One more simple computation can be used to verify that addition is commutative in $\mathbb{Z}[x]_2$. So, $(\mathbb{Z}[x]_2, +)$ is a commutative group.

If we replace "polynomials of degree at most n" by "polynomials of degree exactly n," we do not get a group. Closure is no longer satisfied. For example, $x + 1$ and $-x$ have degree 1, whereas the sum $(x + 1) + (-x) = 1$ has degree 0. □

LEVEL 5

Let $\mathbb{Q}$ be the set of rational numbers.

85. Prove that multiplication is associative in $\mathbb{Q}$.

Proof: Let $\frac{a}{b}, \frac{c}{d}, \frac{e}{f} \in \mathbb{Q}$. Then $a, b, c, d, e, f \in \mathbb{Z}$ and $b, d, f \neq 0$. Since multiplication is associative in $\mathbb{Z}$, we have

$$\left(\frac{a}{b} \cdot \frac{c}{d}\right)\frac{e}{f} = \left(\frac{ac}{bd}\right)\frac{e}{f} = \frac{(ac)e}{(bd)f} = \frac{a(ce)}{b(df)} = \frac{a}{b}\left(\frac{ce}{df}\right) = \frac{a}{b}\left(\frac{c}{d} \cdot \frac{e}{f}\right).$$

Since $\frac{a}{b}, \frac{c}{d}, \frac{e}{f} \in \mathbb{Q}$ were arbitrary, multiplication is associative in $\mathbb{Q}$. □

86. Prove that addition is associative in $\mathbb{Q}$.

Proof: Let $\frac{a}{b}, \frac{c}{d}, \frac{e}{f} \in \mathbb{Q}$. Then $a, b, c, d, e, f \in \mathbb{Z}$ and $b, d, f \neq 0$. Since multiplication and addition are associative in $\mathbb{Z}$, multiplication is (both left and right) distributive over addition in $\mathbb{Z}$, and multiplication is associative in $\mathbb{Z}^*$, we have

$$\left(\frac{a}{b} + \frac{c}{d}\right) + \frac{e}{f} = \frac{ad + bc}{bd} + \frac{e}{f} = \frac{(ad + bc)f + (bd)e}{(bd)f} = \frac{\big((ad)f + (bc)f\big) + (bd)e}{(bd)f}$$

$$= \frac{a(df) + \big(b(cf) + b(de)\big)}{b(df)} = \frac{a(df) + b(cf + de)}{b(df)} = \frac{a}{b} + \frac{cf + de}{df} = \frac{a}{b} + \left(\frac{c}{d} + \frac{e}{f}\right).$$

Since $\frac{a}{b}, \frac{c}{d}, \frac{e}{f} \in \mathbb{Q}$ were arbitrary, addition is associative in $\mathbb{Q}$. □

87. Prove that multiplication is distributive over addition in $\mathbb{Q}$.

Proof: Let $\frac{a}{b}, \frac{c}{d}, \frac{e}{f} \in \mathbb{Q}$. Then $a, b, c, d, e, f \in \mathbb{Z}$ and $b, d, f \neq 0$. Let's start with left distributivity.

$$\frac{a}{b}\left(\frac{c}{d} + \frac{e}{f}\right) = \frac{a}{b}\left(\frac{cf + de}{df}\right) = \frac{a(cf + de)}{b(df)}$$

$$\frac{a}{b} \cdot \frac{c}{d} + \frac{a}{b} \cdot \frac{e}{f} = \frac{ac}{bd} + \frac{ae}{bf} = \frac{(ac)(bf) + (bd)(ae)}{(bd)(bf)}$$

We need to verify that $\frac{(ac)(bf)+(bd)(ae)}{(bd)(bf)} = \frac{a(cf+de)}{b(df)}$.

Since $\mathbb{Z}$ is a ring, $(ac)(bf) + (bd)(ae) = bacf + bade = ba(cf + de)$.

Since multiplication is associative and commutative in $\mathbb{Z}$, we have

$$(bd)(bf) = b\big(d(bf)\big) = b\big((db)f\big) = b\big((bd)f\big) = b\big(b(df)\big).$$

So, $\frac{(ac)(bf)+(bd)(ae)}{(bd)(bf)} = \frac{ba(cf+de)}{b(b(df))} = \frac{a(cf+de)}{b(df)}$.

For right distributivity, we can use left distributivity together with the commutativity of multiplication in $\mathbb{Q}$.

$$(y + z)x = x(y + z) = xy + xz = yx + zx$$ □

88. Prove that $(\mathbb{Q}, +, \cdot)$ is a field.

Proof: We first check that $(\mathbb{Q}, +)$ is a commutative group.

(Closure) Let $x, y \in \mathbb{Q}$. Then there exist $a, c \in \mathbb{Z}$ and $b, d \in \mathbb{Z}^*$ such that $x = \frac{a}{b}$ and $y = \frac{c}{d}$. We have $x + y = \frac{a}{b} + \frac{c}{d} = \frac{ad+bc}{bd}$. Since $\mathbb{Z}$ is closed under multiplication, $ad \in \mathbb{Z}$ and $bc \in \mathbb{Z}$. Since $\mathbb{Z}$ is closed under addition, $ad + bc \in \mathbb{Z}$. Since $\mathbb{Z}^*$ is closed under multiplication, $bd \in \mathbb{Z}^*$. Therefore, $x + y \in \mathbb{Q}$.

(Associativity) This is Problem 86.

(Identity) Let $\overline{0} = \frac{0}{1}$. We show that $\overline{0}$ is an identity for $(\mathbb{Q}, +)$. Let $x \in \mathbb{Q}$. Then there exist $a \in \mathbb{Z}$ and $b \in \mathbb{Z}^*$ such that $x = \frac{a}{b}$. Since 0 is an additive identity for $\mathbb{Z}$, 1 is a multiplicative identity for $\mathbb{Z}$, and $0 \cdot x = x \cdot 0 = 0$ for all $x \in \mathbb{Z}$, we have

$$x + \overline{0} = \frac{a}{b} + \frac{0}{1} = \frac{a \cdot 1 + b \cdot 0}{b \cdot 1} = \frac{a + 0}{b} = \frac{a}{b} = x \text{ and } \overline{0} + x = \frac{0}{1} + \frac{a}{b} = \frac{0b + 1a}{1b} = \frac{0 + a}{b} = \frac{a}{b} = x.$$

(Inverse) Let $x \in \mathbb{Q}$. Then there exist $a \in \mathbb{Z}$ and $b \in \mathbb{Z}^*$ such that $x = \frac{a}{b}$. Let $y = \frac{-1a}{b}$. Since $\mathbb{Z}$ is closed under multiplication, $-1a \in \mathbb{Z}$. So, $y \in \mathbb{Q}$. Since multiplication is associative and commutative in $\mathbb{Z}$ and $(-1)n = -n$ for all $n \in \mathbb{Z}$, we have

$$x + y = \frac{a}{b} + \frac{-1a}{b} = \frac{ab + b(-1a)}{b \cdot b} = \frac{ab + (-1a)b}{b^2} = \frac{ab + (-1)(ab)}{b^2} = \frac{ab - ab}{b^2} = \frac{0}{b^2} = \overline{0}$$

$$y + x = \frac{-1a}{b} + \frac{a}{b} = \frac{(-1a)b + ba}{b \cdot b} = \frac{-1(ab) + ab}{b^2} = \frac{-ab + ab}{b^2} = \frac{0}{b^2} = \overline{0}$$

So, y is the additive inverse of x.

(Commutativity) This is Problem 75.

So, $(\mathbb{Q}, +)$ is a commutative group.

We next check that $(\mathbb{Q} \setminus \{0\}, \cdot)$ is a commutative group.

(Closure) Let $x, y \in \mathbb{Q}^*$. Then there exist $a, b, c, d \in \mathbb{Z}^*$ such that $x = \frac{a}{b}$ and $y = \frac{c}{d}$. We have $xy = \frac{a}{b} \cdot \frac{c}{d} = \frac{ac}{bd}$. Since $\mathbb{Z}^*$ is closed under multiplication, $ac, bd \in \mathbb{Z}^*$. Therefore, $xy \in \mathbb{Q}^*$.

(Associativity) This follows from Problem 85 and the fact that associativity is closed downwards.

(Identity) Let $\overline{1} = \frac{1}{1}$. We show that $\overline{1}$ is an identity for $(\mathbb{Q}^*, \cdot)$. Let $x \in \mathbb{Q}^*$. Then there exist $a, b \in \mathbb{Z}^*$ such that $x = \frac{a}{b}$. Since 1 is an identity for $\mathbb{Z}^*$, we have

$$x \cdot \overline{1} = \frac{a}{b} \cdot \frac{1}{1} = \frac{a \cdot 1}{b \cdot 1} = \frac{a}{b} = x \text{ and } \overline{1}x = \frac{1}{1} \cdot \frac{a}{b} = \frac{1a}{1b} = \frac{a}{b} = x.$$

(Inverse) Let $x \in \mathbb{Q}^*$. Then there exist $a, b \in \mathbb{Z}^*$ such that $x = \frac{a}{b}$. Let $y = \frac{b}{a}$. Then $y \in \mathbb{Q}^*$ (note that $a \neq 0$). Since multiplication is commutative in $\mathbb{Z}^*$, we have

$$xy = \frac{a}{b} \cdot \frac{b}{a} = \frac{ab}{ba} = \frac{ab}{ab} = \frac{1}{1} = \overline{1}.$$

So, y is the multiplicative inverse of x.

(Commutativity) This follows from Problem 74 and the fact that commutativity is closed downwards.

So, $(\mathbb{Q} \setminus \{0\}, \ \cdot)$ is a commutative group.

Now we check that multiplication is distributive over addition in $\mathbb{Q}$.

(Distributivity) This is Problem 87.

So, $(\mathbb{Q}, +, \ \cdot)$ is a field. □

Note: There is actually one more issue here. It's not obvious that the definitions of addition and multiplication are even well-defined. You were asked to prove this in Problem 346 in Problem Set 2. Nonetheless, I will include the details here.

Let's start with multiplication.

Suppose that $\frac{a}{b} = \frac{a'}{b'}$ and $\frac{c}{d} = \frac{c'}{d'}$. We need to check that $\frac{a}{b} \cdot \frac{c}{d} = \frac{a'}{b'} \cdot \frac{c'}{d'}$, or equivalently, $\frac{ac}{bd} = \frac{a'c'}{b'd'}$.

Since $\frac{a}{b} = \frac{a'}{b'}$, we have $ab' = ba'$. Since $\frac{c}{d} = \frac{c'}{d'}$, we have $cd' = dc'$. Now, since $ab' = ba'$, $cd' = dc'$, and multiplication is commutative and associative in $\mathbb{Z}$, we have

$$(ac)(b'd') = (ab')(cd') = (ba')(dc') = (bd)(a'c')$$

Therefore, $\frac{ac}{bd} = \frac{a'c'}{b'd'}$, as desired.

Now let's check that the definition of addition is well-defined.

Suppose that $\frac{a}{b} = \frac{a'}{b'}$ and $\frac{c}{d} = \frac{c'}{d'}$. We need to check that $\frac{a}{b} + \frac{c}{d} = \frac{a'}{b'} + \frac{c'}{d'}$, or equivalently, $\frac{ad+bc}{bd} = \frac{a'd'+b'c'}{b'd'}$.

Since $\frac{a}{b} = \frac{a'}{b'}$, we have $ab' = ba'$. Since $\frac{c}{d} = \frac{c'}{d'}$, we have $cd' = dc'$. Now, since $ab' = ba'$, $cd' = dc'$, multiplication is commutative and associative in $\mathbb{Z}$, and multiplication is distributive over addition in $\mathbb{Z}$, we have

$$\begin{aligned}(ad + bc)(b'd') &= adb'd' + bcb'd' = ab'dd' + cd'bb' = ba'dd' + dc'bb' \\ &= bda'd' + bdb'c' = (bd)(a'd' + b'c').\end{aligned}$$

Therefore, $\frac{ad+bc}{bd} = \frac{a'd'+b'c'}{b'd'}$, as desired.

Let $\mathbb{R}$ be the set of real numbers.

89. Prove that $(\mathbb{R}, +)$ is a commutative group.

Proof:

(Closure) You were asked to prove this in Problem 346 in Problem Set 2.

(Associativity) To see that $+$ is associative in $\mathbb{R}$, we use the associativity of $+$ in $\mathbb{Q}$. If $[(x_n)], [(y_n)], [(z_n)] \in \mathbb{R}$, then

$$([(x_n)] + [(y_n)]) + [(z_n)] = [(x_n + y_n)] + [(z_n)] = \big[\big((x_n + y_n) + z_n\big)\big]$$
$$= \big[\big(x_n + (y_n + z_n)\big)\big] = [(x_n)] + [(y_n + z_n)] = [(x_n)] + ([(y_n)] + [(z_n)]).$$

(Commutativity) To see that $+$ is commutative in $\mathbb{R}$, we use the commutativity of $+$ in $\mathbb{Q}$. If $[(x_n)], [(y_n)] \in \mathbb{R}$, then

$$[(x_n)] + [(y_n)] = [(x_n + y_n)] = [(y_n + x_n)] = [(y_n)] + [(x_n)].$$

(Identity) To see that $[(0)]$ is the additive identity, using the fact that 0 is the additive identity in $\mathbb{Q}$, we have for $[(x_n)] \in \mathbb{R}$,

$$[(0)] + [(x_n)] = [(0 + x_n)] = [(x_n)] \text{ and } [(x_n)] + [(0)] = [(x_n + 0)] = [(x_n)].$$

(Inverse) The inverse of the real number $[(x_n)]$ is $[(-x_n)]$, where for each $n \in \mathbb{N}$, $-x_n$ is the additive inverse of x_n in $\mathbb{Q}$. To see this, simply observe that

$$[(x_n)] + [(-x_n)] = \big[\big(x_n + (-x_n)\big)\big] = [(0)] \text{ and } [(-x_n)] + [(x_n)] = [(-x_n + x_n)] = [(0)]. \quad \square$$

90. Prove that multiplication is associative in $\mathbb{R}$.

To see that $\cdot$ is associative in $\mathbb{R}$, we use the associativity of $\cdot$ in $\mathbb{Q}$. If $[(x_n)], [(y_n)], [(z_n)] \in \mathbb{R}$, then

$$([(x_n)] \cdot [(y_n)]) \cdot [(z_n)] = [(x_n \cdot y_n)] \cdot [(z_n)] = \big[\big((x_n \cdot y_n) \cdot z_n\big)\big]$$
$$= \big[\big(x_n \cdot (y_n \cdot z_n)\big)\big] = [(x_n)] \cdot [(y_n \cdot z_n)] = [(x_n)] \cdot ([(y_n)] \cdot [(z_n)]). \quad \square$$

91. Prove that $\mathbb{R}$ has the multiplicative inverse property.

Proof: We first show that $[(1)]$ is the multiplicative identity for $\mathbb{R}$. Using the fact that 1 is the additive identity in $\mathbb{Q}$, we have for $[(x_n)] \in \mathbb{R}$,

$$[(1)] \cdot [(x_n)] = [(1 \cdot x_n)] = [(x_n)] \text{ and } [(x_n)] \cdot [(1)] = [(x_n \cdot 1)] = [(x_n)].$$

The inverse of the real number $[(x_n)]$ is $[(y_n)]$, where for each $n \in \mathbb{N}$, $y_n = \frac{1}{x_n}$ if $x_n \neq 0$ and $y_n = 0$ if $x_n = 0$. We have that $[(x_n)] \cdot [(y_n)] = [(z_n)]$ and $[(y_n)] \cdot [(x_n)] = [(z_n)]$, where z_n is 0 or 1 for all $n \in \mathbb{N}$. We claim that $[(z_n)] = [(1)]$. To see this, note that since $[(x_n)] \neq [(0)]$, there is a $K > 0$ such that for $n > N$, $x_n \neq 0$. $\square$

92. Prove that multiplication is distributive over addition in $\mathbb{R}$.

Proof:

(Left Distributivity) To see that $\cdot$ is left distributive over $+$ in $\mathbb{R}$, we use the left distributivity of $\cdot$ over $+$ in $\mathbb{Q}$. If $[(x_n)], [(y_n)], [(z_n)] \in \mathbb{R}$, then

$$[(x_n)]([(y_n)] + [(z_n)]) = [(x_n)] \cdot [(y_n + z_n)] = \left[\left(x_n(y_n + z_n)\right)\right]$$
$$= [(x_n y_n + x_n z_n)] = [(x_n y_n)] + [(x_n z_n)] = [(x_n)] \cdot [(y_n)] + [(x_n)] \cdot [(z_n)].$$

Right distributivity is proved similarly. □

93. Prove that $(\mathbb{R}, +, \cdot)$ is a field.

Proof: $(\mathbb{R}, +)$ is a commutative group by Problem 89.

We next check that $(\mathbb{R}^*, \cdot)$ is a commutative group.

(Closure) You were asked to prove this in Problem 346 in Problem Set 2.

(Associativity) This follows from Problem 90 and the fact that associativity is closed downwards.

(Commutativity) To see that $\cdot$ is commutative in $\mathbb{R}^*$, we use the commutativity of $\cdot$ in $\mathbb{Q}$. If $[(x_n)], [(y_n)] \in \mathbb{R}^*$, then $[(x_n)] \cdot [(y_n)] = [(x_n y_n)] = [(y_n x_n)] = [(y_n)] \cdot [(x_n)]$.

(Identity) This is true by Problem 91.

(Inverse) This is also true by Problem 91.

Distributivity is true by Problem 92.

Therefore, $(\mathbb{R}, +, \cdot)$ is a field. □

Let $\mathbb{C}$ be the set of complex numbers.

94. Prove that $(\mathbb{C}, +)$ is a commutative group.

Proof: We will freely use the fact that $(\mathbb{R}, +)$ is a commutative group (see Problem 89 or 93 above).

(Closure) Let $z, w \in \mathbb{C}$. Then there are $a, b, c, d \in \mathbb{R}$ such that $z = a + bi$ and $w = c + di$. By definition, $z + w = (a + bi) + (c + di) = (a + c) + (b + d)i$. Since $\mathbb{R}$ is closed under addition, $a + c \in \mathbb{R}$ and $b + d \in \mathbb{R}$. Therefore, $z + w \in \mathbb{C}$.

(Associativity) Let $z, w, v \in \mathbb{C}$. Then there are $a, b, c, d, e, f \in \mathbb{R}$ such that $z = a + bi$, $w = c + di$, and $v = e + fi$. Since addition is associative in $\mathbb{R}$, we have

$$(z + w) + v = \left((a + bi) + (c + di)\right) + (e + fi) = \left((a + c) + (b + d)i\right) + (e + fi)$$
$$= \left((a + c) + e\right) + \left((b + d) + f\right)i = \left(a + (c + e)\right) + \left(b + (d + f)\right)i$$
$$= (a + bi) + \left((c + e) + (d + f)i\right) = (a + bi) + \left((c + di) + (e + fi)\right) = z + (w + v).$$

(Commutativity) Let $z, w \in \mathbb{C}$. Then there are $a, b, c, d \in \mathbb{R}$ such that $z = a + bi$ and $w = c + di$. Since addition is commutative in $\mathbb{R}$, we have

$$z + w = (a + bi) + (c + di) = (a + c) + (b + d)i = (c + a) + (d + b)i$$
$$= (c + di) + (a + bi) = w + z.$$

(Identity) Let $\overline{0} = 0 + 0i$. We show that $\overline{0}$ is an additive identity for $\mathbb{C}$. Since $0 \in \mathbb{R}$, $\overline{0} \in \mathbb{C}$. Let $z \in \mathbb{C}$. Then there are $a, b \in \mathbb{R}$ such that $z = a + bi$. Since 0 is an additive identity in $\mathbb{R}$, we have

$$\overline{0} + z = (0 + 0i) + (a + bi) = (0 + a) + (0 + b)i = a + bi.$$

$$z + \overline{0} = (a + bi) + (0 + 0i) = (a + 0) + (b + 0)i = a + bi.$$

(Inverse) Let $z \in \mathbb{C}$. Then there are $a, b \in \mathbb{R}$ such that $z = a + bi$. Let $w = -a + (-b)i$. Then

$$z + w = (a + bi) + (-a + (-b)i) = \big(a + (-a)\big) + \big(b + (-b)\big)i = 0 + 0i = \overline{0}.$$

$$w + z = (-a + (-b)i) + (a + bi) = (-a + a) + (-b + b)i = 0 + 0i = \overline{0}$$ □

95. Prove that multiplication is associative in $\mathbb{C}$.

Proof: Let $z, w, v \in \mathbb{C}$. Then there are $a, b, c, d, e, f \in \mathbb{R}$ such that $z = a + bi$, $w = c + di$, and $v = e + fi$. Since addition and multiplication are associative in $\mathbb{R}$, addition is commutative in $\mathbb{R}$, and multiplication is distributive over addition in $\mathbb{R}$ (by Problem 93 above), we have

$$(zw)v = \big((a + bi)(c + di)\big)(e + fi) = \big((ac - bd) + (ad + bc)i\big)(e + fi)$$
$$= [(ac - bd)e - (ad + bc)f] + [(ac - bd)f + (ad + bc)e]i$$
$$= (ace - bde - adf - bcf) + (acf - bdf + ade + bce)i$$
$$= (ace - adf - bcf - bde) + (acf + ade + bce - bdf)i$$
$$= [a(ce - df) - b(cf + de)] + [a(cf + de) + b(ce - df)]i$$
$$= (a + bi)\big((ce - df) + (cf + de)i\big) = (a + bi)\big((c + di)(e + fi)\big) = z(wv).$$ □

96. Prove that $\mathbb{C}$ has the multiplicative inverse property. What is the inverse of the nonzero complex number $a + bi$?

Proof: We first show that $\overline{1} = 1 + 0i$ is a multiplicative identity for $\mathbb{C}$. Since $0, 1 \in \mathbb{R}$, $\overline{1} \in \mathbb{C}$. Let $z \in \mathbb{C}$. Then there are $a, b \in \mathbb{R}$ such that $z = a + bi$. Since 0 is an additive identity in $\mathbb{R}$, 1 is a multiplicative identity in $\mathbb{R}$, and $0 \cdot x = x \cdot 0 = 0$ for all $x \in \mathbb{R}$, we have

$$\overline{1}z = (1 + 0i)(a + bi) = (1a - 0b) + (1b + 0a)i = 1a + 1bi = a + bi.$$

$$z \cdot \overline{1} = (a + bi)(1 + 0i) = (a \cdot 1 - b \cdot 0) + (a \cdot 0 + b \cdot 1)i = a \cdot 1 + b \cdot 1i = a + bi.$$

Now, let $z \in \mathbb{C}^*$. Then there are $a, b \in \mathbb{R}$ such that $z = a + bi$ and $z \neq 0$. Let $w = \frac{a}{a^2+b^2} + \frac{-b}{a^2+b^2}i$. We show that w is the multiplicative inverse of z. We have

$$zw = (a + bi)\left(\frac{a}{a^2 + b^2} + \frac{-b}{a^2 + b^2}i\right)$$
$$= \left(a \cdot \frac{a}{a^2 + b^2} - b \cdot \frac{-b}{a^2 + b^2}\right) + \left(a \cdot \frac{-b}{a^2 + b^2} + b \cdot \frac{a}{a^2 + b^2}\right)i$$
$$= \frac{a^2 + b^2}{a^2 + b^2} + \frac{-ab + ba}{a^2 + b^2}i = 1 + 0i = \overline{1}.$$
$$wz = \left(\frac{a}{a^2 + b^2} + \frac{-b}{a^2 + b^2}i\right)(a + bi)$$
$$= \left(\frac{a}{a^2 + b^2} \cdot a - \frac{-b}{a^2 + b^2} \cdot b\right) + \left(\frac{a}{a^2 + b^2} \cdot b + \frac{-b}{a^2 + b^2} \cdot a\right)i$$
$$= \frac{a^2 + b^2}{a^2 + b^2} + \frac{ab - ba}{a^2 + b^2}i = 1 + 0i = \overline{1}.$$
□

97. Prove that multiplication is distributive over addition in $\mathbb{C}$.

Proof:

(Left Distributivity) Let $z, w, v \in \mathbb{C}$. Then there are $a, b, c, d, e, f \in \mathbb{R}$ such that $z = a + bi$, $w = c + di$, and $v = e + fi$. Since multiplication is left distributive over addition in $\mathbb{R}$, and addition is associative and commutative in $\mathbb{R}$, we have

$$z(w + v) = (a + bi)[(c + di) + (e + fi)] = (a + bi)[(c + e) + (d + f)i]$$
$$= [a(c + e) - b(d + f)] + [a(d + f) + b(c + e)]i$$
$$= (ac + ae - bd - bf) + (ad + af + bc + be)i$$
$$= [(ac - bd) + (ad + bc)i\,] + [(ae - bf) + (af + be)i]$$
$$(a + bi)(c + di) + (a + bi)(e + fi) = zw + zv.$$

(Right Distributivity) Let $z, w, v \in \mathbb{C}$. There are $a, b, c, d, e, f \in \mathbb{R}$ such that $z = a + bi$, $w = c + di$, and $v = e + fi$. Since multiplication is right distributive over addition in $\mathbb{R}$, and addition is associative and commutative in $\mathbb{R}$, we have

$$(w + v)z = [(c + di) + (e + fi)](a + bi) = [(c + e) + (d + f)i](a + bi)$$
$$= [(c + e)a - (d + f)b] + [(c + e)b + (d + f)a]i$$
$$= (ca + ea - db - fb) + (cb + eb + da + fa)i$$
$$= [(ca - db) + (cb + da)i\,] + [(ea - fb) + (eb + fa)i]$$
$$(c + di)(a + bi) + (e + fi)(a + bi) = wz + vz.$$
□

98. Prove that $(\mathbb{C}, +, \cdot)$ is a field.

Proof: We will freely use the fact that $(\mathbb{R}, +, \cdot)$ is a field (see Problem 93 above).

$(\mathbb{C}, +)$ is a commutative group by Problem 94.

We next check that $(\mathbb{C}^*, \cdot)$ is a commutative group.

(Closure) Let $z, w \in \mathbb{C}^*$. Then there are $a, b, c, d \in \mathbb{R}$ such that $z = a + bi$ and $w = c + di$ with $z \neq 0$ and $w \neq 0$. By definition, $zw = (a + bi)(c + di) = (ac - bd) + (ad + bc)i$. Since $\mathbb{R}$ is closed under multiplication, we have $ac, bd, ad, bc \in \mathbb{R}$. Also, $-bd$ is the additive inverse of bd in $\mathbb{R}$. Since $\mathbb{R}$ is closed under addition, we have $ac - bd = ac + (-bd) \in \mathbb{R}$ and $ad + bc \in \mathbb{R}$. Therefore, $zw \in \mathbb{C}$.

We still need to show that $zw \neq 0$. If $zw = 0$, then $ac - bd = 0$ and $ad + bc = 0$. So, $ac = bd$ and $ad = -bc$. Since $w \neq 0$, either $c \neq 0$ or $d \neq 0$. Without loss of generality assume that $c \neq 0$. Multiplying each side of the last equation by c gives us $acd = -bc^2$. Replacing ac with bd on the left gives $bd^2 = -bc^2$, or equivalently, $bd^2 + bc^2 = 0$. So, $b(d^2 + c^2) = 0$. If $d^2 + c^2 = 0$, then $c = 0$ and $d = 0$, contradicting our assumption that $c \neq 0$. So, $b = 0$. Then we must have $a \neq 0$ (otherwise $a + bi) = 0$). Since $ad = -bc = 0$, we have $d = 0$. So, $w = 0$, contrary to our assumption that $w \neq 0$. So, we see that $zw = 0$ is impossible, and so, $zw \in \mathbb{C}^*$.

(Associativity) This follows from Problem 95 and the fact that associativity is closed downwards.

(Commutativity) Let $z, w \in \mathbb{C}^*$. Then there are $a, b, c, d \in \mathbb{R}$ such that $z = a + bi$ and $w = c + di$. Since addition and multiplication are commutative in $\mathbb{R}$, we have

$$zw = (a + bi)(c + di) = (ac - bd) + (ad + bc)i$$
$$= (ca - db) + (cb + da)i = (c + di)(a + bi) = wz$$

(Identity) This is true by Problem 96.

(Inverse) This is also true by Problem 96.

Distributivity is true by Problem 97.

Therefore, $(\mathbb{C}, +, \cdot)$ is field. □

Define a sum and product on $\mathbb{Z}_n$ as follows: For $x, y \in \mathbb{Z}_n$, let $[x]_n + [y]_n = [x + y]_n$ and let $[x]_n \cdot [y]_n = [xy]_n$.

99. Prove that $(\mathbb{Z}_n, +)$ is a commutative group.

Proof: It was already proved that addition is well-defined in Lesson 3.

(Closure) Let $[x]_n, [y]_n \in \mathbb{Z}_n$. Then $[x]_n + [y]_n = [x + y]_n \in \mathbb{Z}_n$.

(Associativity) Let $[x]_n, [y]_n, [z]_n \in \mathbb{Z}_n$. Since addition is associative in $\mathbb{Z}$, we have

$$([x]_n + [y]_n) + [z]_n = [x + y]_n + [z]_n = [(x + y) + z]_n = [x + (y + z)]_n$$
$$= [x]_n + [y + z]_n = [x]_n + ([y]_n + [z]_n).$$

(Commutativity) Let$[x]_n, [y]_n \in \mathbb{Z}_n$. Since addition is commutative in $\mathbb{Z}$, we have

$$[x]_n + [y]_n = [x + y]_n = [y + x]_n = [y]_n + [x]_n.$$

(Identity) We show that $[0]_n$ is the additive identity. Let $[x]_n \in \mathbb{Z}_n$. Since 0 is an additive identity in $\mathbb{Z}$, we have: $[0]_n + [x]_n = [0 + x]_n = [x]_n$ and $[x]_n + [0]_n = [x + 0]_n = [x]_n$.

(Inverse) Let $[x]_n \in \mathbb{Z}_n$. We show that $[-x]_n$ is the additive inverse of $[x]_n$. Since $-x$ is the additive inverse of x in $\mathbb{Z}$, we have

$$[x]_n + [-x]_n = [x + (-x)]_n = [0]_n.$$
$$[-x]_n + [x]_n = [-x + x]_n = [0]_n.$$

□

100. Prove that $(\mathbb{Z}_n, +, \cdot)$ is a commutative ring.

Proof: By Problem 99, $(\mathbb{Z}_n, +)$ is a commutative group.

We next prove that $(\mathbb{Z}_n, \cdot)$ is a commutative monoid. It was already proved that multiplication is well-defined in Lesson 3.

(Closure) Let $[x]_n, [y]_n \in \mathbb{Z}_n$. Then $[x]_n \cdot [y]_n = [xy]_n \in \mathbb{Z}_n$.

(Associativity) Let $[x]_n, [y]_n, [z]_n \in \mathbb{Z}_n$. Since multiplication is associative in $\mathbb{Z}$, we have

$$([x]_n \cdot [y]_n) \cdot [z]_n = [xy]_n \cdot [z]_n = [(xy)z]_n = [x(yz)]_n = [x]_n \cdot [yz]_n = [x]_n \cdot ([y]_n \cdot [z]_n).$$

(Commutativity) Let $[x]_n, [y]_n \in \mathbb{Z}_n$. Since multiplication is commutative in $\mathbb{Z}$, we have

$$[x]_n \cdot [y]_n = [xy]_n = [yx]_n = [y]_n \cdot [x]_n.$$

(Identity) We show that $[1]_n$ is the multiplicative identity. Let $[x]_n \in \mathbb{Z}_n$. Since 1 is a multiplicative identity in $\mathbb{Z}$, we have

$$[1]_n \cdot [x]_n = [1x]_n = [x]_n \quad \text{and} \quad [x]_n \cdot [1]_n = [x \cdot 1]_n = [x]_n.$$

We now prove distributivity. Since $(\mathbb{Z}_n, \cdot)$ is commutative, it suffices to prove left distributivity.

(Left Distributivity) Let $[x]_n, [y]_n, [z]_n \in \mathbb{Z}_n$. Since multiplication is distributive over addition in $\mathbb{Z}$, we have

$$[x]_n([y]_n + [z]_n) = [x]_n \cdot [y + z]_n = [x(y + z)]_n = [xy + xz]_n$$
$$= [xy]_n + [xz]_n = [x]_n \cdot [y]_n + [x]_n \cdot [z]_n.$$

□

Let $S = \{a, b\}$, where $a \neq b$.

101. How many binary operations are there on S?

$2^4 = \mathbf{16}.$

102. Draw the multiplication table for each binary operation on S.

$\star$	a	b
a	a	a
b	a	a

$\star$	a	b
a	b	a
b	a	a

$\star$	a	b
a	a	b
b	a	a

$\star$	a	b
a	a	a
b	b	a

$\star$	a	b
a	a	a
b	a	b

$\star$	a	b
a	b	b
b	a	a

$\star$	a	b
a	b	a
b	b	a

$\star$	a	b
a	b	a
b	a	b

$\star$	a	b
a	a	b
b	b	a

$\star$	a	b
a	a	b
b	a	b

$\star$	a	b
a	a	a
b	b	b

$\star$	a	b
a	b	b
b	b	a

$\star$	a	b
a	b	b
b	a	b

$\star$	a	b
a	b	a
b	b	b

$\star$	a	b
a	a	b
b	b	b

$\star$	a	b
a	b	b
b	b	b

103. How many semigroups are there of the form $(S,\star)$, up to renaming the elements?

Of the 16 binary operations, 8 give rise to semigroups. However, 3 of these are essentially the same as 3 of the others. The 5 circled multiplication tables above represent the 5 semigroups. The 3 tables in rectangles that are crossed out also represent semigroups. However, if you interchange the roles of a and b, you'll see that they are the same as 3 of the others with the names changed (arrows are present to indicate the tables that are essentially the same as these). The other 8 tables represent operations that are not associative (the reader should find a counterexample to associativity for each of these). I leave it to the reader to verify that the 5 circled multiplication tables represent semigroups.

Note: A **magma** is a pair $(M,\star)$, where M is a set and $\star$ is a binary operation on M (and no other conditions). In the solution above we showed that there are 16 magmas of the form $(\{a,b\},\star)$, and of these, 8 are semigroups. However, there are only 5 semigroups up to renaming the elements. Of the 16 magmas, there are only 10 up to renaming the elements. See if you can find the duplicates.

Prove each of the following:

104. There are exactly two groups of order 4, up to renaming the elements.

Proof: Let $G = \{e,a,b,c\}$. We will run through the possible cases.

Case 1: Suppose that $a^2 = e$.

Since $e \star x = x \star e = x$ for all x in the group, we can easily fill out the first row, the first column, and one more entry of the table.

$\star$	e	a	b	c
e	e	a	b	c
a	a	e	$\boxdot$	
b	b	$\boxdot$		
c	c			

Each of the entries labeled with $\boxdot$ cannot be a, e, or b (Why?), and so they must be c. So, we get the following:

$\star$	e	a	b	c
e	e	a	b	c
a	a	e	c	b
b	b	c		
c	c	b		

Now, if $b^2 = e$, the rest of the table is determined:

$\star$	e	a	b	c
e	e	a	b	c
a	a	e	c	b
b	b	c	e	a
c	c	b	a	e

This table gives a group $(G,\star)$ called the **Klein four group**.

If $b^2 = a$, the rest of the table is also determined:

$\star$	e	a	b	c
e	e	a	b	c
a	a	e	c	b
b	b	c	a	e
c	c	b	e	a

This table gives a group $(G,\star)$ called the **Cyclic group of order 4**.

Observe that we cannot have $b^2 = b$ or $b^2 = c$ because b and c already appear in the row (and column) corresponding to b.

$\star$	e	a	b	c
e	e	a	b	c
a	a	e	c	b
b	b	c	~~b, c~~	
c	c	b		

Case 2: Suppose that $a^2 \neq e$.

If $b^2 = e$ or $c^2 = e$, then by renaming elements, we get the same groups in Case 1. So, we may assume that $a^2 \neq e$, $b^2 \neq e$, and $c^2 \neq e$.

So, a, a^2, and a^3 are distinct elements. If $a^3 \neq e$, then $a^4 = e$, and so, $(a^2)^2 = e$. But a^2 must be equal to either b or c. So, $b^2 = e$ or $c^2 = e$, contrary to our assumption.

It follows that $a^3 = e$. So, $a \star a^2 = e$. Therefore, a and a^2 are inverses of each other. If $a^2 = b$, then c must be its own inverse. So, $c^2 = c \star c = e$, contrary to our assumption. Similarly, if $a^2 = c$, then b must be its own inverse. So, $b^2 = b \star b = e$, contrary to our assumption.

It follows that there are exactly 2 groups of order 4, up to renaming the elements. These 2 groups are the **Klein four group** and the **Cyclic group of order 4**. □

105. Let $R[x] = \{a_k x^k + a_{k-1}x^{k-1} + \cdots + a_1 x + a_0 \mid k \in \mathbb{N} \wedge a_0, a_1, \ldots, a_k \in R\}$, where R is a ring. Then $R[x]$ is both a left R-module and a right R-module.

Proof: Addition in $(R[x], +, \cdot)$ is defined by

$$a_k x^k + a_{k-1}x^{k-1} + \cdots + a_1 x + a_0) + (b_k x^k + b_{k-1}x^{k-1} + \cdots + b_1 x + b_0)$$
$$= (a_k + b_k)x^k + (a_{k-1} + b_{k-1})x^{k-1} + \cdots + (a_1 + b_1)x + (a_0 + b_0).$$

Since R is closed under addition, $a_k + b_k, a_{k-1} + b_{k-1}, \ldots, a_1 + b_1, a_0 + b_0 \in R$. Therefore, $R[x]$ is closed under addition.

Simple computations show that addition is associative in $R[x]$, $0 = 0x^k + 0x^{k-1} + \cdots + 0x + 0$ is an additive identity in $R[x]$, and the additive inverse of $a_k x^k + a_{k-1}x^{k-1} + \cdots + a_1 x + a_0$ is $-a_k x^k - a_{k-1}x^{k-1} - \cdots - a_1 x - a_0$. One more simple computation can be used to verify that addition is commutative.

To turn $R[x]$ into a left R-module, we define scalar multiplication as follows:

If $c \in R$ and $a_k x^k + a_{k-1}x^{k-1} + \cdots + a_1 x + a_0 \in R[x]$, then

$$c \cdot (a_k x^k + a_{k-1}x^{k-1} + \cdots + a_1 x + a_0) = (ca_k)x^k + (ca_{k-1})x^{k-1} + \cdots + (ca_1)x + (ca_0).$$

We now prove the remaining left R-module properties:

(Closure under scalar multiplication) Let $c \in R$ and let $a_k x^k + a_{k-1}x^{k-1} + \cdots + a_1 x + a_0 \in R[x]$. Then $a_k, a_{k-1}, \ldots, a_1, a_0 \in R$. Since R is a ring, R is closed under multiplication. Therefore, we have $ca_k, ca_{k-1}, \ldots ca_1, ca_0 \in R$. So, $(ca_k)x^k + (ca_{k-1})x^{k-1} + \cdots + (ca_1)x + (ca_0) \in R[x]$. Therefore, $c \cdot (a_k x^k + a_{k-1}x^{k-1} + \cdots + a_1 x + a_0) \in R[x]$.

(Scalar multiplication identity) Let $a_k x^k + a_{k-1}x^{k-1} + \cdots + a_1 x + a_0 \in R[x]$ and let 1 be the multiplicative identity in R. Then

$$1 \cdot (a_k x^k + a_{k-1}x^{k-1} + \cdots + a_1 x + a_0) = (1a_k)x^k + (1a_{k-1})x^{k-1} + \cdots + (1a_1)x + (1a_0)$$
$$= a_k x^k + a_{k-1}x^{k-1} + \cdots + a_1 x + a_0.$$

(Associativity of scalar multiplication) Let $c, d \in R$ and $a_k x^k + a_{k-1}x^{k-1} + \cdots + a_1 x + a_0 \in R[x]$. Then since multiplication is associative in R, we have

$$\begin{aligned}&(cd)(a_k x^k + a_{k-1}x^{k-1} + \cdots + a_1 x + a_0)\\ &= \big((cd)a_k\big)x^k + \big((cd)a_{k-1}\big)x^{k-1} + \cdots + \big((cd)a_1\big)x + (cd)a_0\\ &= (c(da_k)x^k + \big(c(da_{k-1})\big)x^{k-1} + \cdots + \big(c(da_1)\big)\,x + c(da_0)\\ &= c\Big((da_k)x^k + (da_{k-1})x^{k-1} + \cdots + (da_1)x + da_0\Big)\\ &= c\Big(d(a_k x^k + a_{k-1}x^{k-1} + \cdots + a_1 x + a_0 a_1, a_2, \ldots, a_n)\Big).\end{aligned}$$

(Distributivity of 1 scalar over 2 vectors) Let $c \in R$, $a_k x^k + a_{k-1}x^{k-1} + \cdots + a_1 x + a_0 \in R[x]$, and $b_k x^k + b_{k-1}x^{k-1} + \cdots + b_1 x + b_0 \in R[x]$. Since multiplication is left distributive over addition in R, we have

$$\begin{aligned}&c\Big((a_k x^k + a_{k-1}x^{k-1} + \cdots + a_1 x + a_0) + (b_k x^k + b_{k-1}x^{k-1} + \cdots + b_1 x + b_0)\Big)\\ &= c\Big((a_k + b_k)x^k + (a_{k-1} + b_{k-1})x^{k-1} + \cdots + (a_1 + b_1)x + (a_0 + b_0)\Big)\\ &= [c(a_k + b_k)]x^k + [c(a_{k-1} + b_{k-1})]x^{k-1} + \cdots + [c(a_1 + b_1)]x + c(a_0 + b_0)\\ &= (ca_k + cb_k)x^k + (ca_{k-1} + cb_{k-1})x^{k-1} + \cdots + (ca_1 + cb_1)x + (ca_0 + cb_0)\\ &= \Big((ca_k)x^k + + \cdots + (ca_1)x + (ca_0)\Big) + \Big((cb_k)x^k + \cdots + (cb_1)x + (cb_0)\Big)\\ &= c \cdot (a_k x^k + a_{k-1}x^{k-1} + \cdots + a_1 x + a_0) + c \cdot (b_k x^k + b_{k-1}x^{k-1} + \cdots + b_1 x + b_0).\end{aligned}$$

(Distributivity of 2 scalars over 1 vector) Let $c, d \in R$ and $a_k x^k + a_{k-1}x^{k-1} + \cdots + a_1 x + a_0 \in R[x]$. Since multiplication is distributive over addition in R, we have

$$\begin{aligned}&(c + d)(a_k x^k + a_{k-1}x^{k-1} + \cdots + a_1 x + a_0)\\ &= [(c + d)a_k]x^k + [(c + d)a_{k-1}]x^{k-1} + \cdots + [(c + d)a_1]x + (c + d)a_0\\ &= (ca_k + da_k)x^k + (ca_{k-1} + da_{k-1})x^{k-1} + \cdots + (ca_1 + da_1)x + (ca_0 + da_0)\\ &= \Big((ca_k)x^k + \cdots + (ca_1)x + (ca_0)\Big) + \Big((da_k)x^k + \cdots + (da_1)x + (da_0)\Big)\\ &= c(a_k x^k + a_{k-1}x^{k-1} + \cdots + a_1 x + a_0) + d(a_k x^k + a_{k-1}x^{k-1} + \cdots + a_1 x + a_0).\end{aligned}$$

Similarly, to turn $R[x]$ into a right R-module, we define scalar multiplication as follows: if $c \in R$ and $a_k x^k + a_{k-1}x^{k-1} + \cdots + a_1 x + a_0 \in R[x]$, then

$$(a_k x^k + a_{k-1}x^{k-1} + \cdots + a_1 x + a_0) \cdot c = (a_k c)x^k + (a_{k-1}c)x^{k-1} + \cdots + (a_1 c)x + (a_0 c).$$

The verifications of the right R-module properties are nearly identical to the computations made above. Just replace all multiplications on the left with multiplications on the right. □

Problem Set 4

LEVEL 1

Write each of the following positive integers as a product of prime factors in canonical form:

1. 16

$16 = 2^4$

2. 19

$19 = 19$

3. 35

$35 = 5 \cdot 7$

4. 105

$105 = 3 \cdot 5 \cdot 7$

5. 275

$275 = 5^2 \cdot 11$

Write each of the following sets using the roster method.

6. $\{n \mid n \text{ is an even prime number}\}$

$\{2\}$

7. $\{n \mid n \text{ is a prime number less than } 100\}$

$\{2, 3, 5, 7, 11, 13, 17, 19, 23, 29, 31, 37, 41, 43, 47, 53, 59, 61, 67, 71, 73, 79, 83, 89, 97\}$

8. $\{n \mid n \text{ is one of the first } 25 \text{ composite numbers}\}$

$\{4, 6, 8, 9, 10, 12, 14, 15, 16, 18, 20, 21, 22, 24, 25, 26, 27, 28, 30, 32, 33, 34, 35, 36, 38\}$

Find the gcd and lcm of each of the following sets of numbers:

9. $\{6, 9\}$

$\gcd(6, 9) = \mathbf{3}$; $\text{lcm}(6, 9) = \mathbf{18}$.

10. $\{12, 180\}$

$12|180$. So, $\gcd(12, 180) = \mathbf{12}$; $\text{lcm}(12, 180) = \mathbf{180}$.

11. $\{2,3,5\}$

$\gcd(2,3,5) = \mathbf{1}$; $\operatorname{lcm}(2,3,5) = \mathbf{30}$.

For each of the integers a and k given, find a number b that shows that a is divisible by k.

12. $a = 8, k = 2$

$8 = 2 \cdot 4$. So, $\boldsymbol{b} = \mathbf{4}$.

13. $a = 30, k = 3$

$30 = 3 \cdot 10$. So, $\boldsymbol{b} = \mathbf{10}$.

14. $a = 17, k = 17$

$17 = 17 \cdot 1$. So, $\boldsymbol{b} = \mathbf{1}$.

15. $a = 1006, k = 1$

$1006 = 1 \cdot 1006$. So, $\boldsymbol{b} = \mathbf{1006}$.

LEVEL 2

Use the Principle of Mathematical Induction to prove each of the following:

16. $2^n > n$ for all natural numbers $n \geq 1$.

Proof:

Base Case ($k = 1$): $2^1 = 2 > 1$.

Inductive Step: Let $k \in \mathbb{N}$ with $k \geq 1$ and assume that $2^k > k$. Then we have

$$2^{k+1} = 2^k \cdot 2^1 = 2^k \cdot 2 > k \cdot 2 = 2k = k + k \geq k + 1.$$

Therefore, $2^{k+1} > k + 1$.

By the Principle of Mathematical Induction, $2^n > n$ for all natural numbers $n \geq 1$. □

17. $0 + 1 + 2 + \cdots + n = \frac{n(n+1)}{2}$ for all natural numbers.

Proof:

Base Case ($k = 0$): $0 = \frac{0(0+1)}{2}$.

Inductive Step: Let $k \in \mathbb{N}$ and assume that $0 + 1 + 2 + \cdots + k = \frac{k(k+1)}{2}$. Then we have

$$0 + 1 + 2 + \cdots + k + (k+1) = \frac{k(k+1)}{2} + (k+1) = (k+1)\left(\frac{k}{2} + 1\right) = (k+1)\left(\frac{k}{2} + \frac{2}{2}\right)$$

$$= (k+1)\left(\frac{k+2}{2}\right) = \frac{(k+1)(k+2)}{2} = \frac{(k+1)\big((k+1)+1\big)}{2}$$

By the Principle of Mathematical Induction, $0 + 1 + 2 + \cdots + n = \frac{n(n+1)}{2}$ for all natural numbers n. □

18. $0^2 + 1^2 + 2^2 + \cdots + n^2 = \frac{n(n+1)(2n+1)}{6}$ for all natural numbers.

Proof:

Base Case ($k = 0$): $0^2 = 0$ and $\frac{0(0+1)(2\cdot 0+1)}{6} = 0$.

Inductive Step: Let $k \in \mathbb{N}$ and assume that $0^2 + 1^2 + 2^2 + \cdots + k^2 = \frac{k(k+1)(2k+1)}{6}$. Then we have

$$0^2 + 1^2 + 2^2 + \cdots + k^2 + (k+1)^2 = \frac{k(k+1)(2k+1)}{6} + (k+1)^2$$

$$= (k+1)\left(\frac{k(2k+1)}{6} + (k+1)\right) = (k+1)\left(\frac{2k^2+k}{6} + \frac{6(k+1)}{6}\right) = (k+1)\left(\frac{2k^2+7k+6}{6}\right)$$

$$= (k+1)\left(\frac{(k+2)(2k+3)}{6}\right) = \frac{(k+1)(k+2)(2k+3)}{6} = \frac{(k+1)\big((k+1)+1\big)\big(2(k+1)+1\big)}{6}$$

By the Principle of Mathematical Induction, $0^2 + 1^2 + 2^2 + \cdots + n^2 = \frac{n(n+1)(2n+1)}{6}$ for all natural numbers n. □

19. $n! > 2^n$ for all natural numbers $n \geq 4$ (where $n! = 1 \cdot 2 \cdots n$ for all natural numbers $n \geq 1$).

Proof:

Base Case ($k = 4$): $4! = 1 \cdot 2 \cdot 3 \cdot 4 = 24 > 16 = 2^4$.

Inductive Step: Let $k \in \mathbb{N}$ with $k \geq 4$ and assume that $k! > 2^k$. Then we have

$$(k+1)! = (k+1)k! > (k+1)2^k \geq (4+1) \cdot 2^k = 5 \cdot 2^k \geq 2 \cdot 2^k = 2^1 \cdot 2^k = 2^{1+k} = 2^{k+1}.$$

Therefore, $(k+1)! > 2^{k+1}$.

By the Principle of Mathematical Induction, $n! > 2^n$ for all natural numbers $n \geq 4$. □

20. $2^n \geq n^2$ for all natural numbers $n \geq 4$.

Proof:

Base Case ($k = 4$): $2^4 = 16 = 4^2$. So, $2^4 \geq 4^2$.

Inductive Step: Let $k \in \mathbb{N}$ with $k \geq 4$ and assume that $2^k \geq k^2$. Then we have

$$2^{k+1} = 2^k \cdot 2^1 \geq k^2 \cdot 2 = 2k^2 = k^2 + k^2.$$

By Theorem 4.7, $k^2 > 2k + 1$. So, we have $2^{k+1} > k^2 + 2k + 1 = (k+1)^2$.

Therefore, $2^{k+1} \geq (k+1)^2$.

By the Principle of Mathematical Induction, $2^n \geq n^2$ for all $n \in \mathbb{N}$ with $n \geq 4$. □

Note: $2^0 = 1 \geq 0 = 0^2$. So, the statement in Problem 20 is true for $k = 0$. Also, $2^1 = 2 \geq 1 = 1^2$ and $2^2 = 4 = 2^2$. So, the statement is true for $k = 1$ and $k = 2$. However, $2^3 = 8$ and $3^2 = 9$. So, the statement is false for $k = 3$. It follows that $2^n \geq n^2$ for all natural numbers n except $n = 3$.

21. $n^3 - n$ is divisible by 3 for all natural numbers n.

Proof:

Base Case $(k = 0)$: $0^3 - 0 = 0 = 3 \cdot 0$. So, $0^3 - 0$ is divisible by 3.

Inductive Step: Let $k \in \mathbb{N}$ and assume that $k^3 - k$ is divisible by 3. Then $k^3 - k = 3b$ for some integer b. Now,

$$(k+1)^3 - (k+1) = (k+1)[(k+1)^2 - 1] = (k+1)[(k+1)(k+1) - 1]$$
$$= (k+1)(k^2 + 2k + 1 - 1) = (k+1)(k^2 + 2k) = k^3 + 2k^2 + k^2 + 2k = k^3 + 3k^2 + 2k$$
$$= k^3 - k + k + 3k^2 + 2k = (k^3 - k) + 3k^2 + 3k = 3b + 3(k^2 + k) = 3(b + k^2 + k).$$

Since $\mathbb{Z}$ is closed under addition and multiplication, $b + k^2 + k \in \mathbb{Z}$. Therefore, $(k+1)^3 - (k+1)$ is divisible by 3.

By the Principle of Mathematical Induction, $n^3 - n$ is divisible by 3 for all $n \in \mathbb{N}$. □

Write each of the following positive integers as a product of prime factors in canonical form:

22. 693

$693 = 3^2 \cdot 7 \cdot 11$

23. 67,500

$67{,}500 = 2^2 \cdot 3^3 \cdot 5^4$

24. 384,659

$384{,}659 = 11^3 \cdot 17^2$

25. 9,699,690

$9{,}699{,}690 = 2 \cdot 3 \cdot 5 \cdot 7 \cdot 11 \cdot 13 \cdot 17 \cdot 19$

Determine if each of the following positive integers is prime:

26. 20

$20 = 2 \cdot 10$. So, 20 is **not prime**.

27. 53

$\sqrt{53} < 8$ and 53 is not divisible by 2, 3, 5, and 7. So, 53 is **prime**.

28. 71

$\sqrt{71} < 9$ and 71 is not divisible by 2, 3, 5, and 7. So, 71 is **prime**.

29. 81

$81 = 9 \cdot 9$. So, 81 is **not prime**.

30. 85

$85 = 5 \cdot 17$. So, 85 is **not prime**.

31. 97

$\sqrt{97} < 10$ and 97 is not divisible by 2, 3, 5, and 7. So, 97 is **prime**.

For $n \in \mathbb{Z}^+$, let $M_n = n! + 1$. Determine if M_n is prime for each of the following values of n.

32. $n = 3$

$M_3 = 3! + 1 = 6 + 1 = 7$. So, M_3 is prime.

33. $n = 4$

$M_4 = 4! + 1 = 24 + 1 = 25$. Since $5|25$, M_4 is **not** prime.

34. $n = 5$

$M_5 = 5! + 1 = 120 + 1 = 121$. Since $11|121$, M_5 is **not** prime.

35. $n = 6$

$M_6 = 6! + 1 = 721$. Since $7|721$, M_6 is **not** prime.

36. $n = 7$

$M_7 = 7! + 1 = 5040 + 1 = 5041$. Since $71|5041$, M_7 is **not** prime.

Find the gcd and lcm of each of the following sets of numbers:

37. $\{14, 21, 77\}$

$\gcd(14, 21, 77) = \mathbf{7}$; $\text{lcm}(14, 21, 77) = 2 \cdot 7 \cdot 3 \cdot 11 = \mathbf{462}$.

38. $\{720, 2448, 5400\}$

$720 = 2^4 \cdot 3^2 \cdot 5, 2448 = 2^4 \cdot 3^2 \cdot 17, 5400 = 2^3 \cdot 3^3 \cdot 5^2$.

$\gcd(720, 2448, 5400) = 2^3 \cdot 3^2 = \mathbf{72}$.

$\text{lcm}(720, 2448, 5400) = 2^4 \cdot 3^3 \cdot 5^2 \cdot 17 = \mathbf{183,600}$.

39. $\{2^{17} \cdot 5^4 \cdot 11^9 \cdot 23, 2^5 \cdot 3^2 \cdot 7^4 \cdot 11^3 \cdot 13\}$

$\gcd(2^{17} \cdot 5^4 \cdot 11^9 \cdot 23, 2^5 \cdot 3^2 \cdot 7^4 \cdot 11^3 \cdot 13) = 2^5 \cdot 11^3$.

$\text{lcm}(2^{17} \cdot 5^4 \cdot 11^9 \cdot 23, 2^5 \cdot 3^2 \cdot 7^4 \cdot 11^3 \cdot 13) = 2^{17} \cdot 3^2 \cdot 5^4 \cdot 7^4 \cdot 11^9 \cdot 13 \cdot 23$.

Prove each of the following. (You may assume that $<$ is a strict linear ordering of $\mathbb{N}$.)

40. Addition is commutative in $\mathbb{N}$.

Proof: We first prove by induction on n that $1 + n = n + 1$.

Base Case ($k = 0$): By definition of addition of natural numbers, we have $1 + 0 = 1$. By Theorem 4.4, we have $0 + 1 = 1$. Therefore, $1 + 0 = 0 + 1$.

Inductive Step: Let $k \in \mathbb{N}$ and assume that $1 + k = k + 1$. Then we have

$$1 + (k + 1) = (1 + k) + 1 = (k + 1) + 1.$$

For the first equality, we used the definition of addition of natural numbers. For the second equality, we used the inductive hypothesis.

By the Principle of Mathematical Induction, for all natural numbers n, $1 + n = n + 1$.

We are now ready to use induction to prove the result. Assume that m is a natural number.

Base Case ($k = 0$): By definition of addition of natural numbers, $m + 0 = m$. By Theorem 4.4, $0 + m = m$. Therefore, $m + 0 = 0 + m$.

Inductive Step: Let $k \in \mathbb{N}$ and assume that $m + k = k + m$. Then we have

$$m + (k + 1) = (m + k) + 1 = (k + m) + 1 = k + (m + 1) = k + (1 + m) = (k + 1) + m.$$

For the first and third equalities, we used the definition of addition of natural numbers (or Theorem 4.5). For the second equality, we used the inductive hypothesis. For the fourth equality, we used the preliminary result that we proved above. For the fifth equality, we used Theorem 4.5.

By the Principle of Mathematical Induction, for all natural numbers n, $m + n = n + m$.

Since m was an arbitrary natural number, we have shown that for all natural numbers m and n, we have $m + n = n + m$. □

41. The set of natural numbers is closed under multiplication.

Proof: Assume that m is a natural number.

Base Case ($k = 0$): $m \cdot 0 = 0$, which is a natural number.

Inductive Step: Let k be a natural number and assume that mk is also a natural number. Then $m(k + 1) = mk + m$. Since mk and m are both natural numbers, by Theorem 4.3, $mk + m$ is a natural number.

By the Principle of Mathematical Induction, mn is a natural number for all natural numbers n.

Since m was an arbitrary natural number, we have shown that the product of any two natural numbers is a natural number. □

42. 1 is a multiplicative identity in $\mathbb{N}$.

Proof: Assume that m is a natural number.

We have $m \cdot 1 = m(0 + 1) = m \cdot 0 + m = 0 + m = m$. For the first and fourth equalities, we used Theorem 4.4. For the second and third equalities, we used the definition of multiplication of natural numbers.

We prove that $1 \cdot n = n$ by induction on n.

Base Case ($k = 0$): $1 \cdot 0 = 0$ by the definition of multiplication of natural numbers.

Inductive Step: Let $k \in \mathbb{N}$ and assume that $1 \cdot k = k$. Then

$$1(k + 1) = 1 \cdot k + 1 = k + 1.$$

For the first equality, we used the definition of multiplication of natural numbers. For the second equality, we used the inductive hypothesis.

By the Principle of Mathematical Induction, for all natural numbers n, $1 \cdot n = n$. □

43. Multiplication is distributive over addition in $\mathbb{N}$.

Proof: Let m and n be natural numbers. We first prove that for all $t \in \mathbb{N}$, $(m + n) \cdot t = mt + nt$

(in other words, we first prove that multiplication is **right distributive** over addition in $\mathbb{N}$).

Base Case ($k = 0$): $(m + n) \cdot 0 = 0$ by the definition of multiplication of natural numbers. Similarly, $m \cdot 0 = 0$ and $n \cdot 0 = 0$. So, $m \cdot 0 + n \cdot 0 = 0$. Therefore, $(m + n) \cdot 0 = m \cdot 0 + n \cdot 0$.

Inductive Step: Let $k \in \mathbb{N}$ and assume that $(m + n) \cdot k = mk + nk$. Then

$$(m + n)(k + 1) = (m + n) \cdot k + (m + n) = (mk + nk) + (m + n)$$
$$= (mk + m) + (nk + n) = m(k + 1) + n(k + 1).$$

For the first and fourth equalities, we used the definition of multiplication of natural numbers. For the second equality, we used the inductive hypothesis. For the third equality we used the fact that addition is associative and commutative in $\mathbb{N}$ several times.

By the Principle of Mathematical Induction, for all natural numbers t, $(m+n) \cdot t = mt + nt$.

We next prove that $n \cdot 0 = 0$ and $0 \cdot n = 0$ for all natural numbers n.

$n \cdot 0 = 0$ by the definition of multiplication of natural numbers.

We prove that $0 \cdot n = 0$ by induction on n.

Base Case $(k = 0)$**:** By definition of multiplication of natural numbers, we have $0 \cdot 0 = 0$.

Inductive Step: Let $k \in \mathbb{N}$ and assume that $0 \cdot k = 0$. Then we have

$$0 \cdot (k+1) = 0 \cdot k + 0 = 0 + 0 = 0.$$

For the first equality, we used the definition of multiplication of natural numbers. For the second equality, we used the inductive hypothesis. For the third equality, we used the definition of addition of natural numbers.

By the Principle of Mathematical Induction, for all natural numbers n, $0 \cdot n = n$.

Let m be a natural number. We prove that for all $n \in \mathbb{N}$, $mn = nm$ (multiplication is **commutative** in $\mathbb{N}$).

Base Case $(k = 0)$**:** $m \cdot 0 = 0$ by the definition of multiplication of natural numbers. We just proved that $0 \cdot m = 0$. Therefore, $m \cdot 0 = 0 \cdot m$

Inductive Step: Let $k \in \mathbb{N}$ and assume that $mk = km$. Then

$$m(k+1) = mk + m = km + m = (k+1)m.$$

For the first equality, we used the definition of multiplication of natural numbers. For the second equality, we used the inductive hypothesis. For the third equality we used the fact that multiplication is right distributive over addition in $\mathbb{N}$ (proved above).

By the Principle of Mathematical Induction, for all natural numbers n, $mn = nm$.

Finally, let $m, n, t \in \mathbb{N}$. Then $m(n+t) = (n+t)m = nm + tm = mn + mt$. This shows that multiplication is left distributive over addition in $\mathbb{N}$.

Since multiplication is both right and left distributive over addition in $\mathbb{N}$, multiplication is distributive over addition in $\mathbb{N}$. □

44. Multiplication is associative in $\mathbb{N}$.

Proof: Assume that m and n are natural numbers.

Base Case $(k = 0)$**:** $(mn) \cdot 0 = 0 = m \cdot 0 = m(n \cdot 0)$ by the definition of multiplication of natural numbers.

Inductive Step: Let $k \in \mathbb{N}$ and assume that $(mn)k = m(nk)$. Then

$$(mn)(k+1) = (mn)k + mn = m(nk) + mn = mn + m(nk)$$
$$= m(n + nk) = m(nk + n) = m\big(n(k+1)\big).$$

For the first and sixth equalities, we used the definition of multiplication of natural numbers. For the second equality, we used the inductive hypothesis. For the third and fifth equalities, we used the fact that addition is commutative in $\mathbb{N}$. For the fourth equality, we used the fact that multiplication is distributive over addition in $\mathbb{N}$.

By the Principle of Mathematical Induction, for all natural numbers t, $(mn)t = m(nt)$. This shows that multiplication is associative in $\mathbb{N}$. □

45. Multiplication is commutative in $\mathbb{N}$.

Proof: This was already proved in Problem 43 above. □

46. For all natural numbers m, n, and k, if $m + k = n + k$, then $m = n$.

Proof: Let $m, n \in \mathbb{N}$. We prove by induction on k that $m + k = n + k \to m = n$.

Base Case ($k = 0$ and $k = 1$): If $m + 0 = n + 0$, then since $m + 0 = m$ and $n + 0 = n$ (by definition of addition of natural numbers), $m = n$. Next, suppose $m + 1 = n + 1$. Then $m \cup \{m\} = n \cup \{n\}$. If $n \neq m$, then either $n \in m$ or $m \in n$. Without loss of generality, assume that $n \in m$. Since $\in$ is antisymmetric on $\mathbb{N}$ and $n \neq m$, we must have $m \notin n$. Thus, $m = n$, contrary to our assumption that $n \neq m$. This contradiction shows that $m = n$.

Inductive Step: Let $t \in \mathbb{N}$, assume that $m + t = n + t \to m = n$, and let $m + (t + 1) = n + (t + 1)$. By the definition of addition in $\mathbb{N}$, $(m + t) + 1 = (n + t) + 1$. By the base case, $m + t = n + t$. By the inductive hypothesis, $m = n$.

By the Principle of Mathematical Induction, for all natural numbers m, n, and k, if $m + k = n + k$, then $m = n$. □

47. For all natural numbers m and n, $m < n$ if and only if there is a natural number $k > 0$ such that $n = m + k$.

Proof: First note that for any $m \in \mathbb{N}$, we have $m + 1 = m \cup \{m\} > 0$ because $m \in m \cup \{m\} = m + 1$ Therefore, $m + 1 \neq \emptyset = 0$.

Now, let $m \in \mathbb{N}$. We prove by induction on n that if $m < n$, there is $k > 0$ such that $n = m + k$.

Base Case ($t = 0$): If $m < 0$, then $m \in \emptyset$, which is impossible. So, the conclusion is vacuously true.

Inductive Step: Let $t \in \mathbb{N}$ and assume that if $m < t$, there is $k > 0$ such that $t = m + k$. Assume that $m < t + 1$. Then $m < t$ or $m = t$. If $m < t$, then $t + 1 = (m + k) + 1 = m + (k + 1)$. By the note in the first paragraph above, $k + 1 > 0$. If $m = t$, then $t + 1 = m + 1$. Since $1 = 0 + 1$, again by the note in the first paragraph above, $1 > 0$.

By the Principle of Mathematical Induction, for all $n \in \mathbb{N}$, if $m < n$, there is $k > 0$ so that $n = m + k$.

Now, let $n, m \in \mathbb{N}$. We prove by induction on $k > 0$ that if $n = m + k$, then $m < n$.

Base Case $(t = 1)$: If $n = m + 1 = m \cup \{m\}$, then since $m \in \{m\}$, $m \in n$, and so, $m < n$.

Inductive Step: Assume that for all $m \in \mathbb{N}$, if $n = m + t$, then $m < n$. Let $n = m + (t + 1)$. Then since addition is commutative and associative in $\mathbb{N}$, $n = m + (1 + t) = (m + 1) + t$. By the inductive hypothesis, we have $m + 1 < n$. Since $m < m + 1$ and $<$ is transitive on $\mathbb{N}$ (because $<$ is a strict linear ordering of $\mathbb{N}$), $m < n$.

By the Principle of Mathematical Induction, if $k > 0$ and $n = m + k$, then $m < n$. □

48. For all natural numbers m, n, and k, $m < n$ if and only if $m + k < n + k$.

Proof: Let $m, n, k \in \mathbb{N}$.

By Problem 47, $m < n$ if and only if there is a natural number $t > 0$ such that $n = m + t$. Now, $n + k = (m + t) + k = m + (t + k) = m + (k + t) = (m + k) + t$. So, if $m < n$, then there is a natural number $t > 0$ such that $n + k = (m + k) + t$. Thus, by Problem 47, $m + k < n + k$. Conversely, if $m + k < n + k$, then there $t > 0$ such that $n + k = (m + k) + t = (m + t) + k$. By Problem 46, $n = m + t$. So, by Problem 47 again, $m < n$. □

49. For all natural numbers m and n, if $m > 0$ and $n > 0$, then $mn > 0$.

Proof: Let $m \in \mathbb{N}$ with $m > 0$. We prove by induction on n that $n > 0$ implies $mn > 0$.

Base Case $(k = 1)$: If $k = 1$, then $m \cdot 1 = m \cdot 0 + m = 0 + m = m > 0$.

Inductive Step: Let $k \in \mathbb{N}$ with $k > 0$ and assume that $mk > 0$. Then $m(k + 1) = mk + k$. Since $mk > 0$, by Problem 48, $mk + k > 0 + k = k > 0$. So, $m(k + 1) > 0$.

By the Principle of Mathematical Induction, for all natural numbers m and n, if $m > 0$ and $n > 0$, then $mn > 0$. □

50. For all natural numbers m, n, and k with $k > 0$, if $mk = nk$, then $m = n$.

Proof: First, we prove that for all natural numbers m, n, and k with $k > 0$, $m < n$ implies $mk < nk$. To see this, let $m < n$. By Problem 47, there is a natural number $j > 0$ such that $n = m + j$. It follows from Problem 43 that $nk = (m + j)k = mk + jk$. Since $k > 0$ and $j > 0$, by Problem 49, we have $jk > 0$. By Problem 47, $mk < nk$, as desired.

Now, for the result. Since $(\mathbb{N}, <)$ satisfies trichotomy (because $<$ is a strict linear ordering of $\mathbb{N}$), we have $m < n$, $n < m$, or $m = n$. If $m < n$, then by the last paragraph, $mk < nk$. If $n < m$, then again by the last paragraph, $nk < mk$. Since $<$ is a strict linear ordering of $\mathbb{N}$ and we are given that $mk = nk$, we see that $m < n$ and $n < m$ are both impossible. Therefore, $m = n$. □

51. Every natural number is transitive. (Recall that a set A is **transitive** if $\forall x(x \in A \rightarrow x \subseteq A)$. In words, A is transitive if every element of A is also a subset of A.)

Proof:

Base Case ($k = 0$)**:** $0 = \emptyset$. Since $\emptyset$ has no elements, it is vacuously true that every element of $\emptyset$ is a subset of $\emptyset$.

Inductive Step: Assuming that k is transitive, let $j \in k + 1 = k \cup \{k\}$ and $m \in j$. Then $j \in k$ or $j \in \{k\}$. If $j \in k$, then we have $m \in j \in k$. Since k is transitive, $m \in k$. Therefore, $m \in k \cup \{k\} = k + 1$. If $j \in \{k\}$, then $j = k$. So, $m \in k$, and again, $m \in k \cup \{k\} = k + 1$.

By the Principle of Mathematical Induction, every natural number is transitive. □

LEVEL 3

Determine if each of the following numbers is prime:

52. 101

$\sqrt{101} < 11$ and 101 is not divisible by 2, 3, 5, and 7. So, 101 is **prime**.

53. 399

$399 = 7 \cdot 57$. So, 399 is **not prime**.

54. 1829

$1829 = 31 \cdot 59$. So, 1829 is **not prime**.

55. 1933

$\sqrt{1933} < 44$ and 1933 is not divisible by 2, 3, 5, 7, 11, 13, 17, 19, 23, 29, 31, 37, 41, and 43. So, 1933 is **prime**.

56. 8051

$8051 = 83 \cdot 97$. So, 8051 is **not prime**.

57. 13,873

13,873 is **prime** (check that 13,873 is not divisible by any prime number less than 117).

58. 65,623

$65{,}623 = 137 \cdot 479$. So, $65{,}623$ is **not prime**.

Use the Division Algorithm to find the quotient and remainder when

59. 28 is divided by 3.

$28 = 3 \cdot 9 + 1$. So, **the quotient is $\mathbf{9}$ and the remainder is $\mathbf{1}$**.

60. 522 is divided by 6.

$522 = 6 \cdot 87$. So, **the quotient is $\mathbf{87}$ and the remainder is $\mathbf{0}$**.

61. 723 is divided by 17.

$723 = 17 \cdot 42 + 9$. So, **the quotient is $\mathbf{42}$ and the remainder is $\mathbf{9}$**.

62. 2365 is divided by 71.

$2365 = 71 \cdot 33 + 22$. So, **the quotient is $\mathbf{33}$ and the remainder is $\mathbf{22}$**.

Express k as a linear combination of 70 and 100, or explain why it is not possible.

63. $k = 100$

$100 = 0 \cdot 70 + 1 \cdot 100$.

64. $k = 70$

$70 = 1 \cdot 70 + 0 \cdot 100$.

65. $k = 10$

$10 = 3 \cdot 70 - 2 \cdot 100$.

66. $k = 5$

Not possible because $\gcd(70, 100) = 10$ and $5 < 10$.

67. $k = 1$

Not possible because $\gcd(70, 100) = 10$ and $1 < 10$.

For each of the following, use the Euclidean Algorithm to find $\gcd(a, b)$.

68. $a = 15, b = 40$

$$40 = 15 \cdot 2 + 10$$
$$15 = 10 \cdot 1 + 5$$
$$10 = 5 \cdot 2 + 0$$

So, $\gcd(15, 40) = \mathbf{5}$.

69. $a = 36, b = 120$

$$120 = 36 \cdot 3 + 12$$
$$36 = 12 \cdot 3 + 0$$

So, $\gcd(36, 120) = \mathbf{12}$.

70. $a = 55, b = 300$

$$300 = 55 \cdot 5 + 25$$
$$55 = 25 \cdot 2 + 5$$
$$25 = 5 \cdot 5 + 0$$

So, $\gcd(55, 300) = \mathbf{5}$.

71. $a = 825, b = 2205$

$$2205 = 825 \cdot 2 + 555$$
$$825 = 555 \cdot 1 + 270$$
$$555 = 270 \cdot 2 + 15$$
$$270 = 15 \cdot 18 + \mathbf{0}$$

So, $\gcd(825, 2205) = \mathbf{15}$.

Prove each of the following:

72. If $n \in \mathbb{N}$ and A is a nonempty subset of n, then A has a least element.

Proof by Mathematical Induction:

Base Case ($k = 0$)**:** $0 = \emptyset$. The only subset of $\emptyset$ is $\emptyset$. So, the statement is vacuously true.

Inductive Step: Assume that every nonempty subset of the natural number k has a least element. We will show that every nonempty subset of $k + 1 = k \cup \{k\}$ has a least element.

Let A be a nonempty subset of $k + 1$. Then $A \setminus \{k\} \subseteq k$. If $A \setminus \{k\} \neq \emptyset$, then by the inductive hypothesis, $A \setminus \{k\}$ has a least element, say j. Since $j \in k$, j is the least element of A. If $A \setminus \{k\} = \emptyset$, then k is the only element of A, and therefore, it is the least element of A.

By the Principle of Mathematical Induction, if $n \in \mathbb{N}$ and A is a nonempty subset of n, then A has a least element. □

73. POMI → WOP.

Proof: Assume POMI, let A be a nonempty subset of $\mathbb{N}$, and choose $n \in A$. If $n \cap A = \emptyset$, then n is the least element of A (If $m \in A$ with $m \in n$, then $m \in n \cap A$, contradicting $n \cap A = \emptyset$). Otherwise, $n \cap A$ is a nonempty subset of n, and so, by Problem 72, $n \cap A$ has a least element m. Then m is the least element of A (If $k \in A$ with $k \in m$, then $k \in n$ by Problem 51, and so, m is not the least element of $n \cap A$). □

74. If $a, b, c \in \mathbb{Z}$ with $a|b$ and $b|c$, then $a|c$.

Proof: Let $a, b, c \in \mathbb{Z}$ with $a|b$ and $b|c$. Since $a|b$, there is $j \in \mathbb{Z}$ such that $b = aj$. Since $b|c$, there is $k \in \mathbb{Z}$ such that $c = bk$. It follows that $c = bk = (aj)k = a(jk)$ because multiplication is associative in $\mathbb{Z}$. Since $j, k \in \mathbb{Z}$ and $\mathbb{Z}$ is closed under multiplication, $jk \in \mathbb{Z}$. Therefore, $a|c$. □

75. If $k \in \mathbb{Z}$ with $k > 1$, then $k^3 + 1$ is not prime.

Proof: $k^3 + 1 = (k+1)(k^2 - k + 1)$. Since $k > 1$, $k + 1 > 2$ and $k^2 - k + 1 > 1 - 1 + 1 = 1$. Since neither factor is 1, $k^3 + 1$ is **not** prime. □

76. $\gcd(a, b) \mid \operatorname{lcm}(a, b)$.

Proof: Since $\gcd(a, b)$ is a divisor of a, $\gcd(a, b) \mid a$. Since $\operatorname{lcm}(a, b)$ is a multiple of a, $a \mid \operatorname{lcm}(a, b)$. By the solution to Problem 314 in Problem Set 2, | is a transitive relation. It follows that $\gcd(a, b) \mid \operatorname{lcm}(a, b)$. □

77. If $a, b, c \in \mathbb{Z}$, then $\gcd(a, b) = \gcd(a + bc, b)$.

Proof: Let d be a divisor of a and b. Then there are integers j and k such that $a = dj$ and $b = dk$. It follows that $a + bc = dj + (dk)c = d(j + kc)$, so that d is a divisor of $a + bc$.

Now, let e be a divisor of $a + bc$ and b. Then there are integers j and k such that $a + bc = ej$ and $b = ek$. It follows that $a = (a + bc) - bc = ej - (ek)c = e(j - kc)$, so that e is a divisor of a.

So, a and b have the same common divisors as $a + bc$ and b. Therefore $\gcd(ab) = \gcd(a + bc, b)$. □

78. If $a, b, k, r \in \mathbb{Z}$ with $a = bk + r$, then $\gcd(a, b) = \gcd(r, b)$.

Proof: $\gcd(r, b) = \gcd(a - bk, b) = \gcd(a + b(-k), b) = \gcd(a, b)$ by problem 77. □

LEVEL 4

For each of the following, use your computations from Problems 68 through 71 to express $\gcd(a, b)$ as a linear combination of a and b.

79. $a = 15, b = 40$

$$5 = 15 - 10 \cdot 1 = 15 - (40 - 15 \cdot 2) \cdot 1 = 15 - 40 \cdot 1 + 15 \cdot 2 = \mathbf{3 \cdot 15 - 1 \cdot 40}.$$

80. $a = 36, b = 120$

$$2 = 120 - 36 \cdot 3 = \mathbf{-3 \cdot 36 + 1 \cdot 120}.$$

81. $a = 55, b = 300$

$$5 = 55 - 25 \cdot 2 = 55 - (300 - 55 \cdot 5) \cdot 2 = 55 - 300 \cdot 2 + 55 \cdot 10 = \mathbf{11 \cdot 55 - 2 \cdot 300}.$$

82. $a = 825, b = 2205$

$$15 = 555 - 270 \cdot 2 = 555 - (825 - 555 \cdot 1) \cdot 2 = 3 \cdot 555 - 2 \cdot 825$$
$$= 3(2205 - 825 \cdot 2) - 2 \cdot 825 = \mathbf{-8 \cdot 825 + 3 \cdot 2205}.$$

A **prime pair** is a pair of prime numbers of the form $p, p+2$. For example, $3, 5$ is a prime pair.

83. Is $5, 7$ a prime pair?

Yes.

84. Is $17, 19$ a prime pair?

Yes.

85. Is $1001, 1003$ a prime pair?

No. $1001 = 7 \cdot 11 \cdot 13$

86. Find a prime pair $p, p+2$ such that $p > 2000$.

$\mathbf{2027, 2029}$

87. A **prime triple** is a sequence of three prime numbers of the form p, $p+2$, and $p+4$. For example, $3, 5, 7$ is a prime triple. Prove that there are no other prime triples.

Proof: We will show that if $n \in \mathbb{Z}$, then one of $n, n+2, n+4$ is divisible by 3. By the Division Algorithm, there are unique integers k and r such that $n = 3k + r$ and $r = 0$, 1, or 2. If $r = 0$, then n is divisible by 3. If $r = 1$, then $n + 2 = (3k+1) + 2 = 3k + 3 = 3(k+1)$, and so, $n+2$ is divisible by 3. Finally, if $r = 2$, then $n + 4 = (3k+2) + 4 = 3k + 6 = 3(k+2)$, and so, $n+4$ is divisible by 3.

So, if $p, p+2, p+4$ is a prime triple, then either p, $p+2$, or $p+4$ must be divisible by 3. The only prime number divisible by 3 is 3. So, $p = 3$, $p+2 = 3$, or $p+4 = 3$. If $p = 3$, we get the prime triple $3, 5, 7$. If $p+2 = 3$, then $p = 1$, which is not prime. If $p+4 = 3$, then $p = -1$, again not prime. □

Determine how many factors each of the following positive integers has.

88. 50

$50 = 2 \cdot 5^2$. So, the number of factors is $2 \cdot 3 = \mathbf{6}$.

Notes: (1) When choosing a factor, we first choose either 2^0 or 2^1 (two choices). We then choose $5^0, 5^1$, or 5^2 (three choices).

(2) The complete list of factors of 50 is $1, 2, 5, 10, 25, 50$.

89. 1000

$1000 = 2^3 \cdot 5^3$. So, the number of factors is $4 \cdot 4 = \mathbf{16}$.

90. $7!$

$7! = 1 \cdot 2 \cdot 3 \cdot 4 \cdot 5 \cdot 6 \cdot 7 = 2^4 \cdot 3^2 \cdot 5 \cdot 7$. So, the number of factors is $5 \cdot 3 \cdot 2 \cdot 2 = \mathbf{60}$.

91. $12!$

$12! = 1 \cdot 2 \cdot 3 \cdot 4 \cdot 5 \cdot 6 \cdot 7 \cdot 8 \cdot 9 \cdot 10 \cdot 11 \cdot 12 = 2^{10} \cdot 3^5 \cdot 5^2 \cdot 7 \cdot 11$. So, the number of factors is $11 \cdot 6 \cdot 3 \cdot 2 \cdot 2 = \mathbf{792}$.

Prove each of the following:

92. The Principle of Mathematical Induction is equivalent to the following statement:

($\star$) Let $P(n)$ be a statement and suppose that (i) $P(0)$ is true and (ii) for all $k \in \mathbb{N}$, $P(k) \to P(k+1)$. Then $P(n)$ is true for all $n \in \mathbb{N}$.

Proof: Recall that the Principle of Mathematical Induction says the following: Let S be a set of natural numbers such that (i) $0 \in S$ and (ii) for all $k \in \mathbb{N}$, $k \in S \to k+1 \in S$. Then $S = \mathbb{N}$.

Suppose that the Principle of Mathematical Induction is true and let $P(n)$ be a statement such that $P(0)$ is true, and for all $k \in \mathbb{N}$, $P(k) \to P(k+1)$. Define $S = \{n \mid (P(n)\}$. Since $P(0)$ is true, $0 \in S$. If $k \in S$, then $P(k)$ is true. So, $P(k+1)$ is true, and therefore, $k+1 \in S$. By the Principle of Mathematical Induction, $S = \mathbb{N}$. So, $P(n)$ is true for all $n \in \mathbb{N}$.

Now, suppose that ($\star$) holds, and let S be a set of natural numbers such that $0 \in S$, and for all $k \in \mathbb{N}$, $k \in S \to k+1 \in S$. Let $P(n)$ be the statement $n \in S$. Since $0 \in S$, $P(0)$ is true. If $P(k)$ is true, then $k \in S$. So, $k+1 \in S$, and therefore, $P(k+1)$ is true. By ($\star$), $P(n)$ is true for all n. So, for all $n \in \mathbb{N}$, we have $n \in S$. In other words, $\mathbb{N} \subseteq S$. Since we were given $S \subseteq \mathbb{N}$, we have $S = \mathbb{N}$. □

93. For each $r \in \mathbb{N}$, the Principle of Mathematical Induction implies the following statement:

($\star_r$) Let $P(n)$ be a statement and suppose that (i) $P(r)$ is true and (ii) for all $k \in \mathbb{N}$ with $k \geq r$, $P(k) \to P(k+1)$. Then $P(n)$ is true for all $n \in \mathbb{N}$ with $n \geq r$.

Proof: Suppose that the Principle of Mathematical Induction is true and let $P(n)$ be a statement such that $P(r)$ is true, and for all $k \in \mathbb{N}$ with $k \geq r$, $P(k) \to P(k+1)$. Define $S = \{n \mid (P(n+r)\}$. Since $P(r) = P(0+r)$ is true, $0 \in S$. If $k \in S$, then $P(k+r)$ is true. So, $P\big((k+1)+r\big) = P(k+r+1)$ is true, and therefore, $k+1 \in S$. By the Principle of Mathematical Induction, $S = \mathbb{N}$. So, $P(n+r)$ is true for all $n \in \mathbb{N}$. Therefore, $P(n)$ is true for all $n \in \mathbb{N}$ with $n \geq r$. □

94. The sum of two odd integers is an even integer.

Proof: Let m and n be odd integers. Then there are integers b and c such that $m = 2b+1$ and $n = 2c+1$. So, $m+n = (2b+1)+(2c+1) = 2b+2c+2 = 2(b+c+1)$ because addition is associative and commutative in $\mathbb{Z}$ and multiplication is distributive over addition in $\mathbb{Z}$. Since $\mathbb{Z}$ is closed under addition, $b+c+1 \in \mathbb{Z}$. Thus, $m+n$ is even. □

95. The product of an even integer and any other integer is an even integer.

Proof: Let m be an even integer and let n be any integer. Since m is even, there is an integer b such that $m = 2b$. So, $mn = (2b)n = 2(bn)$ because multiplication is associative in $\mathbb{Z}$. Since $\mathbb{Z}$ is closed under multiplication, $bn \in \mathbb{Z}$. Thus, mn is even. □

96. The sum of two integers that are each divisible by k is also divisible by k.

Proof: Let m and n be integers that are divisible by k. Then there are integers b and c such that $m = kb$ and $n = kc$. So, $m + n = kb + kc = k(b + c)$ because multiplication is distributive over addition in $\mathbb{Z}$. Since $\mathbb{Z}$ is closed under addition, $b + c \in \mathbb{Z}$. Thus, $m + n$ is divisible by k. □

97. The product of an integer divisible by k and any other integer is divisible by k.

Proof: Let m be an integer divisible by k and let n be any integer. Since m is divisible by k, there is an integer b such that $m = kb$. So, $mn = (kb)n = k(bn)$ because multiplication is associative in $\mathbb{Z}$. Since $\mathbb{Z}$ is closed under multiplication, $bn \in \mathbb{Z}$. Thus, mn is divisible by k. □

98. The product of two odd integers is an odd integer.

Proof: Let m and n be odd integers. Then there are integers j and k such that $m = 2j + 1$ and $n = 2k + 1$. So,

$$m \cdot n = (2j + 1) \cdot (2k + 1) = (2j + 1)(2k) + (2j + 1)(1) = (2k)(2j + 1) + (2j + 1)$$
$$= \big((2k)(2j) + 2k\big) + (2j + 1) = \big(2\big(k(2j)\big) + 2k\big) + (2j + 1) = 2(k(2j) + k) + (2j + 1)$$
$$= (2(k(2j) + k) + 2j) + 1 = 2\big((k(2j) + k) + j\big) + 1.$$

Here we used the fact that $(\mathbb{Z}, +, \ \cdot)$ is a ring. Since $\mathbb{Z}$ is closed under addition and multiplication, we have $(k(2j) + k) + j \in \mathbb{Z}$. Therefore, mn is odd. □

Note: We skipped over mentioning every property of the integers we used at each step, and simply mentioned that $(\mathbb{Z}, +, \ \cdot)$ is a ring. The dedicated reader may want to fill in the details.

99. If $a, b, c, d, e \in \mathbb{Z}$ with $a|b$ and $a|c$, then $a|(db + ec)$.

Proof: Let $a, b, c, d, e \in \mathbb{Z}$ with $a|b$ and $a|c$. Since $a|b$, there is $j \in \mathbb{Z}$ such that $b = aj$. Since $a|c$, there is $k \in \mathbb{Z}$ such that $c = ak$. Since $(\mathbb{Z}, +, \ \cdot)$ is a ring, it follows that

$$db + ec = d(aj) + e(ak) = (da)j + (ea)k = (ad)j + (ae)k = a(dj) + a(ek) = a(dj + ek).$$

Since $\mathbb{Z}$ is closed under multiplication, $dj \in \mathbb{Z}$ and $ek \in \mathbb{Z}$. Since $\mathbb{Z}$ is closed under addition, $dj + ek \in \mathbb{Z}$. So, $a|(db + ec)$. □

Note: As in Problem 98, we skipped over mentioning every property of the natural numbers we used at each step, and simply mentioned that $(\mathbb{Z}, +, \ \cdot)$ is a ring. The dedicated reader may want to fill in the details.

100. $3^n - 1$ is even for all natural numbers n.

Proof by Mathematical induction:

Base Case $(k = 0)$**:** $3^0 - 1 = 1 - 1 = 0 = 2 \cdot 0$. So, $3^0 - 1$ is even.

Inductive Step: Let $k \in \mathbb{N}$ and assume that $3^k - 1$ is even. Then $3^k - 1 = 2b$ for some integer b. Now,

$$3^{k+1} - 1 = 3^k \cdot 3^1 - 1 = 3^k \cdot 3 - 1 = 3^k \cdot 3 - 3^k + 3^k - 1 = 3^k(3-1) + (3^k - 1)$$
$$= 3^k \cdot 2 + 2b = 2 \cdot 3^k + 2b = 2(3^k + b).$$

Here we used the fact that $(\mathbb{Z}, +, \cdot)$ is a ring. Since $\mathbb{Z}$ is closed under multiplication, $3^k \in \mathbb{N}$. Since $\mathbb{Z}$ is closed under addition, $3^k + b \in \mathbb{Z}$. Therefore, $3^{k+1} - 1$ is even.

By the Principle of Mathematical Induction, $3^n - 1$ is even for all $n \in \mathbb{N}$. □

Notes: (1) As in Problem 98, we skipped over mentioning every property of the natural numbers we used at each step, and simply mentioned that $(\mathbb{Z}, +, \cdot)$ is a ring. The dedicated reader may want to fill in the details.

(2) Notice our use of SACT in the middle of the first line of the sequence of equations. Recall that SACT says that if you need something to appear, just put it in. Then correct it by performing the opposite of what you just did. We needed $3^k - 1$ to appear, so we added 3^k, and then subtracted 3^k to the left of it.

101. The Euclidean Algorithm:_ Let $a, b \in \mathbb{Z}^+$ with $a \geq b$. Let $r_0 = a$, $r_1 = b$. Apply the Division Algorithm to r_0 and r_1 to find $k_1, r_2 \in \mathbb{Z}^+$ such that $r_0 = r_1 k_1 + r_2$, where $0 \leq r_2 < r_1$. Iterate this process to get $r_j = r_{j+1} k_{j+1} + r_{j+2}$, where $0 \leq r_{j+2} < r_{j+1}$ for $j = 0, 1, \dots, n-1$ so that $r_{n+1} = 0$. Then $\gcd(a, b) = r_n$.

Proof: We use the Division Algorithm to get the following sequence of equations:

$$\begin{aligned} r_0 &= r_1 k_1 + r_2, & 0 \leq r_2 < r_1 \\ r_1 &= r_2 k_2 + r_3, & 0 \leq r_3 < r_2 \\ &\cdots \\ r_{n-2} &= r_{n-1} k_{n-1} + r_n, & 0 \leq r_n < r_{n-1} \\ r_{n-1} &= r_n k_n + 0 \end{aligned}$$

Since each remainder r_i is strictly less than the remainder r_{i-1}, after finitely many iterations of the Division Algorithm, we get a remainder of 0.

By Problem 78, we have $\gcd(a, b) = \gcd(r_0, r_1) = \gcd(r_1, r_2) = \cdots = \gcd(r_n, 0) = r_n$. □

102. If $a|c$ and $b|c$, then $\text{lcm}(a,b) \mid c$.

Proof: Suppose that $a|c$ and $b|c$. By the Division Algorithm, there are integers k and r such that $c = \text{lcm}(a,b) \cdot k + r$ with $0 \leq r < \text{lcm}(a,b)$. Therefore, we have $r = c - \text{lcm}(a,b) \cdot k$. Since $a|c$ and $a|\text{lcm}(a,b)$, it follows from Problem 99 that $a|r$. Similarly, $b|r$. So r is a multiple of both a and b. Since $0 \leq r < \text{lcm}(a,b)$, r must be 0. Thus, $c = \text{lcm}(a,b) \cdot k$ and so, $\text{lcm}(a,b)|c$. □

103. If $a, b \in \mathbb{Z}^+$ with $\gcd(a,b) = 1$ and $c|ab$, then there are integers d and e such that $c = de$, $d|a$, and $e|b$.

Proof: Suppose that $a, b \in \mathbb{Z}^+$, $\gcd(a,b) = 1$, and $c|ab$. Let $a = p_0^{a_0} p_1^{a_1} \cdots p_n^{a_n}$ and $b = p_0^{b_0} p_1^{b_1} \cdots p_n^{b_n}$ be complete prime factorizations of a and b, let $d = \gcd(a,c)$, and let $e = \gcd(b,c)$. If $d = p_0^{d_0} p_1^{d_1} \cdots p_n^{d_n}$ and $e = p_0^{e_0} p_1^{e_1} \cdots p_n^{e_n}$ are complete prime factorizations of d and e, respectively, then we have $d_i \leq a_i$ and $e_i \leq b_i$ for each $i = 1, 2, \dots, n$. It follows that $\gcd(d,e) = 1$, $d|a$, and $e|b$. Every prime power in the prime factorization of c must appear as a factor of d or e. So, $c = de$. □

LEVEL 5

Using the formal definition of $\mathbb{N}$, prove each of the following:

104. $(\mathbb{N}, +, \cdot)$ is a commutative semiring,

Proof: By Theorem 4.3, addition on the set of natural numbers is a binary operation.

By Theorem 4.5, addition is associative in $\mathbb{N}$.

By Problem 40, addition is commutative in $\mathbb{N}$.

By Theorem 4.4 and the notes that follow it, 0 is an additive identity in $\mathbb{N}$.

So, $(\mathbb{N}, +)$ is a commutative monoid.

By Problem 41, multiplication on the set of natural numbers is a binary operation.

By Problem 44, multiplication is associative in $\mathbb{N}$.

By Problem 45, multiplication is commutative in $\mathbb{N}$.

By Problem 42, 1 is a multiplicative identity in $\mathbb{N}$.

So, $(\mathbb{N}, \cdot)$ is a commutative monoid.

By Problem 43, multiplication is distributive over addition in $\mathbb{N}$.

By the definition of multiplication of natural numbers, for all $n \in \mathbb{N}$, $n \cdot 0 = 0$. Since multiplication is commutative in $\mathbb{N}$, $0 \cdot n = n \cdot 0 = 0$.

Therefore, $(\mathbb{N}, +, \cdot)$ is a commutative semiring. □

105. $(\mathbb{N}, +, \cdot)$ is not a ring.

Proof: We show that 1 has no additive inverse in $\mathbb{N}$ (in fact, the only natural number with an additive inverse is 0). Suppose toward contradiction that $n \in \mathbb{N}$ with $n + 1 = 0$. Since $n + 1 = n^+ = n \cup \{n\}$, and $n \in n \cup \{n\}$, we must have $n \in 0 = \emptyset$. But the empty set has no elements. This contradiction proves that there is no natural number n such that $n + 1 = 0$. So, 1 has no additive inverse in $\mathbb{N}$. Therefore, the inverse property fails and $(\mathbb{N}, +)$ is **not** a group. So, $(\mathbb{N}, +, \cdot)$ is **not** a ring. □

The Principle of Strong Induction is the following statement:

> $(\star\star)$ Let $P(n)$ be a statement and suppose that (i) $P(0)$ is true and (ii) for all $k \in \mathbb{N}$, $\forall j \leq k\ (P(j)) \to P(k+1)$. Then $P(n)$ is true for all $n \in \mathbb{N}$.

106. Use the Principle of Mathematical Induction to prove the Principle of Strong Induction.

Proof: Let $P(n)$ be a statement such that $P(0)$ is true, and for all $k \in \mathbb{N}$, $\forall j \leq k\ (P(j)) \to P(k+1)$. Let $Q(n)$ be the statement $\forall j \leq n\ (P(j))$.

Base case: $Q(0) \equiv \forall j \leq 0(P(j)) \equiv P(0)$. Since $P(0)$ is true and $Q(0) \equiv P(0)$, $Q(0)$ is also true.

Inductive step: Suppose that $Q(k)$ is true. Then $\forall j \leq k\ (P(j))$ is true. Therefore, $P(k+1)$ is true. So $Q(k) \wedge P(k+1)$ is true. But notice that

$$Q(k+1) \equiv \forall j \leq k+1(P(j)) \equiv \forall j \leq k(P(j)) \wedge P(k+1) \equiv Q(k) \wedge P(k+1).$$

So, $Q(k+1)$ is true.

By the Principle of Mathematical Induction ($(\star)$ from Problem 92), $Q(n)$ is true for all $n \in \mathbb{N}$. This implies that $P(n)$ is true for all $n \in \mathbb{N}$. □

107. Use the Principle of Strong Induction to prove the Principle of Mathematical Induction.

Proof: Let $P(n)$ be a statement such that (i) $P(0)$ is true, and (ii) for all $k \in \mathbb{N}$, $P(k) \to P(k+1)$. We now prove that for all $k \in \mathbb{N}$, $\forall j \leq k\ (P(j)) \to P(k+1)$.

Base case ($k = 0$): $\forall j \leq 0\ (P(j)) \to P(0+1)$ is equivalent to $P(0) \to P(1)$, which is statement (ii) above with $k = 0$.

Inductive step: Suppose that for fixed $k \in \mathbb{N}$, $\forall j \leq k\ (P(j)) \to P(k+1)$ is true. By statement (ii) above, we have $P(k+1) \to P(k+2)$ is true. By Hypothetical Syllogism (see List 1.27), we see that $\forall j \leq k\ (P(j)) \to P(k+2)$ is true. We also have $\forall j \leq k+1(P(j)) \equiv \forall j \leq k(P(j)) \wedge P(k+1)$ and $\forall j \leq k(P(j)) \wedge P(k+1) \to \forall j \leq k(P(j))$ (by Conjunctive Elimination form List 1.27). So, again by Hypothetical Syllogism, we have $\forall j \leq k+1(P(j)) \to P(k+2)$ is true.

Therefore, (i) and (ii) in the statement of Strong Induction hold. So, by the Principle of Strong Induction, $P(n)$ is true for all $n \in \mathbb{N}$. □

Let $A = \{a, b, c, d\}$, $B = \{X \mid X \subseteq A \wedge d \notin X\}$, and $C = \{X \mid X \subseteq A \wedge d \in X\}$.

108. Show that there is a natural bijection between the elements of B and the elements of C.

Solution: We define the one-to-one correspondence as follows: If $Y \in B$, then Y is a subset of A that does not contain d. Let Y_d be the set that contains the same elements as Y, but with d thrown in (in other words, $Y_d = Y \cup \{d\}$). Then the correspondence $Y \to Y_d$ is a one-to-one correspondence. We can see this correspondence in the table below.

Elements of B	Elements of C
$\emptyset$	$\{d\}$
$\{a\}$	$\{a, d\}$
$\{b\}$	$\{b, d\}$
$\{c\}$	$\{c, d\}$
$\{a, b\}$	$\{a, b, d\}$
$\{a, c\}$	$\{a, c, d\}$
$\{b, c\}$	$\{b, c, d\}$
$\{a, b, c\}$	$\{a, b, c, d\}$

109. Generalize the result in Problem 108 to a set with $n + 1$ elements for $n > 0$.

For the general result, we start with a set A with $n + 1$ elements, and we let d be some element from A. Define B and C the same way as before: $B = \{X \mid X \subseteq A \wedge d \notin X\}$, and $C = \{X \mid X \subseteq A \wedge d \in X\}$. Also, as before, if $Y \in B$, then Y is a subset of A that does not contain d. Let Y_d be the set that contains the same elements as Y, but with d thrown in (in other words, $Y_d = Y \cup \{d\}$). Then the correspondence $Y \to Y_d$ is a one-to-one correspondence.

110. Use the Principle of Mathematical Induction to prove that for every $n \in \mathbb{N}$, if S is a set with $|S| = n$, then S has 2^n subsets.

Proof:

Base Case ($k = 0$)**:** Let S be a set with $|S| = 0$. Then $S = \emptyset$, and the empty set has exactly 1 subset, namely itself. So, the number of subsets of S is $1 = 2^0$.

Inductive Step: Assume that for any set S with $|S| = k$, S has 2^k subsets.

Now, let A be a set with $|A| = k + 1$, let d be any element from A, and let $S = A \setminus \{d\}$ (S is the set consisting of all elements of A except d). $|S| = k$, and so, by the inductive hypothesis, S has 2^k subsets. Let $B = \{X \mid X \subseteq A \wedge d \notin X\}$ and $C = \{X \mid X \subseteq A \wedge d \in X\}$. B is precisely the set of subsets of S, and so $|B| = 2^k$. By Problem 109, $|B| = |C|$ and therefore, $|C| = 2^k$. Also, B and C have no elements in common and every subset of A is in either B or C. So, the number of subsets of A is equal to $|B| + |C| = 2^k + 2^k = 2 \cdot 2^k = 2^1 \cdot 2^k = 2^{1+k} = 2^{k+1}$.

By the Principle of Mathematical Induction, given any $n \in \mathbb{N}$, if S is a set with $|S| = n$, then S has 2^n subsets. □

Notes: (1) Recall from Lesson 2 that $|S| = n$ means that the set S has n elements.

(2) Also, recall from Lesson 2 that if S is a set, then the **power set** of S is the set of subsets of S.

$$\mathcal{P}(S) = \{X \mid X \subseteq S\}$$

In this problem, we proved that a set with n elements has a power set with 2^n elements. Symbolically, we have $|S| = n \to |\mathcal{P}(S)| = 2^n$.

In the following two exercises, you will use the Principle of Mathematical Induction to prove that commutativity implies generalized commutativity.

111. Provide a formal definition of generalized commutativity.

Definition: An operation $\star$ satisfies **generalized commutativity** in a set S if for any $n \in \mathbb{N}^+$, for all $x_1, \ldots, x_n \in S$, and any bijection $f: \{1, 2, \ldots, n\} \to \{1, 2, \ldots, n\}$, we have

$$\prod_{i=1}^{n} x_{f(i)} = \prod_{i=1}^{n} x_i.$$

112. Prove that if $\star$ is a binary operation that is commutative in a set S, then $\star$ satisfies generalized commutativity in S.

Proof: Suppose that $\star$ is commutative in S. We will prove by induction that for all $n \in \mathbb{N}$ with $n \geq 1$, for all $x_1, \ldots, x_n \in S$, and any bijection $f: \{1, 2, \ldots, n\} \to \{1, 2, \ldots, n\}$, we have

$$\prod_{i=1}^{n} x_{f(i)} = \prod_{i=1}^{n} x_i.$$

The base case ($k = 1$) is obvious. So, assume that the statement is true for some $k \geq 1$, and let $x_1, \ldots x_{k+1} \in S$, let $f: \{1, 2, \ldots, k+1\} \to \{1, 2, \ldots, k+1\}$ be a bijection, and suppose $f(j) = k + 1$.

We have

$$\prod_{i=1}^{k+1} x_{f(i)} = \prod_{i=1}^{j-1} x_{f(i)} \star x_{f(j)} \star \prod_{i=1}^{k+1-j} x_{f(j+i)} = \prod_{i=1}^{j-1} x_{f(i)} \star \prod_{i=1}^{k+1-j} x_{f(j+i)} \star x_{f(j)}.$$

Now, define $g: \{1, 2, \ldots, k\} \to \{1, 2, \ldots, k\}$ by $g(i) = f(i)$ if $i < j$ and $g(i) = f(i+1)$ if $i \geq j$. Then we have

$$\prod_{i=1}^{k+1} x_{f(i)} = \prod_{i=1}^{j-1} x_{g(i)} \star \prod_{i=1}^{k+1-j} x_{g(j-1+i)} \star x_{k+1} = \prod_{i=1}^{k} x_{g(i)} \star x_{k+1} = \prod_{i=1}^{k} x_i \star x_{k+1} = \prod_{i=1}^{k+1} x_i.$$

The inductive hypothesis was used in the third equality above.

By the Principle of Mathematical Induction, for all $m, n \in \mathbb{N}^+$, for all $x_1, \dots, x_n \in S$, and any bijection $f: \{1, 2, \dots, n\} \to \{1, 2, \dots, n\}$, we have

$$\prod_{i=1}^{n} x_{f(i)} = \prod_{i=1}^{n} x_i .$$ □

If $a, b \in \mathbb{Z}^+$ and $\gcd(a, b) = 1$, find each of the following:

113. $\gcd(a, a+1)$

Since $a + 1 - a = 1$, we see that 1 can be written as a linear combination of a and $a + 1$. Therefore, $\gcd(a, a+1) = 1$.

114. $\gcd(a, a+2)$

Since $a + 2 - a = 2$, we see that 2 can be written as a linear combination of a and $a + 2$. Therefore, $\gcd(a, a+2) \leq 2$. So, $\gcd(a, a+2) = 1$ or 2.

If a is even, then there is $k \in \mathbb{Z}$ such that $a = 2k$. So, $a + 2 = 2k + 2 = 2(k + 1)$. Thus, we see that $2|a$ and $2|a + 2$. Therefore, if a is even, $\gcd(a, a+2) = 2$.

If a is odd, then 2 does not divide a. So, $\gcd(a, a+2)$ cannot be 2, and therefore, must be 1.

115. $\gcd(3a+2, 5a+3)$

Since $5(3a + 2) - 3(5a + 3) = 15a + 10 - 15a - 9 = 1$, we see that 1 can be written as a linear combination of $3a + 2$ and $5a + 3$. So, $\gcd(3a + 2, 5a + 3) = 1$.

116. $\gcd(a+b, a-b)$

If d divides $a + b$ and $a - b$, then d divides $(a + b) + (a - b) = 2a$ and d divides $(a + b) - (a - b) = 2b$ because by Problem 99, d divides any linear combination of $a + b$ and $a - b$. By Theorem 4.30, $\gcd(2a, 2b)$ can be written as a linear combination of $2a$ and $2b$. So, again by Problem 99, $d | \gcd(2a, 2b) = 2$. So, $d = 1$ or $d = 2$.

If a and b are both odd, then both $a + b$ and $a - b$ are even, and so, $\gcd(a + b, a - b) = 2$.

If a and b do not have the same parity (in other words, one is even and the other is odd), then $a + b$ is odd. So, 2 is not a divisor of $a + b$, and therefore, $\gcd(a + b, a - b) = 1$.

a and b cannot both be even because then $\gcd(a, b) \geq 2$.

117. $\gcd(a+2b, 2a+b)$

If d divides $a + 2b$ and $2a + b$, then d divides $2(a + 2b) - (2a + b) = 3b$ and d divides $2(2a + b) - (a + 2b) = 3a$ because once again by Problem 99 above, d divides any linear combination of $a + 2b$ and $2a + b$. By Theorem 4.30, $\gcd(3a, 3b)$ can be written as a linear combination of $3a$ and $3b$. So, again by Problem 99, $d | \gcd(3a, 3b) = 3$. So, $d = 1$ or $d = 3$.

If we let $a = 2, b = 3$, then $a + 2b = 2 + 2 \cdot 3 = 2 + 6 = 8$ and $2a + b = 2 \cdot 2 + 3 = 4 + 3 = 7$. So, $\gcd(a + 2b, 2a + b) = \gcd(8, 7) = 1$.

If we let $a = 2, b = 5$, then $a + 2b = 2 + 2 \cdot 5 = 2 + 10 = 12$ and $2a + b = 2 \cdot 2 + 5 = 4 + 5 = 9$. So, $\gcd(a + 2b, 2a + b) = \gcd(12, 9) = 3$.

So, both $\gcd(a + 2b, 2a + b) = 1$ and $\gcd(a + 2b, 2a + b) = 3$ can occur.

Problem Set 5

LEVEL 1

Determine if each of the following sets is bounded or unbounded in $\mathbb{Q}$.

1. $\mathbb{Q}$

Unbounded

2. $\mathbb{Z}^-$

Unbounded

3. $\{5, 10, 15, 20, 25, 30\}$

Bounded

4. $\{1, 2, 3, \ldots, 1000\}$

Bounded

Let a and b be the lengths of the legs of a right triangle and let c be the length of the hypotenuse of the right triangle. Find the third side length of the triangle, given the other two.

5. $a = 9, b = 12$

$c^2 = 9^2 + 12^2 = 81 + 144 = 225$. Therefore, $\boldsymbol{c = 15}$.

6. $a = 40, c = 41$

$41^2 = 40^2 + b^2$. So, $1681 = 1600 + b^2$, and thus, $b^2 = 1681 - 1600 = 81$. Therefore, $\boldsymbol{b = 9}$.

7. $a = 1, b = 2$

$c^2 = 1^2 + 2^2 = 1 + 4 = 5$. Therefore, $\boldsymbol{c = \sqrt{5}}$.

8. $b = 17, c = 26$

$26^2 = a^2 + 17^2$. So, $676 = a^2 + 289$, and thus, $a^2 = 676 - 289 = 387$. Therefore, we have $\boldsymbol{a = \sqrt{387} = 3\sqrt{43}}$.

For each of the following sets, find the least upper bound in $\mathbb{Q}$ and greatest lower bound in $\mathbb{Q}$ if they exist.

9. $\{x \in \mathbb{Q} \mid -7 \leq x \leq 28\}$

Least upper bound $= \mathbf{28}$

Greatest lower bound $= -\mathbf{7}$

10. $\{x \in \mathbb{Q} \mid x > 15\}$

Least upper bound **does not exist**.

Greatest lower bound $= \mathbf{15}$

11. $\mathbb{Q}^+$

Least upper bound **does not exist**.

Greatest lower bound $= \mathbf{0}$

12. $\{x \in \mathbb{Q} \mid -2 \leq x^2 \leq 2\}$

Least upper bound **does not exist in** $\mathbb{Q}$ (however, in $\mathbb{R}$ it does exist and is equal to $\sqrt{2}$).

Greatest lower bound **does not exist in** $\mathbb{Q}$ (however, in $\mathbb{R}$ it does exist and is equal to $-\sqrt{2}$).

Let $f: \mathbb{R} \to \mathbb{R}$ be defined by $f(x) = 5x - 1$.

13. Prove that $\lim_{x \to 3} f(x) = 14$.

Proof: Let $\epsilon > 0$ and let $\delta = \frac{\epsilon}{5}$. Suppose that $0 < |x - 3| < \delta$. Then we have

$$|(5x - 1) - 14| = |5x - 15| = 5|x - 3| < 5\delta = 5 \cdot \frac{\epsilon}{5} = \epsilon.$$

So, $\lim_{x \to 3} f(x) = 14$. □

14. Prove that f is continuous on $\mathbb{R}$.

Proof: Let $a \in \mathbb{R}$. We will show that f is continuous at a. Let $\epsilon > 0$, let $\delta = \frac{\epsilon}{5}$, and let $|x - a| < \delta$. Then

$$|(5x - 1) - (5a - 1)| = |5x - 5a| = 5|x - a| < 5\delta = 5 \cdot \frac{\epsilon}{5} = \epsilon.$$

So, f is continuous at a. Since $a \in \mathbb{R}$ was arbitrary, f is continuous on $\mathbb{R}$. □

LEVEL 2

In Problems 49 and 50 below, you will be asked to show that $W\left(\frac{\pi}{3}\right) = \left(\frac{1}{2}, \frac{\sqrt{3}}{2}\right)$ and $W\left(\frac{\pi}{6}\right) = \left(\frac{\sqrt{3}}{2}, \frac{1}{2}\right)$. Use this information to compute the sine, cosine, and tangent of each of the following angles:

15. $\frac{\pi}{6}$

By Problem 50, $W\left(\frac{\pi}{6}\right) = \left(\frac{\sqrt{3}}{2}, \frac{1}{2}\right)$. So, $\cos\frac{\pi}{6} = \boldsymbol{\frac{\sqrt{3}}{2}}$, $\sin\frac{\pi}{6} = \boldsymbol{\frac{1}{2}}$, and $\tan\frac{\pi}{6} = \frac{\sin\frac{\pi}{6}}{\cos\frac{\pi}{6}} = \frac{1}{2} \cdot \frac{2}{\sqrt{3}} = \boldsymbol{\frac{1}{\sqrt{3}}}$.

16. $\frac{\pi}{3}$

By Problem 49, $W\left(\frac{\pi}{3}\right) = \left(\frac{1}{2}, \frac{\sqrt{3}}{2}\right)$. So, $\cos\frac{\pi}{3} = \boldsymbol{\frac{1}{2}}$, $\sin\frac{\pi}{3} = \boldsymbol{\frac{\sqrt{3}}{2}}$, and $\tan\frac{\pi}{3} = \frac{\sin\frac{\pi}{3}}{\cos\frac{\pi}{3}} = \frac{\sqrt{3}}{2} \cdot \frac{2}{1} = \boldsymbol{\sqrt{3}}$.

17. $\frac{2\pi}{3}$

Since $\frac{2\pi}{3} = \pi - \frac{\pi}{3}$, by the symmetry of the unit circle, $W\left(\frac{2\pi}{3}\right) = \left(-\frac{1}{2}, \frac{\sqrt{3}}{2}\right)$. It follows that $\cos\frac{2\pi}{3} = -\boldsymbol{\frac{1}{2}}$, $\sin\frac{2\pi}{3} = \boldsymbol{\frac{\sqrt{3}}{2}}$, and $\tan\frac{2\pi}{3} = \frac{\sin\frac{2\pi}{3}}{\cos\frac{2\pi}{3}} = \frac{\sqrt{3}}{2}\left(\frac{-2}{1}\right) = -\boldsymbol{\sqrt{3}}$.

18. $\frac{5\pi}{6}$

Since $\frac{5\pi}{6} = \pi - \frac{\pi}{6}$, by the symmetry of the unit circle, $W\left(\frac{5\pi}{6}\right) = \left(-\frac{\sqrt{3}}{2}, \frac{1}{2}\right)$. It follows that $\cos\frac{5\pi}{6} = -\boldsymbol{\frac{\sqrt{3}}{2}}$, $\sin\frac{5\pi}{6} = \boldsymbol{\frac{1}{2}}$, and $\tan\frac{5\pi}{6} = \frac{\sin\frac{5\pi}{6}}{\cos\frac{5\pi}{6}} = \frac{1}{2}\left(-\frac{2}{\sqrt{3}}\right) = -\boldsymbol{\frac{1}{\sqrt{3}}}$.

19. $\frac{7\pi}{6}$

Since $\frac{7\pi}{6} = \pi + \frac{\pi}{6}$, by the symmetry of the unit circle, $W\left(\frac{7\pi}{6}\right) = \left(-\frac{\sqrt{3}}{2}, -\frac{1}{2}\right)$. It follows that $\cos\frac{7\pi}{6} = -\boldsymbol{\frac{\sqrt{3}}{2}}$, $\sin\frac{7\pi}{6} = -\boldsymbol{\frac{1}{2}}$, and $\tan\frac{7\pi}{6} = \frac{\sin\frac{7\pi}{6}}{\cos\frac{7\pi}{6}} = -\frac{1}{2}\left(-\frac{2}{\sqrt{3}}\right) = \boldsymbol{\frac{1}{\sqrt{3}}}$.

20. $\frac{4\pi}{3}$

Since $\frac{4\pi}{3} = \pi + \frac{\pi}{3}$, by the symmetry of the unit circle, $W\left(\frac{4\pi}{3}\right) = \left(-\frac{1}{2}, -\frac{\sqrt{3}}{2}\right)$. It follows that $\cos\frac{4\pi}{3} = -\boldsymbol{\frac{1}{2}}$, $\sin\frac{4\pi}{3} = -\boldsymbol{\frac{\sqrt{3}}{2}}$, and $\tan\frac{4\pi}{3} = \frac{\sin\frac{4\pi}{3}}{\cos\frac{4\pi}{3}} = -\frac{\sqrt{3}}{2}\left(\frac{-2}{1}\right) = \boldsymbol{\sqrt{3}}$.

21. $\frac{5\pi}{3}$

Since $\frac{5\pi}{3} = 2\pi - \frac{\pi}{3}$, by the symmetry of the unit circle, $W\left(\frac{5\pi}{3}\right) = \left(\frac{1}{2}, -\frac{\sqrt{3}}{2}\right)$. It follows that $\cos\frac{5\pi}{3} = \mathbf{\frac{1}{2}}$, $\sin\frac{5\pi}{3} = -\mathbf{\frac{\sqrt{3}}{2}}$, and $\tan\frac{5\pi}{3} = \frac{\sin\frac{5\pi}{3}}{\cos\frac{5\pi}{3}} = -\frac{\sqrt{3}}{2}\cdot\frac{2}{1} = -\mathbf{\sqrt{3}}$.

22. $\frac{11\pi}{6}$

Since $\frac{11\pi}{6} = 2\pi - \frac{\pi}{6}$, by the symmetry of the unit circle, $W\left(\frac{11\pi}{6}\right) = \left(\frac{\sqrt{3}}{2}, -\frac{1}{2}\right)$. It follows that $\cos\frac{11\pi}{6} = \mathbf{\frac{\sqrt{3}}{2}}$, $\sin\frac{11\pi}{6} = -\mathbf{\frac{1}{2}}$, and $\tan\frac{11\pi}{6} = \frac{\sin\frac{11\pi}{6}}{\cos\frac{11\pi}{6}} = -\frac{1}{2}\cdot\frac{2}{\sqrt{3}} = -\mathbf{\frac{1}{\sqrt{3}}}$.

Let $x \in \mathbb{R}$ with $x > 0$.

23. Prove that $\frac{1}{2} > 0$.

Proof: Since $2 > 0$, by Theorem 5.4, $\frac{1}{2} > 0$. □

24. Prove that $\frac{1}{2}x > 0$.

Proof: Since $\frac{1}{2} > 0$ (by Problem 23) and $x > 0$ (given), by property (2) of an ordered field, $\frac{1}{2}x > 0$. □

Note: Here we've replaced $\leq$ by $<$ in property (2) of an ordered field. We can always do this because if $xy = 0$, then $y = 1y = (x^{-1}x)y = x^{-1}(xy) = 0$. So, $x, y > 0$ imples $xy \neq 0$. Alternatively, we can just use the fact that every field is an integral domain.

25. Prove that $x > \frac{1}{2}x$.

Proof: $x - \frac{1}{2}x = \left(1 - \frac{1}{2}\right)x = \frac{1}{2}x > 0$ by Problem 24. By property (1) of an ordered field, we have

$$x = x - \frac{1}{2}x + \frac{1}{2}x > 0 + \frac{1}{2}x = \frac{1}{2}x.$$

□

Note: Similar to what we did in Problem 24, here we've replaced $\leq$ by $<$ in property (1) of an ordered field. We can do this for the same reason given in the Note following Problem 24.

26. Prove that there is no smallest positive real number.

Proof: If x is a positive real number, by Problem 24, $\frac{1}{2}x$ is a positive real number, and by Problem 25, $\frac{1}{2}x$ is smaller than x. Since x was an arbitrary positive real number, there is no smallest positive real number. □

Let $a, r \in \mathbb{R}$ with $a \geq 0$.

27. Prove that if $a = 0$, then a is less than every positive real number.

Proof: Let r be a positive real number. By the definition of being positive, $r > 0 = a$. So, $a < 0$. Since r was an arbitrary positive real number, a is less then every positive real number. □

28. Prove that if $a \neq 0$, then there is a positive real number less than a.

Proof: Suppose that $a \neq 0$. Then a is positive. By Problem 26, there is no smallest positive real number. So, there is a positive real number less than a. □

29. Prove that $a = 0$ if and only if a is less than every positive real number.

Proof: If $a = 0$, then by Problem 27, a is less than every positive real number. If $a \neq 0$, then by Problem 28, there is a positive real number less than a. By contrapositive, if there is no positive real number less than a, then $a = 0$. By the trichotomy of $<$, the statement "there is no positive real number less than a" is equivalent to "a is less than or equal to every positive real number." Now, if a is less than every positive real number, then a is less than or equal to every positive real number. It follows that there is no positive real number less than a, and so, $a = 0$. □

Let $x, y \in \mathbb{R}$.

30. Suppose that $x < y$ and let c be a positive irrational number. Prove that there is a rational number q such that $\frac{x}{c} < q < \frac{y}{c}$.

Proof: $\frac{x}{c} < \frac{y}{c}$, and therefore, by the Density Theorem (Theorem 5.12), there is a rational number q such that $\frac{x}{c} < q < \frac{y}{c}$. □

31. Suppose that $x < y$ and let c be a positive irrational number. Prove that there is a **nonzero** rational number q such that $\frac{x}{c} < q < \frac{y}{c}$.

Proof: By Problem 30, there is a rational number s such that $\frac{x}{c} < s < \frac{y}{c}$. If $s \neq 0$, let $q = s$. If $s = 0$, then we apply the Density Theorem (Theorem 5.12) to get a rational number q with $\frac{x}{c} < q < 0$. □

32. Suppose that $x < y$. Prove that there is an irrational number t such that $x < t < y$.

Proof: By Problem 31, there is a nonzero rational number q such that $\frac{x}{c} < q < \frac{y}{c}$. So, $x < cq < y$. If cq were equal to a rational number s, then $c = c \cdot 1 = c\left(q \cdot \frac{1}{q}\right) = (cq)\left(\frac{1}{q}\right) = s\left(\frac{1}{q}\right)$. Since $\mathbb{Q}$ is closed under multiplicative inverses and multiplication, c would be rational, contrary to our assumption that c is irrational. So, $t = cq$ is irrational and $x < t < y$. □

Let $A \subseteq \mathbb{R}$, let $f: A \to \mathbb{R}$, and let $r \in \mathbb{R}$. Prove each of the following:

33. If $A = \mathbb{R}$, $c \in \mathbb{R}$ and $f: \mathbb{R} \to \mathbb{R}$ is defined by $f(x) = c$, then $\lim_{x \to r}[f(x)] = c$.

Proof: Let $\epsilon > 0$ and let $\delta = 1$. If $0 < |x - r| < \delta$, then $|f(x) - c| = |c - c| = |0| = 0 < \epsilon$. Therefore, $\lim_{x \to r}[f(x)] = c$. □

34. If $k \in \mathbb{R}$, and $\lim_{x \to r}[f(x)]$ is a finite real number, then $\lim_{x \to r}[kf(x)] = k \lim_{x \to r}[f(x)]$.

Proof: Suppose that $\lim_{x \to r}[f(x)] = L$ and let $\epsilon > 0$. First assume that $k \neq 0$. Since $\lim_{x \to r}[f(x)] = L$, there is $\delta > 0$ such that $0 < |x - r| < \delta$ implies $|f(x) - L| < \frac{\epsilon}{|k|}$. Suppose that $0 < |x - r| < \delta$. Then

$$|kf(x) - kL| = |k||f(x) - L| < |k|\frac{\epsilon}{|k|} = \epsilon.$$

So, $\lim_{x \to r}[kf(x)] = kL = k \lim_{x \to r}[f(x)]$.

If $k = 0$, let $\delta = 1$. If $0 < |x - r| < \delta$, then

$$|kf(x) - kL| = |0f(x) - 0L| = |0| = 0 < \epsilon.$$

So, in this case, we also have $\lim_{x \to r}[kf(x)] = kL = k \lim_{x \to r}[f(x)]$. □

35. If A is an open interval, then f is continuous at r if and only if $\lim_{x \to r}[f(x)] = f(r)$.

Proof: Let A be an open interval in $\mathbb{R}$ and let $f: A \to \mathbb{R}$. First suppose that f is continuous at r. Let $\epsilon > 0$. Then there is $\delta > 0$ such that $|x - r| < \delta$ implies $|f(x) - f(r)| < \epsilon$. Let $x \in \mathbb{R}$ satisfy $0 < |x - r| < \delta$. Then $|x - r| < \delta$. So, $|f(x) - f(r)| < \epsilon$. Since $\epsilon > 0$ was arbitrary, $\lim_{x \to r}[f(x)] = f(r)$.

Now, suppose that $\lim_{x \to r}[f(x)] = f(r)$. Let $\epsilon > 0$. Then there is $\delta > 0$ such that $0 < |x - r| < \delta$ implies $|f(x) - f(r)| < \epsilon$. Let $x \in \mathbb{R}$ satisfy $|x - r| < \delta$. Then $0 < |x - r| < \delta$ or $x = r$. If $0 < |x - r| < \delta$, then $|f(x) - f(r)| < \epsilon$. If $x = r$, then $|f(x) - f(r)| = |f(r) - f(r)| = |0| = 0 < \epsilon$. Since $\epsilon > 0$ was arbitrary, f is continuous at r. □

Let (s_n) be a real-valued sequence. Prove each of the following:

36. If (s_n) is bounded, then there is $M \in \mathbb{R}^+$ such that for all $m, n \in \mathbb{N}$, $|s_m - s_n| \leq M$.

Proof: Suppose that (s_n) is bounded in $\mathbb{R}$. Then there is $K \in \mathbb{R}^+$ such that $|s_n| \leq K$ for all $n \in \mathbb{N}$. So, for all $m, n \in \mathbb{N}$, we have $|s_m - s_n| \leq |s_m| + |s_n| \leq 2K$ (by the Triangle Inequality). Let $M = 2K$. Then $M \in \mathbb{R}^+$ and for all $m, n \in \mathbb{N}$, we have $|s_m - s_n| \leq M$. □

37. If there is $M \in \mathbb{R}^+$ such that for all $m, n \in \mathbb{N}$, $|s_m - s_n| \leq M$, then (s_n) is bounded.

Proof: Suppose that there is $M \in \mathbb{R}^+$ such that for all $m, n \in \mathbb{N}$, $|s_m - s_n| \leq M$. By the Triangle Inequality (and SACT), we have

$$|s_m| = |(s_m - s_0) + s_0| \leq |s_m - s_0| + |s_0| \leq M + |s_0|.$$

So, if we let $K = M + |s_0|$. Then for all $m \in \mathbb{N}$, $|s_m| \leq K$. So, (s_n) is bounded. □

LEVEL 3

Use the sum identities (Theorem 5.25) to compute the cosine, sine, and tangent of each of the following angles:

38. $\frac{5\pi}{12}$

$$\cos\frac{5\pi}{12} = \cos\left(\frac{\pi}{4} + \frac{\pi}{6}\right) = \cos\frac{\pi}{4}\cos\frac{\pi}{6} - \sin\frac{\pi}{4}\sin\frac{\pi}{6} = \frac{\sqrt{2}}{2} \cdot \frac{\sqrt{3}}{2} - \frac{\sqrt{2}}{2} \cdot \frac{1}{2} = \frac{\sqrt{6} - \sqrt{2}}{4}.$$

$$\sin\frac{5\pi}{12} = \sin\left(\frac{\pi}{4} + \frac{\pi}{6}\right) = \sin\frac{\pi}{4}\cos\frac{\pi}{6} + \cos\frac{\pi}{4}\sin\frac{\pi}{6} = \frac{\sqrt{2}}{2} \cdot \frac{\sqrt{3}}{2} + \frac{\sqrt{2}}{2} \cdot \frac{1}{2} = \frac{\sqrt{6} + \sqrt{2}}{4}.$$

$$\tan\frac{5\pi}{12} = \frac{\sin\frac{5\pi}{12}}{\cos\frac{5\pi}{12}} = \frac{\sqrt{6} + \sqrt{2}}{4} \cdot \frac{4}{\sqrt{6} - \sqrt{2}} = \frac{\sqrt{6} + \sqrt{2}}{\sqrt{6} - \sqrt{2}}.$$

39. $\frac{\pi}{12}$

$$\cos\frac{\pi}{12} = \cos\left(\frac{\pi}{4} - \frac{\pi}{6}\right) = \cos\frac{\pi}{4}\cos\frac{\pi}{6} + \sin\frac{\pi}{4}\sin\frac{\pi}{6} = \frac{\sqrt{2}}{2} \cdot \frac{\sqrt{3}}{2} + \frac{\sqrt{2}}{2} \cdot \frac{1}{2} = \frac{\sqrt{6} + \sqrt{2}}{4}.$$

$$\sin\frac{\pi}{12} = \sin\left(\frac{\pi}{4} - \frac{\pi}{6}\right) = \sin\frac{\pi}{4}\cos\frac{\pi}{6} - \cos\frac{\pi}{4}\sin\frac{\pi}{6} = \frac{\sqrt{2}}{2} \cdot \frac{\sqrt{3}}{2} - \frac{\sqrt{2}}{2} \cdot \frac{1}{2} = \frac{\sqrt{6} - \sqrt{2}}{4}.$$

$$\tan\frac{\pi}{12} = \frac{\sin\frac{\pi}{12}}{\cos\frac{\pi}{12}} = \frac{\sqrt{6} - \sqrt{2}}{4} \cdot \frac{4}{\sqrt{6} + \sqrt{2}} = \frac{\sqrt{6} - \sqrt{2}}{\sqrt{6} + \sqrt{2}}.$$

40. $\frac{11\pi}{12}$

$$\cos\frac{11\pi}{12} = \cos\left(\frac{\pi}{4} + \frac{2\pi}{3}\right) = \cos\frac{\pi}{4}\cos\frac{2\pi}{3} - \sin\frac{\pi}{4}\sin\frac{2\pi}{3} = \frac{\sqrt{2}}{2} \cdot \left(-\frac{1}{2}\right) - \frac{\sqrt{2}}{2} \cdot \frac{\sqrt{3}}{2} = \frac{-\sqrt{2} - \sqrt{6}}{4}.$$

$$\sin\frac{11\pi}{12} = \sin\left(\frac{\pi}{4} + \frac{2\pi}{3}\right) = \sin\frac{\pi}{4}\cos\frac{2\pi}{3} + \cos\frac{\pi}{4}\sin\frac{2\pi}{3} = \frac{\sqrt{2}}{2} \cdot \left(-\frac{1}{2}\right) + \frac{\sqrt{2}}{2} \cdot \frac{\sqrt{3}}{2} = \frac{-\sqrt{2} + \sqrt{6}}{4}.$$

$$\tan\frac{11\pi}{12} = \frac{\sin\frac{11\pi}{12}}{\cos\frac{11\pi}{12}} = \frac{-\sqrt{2} + \sqrt{6}}{4} \cdot \frac{4}{-\sqrt{2} - \sqrt{6}} = \frac{-\sqrt{2} + \sqrt{6}}{-\sqrt{2} - \sqrt{6}} = \frac{\sqrt{2} - \sqrt{6}}{\sqrt{2} + \sqrt{6}}.$$

41. $\frac{19\pi}{12}$

$$\cos\frac{19\pi}{12}=\cos\left(\frac{5\pi}{4}+\frac{\pi}{3}\right)=\cos\frac{5\pi}{4}\cos\frac{\pi}{3}-\sin\frac{5\pi}{4}\sin\frac{\pi}{3}=-\frac{\sqrt{2}}{2}\cdot\frac{1}{2}-\left(-\frac{\sqrt{2}}{2}\right)\cdot\frac{\sqrt{3}}{2}=\frac{-\sqrt{2}+\sqrt{6}}{4}.$$

$$\sin\frac{19\pi}{12}=\sin\left(\frac{5\pi}{4}+\frac{\pi}{3}\right)=\sin\frac{5\pi}{4}\cos\frac{\pi}{3}+\cos\frac{5\pi}{4}\sin\frac{\pi}{3}=-\frac{\sqrt{2}}{2}\cdot\frac{1}{2}+\left(-\frac{\sqrt{2}}{2}\right)\cdot\frac{\sqrt{3}}{2}=\frac{-\sqrt{2}-\sqrt{6}}{4}.$$

$$\tan\frac{19\pi}{12}=\frac{\sin\frac{19\pi}{12}}{\cos\frac{19\pi}{12}}=\frac{-\sqrt{2}-\sqrt{6}}{4}\cdot\frac{4}{-\sqrt{2}+\sqrt{6}}=\frac{-\sqrt{2}-\sqrt{6}}{-\sqrt{2}+\sqrt{6}}=\frac{\sqrt{2}+\sqrt{6}}{\sqrt{2}-\sqrt{6}}.$$

Let $(R,+,\cdot,\leq)$ be a commutative ordered ring. Prove each of the following:

42. If $(R,+,\cdot)$ is a field, $a,b\in R^+$ and $a>b$, then $\frac{1}{a}<\frac{1}{b}$.

Proof: Let $a,b\in R^+$ with $a,b\in R^+$ and $a>b$. Then $a-b=a+(-b)>b+(-b)=0$ (we used property (1) of an ordered ring here). Also, $\frac{1}{b}-\frac{1}{a}=\frac{1}{ab}(a-b)$. Since $a,b\in R^+$, $ab\in R^+$ by property (2) of an ordered ring together with the fact that every field is an integral domain (see Problem 82 in Problem Set 3). So, $\frac{1}{ab}\in R^+$ by Theorem 5.4. Since $\frac{1}{ab}>0$ and $a-b>0$, $\frac{1}{b}-\frac{1}{a}=\frac{1}{ab}(a-b)>0$ (once again, we have used property (2) of an ordered ring together with the fact that every field is an integral domain). So, $\frac{1}{b}>\frac{1}{a}$, or equivalently, $\frac{1}{a}<\frac{1}{b}$. □

43. If $a,b\in R$, then $a\geq b$ if and only if $-a\leq -b$.

Proof: Let $a,b\in R$. Then $a\geq b$ if and only if $a-b\geq 0$ if and only if $-(a-b)\leq 0$ if and only if $-1\big(a+(-b)\big)\leq 0$ if and only if $-1a-1(-b)\leq 0$ if and only if $-a-(-b)\leq 0$ if and only if $-a\leq -b$. □

Note: In the proof of Problem 43, we have used Problem 79 in Problem Set 3 several times.

44. If $a,b\in R^+$, then $a\leq b$ if and only if $a^2\leq b^2$.

Proof: Let $a,b\in R^+$. By property (1) of an ordered ring, $b+a\in R^+$. Therefore, $0\leq b+a$. Also, we have $b^2-a^2=(b-a)(b+a)$.

If $a\leq b$, then $b-a\geq 0$. So, by property (2) of an ordered ring, $0\leq(b-a)(b+a)=b^2-a^2$. Therefore, $a^2\leq b^2$.

If $a\nleq b$, then $b<a$, and so, $0<a-b$. By property (2) of an ordered ring together with the fact that every field is an integral domain (see Problem 82 in Problem Set 3), we have $0<(a-b)(b+a)=a^2-b^2$, and so, $b^2<a^2$. Therefore, $a^2\nleq b^2$. By contrapositive, we have $b^2<a^2$ implies $b<a$. So, $b^2\leq a^2$ implies $b\leq a$. □

Let $g: \mathbb{R} \to \mathbb{R}$ be defined by $g(x) = 2x^2 - 3x + 7$.

45. Prove that $\lim\limits_{x \to 1} g(x) = 6$.

Proof: Let $\epsilon > 0$ and let $\delta = \min\left\{1, \frac{\epsilon}{3}\right\}$. Suppose that $0 < |x - 1| < \delta$. Then we have $|x - 1| < 1$, so that $-1 < x - 1 < 1$. Adding 1, we get $0 < x < 2$. Multiplying by 2, we have $0 < 2x < 4$. Subtracting 1 gives us $-1 < 2x - 1 < 3$. So, $-3 < 2x - 1 < 3$, and therefore, $|2x - 1| < 3$. Now, we have

$$|(2x^2 - 3x + 7) - 6| = |2x^2 - 3x + 1| = |2x - 1||x - 1| < 3\delta \leq 3 \cdot \frac{\epsilon}{3} = \epsilon.$$

So, $\lim\limits_{x \to 1} g(x) = 6$. □

46. Prove that g is continuous on $\mathbb{R}$.

Proof: Let $a \in \mathbb{R}$. We will show that g is continuous at a. Let $\epsilon > 0$ and let $\delta = \min\left\{1, \frac{\epsilon}{M}\right\}$, where $M = \max\{|4a - 8|, |4a - 4|\}$. Suppose that $|x - a| < \delta$. Then we have $|x - a| < 1$, so that $-1 < x - a < 1$. Adding $2a - 3$, we get $2a - 4 < x + a - 3 < 2a - 2$. Multiplying by 2 yields $4a - 8 < 2(x + a - 3) < 4a - 4$. So, $-M < 2(x + a - 3) < M$. Therefore, $|2(x + a - 3)| < M$. Now, we have

$$|(2x^2 - 3x + 7) - (2a^2 - 3a + 7)| = |2(x^2 - a^2) - 3(x - a)| = |x - a||2(x + a - 3)|$$
$$< \delta M \leq \frac{\epsilon}{M} \cdot M = \epsilon.$$

So, g is continuous at a. Since $a \in \mathbb{R}$ was arbitrary, g is continuous on $\mathbb{R}$. □

Let $f: \mathbb{R} \to \mathbb{R}$. Prove each of the following:

47. If f is a polynomial function, then f is continuous on $\mathbb{R}$.

Proof: Let $r \in \mathbb{R}$. We first show by induction that for all $n \in \mathbb{N}$ with $n \geq 1$, $\lim\limits_{x \to r}[x^n] = r^n$.

Base case ($k = 1$): Let $\epsilon > 0$ be given and let $\delta = \epsilon$. Then $0 < |x - r| < \delta$ implies $|x - r| < \delta = \epsilon$. Since $\epsilon > 0$ was arbitrary, $\lim\limits_{x \to r}[x] = r$.

Inductive step: Let $k \in \mathbb{N}$ and assume that $\lim\limits_{x \to r}[x^k] = r^k$. By Theorem 5.40, we have

$$\lim_{x \to r}[x^{k+1}] = \lim_{x \to r}[x^k \cdot x] = \lim_{x \to r}[x^k] \cdot \lim_{x \to r}[x] = r^k \cdot r = r^{k+1}.$$

By the Principle of Mathematical Induction, for all $n \in \mathbb{N}$ with $n \geq 1$, $\lim\limits_{x \to r}[x^n] = r^n$.

Now, let $p: \mathbb{R} \to \mathbb{R}$ be a polynomial, say $p(x) = a_n x^n + a_{n-1}x^{n-1} + \cdots + a_1 x + a_0$. By Problem 33, $\lim\limits_{x \to r}[a_0] = a_0$. By the last paragraph and Problem 34, $\lim\limits_{x \to r}[a_k x^k] = a_k \lim\limits_{x \to r}[x^k] = a_k r^k$. Finally, using Theorem 5.39, we have

$$\begin{aligned}\lim_{x \to r}[p(x)] &= \lim_{x \to r}[a_n x^n + a_{n-1}x^{n-1} + \cdots + a_1 x + a_0] \\ &= \lim_{x \to r}[a_n x^n] + \lim_{x \to r}[a_{n-1}x^{n-1}] + \cdots + \lim_{x \to r}[a_1 x] + \lim_{x \to r}[a_0] \\ &= a_n r^n + a_{n-1}r^{n-1} + \cdots + a_1 r + a_0 = p(r).\end{aligned}$$

By Problem 35, p is continuous at r. Since $r \in \mathbb{R}$ was arbitrary, p is continuous on $\mathbb{R}$. □

48. Suppose that $g: \mathbb{R} \to \mathbb{R}$, $a \in \mathbb{R}$, f is continuous at a, and g is continuous at $f(a)$. Then $g \circ f$ is continuous at a.

Proof: Let $f, g: \mathbb{R} \to \mathbb{R}$, let $a \in \mathbb{R}$, and suppose that f is continuous at a and g is continuous at $f(a)$. Let $\epsilon > 0$. Since g is continuous at $f(a)$, there is $\delta_1 > 0$ such that $|y - f(a)| < \delta_1$ implies $|g(y) - g(f(a))| < \epsilon$. Since f is continuous at a, there is $\delta_2 > 0$ such that $|x - a| < \delta_2$ implies $|f(x) - f(a)| < \delta_1$. Now, suppose that $|x - a| < \delta_2$. Then $|f(x) - f(a)| < \delta_1$. It follows that $|g(f(x)) - g(f(a))| < \epsilon$. Since $\epsilon > 0$ was arbitrary, $g \circ f$ is continuous at a. □

LEVEL 4

For Problems 49 and 50 you may use the following facts about triangles: (i) The interior angle measures of a triangle sum to π radians; (ii) Two sides of a triangle have the same length if and only if the interior angles of the triangle opposite these sides have the same measure; (iii) If two sides of a triangle have the same length, then the line segment beginning at the point of intersection of those two sides and terminating on the opposite base midway between the endpoints of that base is perpendicular to that base.

49. Consider triangle AOP, where $O = (0, 0)$, $A = (1, 0)$, and P is the point on the unit circle so that angle POA has radian measure $\frac{\pi}{3}$. Prove that triangle AOP is equilateral, and then use this to prove that $W\left(\frac{\pi}{3}\right) = \left(\frac{1}{2}, \frac{\sqrt{3}}{2}\right)$.

Proof: Let's start by drawing the unit circle together with triangle AOP. We also draw line segment PE, where E is midway between O and A.

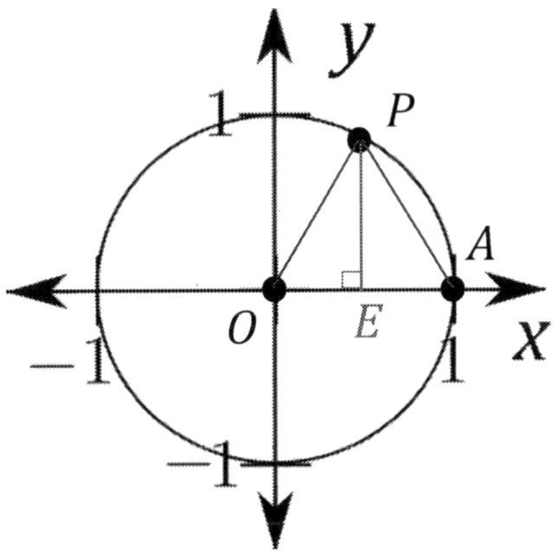

Since OP and OA are both radii of the circle, they have the same length. By (ii), angles OAP and OPA have the same measure. By (i), the sum of these measures is $\pi - \frac{\pi}{3} = \frac{3\pi}{3} - \frac{\pi}{3} = \frac{2\pi}{3}$. So, each of angles OAP and OPA measure $\frac{\pi}{3}$ radians. It follows from (ii) again that triangle AOP is equilateral.

By (iii), PE is perpendicular to OA.

Now, $OP = 1$ because OP is a radius of the unit circle and $OE = \frac{1}{2}$ because OA is a radius of the unit circle and E is midway between O and A. Since triangle OEP is a right triangle with hypotenuse OP, by the Pythagorean Theorem, $PE^2 = OP^2 - OE^2 = 1^2 - \left(\frac{1}{2}\right)^2 = 1 - \frac{1}{4} = \frac{3}{4}$. So, $PE = \sqrt{\frac{3}{4}} = \frac{\sqrt{3}}{\sqrt{4}} = \frac{\sqrt{3}}{2}$. It follows that $W\left(\frac{\pi}{3}\right) = \left(\frac{1}{2}, \frac{\sqrt{3}}{2}\right)$. □

50. Prove that $W\left(\frac{\pi}{6}\right) = \left(\frac{\sqrt{3}}{2}, \frac{1}{2}\right)$.

Proof: Let's start by drawing a picture similar to what we drew in Problem 49. We draw P and Q on the unit circle and A on the positive x-axis so that angle AOP has radian measure $\frac{\pi}{6}$, angle AOQ has radian measure $-\frac{\pi}{6}$, and A is right in the middle of the line segment joining P and Q.

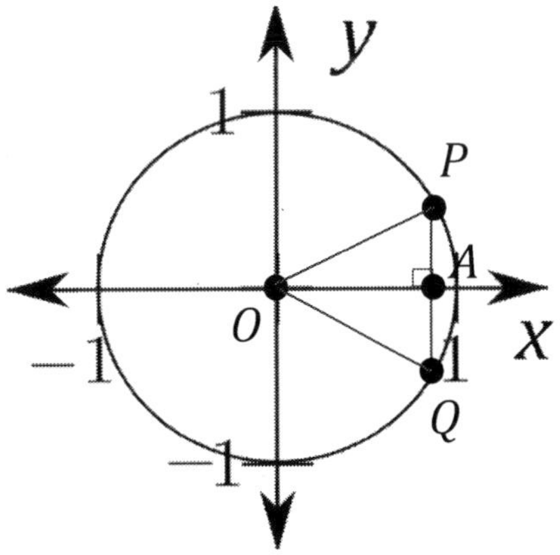

By reasoning similar to what was done in Problem 49, we see that triangle POQ is equilateral and OA is perpendicular to PQ.

Now, $OP = 1$ because OP is a radius of the unit circle and $AP = \frac{1}{2}$ because A is midway between P and Q. Since triangle POA is a right triangle with hypotenuse OP, by the Pythagorean Theorem, $OA^2 = OP^2 - AP^2 = 1^2 - \left(\frac{1}{2}\right)^2 = 1 - \frac{1}{4} = \frac{3}{4}$. Therefore, $OA = \sqrt{\frac{3}{4}} = \frac{\sqrt{3}}{\sqrt{4}} = \frac{\sqrt{3}}{2}$. It follows that $W\left(\frac{\pi}{6}\right) = \left(\frac{\sqrt{3}}{2}, \frac{1}{2}\right)$. □

Let θ and ϕ be the radian measure of angles A and B, respectively. Prove the following identities:

51. $\cos(\theta - \phi) = \cos\theta\cos\phi + \sin\theta\sin\phi$

Proof: Let's draw a picture of the unit circle together with angles θ, ϕ, and $\theta - \phi$ in standard position, and label the corresponding points on the unit circle.

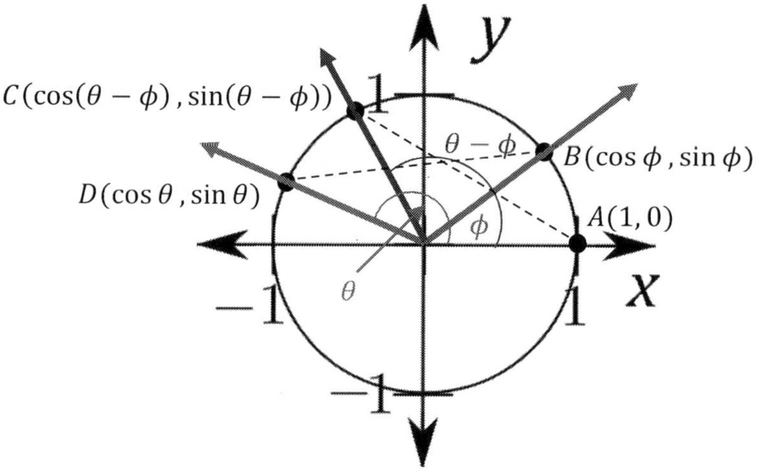

Since the arcs moving counterclockwise from A to C and from B to D both have radian measure $\theta - \phi$, it follows that $AC = BD$, and so, using the Pythagorean Theorem twice, we have

$$(\cos(\theta - \phi) - 1)^2 + (\sin(\theta - \phi) - 0)^2 = (\cos\theta - \cos\phi)^2 + (\sin\theta - \sin\phi)^2$$

The left-hand side of this equation is equal to:

$$(\cos(\theta - \phi) - 1)^2 + (\sin(\theta - \phi) - 0)^2$$
$$= \cos^2(\theta - \phi) - 2\cos(\theta - \phi) + 1 + \sin^2(\theta - \phi)$$
$$= (\cos^2(\theta - \phi) + \sin^2(\theta - \phi)) - 2\cos(\theta - \phi) + 1$$
$$= 1 - 2\cos(\theta - \phi) + 1 \text{ (by the Pythagorean Identity)}$$
$$= 2 - 2\cos(\theta - \phi)$$

The right-hand side of this equation is equal to:

$$(\cos\theta - \cos\phi)^2 + (\sin\theta - \sin\phi)^2$$
$$= \cos^2\theta - 2\cos\theta\cos\phi + \cos^2\phi + \sin^2\theta - 2\sin\theta\sin\phi + \sin^2\phi$$
$$= (\cos^2\theta + \sin^2\theta) + (\cos^2\phi + \sin^2\phi) - 2\cos\theta\cos\phi - 2\sin\theta\sin\phi$$
$$= 1 + 1 - 2\cos\theta\cos\phi - 2\sin\theta\sin\phi$$
$$= 2 - 2\cos\theta\cos\phi - 2\sin\theta\sin\phi$$

Therefore, we have $2 - 2\cos(\theta - \phi) = 2 - 2\cos\theta\cos\phi - 2\sin\theta\sin\phi$. Subtracting 2 from each side of this equation gives us $-2\cos(\theta - \phi) = -2\cos\theta\cos\phi - 2\sin\theta\sin\phi$. Multiplying each side of this last equation by $-\frac{1}{2}$ gives us $\cos(\theta - \phi) = \cos\theta\cos\phi + \sin\theta\sin\phi$, as desired. □

52. $\cos(\theta + \phi) = \cos\theta\cos\phi - \sin\theta\sin\phi$

Proof: $\cos(\theta + \phi) = \cos(\theta - (-\phi)) = \cos\theta\cos(-\phi) + \sin\theta\sin(-\phi)$ (by Problem 51)

$= \cos\theta\cos\phi - \sin\theta\sin\phi$ (by the Negative Identities). □

53. $\cos(\pi - \theta) = -\cos\theta$

Proof: $\cos(\pi - \theta) = \cos\pi\cos\theta + \sin\pi\sin\theta = (-1)\cos\theta + 0 \cdot \sin\theta = -\cos\theta$. □

54. $\cos\left(\frac{\pi}{2} - \theta\right) = \sin\theta$

Proof: $\cos\left(\frac{\pi}{2} - \theta\right) = \cos\frac{\pi}{2}\cos\theta + \sin\frac{\pi}{2}\sin\theta = 0 \cdot \cos\theta + 1 \cdot \sin\theta = \sin\theta$. □

55. $\sin\left(\frac{\pi}{2} - \theta\right) = \cos\theta$

Proof: $\sin\left(\frac{\pi}{2} - \theta\right) = \cos\left(\frac{\pi}{2} - \left(\frac{\pi}{2} - \theta\right)\right) = \cos\left(\frac{\pi}{2} - \frac{\pi}{2} + \theta\right) = \cos\theta$. □

56. $\sin(\theta + \phi) = \sin\theta\cos\phi + \cos\theta\sin\phi$

Proof: $\sin(\theta + \phi) = \cos\left(\frac{\pi}{2} - (\theta + \phi)\right) = \cos\left(\left(\frac{\pi}{2} - \theta\right) - \phi\right)$

$= \cos\left(\frac{\pi}{2} - \theta\right)\cos\phi + \sin\left(\frac{\pi}{2} - \theta\right)\sin\phi = \sin\theta\cos\phi + \cos\theta\sin\phi$. □

57. $\sin(\pi - \theta) = \sin\theta$

Proof: $\sin(\pi - \theta) = \sin(\pi + (-\theta)) = \sin\pi\cos(-\theta) + \cos\pi\sin(-\theta)$

$= 0 \cdot \cos\theta + (-1)(-\sin\theta) = \sin\theta$. □

Let S be a nonempty set of real numbers that is bounded below and let $T = \{-x \mid x \in S\}$.

58. Let K be a lower bound for S. Prove that $-K$ is an upper bound for the set T.

Proof: Let $y \in T$. Then there is $x \in S$ with $y = -x$. Since $x \in S$, $x \geq K$. So, $y = -x \leq -K$ (**Check this!**). Since $y \in T$ was arbitrary, we have shown that $-K$ is an upper bound for the set T. □

59. Explain why T has a least upper bound M.

Proof: By Problem 58, T is bounded above. So, by the Completeness Property of $\mathbb{R}$, T has a least upper bound M. □

60. Let M be the least upper bound of T. Prove that $-M$ is a lower bound of S.

Proof: Let $x \in S$. Then $-x \in T$. Since M is an upper bound for T, $-x \leq M$. So, $x \geq -M$ (**Check this!**). Since $x \in S$ was arbitrary, $-M$ is a lower bound of S. □

61. Let M be the least upper bound of T. Prove that $-M$ is a greatest lower bound of S.

Proof: By Problem 60, $-M$ is a lower bound of S.

Let $B > -M$. Then $-B < M$ (**Check this!**). Since M is the least upper bound for T, there is $y \in T$ with $y > -B$. So, $-y < B$ (**Check this!**). Since $y \in T$, $-y \in S$. Thus, B is not a lower bound of S.

Therefore, $-M$ is a greatest lower bound of S. □

62. Prove that every nonempty set of real numbers that is bounded below has a greatest lower bound in $\mathbb{R}$.

Proof: This follows from Problems 58 – 61. □

Prove each of the following:

63. Every rational number can be written in the form $\frac{m}{n}$, where $m \in \mathbb{Z}$, $n \in \mathbb{Z}^*$, and at least one of m or n is **not** even.

Proof: Let x be a rational number. Then there are $a \in \mathbb{Z}$ and $b \in \mathbb{Z}^*$ such that $x = \frac{a}{b}$. Let j be the largest integer such that 2^j divides a and let k be the largest integer such that 2^k divides b. Since, 2^j divides a, there is $c \in \mathbb{Z}$ such that $a = 2^j c$. Since, 2^k divides b, there is $d \in \mathbb{Z}$ such that $b = 2^k d$.

Observe that c is odd. Indeed, if c were even, then there would be an integer s such that $c = 2s$. But then $a = 2^j c = 2^j(2s) = (2^j \cdot 2)s = (2^j \cdot 2^1)s = 2^{j+1}s$. So, 2^{j+1} divides a, contradicting the maximality of j.

Similarly, d is odd.

So, we have $x = \frac{a}{b} = \frac{2^j c}{2^k d}$.

If $j \geq k$, then, $j - k \geq 0$ and $x = \frac{2^j c}{2^k d} = \frac{2^{j-k} c}{d}$. Let $m = 2^{j-k}c$ and $n = d$. Then $x = \frac{m}{n}$, $m \in \mathbb{Z}$ (because $\mathbb{Z}$ is closed under multiplication), $n \in \mathbb{Z}^*$ (if $n = 0$, then $b = 2^k d = 2^k n = 2^k \cdot 0 = 0$), and $n = d$ is odd.

If $j < k$, then $k - j > 0$ and $x = \frac{2^j c}{2^k d} = \frac{c}{2^{k-j} d}$. Let $m = c$ and $n = 2^{k-j}d$. Then $x = \frac{m}{n}$, $m = c \in \mathbb{Z}$, $n \in \mathbb{Z}^*$ (because $\mathbb{Z}$ is closed under multiplication, and if n were 0, then d would be 0, and then b would be 0), and $m = c$ is odd. □

64. $\lim_{x\to 2}\frac{x^3-4}{x^2+1}=\frac{4}{5}$.

Proof: Let $\epsilon>0$ and let $\delta=\min\left\{1,\frac{2\epsilon}{15}\right\}$. Suppose that $0<|x-2|<\delta$. Then we have $|x-2|<1$, so that $-1<x-2<1$. Adding 2, we get $1<x<3$. So, $23<5x^2+6x+12<75$ and therefore, $-75<5x^2+6x+12<75$. So, $|5x^2+6x+12|<75$. Also, $2<x^2+1<10$. In particular, we have $x^2+1>2$, and so, $\frac{1}{x^2+1}<\frac{1}{2}$. Now, we have

$$\left|\frac{x^3-4}{x^2+1}-\frac{4}{5}\right|=\left|\frac{5(x^3-4)}{5(x^2+1)}-\frac{4(x^2+1)}{5(x^2+1)}\right|=\left|\frac{5x^3-4x^2-24}{5(x^2+1)}\right|$$
$$=\frac{|5x^2+6x+12||x-2|}{5(x^2+1)}<\frac{75\delta}{5\cdot 2}\leq\frac{75}{10}\cdot\frac{2\epsilon}{15}=\epsilon.$$

So, $\lim_{x\to 2}h(x)=\frac{4}{5}$. □

Let $T=\{x\in F\mid -2<x\leq 2\}$. Prove each of the following:

65. $\sup T=2$

Proof: If $x\in T$, then by the definition of T, $x\leq 2$. So, 2 is an upper bound of T.

Now, let $B<2$, and let $z=\max\left\{0,\frac{1}{2}(B+2)\right\}$. Since $B<2$, we have

$$\frac{1}{2}(B+2)<\frac{1}{2}(2+2)=\frac{1}{2}\cdot 4=2.$$

So, if we have $\frac{1}{2}(B+2)>0$, then $\frac{1}{2}(B+2)\in T$. Since $0\in T$, we see that $z\in T$. Also,

$$z\geq\frac{1}{2}(B+2)>\frac{1}{2}(B+B)=\frac{1}{2}(2B)=\left(\frac{1}{2}\cdot 2\right)B=1B=B.$$

So, we see that $z\in T$ and $z>B$. Therefore, B is not an upper bound of T. So, $2=\sup T$. □

66. $\inf T=-2$.

Proof: If $x\in T$, then by the definition of T, $x>-2$. So, -2 is a lower bound of T.

Now, let $C>-2$, and let $w=\min\left\{0,\frac{1}{2}(-2+C)\right\}$. Since $C>-2$, we have

$$\frac{1}{2}(-2+C)>\frac{1}{2}(-2-2)=\frac{1}{2}(-4)=-2.$$

So, if we have $\frac{1}{2}(-2+C)<0$, then $\frac{1}{2}(-2+C)\in T$. Since $0\in T$, we see that $w\in T$. Also,

$$w\leq\frac{1}{2}(-2+C)<\frac{1}{2}(C+C)=\frac{1}{2}(2C)=\left(\frac{1}{2}\cdot 2\right)C=1C=C.$$

So, we see that $w\in T$ and $w<C$. Therefore, C is not a lower bound of T. So, $-2=\inf T$. □

Let $k: (0, \infty) \to \mathbb{R}$ be defined by $k(x) = \sqrt{x}$.

67. Prove that $\lim_{x \to 25} k(x) = 5$.

Proof: Let $\epsilon > 0$ and let $\delta = \min\{1, (5 + \sqrt{24})\epsilon\}$. Suppose that $0 < |x - 25| < \delta$. Then we have $|x - 25| < 1$, so that $-1 < x - 25 < 1$. Adding 25, we get $24 < x < 26$. Taking square roots, we have $\sqrt{24} < \sqrt{x} < \sqrt{26}$. Adding 5 gives us $5 + \sqrt{24} < \sqrt{x} + 5 < 5 + \sqrt{26}$. So, $\frac{1}{\sqrt{x}+5} < \frac{1}{5+\sqrt{24}}$. Now, we have

$$|\sqrt{x} - 5| = \left|\frac{(\sqrt{x} - 5)(\sqrt{x} + 5)}{\sqrt{x} + 5}\right| = \frac{|x - 25|}{\sqrt{x} + 5} < \frac{\delta}{5 + \sqrt{24}} \leq \frac{1}{5 + \sqrt{24}} \cdot (5 + \sqrt{24})\epsilon = \epsilon.$$

So, $\lim_{x \to 25} k(x) = 5$. □

68. Prove that f is continuous on $(0, \infty)$.

Proof: Let $a \in \mathbb{R}$. We will show that f is continuous at a. Let $\epsilon > 0$, let $\delta = \min\{1, \epsilon\sqrt{a}\}$, and let $x \in (0, \infty)$ satisfy $|x - a| < \delta$. Then we have $|x - a| < 1$, so that $-1 < x - a < 1$. Adding a, we get $a - 1 < x < a + 1$. Since $x \in (0, \infty)$, we have $0 < x < a + 1$. Taking square roots, we have $0 < \sqrt{x} < \sqrt{a + 1}$. Adding $\sqrt{a}$ gives us $\sqrt{a} < \sqrt{x} + \sqrt{a} < \sqrt{a + 1} + \sqrt{a}$. Therefore, $\frac{1}{\sqrt{x}+\sqrt{a}} < \frac{1}{\sqrt{a}}$. Now, we have

$$|\sqrt{x} - \sqrt{a}| = \left|\frac{(\sqrt{x} - \sqrt{a})(\sqrt{x} + \sqrt{a})}{\sqrt{x} + \sqrt{a}}\right| = \frac{|x - a|}{\sqrt{x} + \sqrt{a}} < \frac{\delta}{\sqrt{a}} \leq \frac{\epsilon\sqrt{a}}{\sqrt{a}} = \epsilon.$$

So, f is continuous at a. Since $a \in \mathbb{R}$ was arbitrary, f is continuous on $\mathbb{R}$. □

69. Is f uniformly continuous on $(0, \infty)$?

Proof: Let $\epsilon > 0$, let $\delta = \epsilon^2$, and let $x, y \in (0, \infty)$ satisfy $|x - y| < \delta$. Then we have

$$|\sqrt{x} - \sqrt{y}| = \sqrt{(\sqrt{x} - \sqrt{y})^2} \leq \sqrt{|\sqrt{x} - \sqrt{y}||\sqrt{x} + \sqrt{y}|} = \sqrt{|x - y|} < \sqrt{\delta} = \sqrt{\epsilon^2} = \epsilon.$$

So, f is uniformly continuous on $\mathbb{R}$. □

Let $f: \mathbb{R} \to \mathbb{R}$ be defined by $f(x) = x^2$. Prove each of the following:

70. f is continuous on $\mathbb{R}$.

Proof: Let $a \in \mathbb{R}$. We will show that f is continuous at a. Let $\epsilon > 0$ and let $\delta = \min\left\{1, \frac{\epsilon}{M}\right\}$, where $M = \max\{|2a - 1|, |2a + 1|\}$. Suppose that $|x - a| < \delta$. Then $|x - a| < 1$, so that $-1 < x - a < 1$. Adding $2a$, we get $2a - 1 < x + a < 2a + 1$. So, $-M < x + a < M$, or equivalently, $|x + a| < M$. Now, we have

$$|x^2 - a^2| = |x - a||x + a| < \delta \cdot M \leq \frac{\epsilon}{M} \cdot M = \epsilon.$$

So, f is continuous at a. Since $a \in \mathbb{R}$ was arbitrary, f is continuous on $\mathbb{R}$. □

71. f is **not** uniformly continuous on $\mathbb{R}$.

Proof: Let $\epsilon = 1$ and let $\delta > 0$. Let $x = \frac{1}{\delta}$ and $y = \frac{1}{\delta} + \frac{\delta}{2}$. Then we have $|x - y| = \left|\frac{1}{\delta} - \left(\frac{1}{\delta} + \frac{\delta}{2}\right)\right| = \frac{\delta}{2}$, but

$$|f(x) - f(y)| = |x^2 - y^2| = \left|\frac{1}{\delta^2} - \left(\frac{1}{\delta} + \frac{\delta}{2}\right)^2\right| = \left|\frac{1}{\delta^2} - \frac{1}{\delta^2} - 1 - \frac{\delta^2}{4}\right| = 1 + \frac{\delta^2}{4} > 1 = \epsilon.$$

So, f is **not** uniformly continuous on $\mathbb{R}$ (and in fact, not uniformly continuous on $(0, \infty)$ since we only needed positive values of x and y to violate the definition of uniform continuity). □

Let $A \subseteq \mathbb{R}$, let $f: A \to \mathbb{R}$, and let $r \in \mathbb{R}$. Prove each of the following:

72. If $\lim_{x \to r}[f(x)] > 0$, then there is a deleted neighborhood N of r such that $f(x) > 0$ for all $x \in N$.

Proof: Suppose that $\lim_{x \to r}[f(x)] = L$ with $L > 0$. Let $\epsilon = \frac{L}{2}$. There is $\delta > 0$ such that $0 < |x - r| < \delta$ implies $|f(x) - L| < \epsilon$. Consider $N_\delta^\odot(r) = (r - \delta, r) \cup (r, r + \delta)$. Let $x \in N_\delta^\odot(r)$. Then we have $x \in (r - \delta, r) \cup (r, r + \delta)$, so that $0 < |x - r| < \delta$. It follows that $|f(x) - L| < \epsilon = \frac{L}{2}$. So, we have $-\frac{L}{2} < f(x) - L < \frac{L}{2}$, or equivalently, $L - \frac{L}{2} < f(x) < L + \frac{L}{2}$. Since $L - \frac{L}{2} = \frac{L}{2}$ and $L + \frac{L}{2} = \frac{3L}{2}$, we have $\frac{L}{2} < f(x) < \frac{3L}{2}$. In particular, we have $f(x) > \frac{L}{2} > 0$. Since $x \in N_\delta^\odot(r)$ was arbitrary, we have shown that for all $x \in N_\delta^\odot(r)$, $f(x) > 0$. □

73. If $\lim_{x \to r}[f(x)]$ is a finite real number, then there is $M \in \mathbb{R}$ and an open interval (a, b) containing r such that $|f(x)| \leq M$ for all $x \in (a, b) \setminus \{r\}$.

Proof: Let $A \subseteq \mathbb{R}$, $f: A \to \mathbb{R}$, $r \in \mathbb{R}$, $\lim_{x \to r}[f(x)] = L$, and let $\epsilon = 1$. Then there is $\delta > 0$ such that $0 < |x - r| < \delta$ implies $|f(x) - L| < 1$, or $-1 < f(x) - L < 1$, or $L - 1 < f(x) < L + 1$. Let $a = r - \delta$, $b = r + \delta$, and $M = \max\{|L - 1|, |L + 1|\}$. If $x \in (a, b) \setminus \{r\}$, then $r - \delta < x < r + \delta$ and $x \neq r$. So, $0 < |x - r| < \delta$. Therefore, $L - 1 < f(x) < L + 1$. Since $M \geq |L - 1| \geq 1 - L$, we have $-M \leq L - 1$. Also, $M \geq |L + 1| \geq L + 1$. So, we have $-M < f(x) < M$, or equivalently, $|f(x)| < M$. Since $x \in (a, b) \setminus \{r\}$ was arbitrary, $|f(x)| < M$ for all $x \in (a, b) \setminus \{r\}$. □

74. If $g, h: A \to \mathbb{R}$ with $f(x) \leq g(x) \leq h(x)$ for all $x \in A \setminus \{r\}$, and $\lim_{x \to r}[f(x)] = \lim_{x \to r}[h(x)] = L$, then $\lim_{x \to r}[g(x)] = L$. This result is known as the **Squeeze Theorem**.

Proof: Let $\epsilon > 0$. Since $\lim_{x \to r}[f(x)] = L$, there is $\delta_1 > 0$ such that $0 < |x - r| < \delta_1$ implies $|f(x) - L| < \epsilon$. Since $\lim_{x \to r}[h(x)] = L$, there is $\delta_2 > 0$ such that $0 < |x - r| < \delta_2$ implies $|h(x) - L| < \epsilon$. Let $\delta = \min\{\delta_1, \delta_2\}$ and let $0 < |x - r| < \delta$. Then $0 < |x - r| < \delta_1$, so that $|f(x) - L| < \epsilon$, or equivalently, $-\epsilon < f(x) - L < \epsilon$, or $L - \epsilon < f(x) < L + \epsilon$. We will need only that $L - \epsilon < f(x)$. Similarly, we have $0 < |x - r| < \delta_2$, so that $|h(x) - L| < \epsilon$, or equivalently, $-\epsilon < h(x) - L < \epsilon$, or $L - \epsilon < h(x) < L + \epsilon$. We will need only that $h(x) < L + \epsilon$. Now, we have $L - \epsilon < f(x) \leq g(x) \leq h(x) < L + \epsilon$. So, $-\epsilon < g(x) - L < \epsilon$, or equivalently, $|g(x) - L| < \epsilon$. Since $\epsilon > 0$ was arbitrary, $\lim_{x \to r}[g(x)] = L$. □

LEVEL 5

Prove each of the following:

75. $(\mathbb{Q}, +, \cdot, \leq)$ is an ordered field.

Proof: By Problem 88 in Problem Set 3, $(\mathbb{Q}, +, \cdot)$ is a field. By Problem 347 in Problem Set 2 and Theorem 2.39, $(\mathbb{Q}, \leq)$ is a linearly ordered set.

We next prove the order properties.

Let $\frac{a}{b}, \frac{c}{d}, \frac{e}{f} \in \mathbb{Q}$ and assume that $\frac{a}{b} \leq \frac{c}{d}$. Then $ad \leq bc$. Since $\mathbb{Z}$ is an ordered ring, we get

$$(af + be)(df) = afdf + bedf = adff + bedf \leq bcff + bedf = bf(cf + de).$$

Therefore, $\frac{a}{b} + \frac{e}{f} = \frac{af+be}{bf} \leq \frac{cf+de}{df} = \frac{c}{d} + \frac{e}{f}$.

Next, let $\frac{a}{b}, \frac{c}{d} \in \mathbb{Q}$ with $0 \leq \frac{a}{b}$ and $0 \leq \frac{c}{d}$. We may assume that $a, b, c, d \geq 0$. Then $\frac{a}{b} \cdot \frac{c}{d} = \frac{ac}{bd} \geq 0$.

It follows that $(\mathbb{Q}, +, \cdot, \leq)$ is an ordered field. □

76. $(\mathbb{R}, +, \cdot, \leq)$ is an ordered field.

Proof: By Problem 93 in Problem Set 3, $(\mathbb{R}, +, \cdot)$ is a field. By Problem 347 in Problem Set 2 and Theorem 2.39, $(\mathbb{R}, \leq)$ is a linearly ordered set.

We next prove the order properties. Let $[(x_n)], [(y_n)], [(z_n)] \in \mathbb{R}$ and assume that $[(x_n)] \leq [(y_n)]$. Then there is $K \in \mathbb{N}$ such that $n > K$ implies $x_n \leq y_n$. Since $\mathbb{Q}$ is an ordered ring, it follows that $n > K$ implies $x_n + z_n \leq y_n + z_n$. So, $[(x_n)] + [(z_n)] = [(x_n + z_n)] \leq [(y_n + z_n)] = [(y_n)] + [(z_n)]$.

Next, let $[(x_n)], [(y_n)] \in \mathbb{R}$ with $[(0)] \leq [(x_n)]$ and $[(0)] \leq [(y_n)]$. Then there is $K_1 \in \mathbb{N}$ such that $n > K_1$ implies $x_n \geq 0$ and there is $K_2 \in \mathbb{N}$ such that $n > K_2$ implies $y_n \geq 0$. Let $K = \max\{K_1, K_2\}$ and let $n > K$. Since $K \geq K_1$, $x_n \geq 0$. Since $K \geq K_2$, $y_n \geq 0$. Since $\mathbb{Q}$ is an ordered ring, $x_n \cdot y_n \geq 0$. Therefore, we have $[(x_n)] \cdot [(y_n)] \geq [(0)]$.

It follows that $(\mathbb{R}, +, \cdot, \leq)$ is an ordered field. □

77. $(\mathbb{C}, +, \cdot)$ is field that cannot be ordered.

Proof: By Problem 98 in Problem Set 3, $(\mathbb{C}, +, \cdot)$ is a field. By part 3 of Example 5.1, $(\mathbb{C}, +, \cdot)$ cannot be ordered. □

78. Let $A \subseteq \mathbb{R}$, let $f, g: A \to \mathbb{R}$ such that $g(x) \neq 0$ for all $x \in A$, let $r \in \mathbb{R}$, and suppose that $\lim_{x\to r}[f(x)]$ and $\lim_{x\to r}[g(x)]$ are both finite real numbers such that $\lim_{x\to r}[g(x)] \neq 0$. Then $\lim_{x\to r}\left[\frac{f(x)}{g(x)}\right] = \frac{\lim_{x\to r} f(x)}{\lim_{x\to r} g(x)}$.

Proof: Suppose that $\lim_{x\to r}[f(x)] = L$ and $\lim_{x\to r}[g(x)] = K$, and let $\epsilon > 0$. Since $\lim_{x\to r}[g(x)] = K$, there is $\delta_1 > 0$ such that $0 < |x - r| < \delta_1$ implies $|g(x) - K| < \frac{|K|}{2}$. Now, $|g(x) - K| < \frac{|K|}{2}$ is equivalent to $-\frac{|K|}{2} < g(x) - K < \frac{|K|}{2}$, or by adding K, $K - \frac{|K|}{2} < g(x) < K + \frac{|K|}{2}$. If $K > 0$, we have $\frac{K}{2} < g(x) < \frac{3K}{2}$. If $K < 0$, we have $\frac{3K}{2} < g(x) < \frac{K}{2}$. In both cases, we have $\frac{|K|}{2} < |g(x)| < \frac{3|K|}{2}$. Let $M = \frac{|K|}{2}$. Then $|g(x)| > M$, and so, $\frac{1}{|g(x)|} < \frac{1}{M}$.

Now, since $\lim_{x\to r}[f(x)] = L$, there is $\delta_2 > 0$ such that $0 < |x - r| < \delta_2$ implies $|f(x) - L| < \frac{M|K|\epsilon}{|K|+|L|}$. Since $\lim_{x\to r}[g(x)] = K$, there is $\delta_3 > 0$ such that $0 < |x - r| < \delta_3$ implies $|g(x) - K| < \frac{M|K|\epsilon}{|K|+|L|}$. Let $\delta = \min\{\delta_1, \delta_2, \delta_3\}$ and suppose that $0 < |x - r| < \delta$. Then since $\delta \leq \delta_1$, $\frac{1}{|g(x)|} < \frac{1}{M}$. Since $\delta \leq \delta_2$, $|f(x) - L| < \frac{M|K|\epsilon}{|K|+|L|}$. Since $\delta \leq \delta_3$, $|g(x) - K| < \frac{M|K|\epsilon}{|K|+|L|}$. By the Triangle Inequality (and SACT), we have

$$\left|\frac{f(x)}{g(x)} - \frac{L}{K}\right| = \left|\frac{Kf(x) - Lg(x)}{Kg(x)}\right| = \left|\frac{Kf(x) - KL + KL - Lg(x)}{Kg(x)}\right| = \left|\frac{Kf(x) - KL}{Kg(x)} + \frac{KL - Lg(x)}{Kg(x)}\right|$$

$$\leq \left|\frac{Kf(x) - KL}{Kg(x)}\right| + \left|\frac{KL - Lg(x)}{Kg(x)}\right| = \left|\frac{f(x) - L}{g(x)}\right| + \left|\frac{L}{K}\right|\left|\frac{K - g(x)}{g(x)}\right| = \left|\frac{f(x) - L}{g(x)}\right| + \left|\frac{L}{K}\right|\left|\frac{g(x) - K}{g(x)}\right|$$

$$= \frac{1}{|g(x)|}\left(|f(x) - L| + \left|\frac{L}{K}\right||g(x) - K|\right) < \frac{1}{M}\left(\frac{M|K|\epsilon}{|K| + |L|} + \left|\frac{L}{K}\right|\frac{M|K|\epsilon}{|K| + |L|}\right) = \frac{1}{M} \cdot \frac{M|K|\epsilon}{|K| + |L|}\left(1 + \left|\frac{L}{K}\right|\right)$$

$$= \frac{|K|\epsilon}{|K| + |L|}\left(\frac{|K| + |L|}{|K|}\right) = \epsilon.$$

So, $\lim_{x\to r}\left[\frac{f(x)}{g(x)}\right] = \frac{L}{K} = \frac{\lim_{x\to r}[f(x)]}{\lim_{x\to r}[g(x)]}$. □

Give a reasonable equivalent definition for each of the following limits (like what was done in Theorem 5.46). r and L are finite real numbers.

79. $\lim_{x\to r} f(x) = -\infty$

$\lim_{x\to r} f(x) = -\infty$ if and only if $\forall M > 0\ \exists \delta > 0\ (0 < |x - r| < \delta \to f(x) < -M)$.

80. $\lim_{x\to +\infty} f(x) = L$

$\lim_{x\to +\infty} f(x) = L$ if and only if $\forall \epsilon > 0\ \exists K > 0\ (x > K \to |f(x) - L| < \epsilon)$.

81. $\lim_{x\to -\infty} f(x) = L$

$\lim_{x\to -\infty} f(x) = L$ if and only if $\forall \epsilon > 0\ \exists K > 0\ (x < -K \rightarrow |f(x) - L| < \epsilon)$.

82. $\lim_{x\to +\infty} f(x) = +\infty$

$\lim_{x\to +\infty} f(x) = +\infty$ if and only if $\forall M > 0\ \exists K > 0\ (x > K \rightarrow f(x) > M)$.

83. $\lim_{x\to +\infty} f(x) = -\infty$

$\lim_{x\to +\infty} f(x) = -\infty$ if and only if $\forall M > 0\ \exists K > 0\ (x > K \rightarrow f(x) < -M)$.

84. $\lim_{x\to -\infty} f(x) = +\infty$

$\lim_{x\to -\infty} f(x) = +\infty$ if and only if $\forall M > 0\ \exists K > 0\ (x < -K \rightarrow f(x) > M)$.

85. $\lim_{x\to -\infty} f(x) = -\infty$

$\lim_{x\to -\infty} f(x) = -\infty$ if and only if $\forall M > 0\ \exists K > 0\ (x < -K \rightarrow f(x) < -M)$.

Let $f(x) = -x^2 + x + 1$.

86. Use the $M - K$ definition of an infinite limit (that you came up with in Problem 83) to prove $\lim_{x\to +\infty} f(x) = -\infty$.

Proof: Let $M > 0$ and let $K = \frac{1}{2} + \sqrt{M + \frac{5}{4}}$. Suppose that $x > K$. Then $x - \frac{1}{2} > \sqrt{M + \frac{5}{4}}$, and so, $\left(x - \frac{1}{2}\right)^2 > M + \frac{5}{4}$. So, $x^2 - x + \frac{1}{4} > M + \frac{5}{4}$. Thus, $x^2 - x - 1 > M$. So, $-x^2 + x + 1 < -M$. That is, $f(x) < -M$. So, $\lim_{x\to +\infty} f(x) = -\infty$. □

87. Use the $M - K$ definition of an infinite limit (that you came up with in Problem 85) to prove $\lim_{x\to -\infty} f(x) = -\infty$.

Proof: Let $M > 0$ and let $K = \frac{1}{2} + \sqrt{M + \frac{5}{4}}$. Suppose that $x < -K$. Then $x + \frac{1}{2} < -\sqrt{M + \frac{5}{4}}$, and so, $\left(x + \frac{1}{2}\right)^2 > M + \frac{5}{4}$. So, $x^2 + x + \frac{1}{4} > M + \frac{5}{4}$. Thus, $x^2 + x - 1 > M$. So, $-x^2 - x + 1 < -M$. Since $x < 0$, $x < -x$, and so, $f(x) = -x^2 + x + 1 < -x^2 - x + 1 < -M$. So, $\lim_{x\to -\infty} f(x) = -\infty$. □

Give a reasonable definition for each of the following limits (like what was done in Theorem 5.48). r and L are finite real numbers.

88. $\lim_{x \to r^-} f(x) = L$

$\lim_{x \to r^-} f(x) = L$ if and only if $\forall \epsilon > 0\ \exists \delta > 0\ (-\delta < x - r < 0 \to |f(x) - L| < \epsilon)$.

89. $\lim_{x \to r^+} f(x) = +\infty$

$\lim_{x \to r^+} f(x) = +\infty$ if and only if $\forall M > 0\ \exists \delta > 0\ (0 < x - r < \delta \to f(x) > M)$.

90. $\lim_{x \to r^+} f(x) = -\infty$

$\lim_{x \to r^+} f(x) = -\infty$ if and only if $\forall M > 0\ \exists \delta > 0\ (0 < x - r < \delta \to f(x) < -M)$.

91. $\lim_{x \to r^-} f(x) = +\infty$

$\lim_{x \to r^-} f(x) = +\infty$ if and only if $\forall M > 0\ \exists \delta > 0\ (-\delta < x - r < 0 \to f(x) > M)$.

92. $\lim_{x \to r^-} f(x) = -\infty$

$\lim_{x \to r^-} f(x) = -\infty$ if and only if $\forall M > 0\ \exists \delta > 0\ (-\delta < x - r < 0 \to f(x) < -M)$.

Let $f(x) = \frac{1}{x-3}$.

93. Use the $M - \delta$ definition of a one-sided limit (that you came up with in Problem 89) to prove that $\lim_{x \to 3^+} f(x) = +\infty$.

Proof: Let $M > 0$ and let $\delta = \frac{1}{M}$. If $0 < x - 3 < \delta$, then $0 < x - 3 < \frac{1}{M}$, and so, we have $\frac{1}{x-3} > M$. Since $M > 0$ was arbitrary, $\lim_{x \to 3^+} \frac{1}{x-3} = +\infty$. □

94. Use the $M - \delta$ definition of a one-sided limit (that you came up with in Problem 92) to prove that $\lim_{x \to 3^-} f(x) = -\infty$.

Proof: Let $M > 0$ and let $\delta = \frac{1}{M}$. If $-\delta < x - 3 < 0$, then $-\frac{1}{M} < x - 3 < 0$, and so, we have $\frac{1}{x-3} < -M$. Since $M > 0$ was arbitrary, $\lim_{x \to 3^-} \frac{1}{x-3} = -\infty$. □

Let $f(x) = \frac{x+1}{(x-1)^2}$. Prove that

95. $\lim_{x \to +\infty} f(x) = 0$.

Proof: Let $\epsilon > 0$ and let $K = \max\left\{2, 1 + \frac{3}{\epsilon}\right\}$. Let $x > K$. Then $x - 1 > 1 + \frac{3}{\epsilon} - 1 = \frac{3}{\epsilon}$, and therefore, $\frac{1}{x-1} < \frac{\epsilon}{3}$. Also, since $x > 2$, $(x-1)^2 - (x-1) = (x-1)(x-1-1) = (x-1)(x-2) > 0$ (because $x - 1 > 2 - 1 = 1 > 0$ and $x - 2 > 2 - 2 > 0$). Thus, $(x-1)^2 > x - 1$, and so, $\frac{1}{(x-1)^2} < \frac{1}{x-1} < \frac{\epsilon}{3}$. It follows from the Triangle Inequality (and SACT) that

$$\left|\frac{x+1}{(x-1)^2} - 0\right| = \left|\frac{x-1+2}{(x-1)^2}\right| = \left|\frac{x-1}{(x-1)^2} + \frac{2}{(x-1)^2}\right| = \left|\frac{1}{x-1} + \frac{2}{(x-1)^2}\right|$$

$$\leq \left|\frac{1}{x-1}\right| + \left|\frac{2}{(x-1)^2}\right| = \frac{1}{x-1} + 2\frac{1}{(x-1)^2} < \frac{\epsilon}{3} + 2 \cdot \frac{\epsilon}{3} = 3 \cdot \frac{\epsilon}{3} = \epsilon.$$

So, $\lim_{x \to +\infty} f(x) = 0$. □

96. $\lim_{x \to 1^+} f(x) = +\infty$.

Proof: Let $M > 0$ and let $\delta = \min\left\{1, \frac{3}{M}\right\}$. If $0 < x - 1 < \delta$, then $0 < x - 1 < \frac{3}{M}$, and so, we have $\frac{1}{x-1} > \frac{M}{3}$. Since $0 < x - 1 < 1$, $(x-1)^2 < x - 1$, and so, $\frac{1}{(x-1)^2} > \frac{1}{x-1}$. So, we have

$$\frac{x+1}{(x-1)^2} = \frac{x-1+2}{(x-1)^2} = \frac{x-1}{(x-1)^2} + \frac{2}{(x-1)^2} = \frac{1}{x-1} + \frac{2}{(x-1)^2}$$

$$> \frac{1}{x-1} + \frac{2}{x-1} = \frac{3}{x-1} > 3 \cdot \frac{M}{3} = M.$$

So, $\lim_{x \to 1^+} f(x) = +\infty$. □

Let $f: \mathbb{R} \to \mathbb{R}$ be defined by $f(x) = \begin{cases} 0 & \text{if } x \text{ is rational.} \\ 1 & \text{if } x \text{ is irrational.} \end{cases}$

97. Prove that for all $r \in \mathbb{R}$, $\lim_{x \to r}[f(x)] \neq 1$.

Proof: Let $r \in \mathbb{R}$ and suppose toward contradiction that $\lim_{x \to r}[f(x)] = 1$. Let $\epsilon = \frac{1}{2}$. Then there is $\delta > 0$ such that $0 < |x - r| < \delta$ implies $|f(x) - 1| < \frac{1}{2}$. By the Density Theorem (Theorem 5.12), there is a rational number a such that $r < a < r + \delta$. So, we have $0 < |a - r| < \delta$. Therefore, $|f(a) - 1| < \frac{1}{2}$. Since a is a rational number, $f(a) = 0$, and therefore, $|f(a) - 1| = |0 - 1| = 1$. So, $1 < \frac{1}{2}$, which is a contradiction. Therefore, $\lim_{x \to r}[f(x)]$ does not equal 1. □

98. Prove that for all $r \in \mathbb{R}$, $\lim_{x \to r}[f(x)]$ does not exist.

Proof: Let $r, L \in \mathbb{R}$ and suppose toward contradiction that $\lim\limits_{x \to r}[f(x)] = L$. Let $\epsilon = \frac{1}{2}$. Then there is $\delta > 0$ such that $0 < |x - r| < \delta$ implies $|f(x) - L| < \frac{1}{2}$.By the Density Theorem (Theorem 5.12) and Problem 32 above, there is a rational number a and an irrational number b such that $r < a, b < r + \delta$. So, we have $0 < |a - r| < \delta$ and $0 < |b - r| < \delta$. Then $|f(a) - L| < \frac{1}{2}$ and $|f(b) - L| < \frac{1}{2}$. But then by the Triangle Inequality, since $f(a) = 0$ and $f(b) = 1$. we would have

$$1 = |0 - 1| = |f(a) - f(b)| = |f(x) - L + L - f(b)| \leq |f(x) - L| + |L - f(b)| < \frac{1}{2} + \frac{1}{2} = 1.$$

Since $1 < 1$ is false, $\lim\limits_{x \to r}[f(x)]$ does not equal L. Since $L \in \mathbb{R}$ was arbitrary, we see that $\lim\limits_{x \to r}[f(x)]$ does not exist. □

Let $f, g: \mathbb{R} \to \mathbb{R}$ be defined by $f(x) = \cos x$ and $g(x) = \sin x$.

99. Prove that f and g are uniformly continuous on $\mathbb{R}$. Hint: Use the fact that the least distance between two points is a straight line.

Proof: Let $\epsilon > 0$ and let $\delta = \min\{\epsilon, 2\pi\}$. Let $x, y \in \mathbb{R}$ with $|x - y| < \delta$. Suppose that $W(x) = (a, b)$ and $W(y) = (c, d)$. The arc length along the unit circle between (a, b) and (c, d) is $|x - y|$ and the straight-line distance between (a, b) and (c, d) is $\sqrt{(a-c)^2 + (b-d)^2}$. Thus,

$$|\cos x - \cos y| = |a - c| \leq \sqrt{(a-c)^2 + (b-d)^2} \leq |x - y|.$$

$$|\sin x - \sin y| = |b - d| \leq \sqrt{(a-c)^2 + (b-d)^2} \leq |x - y|.$$

Therefore, we have $|\cos x - \cos y| \leq |x - y| < \delta \leq \epsilon$ and $|\sin x - \sin y| \leq |x - y| < \delta \leq \epsilon$. It follows that f and g are uniformly continuous on $\mathbb{R}$. □

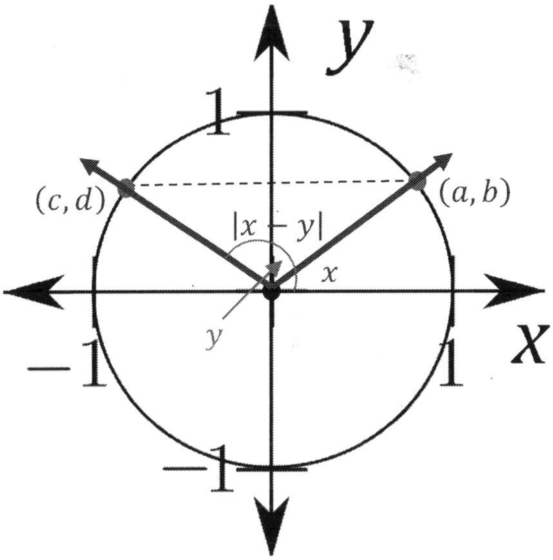

100. Prove that $\lim\limits_{x \to 0} \frac{g(x)}{x} = 1$. You may use Problem 108 below.

Proof: By Problem 108, for $x \in \left(-\frac{\pi}{2}, 0\right) \cup \left(0, \frac{\pi}{2}\right)$, $\cos x \leq \frac{\sin x}{x} \leq 1$. By Problem 99, $\cos x$ is continuous at $x = 0$, and so, by Problem 35, $\lim\limits_{x \to 0} \cos x = \cos 0 = 1$. By Problem 33, $\lim\limits_{x \to 0} 1 = 1$. By the Squeeze Theorem (Problem 74), $\lim\limits_{x \to 0} \frac{\sin x}{x} = 1$. □

101. Prove that $\lim\limits_{x \to 0} \frac{f(x) - 1}{x} = 0$.

Proof: $\lim\limits_{x \to 0} \frac{\cos x - 1}{x} = \lim\limits_{x \to 0} \frac{(\cos x - 1)(\cos x + 1)}{x(\cos x + 1)} = \lim\limits_{x \to 0} \frac{\cos^2 x - 1}{x(\cos x + 1)} = \lim\limits_{x \to 0} \frac{-\sin^2 x}{x(\cos x + 1)} = -\lim\limits_{x \to 0}\left(\frac{\sin x}{x} \cdot \frac{\sin x}{\cos x + 1}\right)$. By Theorem 5.40, this is equal to $-\left(\lim\limits_{x \to 0} \frac{\sin x}{x} \cdot \lim\limits_{x \to 0} \frac{\sin x}{\cos x + 1}\right)$. By Problem 100, this is equal to $-\left(1 \cdot \lim\limits_{x \to 0} \frac{0}{2}\right) = -(1 \cdot 0) = 0$. □

Problem Set 6

LEVEL 1

Determine if each of the following sets is open, closed, both, or neither.

1. $(5,11]$

Neither

2. $\{x \in \mathbb{R} \mid x < -103\}$

Open

3. $(12,17) \cup (26,\infty)$

Open

4. $(0,5) \cap (2,6]$

Open: $(0,5) \cap (2,6] = (2,5)$

5. $(0,5) \cap [2,6)$

Neither: $(0,5) \cap [2,6) = [2,5)$

6. $(0,5] \cap [2,\infty)$

Closed: $(0,5] \cap [2,\infty) = [2,5]$

7. $(0,1) \cap (1,2)$

Both: $(0,1) \cap (1,2) = \emptyset$

Find the accumulation points of each of the following sets:

8. $\left\{\frac{1}{2}, \frac{2}{3}, \frac{3}{4} \dots\right\}$

1

9. $[0,1)$

The set of accumulation points is $[\mathbf{0}, \mathbf{1}]$.

10. $\mathbb{N}$

No accumulation points

11. $\mathbb{R} \setminus \mathbb{Q}$

The set of accumulation points is $\mathbb{R}$.

12. $\{(-1)^n \mid n \in \mathbb{Z}^+\}$

This set is equal to $\{1, -1\}$. It has no accumulation points.

13. $\{x \mid |x| < 1\}$

The set of accumulation points is $\{x \mid |x| \leq 1\} = [-1, 1]$.

14. $\{x \mid 0 < |x - 2| \leq 3\}$

The set of accumulation points is $\{x \mid |x - 2| \leq 3\} = [-1, 5]$.

15. Define a set of real numbers with exactly two accumulation points.

Let $S = \left\{(-1)^n \left(1 + \frac{1}{n}\right) \middle| n \in \mathbb{N}\right\}$. Then S has exactly two accumulation points: 1 and -1.

Let $A \subseteq \mathbb{R}$ and let $f: A \to \mathbb{R}$. Provide a counterexample showing that each of the following is false.

16. If A is closed and f is continuous, then f is bounded.

Let $A = [0, \infty)$ and define $f: A \to \mathbb{R}$ by $f(x) = x$. Then A is closed and f is continuous, but f is not bounded.

17. If A is bounded and f is continuous, then f is bounded.

Let $A = (0, 1)$ and define $f: A \to \mathbb{R}$ by $f(x) = \frac{1}{x}$. Then A is bounded and f is continuous, but f is not bounded.

18. If A is closed and bounded, then f is bounded.

Let $A = [0, 1]$ and define $f: A \to \mathbb{R}$ by $f(x) = \begin{cases} 1 & \text{if } x = 0. \\ \frac{1}{x} & \text{if } x \neq 0. \end{cases}$ Then A is closed and bounded, but f is not bounded.

19. If A is a closed and f is continuous then $f[A]$ is closed.

Let $A = [0, \infty)$ and define $f: A \to \mathbb{R}$ by $f(x) = \frac{1}{x^2+1}$. Then A is closed and f is continuous on A, but $f[A] = (0, 1]$, which is not closed.

20. If A is open and f is continuous, then $f[A]$ is open.

Let $A = (-1, 1)$ and define $f: A \to \mathbb{R}$ by $f(x) = \frac{1}{x^2+1}$. Then A is open and f is continuous on A, but $f[A] = \left(\frac{1}{2}, 1\right]$, which is not open.

LEVEL 2

For each of the following, compute $\bigcup \boldsymbol{X}$ and $\bigcap \boldsymbol{X}$.

21. $\boldsymbol{X} = \{(0, q] \mid q \in \mathbb{Q}^+\}$

$\bigcup \boldsymbol{X} = (\mathbf{0}, \infty)$ $\qquad$ $\bigcap \boldsymbol{X} = \emptyset$

22. $\boldsymbol{X} = \left\{\left(\frac{1}{n}, 1\right) \middle| n \in \mathbb{Z}^+\right\}$

$\bigcup \boldsymbol{X} = (\mathbf{0}, \mathbf{1})$ $\qquad$ $\bigcap \boldsymbol{X} = \emptyset$

23. $\boldsymbol{X} = \{(n, n+3) \mid n \in \mathbb{Z}\}$

$\bigcup \boldsymbol{X} = \mathbb{R}$ $\qquad$ $\bigcap \boldsymbol{X} = \emptyset$

24. $\boldsymbol{X} = \left\{\left(0, 1 + \frac{1}{n}\right) \mid n \in \mathbb{Z}^+\right\}$

$\bigcup \boldsymbol{X} = (\mathbf{0}, \mathbf{2})$ $\qquad$ $\bigcap \boldsymbol{X} = (\mathbf{0}, \mathbf{1}]$

25. $\boldsymbol{X} = \left\{\left(-\infty, \frac{1}{n}\right] \middle| n \in \mathbb{Z}^+\right\}$

$\bigcup \boldsymbol{X} = (-\infty, \mathbf{1}]$ $\qquad$ $\bigcap \boldsymbol{X} = (-\infty, \mathbf{0}]$

Determine if each of the following statements is true or false. If true, provide a proof. If false, provide a counterexample.

26. Let I be an open or half-open interval of real numbers. Then I is not compact.

True

Proof: If I is unbounded, then $C = \{(n-1, n+1) \mid n \in \mathbb{Z}^+\}$ is an open covering of I with no finite subcover. So, assume that I is bounded.

Let $I = (a, b)$. The collection of open sets $C = \left\{\left(a + \frac{1}{n}, b\right) \middle| n \in \mathbb{Z}^+\right\}$ is an open covering of I with no finite subcover.

Let $I = (a, b]$. The collection $C = \left\{\left(a + \frac{1}{n}, b + 1\right) \middle| n \in \mathbb{Z}^+\right\}$ is an open covering of I with no finite subcover.

Let $I = [a, b)$. The collection $C = \left\{\left(a - 1, b - \frac{1}{n}\right) \middle| n \in \mathbb{Z}^+\right\}$ is an open covering of I with no finite subcover. □

27. A finite union of compact subsets of $\mathbb{R}$ is compact.

True

Proof: For each $k = 0, 1, \dots, n$, let A_k be a compact subset of $\mathbb{R}$. Let $A = \bigcup\{A_k \mid k = 0, 1, \dots, n\}$. We will show that A is compact. To this end, let $\mathcal{C}$ be an open covering of A. Then for each $k = 0, 1, \dots, n$, $\mathcal{C}$ is an open covering of A_k. Since A_k is compact, there is a finite subcollection $\mathcal{C}_k$ such that $\mathcal{C}_k$ still covers A_k. Then $\bigcup\{\mathcal{C}_k \mid k = 0, 1, \dots, n\}$ is a finite subcollection of $\mathcal{C}$ that covers A. □

28. A countable union of compact subsets of $\mathbb{R}$ is compact.

False

Let $X = \{[n, n+1] \mid n \in \mathbb{Z}\}$. Each set in X is compact, but $\bigcup X = \mathbb{R}$, which is not compact.

29. The Cantor set is compact.

True

Proof: Let C be the Cantor set. We saw in part 2 of Example 6.22 that C is closed. Since $C \subseteq [0, 1]$, C is bounded. So, by the Heine-Borel Theorem (Theorem 6.28), C is compact. □

30. Let $A \subseteq \overline{A} \subseteq B \subseteq \mathbb{R}$ and let $f, g: B \to \mathbb{R}$ be continuous. Suppose that $f(x) = g(x)$ for all $x \in A$. Then $f(x) = g(x)$ for all $x \in \overline{A}$.

True

Proof: Let $x \in \overline{A}$. Then there is a sequence (x_n) in A such that $x_n \to x$ (simply choose $x_n \in N_{\frac{1}{n}}(x) \cap A$ for each $n \in \mathbb{N}$). Since f is continuous, $f(x_n) \to f(x)$. Since g is continuous, $f(x_n) = g(x_n) \to g(x)$. So, $f(x_n)$ converges to both $f(x)$ and $g(x)$. By the Note following Example 5.49, $f(x) = g(x)$. □

LEVEL 3

Determine if each of the following sets is open, closed, both, or neither.

31. $\mathbb{Q}^+$

Neither

32. $\mathbb{Z}$

Closed

33. $\mathbb{R} \setminus \mathbb{N}$

Open

34. $\bigcup\left\{\left(0, 1+\frac{1}{n}\right) \mid n \in \mathbb{Z}^+\right\}$

Open

35. $\bigcap\left\{\left[0, 1-\frac{1}{n}\right] \mid n \in \mathbb{Z}^+\right\}$

Closed

Prove each of the following:

36. For all $b \in \mathbb{R}$, the infinite interval $(-\infty, b)$ is open in $\mathbb{R}$.

Proof: Let $x \in (-\infty, b)$ and let $a = x - 1$. Since $x \in (-\infty, b)$, $x < b$. Since $x - (x-1) = 1 > 0$, we have $x > x - 1 = a$. So, we have $a < x < b$. That is, $x \in (a, b)$. Also, $(a, b) \subseteq (-\infty, b)$. Since $x \in (-\infty, b)$ was arbitrary, $(-\infty, b)$ is an open set. □

37. The intersection of two open intervals in $\mathbb{R}$ is either empty or an open interval in $\mathbb{R}$.

Proof: Let (a, b) and (c, d) be open intervals in $\mathbb{R}$ (a and c can be $-\infty$, and b and d can be ∞, where $-\infty$ is less than any real number and ∞, and ∞ is greater than any real number and $-\infty$). Without loss of generality, we may assume that $a \leq c$. If $b \leq c$, then we have $(a, b) \cap (c, d) = \emptyset$ because if $a < x < b$ and $c < x < d$, then $x < b \leq c < x$, and so, $x < x$, which is impossible.

So, we may assume that $c < b$. Let $e = \min\{b, d\}$. We claim that $(a, b) \cap (c, d) = (c, e)$.

Let $x \in (a, b) \cap (c, d)$. Then $x \in (a, b)$ and $x \in (c, d)$. So, $a < x < b$ and $c < x < d$. In particular, $x > c$, $x < b$, and $x < d$. Since $x < b$ and $x < d$, $x < e$. So, $x \in (c, e)$. Since $x \in (a, b) \cap (c, d)$ was arbitrary, we have shown that $(a, b) \cap (c, d) \subseteq (c, e)$.

Now, let $x \in (c, e)$. Then $c < x < e$. We are assuming that $a \leq c$. We also have $e \leq b$. So, $a \leq c < x < e \leq b$. Therefore, $x \in (a, b)$. We also have $e \leq d$. So, $c < x < e \leq d$, and therefore, $x \in (c, d)$. Since $x \in (a, b)$ and $x \in (c, d)$, we have $x \in (a, b) \cap (c, d)$. Since $x \in (c, e)$ was arbitrary, we have shown that $(c, e) \subseteq (a, b) \cap (c, d)$.

Finally, since we have shown $(a, b) \cap (c, d) \subseteq (c, e)$ and $(c, e) \subseteq (a, b) \cap (c, d)$, we have $(a, b) \cap (c, d) = (c, e)$.

Therefore, the intersection of two open intervals in $\mathbb{R}$ is either empty or an open interval in $\mathbb{R}$. □

38. The intersection of finitely many open sets in $\mathbb{R}$ is an open set in $\mathbb{R}$.

Proof: The intersection of a single set with itself is just that set itself, and so, the result holds trivially for one open set.

So, we will prove the following statement: "The intersection of a set of finitely many open sets in $\mathbb{R}$ consisting of at least 2 sets is an open set in $\mathbb{R}$." We will prove this by induction on the number of open sets we are taking the intersection of. Theorem 6.19 is the base case $n = 2$.

For the inductive step, assume that the intersection of any k nonempty open sets in $\mathbb{R}$ is always open, and let $\boldsymbol{X}$ be a set of $k+1$ open sets. Let $A \in \boldsymbol{X}$ and let B be the intersection of all the sets in $\boldsymbol{X}$ except A. By the induction hypotheses, B is open. Therefore, $\bigcap \boldsymbol{X} = A \cap B$ is open by Theorem 6.19.

By the Principle of Mathematical Induction, we have shown that the intersection of a set of finitely many open sets in $\mathbb{R}$ consisting of at least 2 sets is an open set in $\mathbb{R}$. □

39. The Triangle Inequality (Theorem 6.4).

Proof: $|x+y|^2 = (x+y)(x+y) = x^2 + 2xy + y^2 \leq |x|^2 + 2|x||y| + |y|^2 = (|x|+|y|)^2$. Since $|x+y|$ and $|x|+|y|$ are nonnegative, by Problem 44 from Problem Set 5, $|x+y| \leq |x| + |y|$. □

40. If x and y are real numbers, then $\big||x|-|y|\big| \leq |x \pm y| \leq |x|+|y|$.

Proof: $|x| = |(x+y)+(-y)| \leq |x+y| + |-y| = |x+y| + |y|$. So, $|x+y| \geq |x| - |y|$.

$|y| = |(x+y)+(-x)| \leq |x+y| + |-x| = |x+y| + |x|$. So, $|x+y| \geq |y| - |x| = -(|x|-|y|)$.

Since for all $x, y \in \mathbb{R}$, we have $\big||x|-|y|\big| = |x|-|y|$ or $\big||x|-|y|\big| = -(|x|-|y|)$, it follows that $\big||x|-|y|\big| \leq |x+y|$.

Combining this result with the Triangle Inequality, gives us $\big||x|-|y|\big| \leq |x+y| \leq |x|+|y|$.

Now, by the Triangle Inequality we have $|x-y| = |x+(-y)| \leq |x| + |-y| = |x| + |y|$.

Finally, by the third paragraph, we have $|x-y| = |x+(-y)| \geq \big||x|-|-y|\big| = \big||x|-|y|\big|$. □

41. If $\boldsymbol{X}$ is a nonempty set of closed subsets of $\mathbb{R}$, then $\bigcap \boldsymbol{X}$ is closed.

Proof: Let $\boldsymbol{X}$ be a nonempty set of closed subsets of $\mathbb{R}$. Then for each $A \in \boldsymbol{X}$, $\mathbb{R} \setminus A$ is an open set in $\mathbb{R}$. By Theorem 6.15, $\bigcup\{\mathbb{R} \setminus A \mid A \in \boldsymbol{X}\}$ is open in $\mathbb{R}$. Therefore, $\mathbb{R} \setminus \bigcup\{\mathbb{R} \setminus A \mid A \in \boldsymbol{X}\}$ is closed in $\mathbb{R}$. So, it suffices to show that $\bigcap \boldsymbol{X} = \mathbb{R} \setminus \bigcup\{\mathbb{R} \setminus A \mid A \in \boldsymbol{X}\}$. Well, $x \in \bigcap \boldsymbol{X}$ if and only if for all $A \in \boldsymbol{X}$, $x \in A$ if and only if for all $A \in \boldsymbol{X}$, $x \notin \mathbb{R} \setminus A$ if and only if $x \notin \bigcup\{\mathbb{R} \setminus A \mid A \in \boldsymbol{X}\}$ if and only if $x \in \mathbb{R} \setminus \bigcup\{\mathbb{R} \setminus A \mid A \in \boldsymbol{X}\}$. So, $\bigcap \boldsymbol{X} = \mathbb{R} \setminus \bigcup\{\mathbb{R} \setminus A \mid A \in \boldsymbol{X}\}$, completing the proof. □

42. Let S be a set of real numbers and let S' be the set of accumulation points of S. Then S' is closed in $\mathbb{R}$.

Proof: Let S be a set of real numbers, let S' be the set of accumulation points of S, and let x be an accumulation point of S'. We will show that x is an accumulation point of S. To see this, let (a, b) be an open interval containing x. Since x is an accumulation point of S', there is $y \in S'$ with $y \in (a, b)$ and $y \neq x$. Without loss of generality, assume that $x < y$. Since $y \in S'$, y is an accumulation point of S. So, there is $z \in S$ with $z \in (x, b)$. Thus, we have found an element of S in (a, b) different from x. So, x is an accumulation point of S, and therefore, $x \in S'$. Since x was an arbitrary accumulation point of S', it follows that S' contains each of its accumulation points. By Theorem 6.24, S' is closed in $\mathbb{R}$. □

43. If $x \in \mathbb{R}$ is an accumulation point of a set S, then every open interval containing x contains infinitely many points of S.

Proof: Let $x \in \mathbb{R}$ be an accumulation point of a set S, let (a, b) be an open interval containing x, and let $X = \{x_1, x_2, \dots, x_n\}$ be an arbitrary subset of $(a, b) \cap S \cap (\mathbb{R} \setminus \{x\})$. Let $A = \{x_i \mid x_i < x\} \cup \{a\}$ and let $B = \{x_i \mid x_i > x\} \cup \{b\}$. Now, let c be the largest element of A and let d be the smallest element of B. Then $c < x < d$, and so, (c, d) is an open interval containing x. Since x is an accumulation point of S, there is $y \in (c, d)$ with $y \in S$ and $y \neq x$. Since $a \leq c$ and $d \leq b$, $(c, d) \subseteq (a, b)$. Therefore, $y \in (a, b)$. Also, $y \notin X$. So, given any finite collection of elements of $Y = (a, b) \cap S \cap (\mathbb{R} \setminus \{x\})$, we can always find another element of Y that is not in that finite collection. □

44. A closed subset of a compact set is compact.

Proof: Let $A \subseteq B \subseteq \mathbb{R}$ with A closed and B compact. Let $\mathcal{C}$ be an open covering of A. Then we have that $\mathcal{D} = \mathcal{C} \cup \{\mathbb{R} \setminus A\}$ is an open covering of B. By the compactness of B, there is a finite subcollection $\mathcal{E}$ of $\mathcal{D}$ such that $\mathcal{E}$ is an open covering of B. It follows that $\mathcal{E} \setminus \{\mathbb{R} \setminus A\}$ is an open covering of A. So, A is compact. □

45. Let I be a closed bounded interval and let J be an open or half-open interval. Then I and J are not homeomorphic.

Proof: Suppose toward contradiction that $f: I \to J$ is a homeomorphism. By the Heine-Borel Theorem, I is compact. By Theorem 6.39, $f[I] = J$ is compact. But then, by the Heine-Borel Theorem again, J is closed, which is a contradiction. □

46. Any two open intervals are homeomorphic. (You may exclude the interval $(-\infty, \infty)$.)

Proof: In the solution to Problem 294 in Problem Set 2, we saw that if $a, b \in \mathbb{R}$ with $a < b$, then the function $h: (0, 1) \to (a, b)$ defined by $h(x) = (b - a)x + a$ is a bijection. By Problem 47 in Problem Set 5, h is continuous. Also, $h^{-1}(x) = \frac{1}{b-a}(x - a)$ is continuous for the same reason. It follows that h is a homeomorphism. Since compositions of bijective functions are bijective (Corollary 2.59) and compositions of continuous functions are continuous (Problem 48 from Problem Set 5), it follows that any two bounded open intervals are homeomorphic.

Again, in Problem 294 in Problem Set 2, we saw that the function $g: (0, \infty) \to (0, 1)$ defined by $g(x) = \frac{1}{x^2+1}$ is a bijection. If $a \in (0, \infty)$, then by Problems 47 and 78 in Problem Set 5, $\lim_{x \to a} g(x) = \frac{1}{a^2+1} = g(a)$. By Problem 35 in Problem Set 5, g is continuous. Also, $g^{-1}(x) = \sqrt{\frac{1}{x} - 1}$. By Problem 68 in Problem Set 5, as well as other results already mentioned, g^{-1} is continuous. It follows that g is a homeomorphism.

The function $k: (0, \infty) \to (-\infty, 0)$ defined by $f(x) = -x$ is easily seen to be a homeomorphism. Using the fact that compositions of bijective functions are bijective (Corollary 2.59) and compositions of continuous functions are continuous (Problem 48 from Problem Set 5), we now have that any two open intervals I and J are homeomorphic, as long as $I \neq (-\infty, \infty)$ and $J \neq (-\infty, \infty)$. □

Note: It is true that $\mathbb{R} = (-\infty, \infty)$ is homeomorphic to every other open interval as well. For example, the function $f: \mathbb{R} \to (0, \infty)$ defined by $f(x) = 2^x$ is a homeomorphism. However, to prove this requires a formal definition of 2^x. See Lesson 13 for a formal definition of this function and an explanation as to why it is a continuous bijection with a continuous inverse.

LEVEL 4

Give an example of each of the following and provide a proof for each example:

47. An infinite collection of open sets in $\mathbb{R}$ whose intersection is not open in $\mathbb{R}$

Let $\boldsymbol{X} = \left\{\left(0, 1 + \frac{1}{n}\right) \mid n \in \mathbb{Z}^+\right\}$. Each set in $\boldsymbol{X}$ is an open interval, and therefore, open. We will show that $\bigcap \boldsymbol{X} = (0, 1]$. Note that $x \in \bigcap \boldsymbol{X}$ if and only if for all $n \in \mathbb{Z}^+$, $x \in \left(0, 1 + \frac{1}{n}\right)$ if and only if for all $n \in \mathbb{Z}^+, 0 < x < 1 + \frac{1}{n}$. We need to show that $x \leq 1$ is equivalent to $\forall n \in \mathbb{Z}^+ \left(x < 1 + \frac{1}{n}\right)$.

Suppose that $x \leq 1$. Let $n \in \mathbb{Z}^+$. By Theorem 5.4, $\frac{1}{n} > 0$. So, $1 + \frac{1}{n} - 1 > 0$ (SACT). Thus, $1 + \frac{1}{n} > 1$. So, we have $x \leq 1 < 1 + \frac{1}{n}$, and therefore, $x < 1 + \frac{1}{n}$. Since $n \in \mathbb{Z}^+$ was arbitrary, we have shown that $x \leq 1$ implies $\forall n \in \mathbb{Z}^+ \left(x < 1 + \frac{1}{n}\right)$.

Now, suppose $x > 1$ (proof by contrapositive). Then $x - 1 > 0$. Since there is no smallest positive real number (by Problem 26 in Problem Set 5), there is a real number $\epsilon > 0$ with $x - 1 > \epsilon$. By the Archimedean Property of the reals (Theorem 5.11), there is a natural number n with $n > \frac{1}{\epsilon}$. So, $\frac{1}{n} < \epsilon$, or equivalently, $\epsilon > \frac{1}{n}$. Thus, $x - 1 > \frac{1}{n}$, and so, $x > 1 + \frac{1}{n}$. We have shown that there is $n \in \mathbb{Z}^+$ such that $x \geq 1 + \frac{1}{n}$. So, $\forall n \in \mathbb{Z}^+ \left(x < 1 + \frac{1}{n}\right)$ is false.

This equivalence proves that $\bigcap \boldsymbol{X} = (0, 1]$, an interval that is **not** an open set.

48. An infinite collection of closed sets in $\mathbb{R}$ whose union is not closed in $\mathbb{R}$.

Let $\boldsymbol{Y} = \left\{\left[0, 1 - \frac{1}{n}\right] \mid n \in \mathbb{Z}^+\right\}$. Each set in Y is a closed interval, and therefore, closed. We will show that $\bigcup \boldsymbol{Y} = [0, 1)$. Note that $x \in \bigcup \boldsymbol{Y}$ if and only if there is $n \in \mathbb{Z}^+$ such that $x \in \left[0, 1 - \frac{1}{n}\right]$ if and only if there is $n \in \mathbb{Z}^+$ such that $0 \leq x \leq 1 - \frac{1}{n}$. We need to show that $x < 1$ is equivalent to $\exists n \in \mathbb{Z}^+ \left(x \leq 1 - \frac{1}{n}\right)$ (where $\exists$ is read "there exists" or "there is").

Suppose $x < 1$. Then $1 - x > 0$. Since there is no smallest positive real number, there is a real number $\epsilon > 0$ with $1 - x > \epsilon$. By the Archimedean Property of the reals (Theorem 5.11), there is a natural number n with $n > \frac{1}{\epsilon}$. So, $\frac{1}{n} < \epsilon$, or equivalently, $\epsilon > \frac{1}{n}$. Thus, $1 - x > \frac{1}{n}$, and so, $x < 1 - \frac{1}{n}$. We have shown that there is $n \in \mathbb{Z}^+$ such that $x \leq 1 - \frac{1}{n}$. So, $\exists n \in \mathbb{Z}^+ \left(x \leq 1 - \frac{1}{n}\right)$.

Now, suppose $x \geq 1$ (proof by contrapositive). Let $n \in \mathbb{Z}^+$. By Theorem 5.4, $\frac{1}{n} > 0$. So, $1 - 1 + \frac{1}{n} > 0$ (SACT). So, $1 > 1 - \frac{1}{n}$. It follows that $x > 1 - \frac{1}{n}$. Since $n \in \mathbb{Z}^+$ was arbitrary, $\forall n \in \mathbb{Z}^+ \left(x > 1 - \frac{1}{n}\right)$. It follows that $\exists n \in \mathbb{Z}^+ \left(x \leq 1 - \frac{1}{n}\right)$ is false.

This equivalence proves that $\bigcup \boldsymbol{Y} = [0, 1)$, an interval that is **not** a closed set.

Determine if each of the following subsets of $\mathbb{R}$ is open, closed, both, or neither. Give a proof in each case.

49. $\{x \in \mathbb{R} \mid |x| > 1\}$

$S = \{x \in \mathbb{R} \mid |x| > 1\}$ is **open**.

Proof: Let $y \in S$ and let $r = |y| - 1$. We will show that $N_r(y) = \{x \in \mathbb{R} \mid |x - y| < r\}) \subseteq S$. Let $x \in N_r(y)$. Then we have $|x - y| < r = |y| - 1$. So, using the Triangle Inequality, we have

$$|y| = |(y - x) + x| \leq |y - x| + |x| = |x - y| + |x| < |y| - 1 + |x|.$$

Thus, $|y| < |y| - 1 + |x|$, and therefore, $|x| > 1$. So, $x \in S$. Since $x \in N_r(y)$ was arbitrary, we have shown that $N_r(y) \subseteq S$. So, S is open. □

$S = \{x \in \mathbb{R} \mid |x| > 1\}$ is **not closed** because $\mathbb{R} \setminus S = \{x \in \mathbb{R} \mid |x| \leq 1\}$ is not open.

50. $\{(-1)^n \mid n \in \mathbb{Z}^+\}$

Note that $S = \{(-1)^n \mid n \in \mathbb{Z}^+\}$ is a finite set consisting of just two real numbers. Indeed, $S = \{1, -1\}$.

S is **not open**.

Proof: Let $N_r(1)$ be an arbitrary r-neighborhood of 1. Then $1 + \frac{r}{2} \in N_r(1)$ because $\left|\left(1 + \frac{r}{2}\right) - 1\right| = \left|\frac{r}{2}\right| = \frac{r}{2} < r$, but $1 + \frac{r}{2} \notin S$ because $1 + \frac{r}{2}$ is not equal to 1 or -1. □

S is **closed**.

Proof: We show that $T = \mathbb{R} \setminus S$ is open.

Let $y \in T$ and let $r = \min\{|y - 1|, |y + 1|\}$.

We now show that $N_r(y) \subseteq T$. Since $r \leq |y - 1|$, $1 \notin N_r(y)$ (otherwise, $|y - 1| < r$). Similarly, $-1 \notin N_r(y)$. So, if $x \in N_r(y)$, then $x \notin S$, and so, $x \in T$.

Thus, $T = \mathbb{R} \setminus S$ is open. Therefore, S is closed. □

51. $\{x \in \mathbb{R} \mid 2 < |x - 2| < 4\}$

$S = \{x \in \mathbb{R} \mid 2 < |x - 2| < 4\}$ is **open** and **not closed**.

Proof: To see that S is open, let $x \in S$ and let $r = \min\{|x-2|-2, 4-|x-2|\}$. We show that $N_r(x) \subseteq S$. Let $y \in N_r(x)$. Then $|y-x| < r$. So, $|y-x| < 4-|x-2|$. Therefore, we have $|y-x|+|x-2| < 4$, and so, $|y-2| = |y-x+x-2| \leq |y-x|+|x-2| < 4$.

Also, $|y-x| < |x-2|-2$. So, we have

$$2 < |x-2|-|y-x| = |x-y+y-2|-|y-x| \leq |x-y|+|y-2|-|y-x|$$
$$= |y-x|+|y-2|-|y-x| = |y-2|.$$

So, $2 < |y-2| < 4$, and therefore, $y \in S$.

Since $x \in S$ was arbitrary, S is open.

To see that S is not closed, we show that $\mathbb{R} \setminus S = \{x \in \mathbb{R} \mid |x-2| \leq 2 \text{ or } |x-2| \geq 4\}$ is not open. To see this, first note that $|6-2| = |4| = 4 \geq 4$, and so, $6 \in \mathbb{R} \setminus S$. Let $N_r(6)$ be an r-neighborhood of 6 and let $k = \min\left\{1, \frac{r}{2}\right\}$. Then we have $6-k \in N_r(6)$ because $|(6-k)-6| = |-k| = k \leq \frac{r}{2} < r$. However, $6-k \notin \mathbb{R} \setminus S$. To see this, first observe that $|(6-k)-2| = |4-k|$. If $k=1$, then $|4-k| = |4-1| = |3| = 3$ and it follows that $6-k \notin \mathbb{R} \setminus S$. If $k = \frac{r}{2}$, then $0 < \frac{r}{2} \leq 1$, so that $-1 \leq -\frac{r}{2} < 0$, and thus, $3 < 4-\frac{r}{2} < 4$. So, $3 < \left|4-\frac{r}{2}\right| < 4$ and once again, $6-k \notin \mathbb{R} \setminus S$. So, $\mathbb{R} \setminus S$ is not open. Therefore, S is not closed. □

Prove each of the following:

52. Let $A \subseteq \mathbb{R}$ and let $f: A \to \mathbb{R}$. f is continuous on A if and only if for each closed set $C \subseteq \mathbb{R}$, there is a closed set $B \subseteq \mathbb{R}$ such that $f^{-1}[C] = B \cap A$.

Proof: Let $f: A \to \mathbb{R}$. First assume that f is continuous on A and let $C \subseteq \mathbb{R}$ be closed. Then $\mathbb{R} \setminus C$ is open. Since f is continuous on A, by Theorem 6.38, there is an open set $U \subseteq \mathbb{R}$ such that $f^{-1}[\mathbb{R} \setminus C] = U \cap A$. We first show that $f^{-1}[\mathbb{R} \setminus C] = A \setminus f^{-1}[C]$. To see this, note that $x \in f^{-1}[\mathbb{R} \setminus C]$ if and only if $f(x) \in \mathbb{R} \setminus C$ if and only if $f(x) \in \mathbb{R}$ and $f(x) \notin C$ if and only if $x \in A$ and $x \notin f^{-1}[C]$ if and only if $x \in A \setminus f^{-1}[C]$. It follows that $A \setminus f^{-1}[C] = U \cap A$. Let $B = \mathbb{R} \setminus U$. Then B is closed. We complete the argument by showing that $f^{-1}[C] = B \cap A$. To see this, note that $x \in f^{-1}[C]$ if and only if $x \notin A \setminus f^{-1}[C] = U \cap A$. Since x must be in A, $x \in f^{-1}[C]$ if and only if $x \notin U$ and $x \in A$ if and only if $x \in B$ and $x \in A$ if and only if $x \in B \cap A$.

Conversely, assume that for each closed $C \subseteq \mathbb{R}$, there is a closed set $B \subseteq \mathbb{R}$ such that $f^{-1}[C] = B \cap A$. Let $U \subseteq \mathbb{R}$ be open. Then $\mathbb{R} \setminus U$ is closed. By our assumption, there is a closed set $B \subseteq \mathbb{R}$ such that $f^{-1}[\mathbb{R} \setminus U] = B \cap A$. By the same argument given in the first paragraph above, we have $f^{-1}[\mathbb{R} \setminus U] = A \setminus f^{-1}[U]$. It follows that $A \setminus f^{-1}[U] = B \cap A$. Let $V = \mathbb{R} \setminus B$. Then V is open. We complete the argument by showing that $f^{-1}[U] = V \cap A$. To see this, note that $x \in f^{-1}[U]$ if and only if $x \notin A \setminus f^{-1}[U] = B \cap A$. Since x must be in A, $x \in f^{-1}[U]$ if and only if $x \notin B$ and $x \in A$ if and only if $x \in V$ and $x \in A$ if and only if $x \in V \cap A$. □

53. The image of a Cauchy sequence under a uniformly continuous function is a Cauchy sequence. If we replace "uniformly continuous" by "continuous," is the result still true?

Proof: Let $A \subseteq \mathbb{R}$, let $f: A \to B$ be uniformly continuous, let (x_n) be a Cauchy sequence in A, and let $\epsilon > 0$. Since f is uniformly continuous, there is $\delta > 0$ such that for all $x, y \in A$, $|x - y| < \delta \to |f(x) - f(y)| < \epsilon$. Since (x_n) is a Cauchy sequence, there is $K \in \mathbb{N}$ such that $m \geq n > K$ implies $|x_m - x_n| < \delta$. Now, assume that $m \geq n > K$. Then we have $|x_m - x_n| < \delta$. Therefore, $|f(x_m) - f(x_n)| < \epsilon$. It follows that $\big(f(x_n)\big)$ is a Cauchy sequence in $\mathbb{R}$. □

If we replace "uniformly continuous" by "continuous," the result is false. Define $f: (0,1) \to \mathbb{R}$ by $f(x) = \frac{1}{x}$. Then $\left(\frac{1}{n}\right)$ is a Cauchy sequence in $(0,1)$, but $f\left(\frac{1}{n}\right) = n$, and (n) is not a Cauchy sequence.

54. Let $S \subseteq \mathbb{R}$, let A and B be disjoint closed subsets of S, and let a and b be real numbers with $a < b$. There is a continuous function $f: S \to [a, b]$ such that $f[A] = \{a\}$ and $f[B] = \{b\}$. You may use Urysohn's Lemma (see Problem 69 below) to prove this.

Proof: Let $S \subseteq \mathbb{R}$, let A and B be disjoint closed subsets of S, and let a and b be real numbers with $a < b$. By Urysohn's Lemma (Problem 69), there is a continuous function $g: S \to [0,1]$ such that $g[A] = \{0\}$ and $g[B] = \{1\}$. Define $f: S \to [a, b]$ by $f(x) = (b - a)g(x) + a$. Then we have $f[A] = (b - a)g[A] + a = (b - a)(0) + a = a$, $f[B] = (b - a)g[B] + a = (b - a)(1) + a = b$, and f is continuous by a standard continuity argument. □

55. If $A \subseteq \mathbb{R}$ and $f: A \to \mathbb{R}$, then f is continuous on A if and only if for each $X \subseteq A$, $f\left[\overline{X} \cap A\right] \subseteq \overline{f[X]}$.

Proof: Let $A \subseteq \mathbb{R}$ and let $f: A \to \mathbb{R}$. First assume that f is continuous on A. Let $X \subseteq A$ and let $y \in f\left[\overline{X} \cap A\right]$. Then there is $x \in \overline{X} \cap A$ with $y = f(x)$. Now, let $V \subseteq \mathbb{R}$ be an open interval containing $f(x)$. By Theorem 6.36 ($1 \to 4$), there is an open set $U \subseteq \mathbb{R}$ with $x \in U$ such that $f[U \cap A] \subseteq V$. By part 5 of Theorem 6.25, there is $z \in U \cap X$ (Check this carefully!). Since $X \subseteq A$, $f[U \cap X] \subseteq f[U \cap A]$. Therefore, $f(z) \in f[U \cap X] \subseteq f[U \cap A] \subseteq V$. Also, since $z \in X$, we have $f(z) \in f[X]$. Once again, by part 5 of Theorem 6.25, $y = f(x) \in \overline{f[X]}$. Since $y \in f\left[\overline{X} \cap A\right]$ was arbitrary, $f\left[\overline{X} \cap A\right] \subseteq \overline{f[X]}$.

Conversely, assume that for each $X \subseteq A$, $f\left[\overline{X} \cap A\right] \subseteq \overline{f[X]}$. Let C be a closed set in $\mathbb{R}$ and let $D = f^{-1}[C]$. Then $f[D] = f\left[f^{-1}[C]\right] \subseteq C$ (see part 3 of Example 2.61). Now, if $x \in \overline{D} \cap A$, then we have $f(x) \in f\left[\overline{D} \cap A\right]$. By our assumption, $f\left[\overline{D} \cap A\right] \subseteq \overline{f[D]}$. So, $f(x) \in \overline{f[D]} \subseteq \overline{C} = C$ (because C is closed in $\mathbb{R}$). So, $f(x) \in C$, and therefore, $x \in f^{-1}[C] = D$. Since $x \in \overline{D} \cap A$ was arbitrary, $\overline{D} \cap A \subseteq D$. Since $D \subseteq \overline{D}$ is always true (by part 1 of Theorem 6.25) and $D \subseteq A$ (because $x \in D = f^{-1}[C]$ implies that $f(x) \in C$, which implies that $x \in \operatorname{dom} f = A$), we have $D \subseteq \overline{D} \cap A$. Since both $\overline{D} \cap A \subseteq D$ and $D \subseteq \overline{D} \cap A$, it follows that $f^{-1}[C] = D = \overline{D} \cap A$. Since $\overline{D}$ is closed in $\mathbb{R}$, by problem 52, f is continuous on A. □

56. If $A \subseteq \mathbb{R}$ is compact and $f: A \to \mathbb{R}$ is continuous and injective, then f is a homeomorphism.

Proof: We need to show that $f^{-1}: f[A] \to A$ is continuous. Let $V \subseteq \mathbb{R}$ be open. Then $\mathbb{R} \setminus V$ is closed. Since A is compact, by the Heine-Borel Theorem, A is closed. Therefore, $A \cap (\mathbb{R} \setminus V)$ is closed. By Problem 44, $A \cap (\mathbb{R} \setminus V) = A \setminus V$ is compact. By Theorem 6.39, $f[A \setminus V]$ is compact. Once again, the Heine-Borel Theorem tells us that $f[A \setminus V]$ is closed. Since f is injective, $f[A \setminus V] = f[A] \setminus f[V]$ (Check this carefully). So, $\mathbb{R} \setminus (f[A] \setminus f[V])$ is open. We show $f[V] = \big(\mathbb{R} \setminus (f[A] \setminus f[V])\big) \cap f[A]$. To see this, note that $x \in \big(\mathbb{R} \setminus (f[A] \setminus f[V])\big) \cap f[A]$ if and only if $x \notin f[A] \setminus f[V]$ and $x \in f[A]$ if and only if we have $x \in f[V]$ or $x \notin f[A]$ as well as $x \in f[A]$ if and only if $x \in f[V] \cap f[A] = f[V]$. By Theorem 6.38, f^{-1} is continuous. □

LEVEL 5

Prove each of the following:

57. Every closed set A in $\mathbb{R}$ with $A \neq \mathbb{R}$ can be written as an intersection $\bigcap \boldsymbol{X}$, where each element of $\boldsymbol{X}$ is a union of two closed intervals.

Proof: Let A be a closed set in $\mathbb{R}$ with $A \neq \mathbb{R}$. Then $\mathbb{R} \setminus A$ is a nonempty open set in $\mathbb{R}$. By Theorem 6.17, $\mathbb{R} \setminus A$ can be expressed as $\bigcup \boldsymbol{X}$, where $\boldsymbol{X}$ is a set of bounded open intervals. For each B in $\boldsymbol{X}$, $\mathbb{R} \setminus B$ is a union of two closed intervals (if $B = (a, b)$, then $\mathbb{R} \setminus B = (-\infty, a] \cup [b, \infty)$). Now, by Problem 307 from Problem Set 2, we have $A = \mathbb{R} \setminus (\mathbb{R} \setminus A) = \mathbb{R} \setminus \bigcup \boldsymbol{X} = \bigcap\{\mathbb{R} \setminus B \mid B \in \boldsymbol{X}\}$. □

58. If X, Y, and Z are sets of real numbers with $Z = X \cup Y$, then $\overline{Z} = \overline{X} \cup \overline{Y}$.

Proof: Let X, Y, and Z be sets of real numbers with $Z = X \cup Y$. Let $x \in \overline{Z}$. By part 3 of Theorem 6.25, $x \in Z$ or x is an accumulation point of Z. If $x \in Z$, then $x \in X$ or $x \in Y$. By part 1 of Theorem 6.25, $X \subseteq \overline{X}$ and $Y \subseteq \overline{Y}$. So, $x \in \overline{X}$ or $x \in \overline{Y}$. Thus, $x \in \overline{X} \cup \overline{Y}$. Now, suppose that x is an accumulation point of Z. Without loss of generality, assume that x is **not** an accumulation point of Y. Then there is an open interval (c, d) with $x \in (c, d)$ and such that if $y \in Y$ with $y \neq x$, then $y \notin (c, d)$. Let (a, b) be any open interval containing x and let $U = (a, b) \cap (c, d)$. Since $x \in U$, U is a nonempty open interval. Since x is an accumulation point of Z, there is $y \in Z$ with $y \in U$ and $y \neq x$. Since $U \subseteq (c, d)$, $y \notin Y$. Therefore, $y \in X$. Since $U \subseteq (a, b)$, we have $y \in (a, b)$. It follows that x is an accumulation point of X. So, $x \in \overline{X}$. Therefore, $x \in \overline{X}$ or $x \in \overline{Y}$, and it follows that $x \in \overline{X} \cup \overline{Y}$. Since $x \in \overline{Z}$ was arbitrary, $\overline{Z} \subseteq \overline{X} \cup \overline{Y}$.

Conversely, let $x \in \overline{X} \cup \overline{Y}$. Then $x \in \overline{X}$ or $x \in \overline{Y}$. Without loss of generality, assume that $x \in \overline{X}$. By part 3 of Theorem 6.25, $x \in X$ or x is an accumulation point of X. If $x \in X$, then $x \in X$ or $x \in Y$. So, $x \in X \cup Y = Z$. By part 1 of Theorem 6.25, $Z \subseteq \overline{Z}$. So, $x \in \overline{Z}$. Now, suppose that x is an accumulation point of X and let (a, b) be any open interval containing x. Then there is $y \in X$ with $y \in (a, b)$ and $y \neq x$. So, $y \in X$ or $y \in Y$, and thus, $y \in X \cup Y = Z$. It follows that x is an accumulation point of Z. Therefore, $x \in \overline{Z}$. Since $x \in \overline{X} \cup \overline{Y}$ was arbitrary, $\overline{X} \cup \overline{Y} \subseteq \overline{Z}$.

Since $\overline{Z} \subseteq \overline{X} \cup \overline{Y}$ and $\overline{X} \cup \overline{Y} \subseteq \overline{Z}$, we have $\overline{Z} = \overline{X} \cup \overline{Y}$. □

59. If $A \subseteq \mathbb{R}$ is compact and $f: A \to \mathbb{R}$ is continuous on A, then f is uniformly continuous on A.

Proof: Let $A \subseteq \mathbb{R}$ be compact, let $f: A \to \mathbb{R}$ be continuous on A, and let $\epsilon > 0$. By the continuity of f, for each $a \in A$, there is a positive number δ_a such that $\forall b \in A\left(|a-b| < \delta_a \to |f(a) - f(b)| < \frac{\epsilon}{2}\right)$. For each $a \in A$, let $B_a = \left\{b \in A \,\middle|\, |a-b| < \frac{\delta_a}{2}\right\}$. Then $\mathcal{C} = \{B_a \,|\, a \in A\}$ is an open cover of A. Since A is compact, there is a finite subcover $\mathcal{D} \subseteq \mathcal{C}$. Let $\delta = \min\left\{\frac{\delta_a}{2} \,\middle|\, B_a \in \mathcal{D}\right\}$. Since $\mathcal{D}$ is finite, $\delta > 0$. Let $a, b \in A$ with $|a-b| < \delta$. Let $B_c \in \mathcal{D}$ with $a \in B_c$. Then $|a-c| < \frac{\delta_c}{2}$. Also,

$$|b-c| \leq |b-a+a-c| \leq |b-a| + |a-c| < \delta + \frac{\delta_c}{2} \leq \delta_c.$$

So, $|f(a) - f(b)| = |f(a) - f(c) + f(c) - f(b)| \leq |f(a) - f(c)| + |f(c) - f(b)| < \frac{\epsilon}{2} + \frac{\epsilon}{2} < \epsilon$. □

60. If $C \subseteq A \subseteq \mathbb{R}$ with C dense in A and $f: C \to \mathbb{R}$ is uniformly continuous, then f can be extended uniquely to a uniformly continuous function $g: A \to \mathbb{R}$. (If $X \subseteq Y$, then X is said to be dense in Y if $\overline{X} = Y$.)

Proof: Let $C \subseteq A \subseteq \mathbb{R}$ with C dense in A and let $f: C \to \mathbb{R}$ be uniformly continuous. If $C = A$, we can let $g = f$. So, assume that $C \neq A$. If $x \in C$, let $g(x) = f(x)$. Now, let $x \in A \setminus C$. Since C is dense in A, $\overline{C} = A$. Therefore, there is a sequence (x_n) in C such that $x_n \to x$. By Theorem 5.51, (x_n) is a Cauchy sequence in C. By Problem 53 above, $(f(x_n))$ is a Cauchy sequence in $\mathbb{R}$. By Theorem 5.53, $(f(x_n))$ converges to some point y in $\mathbb{R}$. Let $g(x) = y$. If (z_n) is another Cauchy sequence in C such that $z_n \to x$, then $|x_n - z_n| \to 0$. Since f is uniformly continuous, $|f(x_n) - f(z_n)| \to 0$. Therefore, $f(z_n) \to y = g(x)$.

We need to show that g is uniformly continuous. To this end, let $\epsilon > 0$. Since f is uniformly continuous, there is $\delta > 0$ such that for all $x, z \in C$, $|x-z| < \delta \to |f(x) - f(z)| < \frac{\epsilon}{2}$. Suppose that $a, b \in A$ with $|a-b| < \delta$. Let $(a_n), (b_n)$ be sequences in C such that $a_n \to a$ and $b_n \to b$. We then have that $|a_n - b_n| = |a_n - a + a - b + b - b_n| \leq |a_n - a| + |a-b| + |b - b_n|$. Since $|a_n - a| \to 0$, $|b - b_n| \to 0$, and $|a-b| < \delta$, we can find $K \in \mathbb{N}$ such that for all $n > K$, $|a_n - b_n| < \delta$. So, $n > K \to |f(a_n) - f(b_n)| < \frac{\epsilon}{2}$. From this it is easy to show that $|g(a) - g(b)| \leq \frac{\epsilon}{2} < \epsilon$.

The proof that g is unique is straightforward. □

A real number x is an **interior point** of a set S of real numbers if there is a neighborhood of x that contains only points in S, whereas x is a **boundary point** of S if each neighborhood of x contains at least one point in S and one point not in S. Prove each of the following:

61. A set of real numbers is open in $\mathbb{R}$ if and only if each point in the set is an interior point of the set.

Proof: Let S be a set of real numbers. Then S is open if and only if for every real number $x \in S$, there is an open interval (a, b) with $x \in (a, b)$ and $(a, b) \subseteq S$ if and only if for every real number $x \in S$, there is a neighborhood of x that contains only points in S if and only if every real number in S is an interior point of S. □

62. A set of real numbers is open in $\mathbb{R}$ if and only if it contains none of its boundary points.

Proof: Suppose that S is an open set of real numbers and let $x \in S$. By Problem 61, x is an interior point of S. So, there is a neighborhood of x containing only points of S. So, x is **not** a boundary point of S. Since $x \in S$ was arbitrary, S contains none of its boundary points.

We now prove that if S contains none of its boundary points, then S is open by contrapositive. Suppose S is not open. By Problem 61, there is $x \in S$ such that x is **not** an interior point. Let N be a neighborhood of x. Since $x \in S$, N contains a point in S (namely, x). Since x is not an interior point of S, N contains a point not in S. So, x is a boundary point of S. Thus, S contains at least one of its boundary points. □

63. A set of real numbers is closed in $\mathbb{R}$ if and only if it contains all its boundary points.

Proof: First note that a real number x is a boundary point of S if and only if x is a boundary point of $\mathbb{R} \setminus S$ (because $x \in S$ if and only if $x \notin \mathbb{R} \setminus S$, and vice versa).

Let S be a set of real numbers. Then S is closed if and only if $\mathbb{R} \setminus S$ is open if and only if $\mathbb{R} \setminus S$ contains none of its boundary points (by Problem 62) if and only if $S = \mathbb{R} \setminus (\mathbb{R} \setminus S)$ contains all its boundary points. □

Let C be the Cantor set. Prove each of the following:

64. $\frac{1}{4} \in C$.

Proof: Recall that $C = \bigcap\{C_n \mid n \in \mathbb{N}\}$. First note that if $x \in C_n$, then $\frac{1}{3}x \in C_{n+1}$. Also, note that $x \in C_n$ if and only if $1 - x \in C_n$. Now, we prove by induction that for all $n \in \mathbb{N}$, $\frac{1}{4} \in C_n$ and $\frac{3}{4} \in C_n$.

Base case ($k = 0$): Clearly $\frac{1}{4}, \frac{3}{4} \in C_0 = [0, 1]$.

Inductive step: Assume that $\frac{1}{4}, \frac{3}{4} \in C_k$. Since $\frac{3}{4} \in C_k$, $\frac{1}{4} = \frac{1}{3} \cdot \frac{3}{4} \in C_{k+1}$. By the symmetry of C_{k+1}, $\frac{3}{4} = 1 - \frac{1}{4} \in C_{k+1}$.

By the Principle of Mathematical Induction, for all $n \in \mathbb{N}$, $\frac{1}{4} \in C_n$ and $\frac{3}{4} \in C_n$. Therefore, $\frac{1}{4}, \frac{3}{4} \in \bigcap\{C_n \mid n \in \mathbb{N}\} = C$. □

65. C is uncountable.

Proof: We define an injection $f: [0, 1) \to C$ as follows. For $x \in [0, 1)$, let $x = 0.x_0x_1x_2 \ldots$ be the binary expansion of x (see Note 2 following Example 2.73). For each $n \in \mathbb{N}$, let $z_n = 2x_n$ and consider $z = 0.z_0z_1z_2 \ldots$ as the ternary expansion of a real number. Observe that for all $n \in \mathbb{N}$, $z_n = 0$ or $z_n = 2$. Since $z_0 \neq 1$, $z \notin \left[\frac{1}{3}, \frac{2}{3}\right)$. So $z \in C_1$. Since $z_1 \neq 1$, $z \notin \left[\frac{1}{9}, \frac{2}{9}\right)$ and $z \notin \left[\frac{7}{9}, \frac{8}{9}\right)$. So, $z \in C_2$. By a continuation of this reasoning, we see that for all $n \in \mathbb{N}$, $z \in C_n$. Therefore, $z \in \bigcap\{C_n \mid n \in \mathbb{N}\} = C$. So, if we let $f(x) = z$, then $f: [0, 1) \to C$. To see that f is injective, observe that if $x \neq y$, then for some $n \in \mathbb{N}$, $x_n \neq y_n$. Without loss of generality, assume that $x_n = 0$ and $y_n = 1$. So, if $f(x) = z$ and $f(y) = w$, then $z_n = 0$ and $w_n = 2$. It follows that $f(x) = y \neq w = f(y)$. □

Problem Set 7

LEVEL 1

Let $z = 1 + i$ and $w = 2 - 3i$. Compute each of the following:

1. Re z

1

2. Im z

1

3. Re w

2

4. Im w

– 3

5. $z + w$

$\mathbf{3 - 2i}$

6. $z - w$

$\mathbf{-1 + 4i}$

7. $2z$

$\mathbf{2 + 2i}$

8. $5w$

$\mathbf{10 - 15i}$

9. $2z + 5w$

$\mathbf{12 - 13i}$

10. $w - 5z$

$(2 - 3i) - 5(1 + i) = (2 - 3i) - (5 + 5i) = \mathbf{-3 - 8i}$

11. $\overline{z}$

$\mathbf{1 - i}$

12. $\overline{w}$

$\mathbf{2 + 3i}$

13. $\overline{z} + \overline{w}$

$\mathbf{3 + 2i}$

14. $\overline{z + w}$

$\mathbf{3 + 2i}$

LEVEL 2

Find all square roots of each of the following complex numbers:

15. 4

We are looking for a complex number $a + bi$ such that $(a + bi)^2 = 4$. In other words, we need $(a + bi)(a + bi) = (a^2 - b^2) + 2abi$ to equal $4 = 4 + 0i$. So, we have $a^2 - b^2 = 4$ and $2ab = 0$. The second equation tells us that $a = 0$ or $b = 0$.

If $a = 0$, then $-b^2 = 4$, or equivalently, $b^2 = -4$. Since the square of a real number must be positive, this is impossible.

If $b = 0$, then $a^2 = 4$, and so, $a = \pm 2$.

It follows that the two square roots of 4 are $\mathbf{2}$ and $-\mathbf{2}$.

16. $2i$

We are looking for a complex number $a + bi$ such that $(a + bi)^2 = 2i$. In other words, we need $(a + bi)(a + bi) = (a^2 - b^2) + 2abi$ to equal $2i = 0 + 2i$. So, we have $a^2 - b^2 = 0$ and $2ab = 2$, or equivalently, $ab = 1$. The first equation is equivalent to $(a + b)(a - b) = 0$, so that $a + b = 0$ or $a - b = 0$. It follows that $a = -b$ or $a = b$.

If $a = -b$, then $ab = (-b)b = -b^2$. So, $-b^2 = 1$, or equivalently, $b^2 = -1$. Since the square of a real number must be positive, this is impossible.

If $a = b$, then $ab = b^2$. So, $b^2 = 1$. Therefore, $b = \pm 1$. So, $a = \pm 1$.

It follows that the two square roots of i are $\mathbf{1 + i}$ and $-\mathbf{1 - i}$.

Let $z = -4 - i$ and $w = 3 - 5i$. Compute each of the following:

17. zw

$$zw = (-4 - i)(3 - 5i) = (-12 - 5) + (20 - 3)i = \mathbf{-17 + 17i}.$$

18. $\frac{z}{w}$

$$\frac{z}{w}=\frac{-4-i}{3-5i}=\frac{(-4-i)(3+5i)}{(3-5i)(3+5i)}=\frac{(-12+5)+(-20-3)i}{3^2+5^2}=\frac{-7-23i}{9+25}=-\mathbf{\frac{7}{34}}-\mathbf{\frac{23}{34}}\boldsymbol{i}.$$

19. $|z|$

$$|z|=\sqrt{(-4)^2+(-1)^2}=\sqrt{16+1}=\sqrt{\mathbf{17}}.$$

20. $|w|$

$$|w|=\sqrt{3^2+(-5)^2}=\sqrt{9+25}=\sqrt{\mathbf{34}}.$$

21. $|zw|$

$|zw|=|-17+17i|=\sqrt{(-17)^2+17^2}=\sqrt{17^2\cdot 2}=\mathbf{17\sqrt{2}}.$

22. the distance between z and w.

$$|z-w|=|(-4-i)-(3-5i)|=|(-4-3)+(-1+5)i|=|-7+4i|=\sqrt{(-7)^2+4^2}$$
$$=\sqrt{49+16}=\sqrt{\mathbf{65}}.$$

Each of the following complex numbers is written in exponential form. Rewrite each complex number in standard form:

23. $e^{\pi i}$

$e^{\pi i}=\cos\pi+i\sin\pi=-1+0i=-\mathbf{1}.$

24. $e^{-\frac{5\pi}{2}i}$

$e^{-\frac{5\pi}{2}i}=\cos\left(-\frac{5\pi}{2}\right)+i\sin\left(-\frac{5\pi}{2}\right)=\cos\frac{5\pi}{2}-i\sin\frac{5\pi}{2}=0-1i=-\boldsymbol{i}.$

25. $3e^{\frac{\pi}{4}i}$

$3e^{\frac{\pi}{4}i}=3\left(\cos\frac{\pi}{4}+i\sin\frac{\pi}{4}\right)=3\left(\frac{\sqrt{2}}{2}+\frac{\sqrt{2}}{2}i\right)=\mathbf{\frac{3\sqrt{2}}{2}}+\mathbf{\frac{3\sqrt{2}}{2}}\boldsymbol{i}.$

26. $2e^{\frac{\pi}{3}i}$

$2e^{\frac{\pi}{3}i}=2\left(\cos\frac{\pi}{3}+i\sin\frac{\pi}{3}\right)=2\left(\frac{1}{2}+\frac{\sqrt{3}}{2}i\right)=\mathbf{1}+\sqrt{\mathbf{3}}\boldsymbol{i}.$

27. $\sqrt{2}e^{\frac{7\pi}{6}i}$

$\sqrt{2}e^{\frac{7\pi}{6}i}=\sqrt{2}\left(\cos\frac{7\pi}{6}+i\sin\frac{7\pi}{6}\right)=\sqrt{2}\left(-\frac{\sqrt{3}}{2}-\frac{1}{2}i\right)=-\mathbf{\frac{\sqrt{6}}{2}}-\mathbf{\frac{\sqrt{2}}{2}}\boldsymbol{i}.$

28. $\pi e^{-\frac{5\pi}{4}i}$

$\pi e^{-\frac{5\pi}{4}i} = \pi\left(\cos\left(-\frac{5\pi}{4}\right) + i\sin\left(-\frac{5\pi}{4}\right)\right) = \pi\left(\cos\frac{5\pi}{4} - i\sin\frac{5\pi}{4}\right) = -\frac{\pi\sqrt{2}}{2} + \frac{\pi\sqrt{2}}{2}i.$

29. $e^{\frac{19\pi}{12}}$

$e^{\frac{19\pi}{12}} = \cos\frac{19\pi}{12} + i\sin\frac{19\pi}{12} = \frac{-\sqrt{2}+\sqrt{6}}{4} + \frac{-\sqrt{2}-\sqrt{6}}{4}i$ (see Problem 41 from Problem Set 5).

Each of the following complex numbers is written in standard form. Rewrite each complex number in exponential form:

30. $-1 - i$

$r^2 = (-1)^2 + (-1)^2 = 1 + 1 = 2$. So, $r = \sqrt{2}$. $\tan\theta = \frac{-1}{-1} = 1$. So, $\theta = \pi + \frac{\pi}{4} = \frac{5\pi}{4}$. Therefore, $-1 - i = \sqrt{2}e^{\frac{5\pi}{4}i} = \sqrt{2}e^{-\frac{3\pi}{4}i}$.

31. $\sqrt{3} + i$

$r^2 = \left(\sqrt{3}\right)^2 + 1^2 = 3 + 1 = 4$. So, $r = 2$. $\tan\theta = \frac{1}{\sqrt{3}}$. So, $\theta = \frac{\pi}{6}$. Therefore, we have $\sqrt{3} + i = 2e^{\frac{\pi}{6}i}$.

32. $1 - \sqrt{3}i$

$r^2 = 1^2 + \left(-\sqrt{3}\right)^2 = 1 + 3 = 4$. So, $r = 2$. $\tan\theta = \frac{-\sqrt{3}}{1}$. So, $\theta = -\frac{\pi}{3}$. Therefore, we have $1 - \sqrt{3}i = 2e^{-\frac{\pi}{3}i}$.

33. $\left(\frac{\sqrt{6}+\sqrt{2}}{4}\right) + \left(\frac{\sqrt{6}-\sqrt{2}}{4}\right)i$

$r^2 = \left(\frac{\sqrt{6}+\sqrt{2}}{4}\right)^2 + \left(\frac{\sqrt{6}-\sqrt{2}}{4}\right)^2 = \frac{6+2+2\sqrt{12}}{16} + \frac{6+2-2\sqrt{12}}{16} = \frac{16}{16} = 1$. So, $r = 1$. By Problem 39 from Problem Set 5, $\theta = \frac{\pi}{12}$. Therefore, $\left(\frac{\sqrt{6}+\sqrt{2}}{4}\right) + \left(\frac{\sqrt{6}-\sqrt{2}}{4}\right)i = 1e^{\frac{\pi}{12}i} = e^{\frac{\pi}{12}i}$.

Write the following complex numbers in standard form:

34. $\left(\frac{\sqrt{2}}{2} + \frac{\sqrt{2}}{2}i\right)^4$

If $z = \frac{\sqrt{2}}{2} + \frac{\sqrt{2}}{2}i$, then $r = \sqrt{\left(\frac{\sqrt{2}}{2}\right)^2 + \left(\frac{\sqrt{2}}{2}\right)^2} = \sqrt{\frac{2}{4} + \frac{2}{4}} = 1$ and $\tan\theta = \frac{\frac{\sqrt{2}}{2}}{\frac{\sqrt{2}}{2}} = 1$, so that $\theta = \frac{\pi}{4}$. So, in exponential form, $z = e^{\frac{\pi}{4}i}$. Therefore, $\left(\frac{\sqrt{2}}{2} + \frac{\sqrt{2}}{2}i\right)^4 = \left(e^{\frac{\pi}{4}i}\right)^4 = e^{\pi i} = -1$.

35. $(1+\sqrt{3}i)^5$

If $z = 1+\sqrt{3}i$, then $r = \sqrt{1^2 + (\sqrt{3})^2} = \sqrt{1+3} = \sqrt{4} = 2$ and $\tan\theta = \frac{\sqrt{3}}{1} = \sqrt{3}$, so that $\theta = \frac{\pi}{3}$. So, in exponential form, $z = 2e^{\frac{\pi}{3}i}$. So, $(1+\sqrt{3}i)^5 = \left(2e^{\frac{\pi}{3}i}\right)^5 = 2^5 e^{\frac{5\pi}{3}i} = \mathbf{16 - 16\sqrt{3}i}$.

LEVEL 3

Determine the accumulation points of each of the following sets:

36. $\left\{\frac{1}{n} \,\middle|\, n \in \mathbb{Z}^+\right\}$

$\mathbf{0}$ is the only accumulation point of this set.

37. $\left\{\frac{i}{n} \,\middle|\, n \in \mathbb{Z}^+\right\}$

$\mathbf{0}$ is the only accumulation point of this set.

38. $\{i^n \mid n \in \mathbb{Z}^+\}$

This set is equal to $\{1, -1, i, -i\}$. It has **no accumulation points**.

39. $\left\{\frac{i^n}{n} \,\middle|\, n \in \mathbb{Z}^+\right\}$

$\mathbf{0}$ is the only accumulation point of this set.

40. $\{z \mid |z| < 1\}$

The set of accumulation points of the set $\{z \mid |z| < 1\}$ is the set $\{\mathbf{z} \mid |\mathbf{z}| \leq \mathbf{1}\}$.

41. $\{z \mid 0 < |z-2| \leq 3\}$

The set of accumulation points of the set $\{z \mid 0 < |z-2| \leq 3\}$ is the set $\{z \mid |z-2| \leq 3\}$.

42. $\{a+bi \mid a, b \in \mathbb{Z}\}$

The set of accumulation points of the set $\{a+bi \mid a, b \in \mathbb{Z}\}$ is $\emptyset$.

43. $\{a+bi \mid a, b \in \mathbb{Q}\}$

The set of accumulation points of the set $\{a+bi \mid a, b \in \mathbb{Q}\}$ is $\mathbb{C}$, the entire set of complex numbers.

Use De Moivre's Theorem to prove the following identities:

44. $\cos 2\theta = \cos^2\theta - \sin^2\theta$

Proof: By De Moivre's Theorem, $\left(e^{i\theta}\right)^2 = e^{i(2\theta)}$, so that $(\cos\theta + i\sin\theta)^2 = \cos 2\theta + i\sin 2\theta$. Multiplying the left-hand side gives us

$$(\cos\theta + i\sin\theta)^2 = (\cos\theta + i\sin\theta)(\cos\theta + i\sin\theta) = \cos^2\theta - \sin^2\theta + i(2\sin\theta\cos\theta).$$

So, $\cos^2\theta - \sin^2\theta + i(2\sin\theta\cos\theta) = \cos 2\theta + i\sin 2\theta$. Equating the real parts of this equation gives us $\cos^2\theta - \sin^2\theta = \cos 2\theta$. □

45. $\sin 2\theta = 2\sin\theta\cos\theta$

Proof: By Problem 44, we have $\cos^2\theta - \sin^2\theta + i(2\sin\theta\cos\theta) = \cos 2\theta + i\sin 2\theta$. Equating the imaginary parts of this equation gives us $2\sin\theta\cos\theta = \sin 2\theta$. □

46. $\cos 3\theta = \cos^3\theta - 3\cos\theta\sin^2\theta$

Proof: By De Moivre's Theorem, $\left(e^{i\theta}\right)^3 = e^{i(3\theta)}$, so that $(\cos\theta + i\sin\theta)^3 = \cos 3\theta + i\sin 3\theta$. Multiplying the left-hand side and using the computation from Problem 44 gives us

$$(\cos\theta + i\sin\theta)^3 = \left(\cos^2\theta - \sin^2\theta + i(2\sin\theta\cos\theta)\right)(\cos\theta + i\sin\theta)$$
$$= (\cos^3\theta - \cos\theta\sin^2\theta - 2\cos\theta\sin^2\theta) + i(\cos^2\theta\sin\theta - \sin^3\theta + 2\cos^2\theta\sin\theta)$$
$$= (\cos^3\theta - 3\cos\theta\sin^2\theta) + i(3\cos^2\theta\sin\theta - \sin^3\theta)$$

So, $(\cos^3\theta - 3\cos\theta\sin^2\theta) + i(3\cos^2\theta\sin\theta - \sin^3\theta) = \cos 3\theta + i\sin 3\theta$. Equating the real parts of this equation gives us $\cos^3\theta - 3\cos\theta\sin^2\theta = \cos 3\theta$. □

Note: Equating imaginary parts gives us one more trigonometric identity:

$$\sin 3\theta = 3\cos^2\theta\sin\theta - \sin^3\theta$$

Suppose that $z = re^{i\theta}$ and $w = se^{i\phi}$ are complex numbers written in exponential form. Express each of the following in exponential form. Provide a proof in each case:

47. zw

$$zw = re^{i\theta}se^{i\phi} = rse^{i\theta}e^{i\phi} = rs(\cos\theta + i\sin\theta)(\cos\phi + i\sin\phi)$$
$$= rs[(\cos\theta\cos\phi - \sin\theta\sin\phi) + i(\sin\theta\cos\phi + \cos\theta\sin\phi)]$$
$$= rs[\cos(\theta+\phi) + i\sin(\theta+\phi)] = \boldsymbol{rse^{(\theta+\phi)i}}.$$

48. $\frac{z}{w}$

$$\frac{z}{w} = \frac{re^{i\theta}}{se^{i\phi}} = \frac{r}{s}\cdot\frac{e^{i\theta}e^{-i\phi}}{e^{i\phi}e^{-i\phi}} = \frac{r}{s}\cdot\frac{e^{i(\theta-\phi)}}{e^{i(\phi-\phi)}} = \boldsymbol{\frac{r}{s}e^{i(\theta-\phi)}}$$

Write each function in the form $f(z) = u(x, y) + iv(x, y)$ and $f(z) = u(r, \theta) + iv(r, \theta)$:

49. $f(z) = 2z^2 - 5$

$$f(z) = 2z^2 - 5 = 2(x + yi)^2 - 5 = 2(x^2 - y^2 + 2xyi) - 5 = \mathbf{(2x^2 - 2y^2 - 5) + 4xyi}$$

$$f(z) = 2z^2 - 5 = 2\left(re^{i\theta}\right)^2 - 5 = 2\left(r^2 e^{i(2\theta)}\right) - 5 = 2r^2 e^{i(2\theta)} - 5$$

$$= 2r^2(\cos 2\theta + i \sin 2\theta) - 5 = \mathbf{(2r^2 \cos(2\theta) - 5) + i(2r^2 \sin 2\theta)}$$

50. $f(z) = \frac{1}{z}$

$$f(z) = \frac{1}{z} = \frac{1}{z} \cdot \frac{\bar{z}}{\bar{z}} = \frac{\bar{z}}{z\bar{z}} = \frac{x - yi}{x^2 + y^2} = \frac{\boldsymbol{x}}{\boldsymbol{x^2 + y^2}} - \frac{\boldsymbol{y}}{\boldsymbol{x^2 + y^2}}\boldsymbol{i}$$

$$f(z) = \frac{1}{z} = \frac{1}{re^{i\theta}} = \frac{1}{r}e^{-i\theta} = \frac{1}{r}(\cos(-\theta) + i \sin(-\theta)) = \frac{1}{r}(\cos \theta - i \sin \theta)$$

$$= \boldsymbol{\frac{1}{r} \cos \theta + i\left(\frac{1}{r} \sin \theta\right)}$$

51. $f(z) = z^3 + z^2 + z + 1$

$$f(z) = z^3 + z^2 + z + 1 = (x + yi)(x^2 - y^2 + 2xyi) + (x^2 - y^2 + 2xyi) + (x + yi) + 1$$

$$= x^3 - xy^2 - 2xy^2 + (x^2y - y^3 + 2x^2y)i + (x^2 - y^2 + x + 1) + (2xy + y)i$$

$$= (x^3 - 3xy^2) + (3x^2y - y^3)i + (x^2 - y^2 + x + 1) + (2xy + y)i$$

$$= \mathbf{(x^3 - 3xy^2 + x^2 - y^2 + x + 1) + (3x^2y - y^3 + 2xy + y)i}$$

$$f(z) = z^3 + z^2 + z + 1 = \left(re^{i\theta}\right)^3 + \left(re^{i\theta}\right)^2 + re^{i\theta} + 1 = r^3 e^{i(3\theta)} + r^2 e^{i(2\theta)} + re^{i\theta} + 1$$

$$= r^3(\cos 3\theta + i \sin 3\theta) + r^2(\cos 2\theta + i \sin 2\theta) + r(\cos \theta + i \sin \theta) + 1$$

$$= \boldsymbol{(r^3 \cos(3\theta) + r^2 \cos(2\theta) + r \cos \theta + 1) + i(r^3 \sin(3\theta) + r^2 \sin(2\theta) + r \sin \theta)}$$

52. Let $f(z) = x^2 - y^2 - 2x + 2y(x + 1)i$. Rewrite $f(z)$ in terms of z.

$f(z) = x^2 - y^2 - 2x + 2y(x + 1)i = x^2 - y^2 + 2xyi - 2(x - yi) = z^2 - 2\bar{z}$.

Note: We can also do this formally by replacing $x = \text{Re } z$ by $\frac{z+\bar{z}}{2}$ and $y = \text{Im } z$ by $\frac{z-\bar{z}}{2i}$. In this case, we get the following:

$$f(z) = \left(\frac{z+\overline{z}}{2}\right)^2 - \left(\frac{z-\overline{z}}{2i}\right)^2 - 2\left(\frac{z+\overline{z}}{2}\right) + 2\left(\frac{z-\overline{z}}{2i}\right)\left(\frac{z+\overline{z}}{2}+1\right)i$$
$$= \frac{z^2+2z\overline{z}+\overline{z}^2}{4} - \frac{z^2-2z\overline{z}+\overline{z}^2}{-4} - (z+\overline{z}) + (z-\overline{z})\left(\frac{z+\overline{z}}{2}+1\right)$$
$$= \frac{z^2+2z\overline{z}+\overline{z}^2}{4} + \frac{z^2-2z\overline{z}+\overline{z}^2}{4} - z - \overline{z} + (z-\overline{z})\left(\frac{z+\overline{z}}{2}\right) + (z-\overline{z})$$
$$= \frac{2z^2+2\overline{z}^2}{4} - z - \overline{z} + \frac{1}{2}(z^2-\overline{z}^2) + z - \overline{z}$$
$$= \frac{1}{2}(z^2+\overline{z}^2) - z - \overline{z} + \frac{1}{2}(z^2-\overline{z}^2) + z - \overline{z}$$
$$= \frac{1}{2}(z^2+\overline{z}^2) + \frac{1}{2}(z^2-\overline{z}^2) - 2\overline{z}$$
$$= z^2 - 2\overline{z}.$$

Find all complex numbers that satisfy the given equation:

53. $z^6 - 1 = 0$

We are looking for the sixth roots of unity. So, $z = \sqrt[6]{1}e^{i\left(\frac{0}{6}+\frac{2k\pi}{6}\right)} = e^{\frac{k\pi}{3}i}$ for $k = 0, 1, 2, 3, 4, 5$. Substituting each of these values for k into the expression $e^{\frac{k\pi}{3}i}$ gives us the following 6 sixth roots of unity.

$$\mathbf{1}, \frac{\mathbf{1}}{\mathbf{2}} + \frac{\sqrt{\mathbf{3}}}{\mathbf{2}}\boldsymbol{i}, -\frac{\mathbf{1}}{\mathbf{2}} + \frac{\sqrt{\mathbf{3}}}{\mathbf{2}}\boldsymbol{i}, -\mathbf{1}, -\frac{\mathbf{1}}{\mathbf{2}} - \frac{\sqrt{\mathbf{3}}}{\mathbf{2}}\boldsymbol{i}, \frac{\mathbf{1}}{\mathbf{2}} - \frac{\sqrt{\mathbf{3}}}{\mathbf{2}}\boldsymbol{i}$$

54. $z^4 + 4 = 0$

We are looking for the fourth roots of -4. So, $z = \sqrt[4]{4}e^{i\left(\frac{\pi}{4}+\frac{2k\pi}{4}\right)} = \sqrt{2}e^{\frac{(2k+1)\pi}{4}i}$ for $k = 0, 1, 2, 3$. Substituting each of these values for k into the expression $e^{\frac{k\pi}{3}i}$ gives us the following 4 fourth roots of -4.

$$\mathbf{1} + \boldsymbol{i}, -\mathbf{1} + \boldsymbol{i}, -\mathbf{1} - \boldsymbol{i}, \mathbf{1} - \boldsymbol{i}$$

Let z and w be complex numbers. Prove each of the following:

55. $\operatorname{Re} z = \frac{z+\overline{z}}{2}$

Proof: Let $z = a + bi$. Then $\overline{z} = a - bi$, and so, $\frac{z+\overline{z}}{2} = \frac{(a+bi)+(a-bi)}{2} = \frac{2a+bi-bi}{2} = \frac{2a}{2} = a = \operatorname{Re} z$. □

56. $\operatorname{Im} z = \frac{z-\overline{z}}{2i}$

Proof: Let $z = a + bi$. Then $\frac{z-\overline{z}}{2i} = \frac{(a+bi)-(a-bi)}{2i} = \frac{a-a+bi+bi}{2i} = \frac{2bi}{2i} = b = \operatorname{Im} z$. □

57. $\overline{z+w} = \overline{z} + \overline{w}$

Proof: Let $z = a + bi$ and $w = c + di$. Then we have

$$\overline{z+w} = \overline{(a+bi)+(c+di)} = \overline{(a+c)+(b+d)i} = (a+c)-(b+d)i$$
$$= (a-bi)+(c-di) = \overline{a+bi} + \overline{c+di} = \overline{z} + \overline{w}.$$ □

58. $\overline{zw} = \overline{z} \cdot \overline{w}$

Proof: Let $z = a + bi$ and $w = c + di$. Then we have

$$\overline{zw} = \overline{(a+bi)(c+di)} = \overline{(ac-bd)+(ad+bc)i} = (ac-bd)-(ad+bc)i$$
$$= (a-bi)(c-di) = \left(\overline{a+bi}\right)\left(\overline{c+di}\right) = \overline{z} \cdot \overline{w}.$$ □

59. $\overline{\left(\frac{z}{w}\right)} = \frac{\overline{z}}{\overline{w}}$

Proof: Let $z = a + bi$ and $w = c + di$. Then we have

$$\overline{\left(\frac{z}{w}\right)} = \overline{\frac{(a+bi)}{(c+di)}} = \overline{\frac{(a+bi)}{(c+di)} \cdot \frac{(c-di)}{(c-di)}} = \overline{\frac{(a+bi)(c-di)}{(c+di)(c-di)}} = \overline{\frac{(ac+bd)+(-ad+bc)i}{c^2+d^2}}$$
$$= \overline{\frac{(ac+bd)}{c^2+d^2} + \frac{(-ad+bc)}{c^2+d^2}i} = \frac{(ac+bd)}{c^2+d^2} - \frac{(-ad+bc)}{c^2+d^2}i = \frac{(ac+bd)+(ad-bc)i}{c^2+d^2}$$
$$= \frac{(a-bi)(c+di)}{(c-di)(c+di)} = \frac{(a-bi)}{(c-di)} \cdot \frac{(c+di)}{(c+di)} = \frac{a-bi}{c-di} = \frac{\overline{a+bi}}{\overline{c+di}} = \frac{\overline{z}}{\overline{w}}.$$ □

60. $z\overline{z} = |z|^2$

Proof: Let $z = a + bi$. Then $z\overline{z} = (a+bi)(a-bi) = a^2 + b^2 = |z|^2$. □

61. $|zw| = |z||w|$

Proof: $|zw|^2 = (zw)(\overline{zw}) = (zw)(\overline{z} \cdot \overline{w}) = z\overline{z}w\overline{w} = |z|^2|w|^2 = (|z||w|)^2$. Since $|z|$, $|w|$, and $|zw|$ are nonnegative, $|zw| = |z||w|$. □

62. If $w \neq 0$, then $\left|\frac{z}{w}\right| = \frac{|z|}{|w|}$

Proof: $\left|\frac{z}{w}\right|^2 = \left(\frac{z}{w}\right)\overline{\left(\frac{z}{w}\right)} = \left(\frac{z}{w}\right)\left(\frac{\overline{z}}{\overline{w}}\right) = \frac{z\overline{z}}{w\overline{w}} = \frac{|z|^2}{|w|^2} = \left(\frac{|z|}{|w|}\right)^2$. Since $|z|$, $|w|$, and $|zw|$ are nonnegative, $\left|\frac{z}{w}\right| = \frac{|z|}{|w|}$.

63. $\text{Re } z \leq |\text{Re } z| \leq |z|$

Proof: We have $|z|^2 = (\text{Re } z)^2 + (\text{Im } z)^2$. Therefore, $(\text{Re } z)^2 \leq |z|^2$, and so, $|\text{Re } z| \leq |z|$. It follows that $\text{Re } z \leq |\text{Re } z| \leq |z|$. □

64. $\text{Im } z \leq |\text{Im } z| \leq |z|$

Proof: We have $|z|^2 = (\text{Re } z)^2 + (\text{Im } z)^2$. Therefore, $(\text{Im } z)^2 \leq |z|^2$, and so, $|\text{Im } z| \leq |z|$. It follows that $\text{Im } z \leq |\text{Im } z| \leq |z|$. □

65. $|z + w| \leq |z| + |w|$ (This is the **Triangle Inequality**.)

$$|z+w|^2 = (z+w)\left(\overline{z+w}\right) = (z+w)(\overline{z}+\overline{w}) = z\overline{z} + z\overline{w} + w\overline{z} + w\overline{w}$$

$$= z\overline{z} + z\overline{w} + \overline{z\overline{w}} + w\overline{w} = z\overline{z} + 2\text{Re}(z\overline{w}) + w\overline{w} \leq z\overline{z} + 2|z\overline{w}| + w\overline{w} = z\overline{z} + 2|z||\overline{w}| + w\overline{w}$$

$$= |z|^2 + 2|z||w| + |w|^2 = (|z| + |w|)^2.$$

Since $|z + w|$ and $|z| + |w|$ are nonnegative, $|z + w| \leq |z| + |w|$. □

66. $\big||z| - |w|\big| \leq |z \pm w| \leq |z| + |w|$.

Proof: $|z| = |(z + w) + (-w)| \leq |z + w| + |-w| = |z + w| + |w|$. So, $|z + w| \geq |z| - |w|$.

$|w| = |(z + w) + (-z)| \leq |z + w| + |-z| = |z + w| + |z|$. So, $|z + w| \geq |w| - |z| = -(|z| - |w|)$.

Since for all $w, z \in \mathbb{C}$, we have $\big||z| - |w|\big| = |z| - |w|$ or $\big||z| - |w|\big| = -(|z| - |w|)$, it follows that $\big||z| - |w|\big| \leq |z + w|$.

Combining this result with the Triangle Inequality (Problem 65 above), gives us

$$\big||z| - |w|\big| \leq |z + w| \leq |z| + |w|.$$

Now, by the Triangle Inequality we have $|z - w| = |z + (-w)| \leq |z| + |-w| = |z| + |w|$.

Finally, by the third paragraph, we have $|z - w| = |z + (-w)| \geq \big||z| - |-w|\big| = \big||z| - |w|\big|$. □

LEVEL 4

Let $c = 3 - i$ and $r = \sqrt{5}$. Use set-builder notation to describe each of the following:

67. The circle with center c and radius r.

$\{\boldsymbol{z} \in \mathbb{C} \mid |\boldsymbol{z} - (\mathbf{3} - \boldsymbol{i})| = \sqrt{\mathbf{5}}\}$ or $\{\boldsymbol{x} + \boldsymbol{yi} \in \mathbb{C} \mid (\boldsymbol{x} - \mathbf{3})^2 + (\boldsymbol{y} + \mathbf{1})^2 = \mathbf{5}\}$

68. The open disk with center c and radius r.

$\{\boldsymbol{z} \in \mathbb{C} \mid |\boldsymbol{z} - (\mathbf{3} - \boldsymbol{i})| < \sqrt{\mathbf{5}}\}$ or $\{\boldsymbol{x} + \boldsymbol{yi} \in \mathbb{C} \mid (\boldsymbol{x} - \mathbf{3})^2 + (\boldsymbol{y} + \mathbf{1})^2 < \mathbf{5}\}$

69. The closed disk with center c and radius r.

$\{\boldsymbol{z} \in \mathbb{C} \mid |\boldsymbol{z} - (\mathbf{3} - \boldsymbol{i})| \leq \sqrt{\mathbf{5}}\}$ or $\{\boldsymbol{x} + \boldsymbol{yi} \in \mathbb{C} \mid (\boldsymbol{x} - \mathbf{3})^2 + (\boldsymbol{y} + \mathbf{1})^2 \leq \mathbf{5}\}$

70. The punctured open disk with center c and radius r.

$\{\boldsymbol{z} \in \mathbb{C} \mid \boldsymbol{0} < |\boldsymbol{z} - (\boldsymbol{3} - \boldsymbol{i})| < \sqrt{\boldsymbol{5}}\}$ or $\{\boldsymbol{x} + \boldsymbol{y}\boldsymbol{i} \in \mathbb{C} \mid \boldsymbol{0} < (\boldsymbol{x} - \boldsymbol{3})^2 + (\boldsymbol{y} + \boldsymbol{1})^2 < \boldsymbol{5}\}$

Determine if each of the following sets is open, closed, both, or neither. Give a proof in each case.

71. $\emptyset$

$\emptyset$ is **open and closed**. The statement that $\emptyset$ is open is vacuously true (since $\emptyset$ has no elements, there is nothing to check). $\emptyset$ is closed because $\mathbb{C} \setminus \emptyset = \mathbb{C}$ is open (proof below). □

72. $\mathbb{C}$

$\mathbb{C}$ is **open and closed**. $\mathbb{C}$ is closed because $\mathbb{C} \setminus \mathbb{C} = \emptyset$ is open (see Problem 71). To see that $\mathbb{C}$ is open. Let $a \in \mathbb{C}$, and let $D = \{z \in \mathbb{C} \mid |z - a| < 1\}$ be the open disk with center a and radius 1. Since $|a - a| = 0 < 1$, $a \in D$, and since every element of D is a complex number, $D \subseteq \mathbb{C}$. It follows that $\mathbb{C}$ is open. □

73. $\{z \in \mathbb{C} \mid |z| > 1\}$

$S = \{z \in \mathbb{C} \mid |z| > 1\}$ is **open**. To see this, let $w \in S$ and let $r = |w| - 1$. We will show that $N_r(w) \subseteq S$ (recall that $N_r(w) = \{z \in \mathbb{C} \mid |z - w| < r\}$). Let $z \in N_r(w)$. Then we have $|z - w| < r = |w| - 1$. So, using the Triangle Inequality, we have

$$|w| = |(w - z) + z| \leq |w - z| + |z| = |z - w| + |z| < |w| - 1 + |z|.$$

Thus, $|w| < |w| - 1 + |z|$, and therefore, $|z| > 1$. So, $z \in S$. Since $z \in N_r(w)$ was arbitrary, we have shown that $N_r(w) \subseteq S$. So, S is open. □

$S = \{z \in \mathbb{C} \mid |z| > 1\}$ is **not closed** because $\mathbb{C} \setminus S = \{z \in \mathbb{C} \mid |z| \leq 1\}$ is not open.

74. $\{z \in \mathbb{C} \mid \operatorname{Im} z \leq -2\}$

$S = \{z \in \mathbb{C} \mid \operatorname{Im} z \leq -2\}$ is **not open**. To see this, first note that $\operatorname{Im}(-2i) = -2$, and so, $-2i \in S$. If $N_r(-2i)$ is an r-neighborhood of $-2i$, then $\left|\frac{-4+r}{2}i - (-2i)\right| = \left|\left(\frac{-4+r}{2} + \frac{4}{2}\right)i\right| = \left|\frac{r}{2}i\right| = \frac{r}{2} < r$. Therefore, $\frac{-4+r}{2}i \in N_r(-2i)$. Now, $\frac{-4+r}{2}i \notin S$ because $\operatorname{Im}\left(\frac{-4+r}{2}i\right) = \frac{-4+r}{2}$, and $\frac{-4+r}{2} > -\frac{4}{2} = -2$. So, for all $r > 0$, $N_r(-2i) \nsubseteq S$, showing that S is not open. □

S is **closed**. To see this, we show that $T = \mathbb{C} \setminus S = \{z \in \mathbb{C} \mid \operatorname{Im} z > -2\}$ is open.

Let $w \in T$ and let $r = 2 + \operatorname{Im} w$. We will show that $N_r(w) \subseteq T$. Let $z \in N_r(w)$. Then we have $\operatorname{Im} w - \operatorname{Im} z = \operatorname{Im}(w - z) \leq |w - z| = |z - w| < r = 2 + \operatorname{Im} w$. So, $-\operatorname{Im} z < 2$, and therefore, $\operatorname{Im} z > -2$. So, $z \in T$. Since $z \in N_r(w)$ was arbitrary, we have shown that $N_r(w) \subseteq T$. So, T is open, and therefore, $S = \mathbb{C} \setminus T$ is closed. □

75. $\{i^n \mid n \in \mathbb{Z}^+\}$

Note that $S = \{i^n \mid n \in \mathbb{Z}^+\}$ is a finite set consisting of just four complex numbers. Indeed, $S = \{1, -1, i, -i\}$.

S is **not open**. To see this, let $N_r(i)$ be an arbitrary r-neighborhood of i. Then $i + \frac{r}{2} \in N_r(i)$ because $\left|\left(i + \frac{r}{2}\right) - i\right| = \left|\frac{r}{2}\right| = \frac{r}{2} < r$, but $i + \frac{r}{2} \notin S$ because $i + \frac{r}{2}$ is not equal to $1, -1, i$, or $-i$. □

S is **closed**. To see this, we show that $T = \mathbb{C} \setminus S$ is open.

Let $w \in T$ and let $r = \min\{|w-1|, |w+1|, |w-i|, |w+i|\}$.

We now show that $N_r(w) \subseteq T$. Since $r \leq |w-1|$, $1 \notin N_r(w)$ (otherwise, $|w-1| < r$). Similarly, $-1, i$, and $-i \notin N_r(w)$. So, if $z \in N_r(w)$, then $z \notin S$, and so, $z \in T$.

Therefore, $\mathbb{C} \setminus S$ is open, and so, S is closed. □

76. $\{z \in \mathbb{C} \mid 2 < |z-2| < 4\}$

$S = \{z \in \mathbb{C} \mid 2 < |z-2| < 4\}$ is **open** and **not closed**.

To see that S is open, let $z \in S$ and let $r = \min\{|z-2|-2, 4-|z-2|\}$. We show that $D_r(z) \subseteq S$. Let $w \in D_r(z)$. Then $|w-z| < r$. So, $|w-z| < 4 - |z-2|$. So, we have $|w-z| + |z-2| < 4$, and so, $|w-2| = |w-z+z-2| \leq |w-z| + |z-2| < 4$.

Also, $|w-z| < |z-2| - 2$. So, we have

$$2 < |z-2| - |w-z| = |z-w+w-2| - |w-z| \leq |z-w| + |w-2| - |w-z|$$
$$= |w-z| + |w-2| - |w-z| = |w-2|.$$

So, $2 < |w-2| < 4$, and therefore, $w \in S$.

Since $z \in S$ was arbitrary, S is open.

To see that S is not closed, we show that $\mathbb{C} \setminus S = \{z \in \mathbb{C} \mid |z-2| \leq 2 \text{ or } |z-2| \geq 4\}$ is not open. To see this, first note that $|6-2| = |4| = 4 \geq 4$, and so, $6 \in \mathbb{C} \setminus S$. Let $N_r(6)$ be an r-neighborhood of 6 and let $k = \min\left\{1, \frac{r}{2}\right\}$. Then we have $6 - k \in N_r(6)$ because $|(6-k)-6| = |-k| = k \leq \frac{r}{2} < r$. However, $6 - k \notin \mathbb{C} \setminus S$. To see this, first observe that $|(6-k)-2| = |4-k|$. If $k = 1$, then $|4-k| = |4-1| = |3| = 3$ and it follows that $6-k \notin \mathbb{C} \setminus S$. If $k = \frac{r}{2}$, then $0 < \frac{r}{2} \leq 1$, so that $-1 \leq -\frac{r}{2} < 0$, and thus, $3 < 4 - \frac{r}{2} < 4$. So, $3 < \left|4 - \frac{r}{2}\right| < 4$ and once again, $6 - k \notin \mathbb{C} \setminus S$. So, $\mathbb{C} \setminus S$ is not open. Therefore, S is not closed. □

Prove each of the following:

77. An arbitrary union of open sets in $\mathbb{C}$ is an open set in $\mathbb{C}$.

Proof: Let $\boldsymbol{X}$ be a set of open subsets of $\mathbb{C}$ and let $z \in \bigcup\boldsymbol{X}$. Then $z \in A$ for some $A \in \boldsymbol{X}$. Since A is open in $\mathbb{C}$, there is an open disk D with $z \in D$ and $D \subseteq A$. By Problem 303 from Problem Set 2, $A \subseteq \bigcup\boldsymbol{X}$. Since $\subseteq$ is transitive (Theorem 2.15), $D \subseteq \bigcup\boldsymbol{X}$. Therefore, $\bigcup\boldsymbol{X}$ is open. □

78. A finite intersection of open sets in $\mathbb{C}$ is an open set in $\mathbb{C}$.

Proof: Let $\boldsymbol{X}$ be a finite set of open sets in $\mathbb{C}$. If $\bigcap \boldsymbol{X} = \emptyset$, then $\bigcap \boldsymbol{X}$ is open by the proof of Problem 71. Otherwise, let $z \in \bigcap \boldsymbol{X}$. By Theorem 7.18, for each A in $\boldsymbol{X}$, there is an open disk D_A with center z and radius r_A such that $z \in D_A$ and $D_A \subseteq A$. Let $r = \min\{r_A \mid A \in \boldsymbol{X}\}$ and let D be the open disk with center z and radius r. Since $D = D_A$ for some $A \in \boldsymbol{X}$, $z \in D$. Let $w \in D$ and let $B \in \boldsymbol{X}$. Then $|z - w| < r \leq r_B$. So, $w \in D_B$. Since $B \in \boldsymbol{X}$ was arbitrary, $w \in \bigcap \boldsymbol{X}$. Therefore, $D \subseteq \bigcap \boldsymbol{X}$, and so, $\bigcap \boldsymbol{X}$ is open. □

79. An arbitrary intersection of closed sets in $\mathbb{C}$ is a closed set in $\mathbb{C}$.

Proof: Let $\boldsymbol{X}$ be a nonempty set of closed sets in $\mathbb{C}$. Then for each $A \in \boldsymbol{X}$, $\mathbb{C} \setminus A$ is an open set in $\mathbb{C}$. By Problem 77, $\bigcup\{\mathbb{C} \setminus A \mid A \in \boldsymbol{X}\}$ is open in $\mathbb{C}$. Therefore, $\mathbb{C} \setminus \bigcup\{\mathbb{C} \setminus A \mid A \in \boldsymbol{X}\}$ is closed in $\mathbb{C}$. So, it suffices to show that $\bigcap \boldsymbol{X} = \mathbb{C} \setminus \bigcup\{\mathbb{C} \setminus A \mid A \in \boldsymbol{X}\}$. Well, $x \in \bigcap \boldsymbol{X}$ if and only if for all $A \in \boldsymbol{X}$, $x \in A$ if and only if for all $A \in \boldsymbol{X}$, $x \notin \mathbb{C} \setminus A$ if and only if $x \notin \bigcup\{\mathbb{C} \setminus A \mid A \in \boldsymbol{X}\}$ if and only if $x \in \mathbb{C} \setminus \bigcup\{\mathbb{C} \setminus A \mid A \in \boldsymbol{X}\}$. So, $\bigcap \boldsymbol{X} = \mathbb{C} \setminus \bigcup\{\mathbb{C} \setminus A \mid A \in \boldsymbol{X}\}$, completing the proof. □

80. A finite union of closed sets in $\mathbb{C}$ is a closed set in $\mathbb{C}$.

Proof: Let $\boldsymbol{X}$ be a finite set of closed subsets of $\mathbb{C}$. Then for each $A \in \boldsymbol{X}$, $\mathbb{C} \setminus A$ is an open set in $\mathbb{C}$. By Problem 78, $\bigcap\{\mathbb{C} \setminus A \mid A \in \boldsymbol{X}\}$ is open in $\mathbb{C}$. Therefore, $\mathbb{C} \setminus \bigcap\{\mathbb{C} \setminus A \mid A \in \boldsymbol{X}\}$ is closed in $\mathbb{C}$. So, it suffices to show that $\bigcup \boldsymbol{X} = \mathbb{C} \setminus \bigcap\{\mathbb{C} \setminus A \mid A \in \boldsymbol{X}\}$. Well, $x \in \bigcup \boldsymbol{X}$ if and only if there is an $A \in \boldsymbol{X}$ such that $x \in A$ if and only if there is an $A \in \boldsymbol{X}$ such that $x \notin \mathbb{C} \setminus A$ if and only if $x \notin \bigcap\{\mathbb{C} \setminus A \mid A \in \boldsymbol{X}\}$ if and only if $x \in \mathbb{C} \setminus \bigcap\{\mathbb{C} \setminus A \mid A \in \boldsymbol{X}\}$. Therefore, we have $\bigcup \boldsymbol{X} = \mathbb{C} \setminus \bigcap\{\mathbb{C} \setminus A \mid A \in \boldsymbol{X}\}$, completing the proof. □

81. Every open set in $\mathbb{C}$ can be expressed as a union of open disks.

Proof: Let X be an open set in $\mathbb{C}$. Since X is open, for each $z \in X$, there is an open disk D_z with $z \in D_z$ and $D_z \subseteq X$. Let $Y = \{D_z \mid z \in X\}$. We will show that $X = \bigcup Y$.

First, let $z \in X$. Then $z \in D_z$. Since $D_z \in Y$, $z \in \bigcup Y$. Since $z \in X$ was arbitrary, $X \subseteq \bigcup Y$.

Now, let $z \in \bigcup Y$. Then there is $w \in X$ with $z \in D_w$. Since $D_w \subseteq X$, $z \in X$. Since $z \in \bigcup Y$ was arbitrary, $\bigcup Y \subseteq X$.

Since $X \subseteq \bigcup Y$ and $\bigcup Y \subseteq X$, it follows that $X = \bigcup Y$. □

82. Let $z, w \in \mathbb{C}$. Then $\arg zw = \arg z + \arg w$ in the sense that if two of the three terms in the equation are specified, then there is a value for the third term so that the equation holds. Provide an example to show that the corresponding equation is false if we replace "arg" by "Arg."

Proof: Let θ and ϕ be any values of $\arg z$ and $\arg w$, respectively. Then there are positive real numbers r and s such that $z = re^{i\theta}$ and $w = se^{i\phi}$. By the solution to Problem 47, $zw = rse^{i(\theta+\phi)}$. So, $\theta + \phi$ is a value of $\arg zw$. Next, without loss of generality, choose values for $\arg zw$ and $\arg z$, so that for some $j, k \in \mathbb{Z}$, $\arg zw = (\theta + \phi) + 2j\pi$ and $\arg z = \theta + 2k\pi$. If we use $\phi + 2(j - k)\pi$ for $\arg w$, we have

$$\arg zw = (\theta + \phi) + 2j\pi = (\theta + 2k\pi) + (\phi + 2(j - k)\pi) = \arg z + \arg w.$$

To see that the equation $\operatorname{Arg} zw = \operatorname{Arg} z + \operatorname{Arg} w$ can be false, let $z = e^{\frac{3\pi}{4}i}$ and $w = e^{\frac{\pi}{2}i}$. Then we have $zw = e^{\frac{3\pi}{4}i}e^{\frac{\pi}{2}i} = e^{\left(\frac{3\pi}{4}+\frac{\pi}{2}\right)i} = e^{\frac{5\pi}{4}i} = e^{-\frac{3\pi}{4}i}$. So, $\operatorname{Arg} zw = -\frac{3\pi}{4}$, whereas $\operatorname{Arg} z + \operatorname{Arg} w = \frac{3\pi}{4} + \frac{\pi}{2} = \frac{5\pi}{4}$. □

83. Let $z, w \in \mathbb{C}$. Then $\arg\frac{z}{w} = \arg z - \arg w$ in the sense that if two of the three terms in the equation are specified, then there is a value for the third term so that the equation holds. Provide an example to show that the corresponding equation is false if we replace "arg" by "Arg."

Proof: Let θ and ϕ be any values of $\arg z$ and $\arg w$, respectively. Then there are positive real numbers r and s such that $z = re^{i\theta}$ and $w = se^{i\phi}$. By the solution to Problem 48, $\frac{z}{w} = \frac{r}{s}e^{i(\theta-\phi)}$. So, $\theta - \phi$ is a value of $\arg\frac{z}{w}$. Next, without loss of generality, choose values for $\arg\frac{z}{w}$ and $\arg z$, so that for some $j, k \in \mathbb{Z}$, $\arg\frac{z}{w} = (\theta - \phi) + 2j\pi$ and $\arg z = \theta + 2k\pi$. If we use $\phi + 2(k-j)\pi$ for $\arg w$, we have

$$\arg\frac{z}{w} = (\theta - \phi) + 2j\pi = (\theta + 2k\pi) - (\phi + 2(k-j)\pi) = \arg z - \arg w.$$

To see that the equation $\operatorname{Arg}\frac{z}{w} = \operatorname{Arg} z - \operatorname{Arg} w$ can be false, let $z = e^{\frac{3\pi}{4}i}$ and $w = e^{-\frac{\pi}{2}i}$. Then we have $\frac{z}{w} = \frac{e^{\frac{3\pi}{4}i}}{e^{-\frac{\pi}{2}i}} = e^{\left(\frac{3\pi}{4}+\frac{\pi}{2}\right)i} = e^{\frac{5\pi}{4}i} = e^{-\frac{3\pi}{4}i}$. So, $\operatorname{Arg}\frac{z}{w} = -\frac{3\pi}{4}$, whereas $\operatorname{Arg} z - \operatorname{Arg} w = \frac{3\pi}{4} - \left(-\frac{\pi}{2}\right) = \frac{5\pi}{4}$. □

LEVEL 5

Define the function $f: \mathbb{C} \to \mathbb{C}$ by $f(z) = z^2$. Determine the images under f of each of the following sets:

84. $A = \{x + yi \mid x^2 - y^2 = 1\}$

$f(z) = z^2 = x^2 - y^2 + 2xyi$. So, $u(x,y) = x^2 - y^2$ and $v(x,y) = 2xy$. When $x^2 - y^2 = 1$, $u(x,y) = 1$. It follows that $f[A] \subseteq \{u + vi \mid u = 1\}$. Now, $x^2 - y^2 = 1$ is equivalent to $x^2 = y^2 + 1$ or $x = \pm\sqrt{y^2+1}$. When $x = \sqrt{y^2+1}$, $v(x,y) = 2y\sqrt{y^2+1}$. v is a continuous function with $\lim_{y\to-\infty} v(x,y) = -\infty$ and $\lim_{y\to+\infty} v(x,y) = +\infty$. It follows that the image of A under the function f defined by $f(z) = z^2$ is the entire vertical line $u = 1$. In other words, we have $f[A] = \{u + vi \mid u = 1\}$.

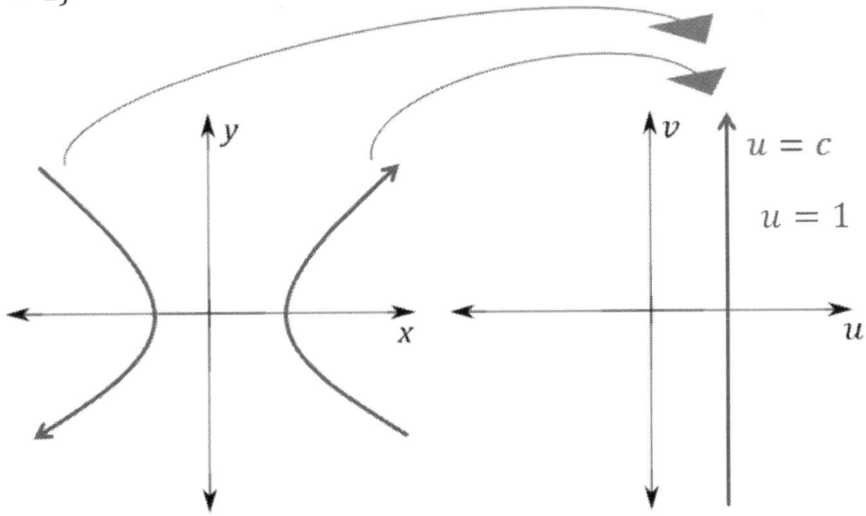

85. $B = \{x + yi \mid x > 0 \wedge y > 0 \wedge xy < 1\}$

As in Problem 84, we let $f(z) = z^2 = x^2 - y^2 + 2xyi$. So, $u(x,y) = x^2 - y^2$ and $v(x,y) = 2xy$. When $0 < xy < 1$, $0 < v(x,y) < 2$. It follows that $f[B] \subseteq \{u + vi \mid 0 < v < 2\}$. Let's choose an arbitrary but specific real number a between 0 and 2 and consider $v(x,y) = a$. Then we have $2xy = a$, or equivalently, $y = \frac{a}{2x}$. So, $u(x,y) = x^2 - \left(\frac{a}{2x}\right)^2 = x^2 - \frac{a^2}{4x^2}$. Now, u is continuous on $(0, +\infty)$ with $\lim\limits_{x \to 0^+} u(x,y) = -\infty$ and $\lim\limits_{x \to +\infty} u(x,y) = +\infty$. It follows that the image of B under the function f defined by $f(z) = z^2$ is the entire horizontal strip $0 < v < 2$. In other words, we have $f[B] = \{u + vi \mid 0 < v < 2\}$.

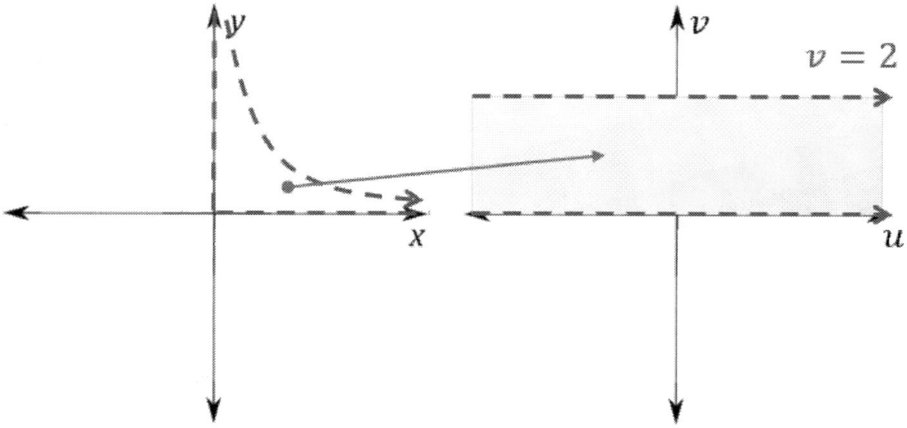

86. $C = \{x + yi \mid x \geq 0 \wedge y \geq 0\}$

As we saw in part 4 of Example 7.24, $f(z) = r^2 e^{i(2\theta)} = r^2(\cos 2\theta + i \sin 2\theta)$. So, $u(r,\theta) = r^2 \cos 2\theta$ and $v(r,\theta) = r^2 \sin 2\theta$. Now, if $x \geq 0$ and $y \geq 0$, then $0 \leq \theta \leq \frac{\pi}{2}$. It follows that $0 \leq 2\theta \leq \pi$. So, $0 \leq \sin 2\theta \leq 1$, and therefore, $0 \leq r^2 \sin 2\theta \leq r^2$ as r ranges over all possible nonnegative real numbers. Thus, $v \geq 0$. So, $f[C] \subseteq \{u + vi \mid v \geq 0\}$. Now, if $u + vi$ is an arbitrary complex number with $v \geq 0$, let $u = r\cos\phi$ and $v = r \sin \phi$ with $r \geq 0$ and $0 \leq \phi \leq \pi$. Let $z = \sqrt{r} e^{\frac{\phi}{2} i}$. Then $z \in C$ and $f(z) = re^{\phi i}$. It follows that the image under C is the entire half plane $v \geq 0$. In other words, we have $f[C] = \{u + vi \mid v \geq 0\}$.

87. $D = \{x + yi \mid y \geq 0\}$

As in problem 86, we have $f(z) = r^2 e^{i(2\theta)} = r^2(\cos 2\theta + i \sin 2\theta)$, so that $u(r,\theta) = r^2 \cos 2\theta$ and $v(r,\theta) = r^2 \sin 2\theta$. Now, if $y \geq 0$, then $0 \leq \theta \leq \pi$. We will show that $f[D] = \mathbb{C}$. If $u + vi$ is an arbitrary complex number, let $u = r\cos\phi$ and $v = r\sin\phi$ with $r \geq 0$ and $0 \leq \phi < 2\pi$. Let $z = \sqrt{r} e^{\frac{\phi}{2} i}$. Then $z \in D$ and $f(z) = re^{\phi i}$. It follows that the image under D is the entire complex plane. In other words, we have $f[D] = \mathbb{C}$.

Give a reasonable definition for each of the following limits (like what was done right before Theorem 7.30). L is a finite real number.

88. $\lim_{z\to\infty} f(z) = L$

$\lim_{z\to\infty} f(z) = L$ if and only if $\forall\epsilon > 0\ \exists\delta > 0\ \left(|z| > \frac{1}{\delta} \to |f(z) - L| < \epsilon\right)$.

89. $\lim_{z\to\infty} f(z) = \infty$

$\lim_{z\to\infty} f(z) = \infty$ if and only if $\forall\epsilon > 0\ \exists\delta > 0\ \left(|z| > \frac{1}{\delta} \to |f(z)| > \frac{1}{\epsilon}\right)$.

A complex number z is an **interior point** of a set S of complex numbers if there is a neighborhood of z that contains only points in S, whereas w is a **boundary point** of S if each neighborhood of w contains at least one point in S and one point not in S. Show the following:

90. A set of complex numbers is open if and only if each point in S is an interior point of S.

Proof: Let S be a set of complex numbers. Then S is open if and only if for every complex number $z \in S$, there is an open disk D with $z \in D$ and $D \subseteq S$ if and only if for every complex number $z \in S$, there is a neighborhood of z that contains only points in S if and only if every complex number in S is an interior point of S. □

91. A set of complex numbers is open if and only if it contains none of its boundary points.

Proof: Suppose that S is an open set of complex numbers and let $z \in S$. By Problem 90, z is an interior point of S. So, there is a neighborhood of z containing only points of S. So, z is **not** a boundary point of S. Since $z \in S$ was arbitrary, S contains none of its boundary points.

We now prove that if S contains none of its boundary points, then S is open by contrapositive. Suppose S is not open. By Problem 90, there is $z \in S$ such that z is **not** an interior point. Let N be a neighborhood of z. Since $z \in S$, N contains a point in S (namely, z). Since z is not an interior point of S, N contains a point not in S. So, z is a boundary point of S. So, S contains at least one of its boundary points. □

92. A set of complex numbers is closed if and only if it contains all its boundary points.

Proof: First note that a complex number z is a boundary point of S if and only if z is a boundary point of $\mathbb{C} \setminus S$ (because $z \in S$ if and only if $z \notin \mathbb{C} \setminus S$, and vice versa).

Let S be a set of complex numbers. Then S is closed if and only if $\mathbb{C} \setminus S$ is open if and only if $\mathbb{C} \setminus S$ contains none of its boundary points (by Problem 91) if and only if $S = \mathbb{C} \setminus (\mathbb{C} \setminus S)$ contains all its boundary points. □

Prove each of the following:

93. Let $A \subseteq \mathbb{C}$, let $f: A \to \mathbb{C}$, let $L = j + ki \in \mathbb{C}$, and let $a = b + ci \in \mathbb{C}$ be a point such that A contains some deleted neighborhood of a. Suppose that $f(x + yi) = u(x, y) + iv(x, y)$. Then $\lim_{z \to a} f(z) = L$ if and only if $\lim_{(x,y)\to(b,c)} u(x, y) = j$ and $\lim_{(x,y)\to(b,c)} v(x, y) = k$.

Proof: Suppose that $\lim_{z \to a} f(z) = L$ and let $\epsilon > 0$. Then there is $\delta > 0$ such that $0 < |z - a| < \delta$ implies $|f(z) - L| < \epsilon$. Now,

$$|z - a| = |(x + yi) - (b + ci)| = |(x - b) + (y - c)i| = \sqrt{(x - b)^2 + (y - c)^2}$$

Also, $|f(z) - L| = \left|\big(u(x, y) + iv(x, y)\big) - (j + ki)\right| = |(u(x, y) - j) + (v(x, y) - k)i|$.

So, if $0 < \sqrt{(x - b)^2 + (y - c)^2} < \delta$, then $0 < |z - a| < \delta$, and therefore, $|f(z) - L| < \epsilon$. By Problem 63,

$$|u(x, y) - j| \leq |(u(x, y) - j) + (v(x, y) - k)i| = |f(z) - L| < \epsilon.$$

By Problem 64,

$$|v(x, y) - k| \leq |(u(x, y) - j) + (v(x, y) - k)i| = |f(z) - L| < \epsilon$$

Therefore, $\lim_{(x,y)\to(b,c)} u(x, y) = j$ and $\lim_{(x,y)\to(b,c)} v(x, y) = k$.

Conversely, suppose that $\lim_{(x,y)\to(b,c)} u(x, y) = j$ and $\lim_{(x,y)\to(b,c)} v(x, y) = k$ and let $\epsilon > 0$. Then there are $\delta_1, \delta_2 > 0$ such that

$$0 < \sqrt{(x - b)^2 + (y - c)^2} < \delta_1 \text{ implies } |u(x, y) - j| < \tfrac{\epsilon}{2}$$

and

$$0 < \sqrt{(x - b)^2 + (y - c)^2} < \delta_2 \text{ implies } |v(x, y) - k| < \tfrac{\epsilon}{2}.$$

Let $\delta = \min\{\delta_1, \delta_2\}$ and assume that $0 < |z - a| < \delta$. Since $|z - a| = \sqrt{(x - b)^2 + (y - c)^2}$ and $\delta \leq \delta_1, \delta_2$, we have $|u(x, y) - j| < \frac{\epsilon}{2}$ and $|v(x, y) - k| < \frac{\epsilon}{2}$. It follows that

$$|f(z) - L| = \left|\big(u(x, y) + iv(x, y)\big) - (j + ki)\right| = |(u(x, y) - j) + (v(x, y) - k)i|$$
$$\leq |u(x, y) - j| + |v(x, y) - k| < \frac{\epsilon}{2} + \frac{\epsilon}{2} = \epsilon.$$

Therefore, $\lim_{z \to a} f(z) = L$. □

94. $\lim_{z\to\infty} f(z) = L$ if and only $\lim_{z\to 0} f\left(\frac{1}{z}\right) = L$.

Proof: Suppose that $\lim_{z\to\infty} f(z) = L$ and let $\epsilon > 0$. There is $\delta > 0$ so that $|z| > \frac{1}{\delta} \to |f(z) - L| < \epsilon$. If we let $w = \frac{1}{z}$, we have $z = \frac{1}{w}$, and therefore, $\left|\frac{1}{w}\right| > \frac{1}{\delta} \to \left|f\left(\frac{1}{w}\right) - L\right| < \epsilon$. But $\left|\frac{1}{w}\right| > \frac{1}{\delta}$ is equivalent to $0 < |w| < \delta$. So, $0 < |w - 0| < \delta \to \left|f\left(\frac{1}{w}\right) - L\right| < \epsilon$. Thus, $\lim_{w\to 0} f\left(\frac{1}{w}\right) = L$. This is equivalent to $\lim_{z\to 0} f\left(\frac{1}{z}\right) = L$.

Conversely, suppose that $\lim_{z\to 0} f\left(\frac{1}{z}\right) = L$ and let $\epsilon > 0$. Then there is $\delta > 0$ so that $0 < |z - 0| < \delta \to \left|f\left(\frac{1}{z}\right) - L\right| < \epsilon$. If we let $w = \frac{1}{z}$, then $z = \frac{1}{w}$, and therefore, we have $0 < \left|\frac{1}{w}\right| < \delta \to |f(w) - L| < \epsilon$. Now, $0 < \left|\frac{1}{w}\right| < \delta$ is equivalent to $|w| > \frac{1}{\delta}$. So, we have $|w| > \frac{1}{\delta} \to |f(w) - L| < \epsilon$. Therefore, $\lim_{w\to\infty} f(w) = L$, or equivalently, $\lim_{z\to\infty} f(z) = L$. □

95. $\lim_{z\to\infty} f(z) = \infty$ if and only $\lim_{z\to 0} \frac{1}{f\left(\frac{1}{z}\right)} = 0$.

Proof: Suppose that $\lim_{z\to\infty} f(z) = \infty$ and let $\epsilon > 0$. There is $\delta > 0$ so that $|z| > \frac{1}{\delta} \to |f(z)| > \frac{1}{\epsilon}$. If we let $w = \frac{1}{z}$, we have $z = \frac{1}{w}$, and therefore, $\left|\frac{1}{w}\right| > \frac{1}{\delta} \to \left|f\left(\frac{1}{w}\right)\right| > \frac{1}{\epsilon}$. But, $\left|\frac{1}{w}\right| > \frac{1}{\delta}$ is equivalent to $0 < |w| < \delta$ and $\left|f\left(\frac{1}{w}\right)\right| > \frac{1}{\epsilon}$ is equivalent to $\left|\frac{1}{f\left(\frac{1}{w}\right)}\right| < \epsilon$. So, $0 < |w| < \delta \to \left|\frac{1}{f\left(\frac{1}{w}\right)}\right| < \epsilon$, and therefore, $\lim_{w\to 0} \frac{1}{f\left(\frac{1}{w}\right)} = 0$. This is equivalent to $\lim_{z\to 0} \frac{1}{f\left(\frac{1}{z}\right)} = 0$.

Now, let $\lim_{z\to 0} \frac{1}{f\left(\frac{1}{z}\right)} = 0$ and let $\epsilon > 0$. There is $\delta > 0$ so that $0 < |z - 0| < \delta \to \left|\frac{1}{f\left(\frac{1}{z}\right)} - 0\right| < \epsilon$. If we let $w = \frac{1}{z}$, we have $z = \frac{1}{w}$, and therefore, $0 < |z - 0| < \delta$ is equivalent to $0 < \left|\frac{1}{w}\right| < \delta$, or equivalently, $|w| > \frac{1}{\delta}$. Also, $\left|\frac{1}{f\left(\frac{1}{z}\right)} - 0\right| < \epsilon$ is equivalent to $\left|\frac{1}{f(w)}\right| < \epsilon$, which in turn is equivalent to $|f(w)| > \frac{1}{\epsilon}$. So, $|w| > \frac{1}{\delta} \to |f(w)| > \frac{1}{\epsilon}$. Thus, $\lim_{w\to\infty} f(w) = \infty$. This is equivalent to $\lim_{z\to\infty} f(z) = \infty$. □

96. Let $D = \{z \in \mathbb{C} \mid |z| \le 1\}$ be the closed unit disk and let S be a subset of D that includes the interior of the disk but is missing at least one point on the bounding circle of the disk. Then S is not a closed set.

Proof: Let S be a set of complex numbers such that $S \subseteq D$, where $D = \{z \in \mathbb{C} \mid |z| \le 1\}$, such that S contains $\{z \in \mathbb{C} \mid |z| < 1\}$, but is missing some point w with $|w| = 1$. We will show that w is a boundary point of S. To see this, let N be a neighborhood of w with radius r. w is a point in N that is not in S. We need to find a point in N that is in S.

If $r > 1$, let $z = 0$. Since $0 < 1$, $z \in S$. Also, $|z - w| = |0 - w| = |-w| = |w| = 1 < r$. So, $z \in N$.

If $r \leq 1$, let $z = \frac{2-r}{2}w$. Then $z \in N$ because we have

$$|z - w| = \left|\frac{2-r}{2}w - w\right| = \left|\frac{2-r}{2}w - \frac{2}{2}w\right| = \left|\frac{2-r-2}{2}w\right| = \left|-\frac{r}{2}w\right| = \left|-\frac{r}{2}\right||w| = \frac{r}{2} \cdot 1 = \frac{r}{2} < r.$$

Also, we have $z \in S$ because $|z| = \left|\frac{2-r}{2}w\right| = \left|\frac{2-r}{2}\right||w| = \left|1 - \frac{r}{2}\right|(1) = 1 - \frac{r}{2} < 1.$

So, we have found a boundary point of S that is not in S. By Problem 92, S is **not** closed. □

97. A set of complex numbers is closed if and only if it contains all its accumulation points. (This is Theorem 7.21.)

Proof: Suppose that S is a closed set of complex numbers and let a be an accumulation point of S. Assume toward contradiction that $a \notin S$, and let N be a neighborhood of a. Since a is an accumulation point, N contains a point in S. Since $a \notin S$, N contains a point not in S (namely, a). So, a is a boundary point of S. Since S is closed, by Problem 92, $a \in S$, contradicting our assumption that $a \notin S$. So, we must have $a \in S$. Since a was an arbitrary accumulation point of S, we see that S contains all its accumulation points.

Now, suppose that S contains all its accumulation points, and let a be a boundary point of S. Assume toward contradiction that $a \notin S$. Then each neighborhood of a contains a point in S that is not equal to a. So, each deleted neighborhood of a contains a point in S. So, a is an accumulation point of S, and therefore, by our assumption that S contains all its accumulation points, $a \in S$. This contradicts our assumption that $a \notin S$. So, we must have $a \in S$. Since a was an arbitrary boundary point of S, we see that S contains all its boundary points. By Problem 92, S is closed. □

98. A set consisting of finitely many complex numbers is a closed set in $\mathbb{C}$. (Hint: Show that a finite set has no accumulation points.)

Proof: Let S be a set consisting of finitely many points. We will show that S has no accumulation points. Let $a \in \mathbb{C}$ and let $r = \min\{|w - a| \mid w \in S \wedge w \neq a\}$. Suppose toward contradiction that the deleted neighborhood $N_r^{\odot}(a) = \{z \mid 0 < |z - a| < r\}$ contains a complex number in S. Let's call this complex number w. Since $w \in S$ and $w \neq a$, by the definition of r, we have $|w - a| \geq r$. Since $w \in N_r^{\odot}(a)$, we have $|w - a| < r$. So, $r \leq |w - a| < r$, and therefore, $r < r$, a contradiction. Therefore, a is not an accumulation point of S. Since $a \in \mathbb{C}$ was arbitrary, we have shown that S has no accumulation points.

Since S has no accumulation points, the statement "S contains all its accumulation points" is vacuously true. By Problem 97, S is closed. □

99. $\mathbb{C}$ has a countable dense subset. (If $X \subseteq Y$, then X is said to be **dense** in Y if $\overline{X} = Y$.) Does $\mathbb{R}^n$ have a countable dense subset for each $n \in \mathbb{N}$?

Proof: Let $S = \{a + bi \mid a, b \in \mathbb{Q}\}$, let $z = c + di \in \mathbb{C}$, and let $N_r(z)$ be an r-neighborhood of z. By the Density Theorem, there is $q \in \mathbb{Q}$ with $c < q < c + r$. So, $q - c < r$, and so, $r^2 - (q - c)^2 > 0$. It follows that $d < d + \sqrt{r^2 - (q - c)^2}$. Once again, by the Density Theorem, there is $p \in \mathbb{Q}$ with $d < p < d + \sqrt{r^2 - (q - c)^2}$. Since $q, p \in \mathbb{Q}$, $q + pi \in S$. Since $c < q$, $q + pi \neq z$. Also,

$$|(q + pi) - (c + di)| = |(q - c) + (p - d)i| = \sqrt{(q - c)^2 + (p - d)^2}$$

$$< \sqrt{(q - c)^2 + \left(\sqrt{r^2 - (q - c)^2}\right)^2} = \sqrt{(q - c)^2 + r^2 - (q - c)^2} = \sqrt{r^2} = r.$$

So, z is an accumulation point of S. Therefore, $z \in \overline{S}$. Since $z \in \mathbb{C}$ was arbitrary, $\mathbb{C} \subseteq \overline{S}$. It follows that $\overline{S} = \mathbb{C}$, and so, S is a dense subset of $\mathbb{C}$.

Since $\mathbb{Q} \sim \mathbb{N}$, it follows that $\mathbb{Q} \times \mathbb{Q} \sim \mathbb{N} \times \mathbb{N} \sim \mathbb{N}$. Therefore, S is a countable dense subset of $\mathbb{C}$.

Similarly, $\mathbb{Q}^n$ is a countable dense subset of $\mathbb{R}^n$. □

100. Let $f: \mathbb{R}^2 \to \mathbb{R}$ be defined by $f(x, y) = xy$. Prove that f is continuous. Then generalize this result to the function $f: \mathbb{R}^n \to \mathbb{R}$ defined by $f(\mathbf{x}) = x_1 \cdot x_2 \cdots x_n$.

Proof: Let $(a, b) \in \mathbb{R}^2$. We will show that f is continuous at (a, b). Let $\epsilon > 0$ and let $\delta = \min\left\{1, \frac{\epsilon}{1+|a|+|b|}\right\}$. Suppose that

$$|(x, y) - (a, b)| < \delta, \text{ or equivalently, } \sqrt{(x - a)^2 + (y - b)^2} < \delta.$$

Then $|x| = |(x - a) + a| \leq |x - a| + |a| \leq \sqrt{(x - a)^2 + (y - b)^2} + |a| < \delta + |a| \leq 1 + |a|$. So,

$$|xy - ab| = |xy - xb + xb - ab| \leq |xy - xb| + |xb - ab| = |x||y - b| + |b||x - a|$$

$$\leq (1 + |a|)\sqrt{(x - a)^2 + (y - b)^2} + |b|\sqrt{(x - a)^2 + (y - b)^2} < (1 + |a| + |b|)\delta$$

$$\leq (1 + |a| + |b|) \cdot \frac{\epsilon}{(1 + |a| + |b|)} = \epsilon.$$

So, f is continuous at (a, b). Since $(a, b) \in \mathbb{R}^2$ was arbitrary, f is continuous.

We now prove that if $h_1, h_2, \dots, h_m: \mathbb{R}^n \to \mathbb{R}$ are continuous and $g: \mathbb{R}^n \to \mathbb{R}^m$ is defined by $g(x_1, x_2, \dots, x_n) = \left(h_1(x_1, \dots, x_n), h_2(x_1, \dots, x_n), \dots, h_m(x_1, \dots, x_n)\right)$, then g is continuous. To see this, let $(a_1, \dots, a_n) \in \mathbb{R}^n$ and let $\epsilon > 0$. Since each h_j is continuous, for each $j = 1, 2, \dots, m$, we can find $\delta_j > 0$ such that

$$0 < \sqrt{(x_1 - a_1)^2 + \cdots + (x_n - a_n)^2} < \delta_j \text{ implies } \left|h_j(x_1, \dots, x_n) - h_j(a_1, \dots, a_n)\right| < \frac{\epsilon}{\sqrt{m}}.$$

Let $\delta = \min\{\delta_1, \dots, \delta_m\}$. Suppose that $0 < \sqrt{(x_1 - a_1)^2 + \cdot + (x_n - a_n)^2} < \delta$. Then we have

$$\sqrt{\left(h_1(x_1, \dots, x_n) - h_1(a_1, \dots, a_n)\right)^2 + \dots + \left(h_m(x_1, \dots, x_n) - h_m(a_1, \dots, a_n)\right)^2}$$

$$< \sqrt{\left(\frac{\epsilon}{\sqrt{m}}\right)^2 + \dots + \left(\frac{\epsilon}{\sqrt{m}}\right)^2} = \sqrt{m \cdot \frac{\epsilon^2}{m}} = \sqrt{\epsilon^2} = \epsilon.$$

Therefore, g is continuous at $(a_1, \dots, a_n)$. Since $(a_1, \dots, a_n) \in \mathbb{R}^n$ was arbitrary, g is continuous.

We now prove that if $h: \mathbb{R}^m \to \mathbb{R}$ and $g: \mathbb{R}^n \to \mathbb{R}^m$ are continuous, then $h \circ g: \mathbb{R}^n \to \mathbb{R}$ is continuous. To see this, let $(a_1, \dots, a_n) \in \mathbb{R}^n$ and let $\epsilon > 0$. Since h is continuous at $g(a_1, \dots, a_n)$, there is $\delta_1 > 0$ such that $|(y_1, \dots, y_m) - g(a_1, \dots, a_n)| < \delta_1$ implies $\left|h(y_1, \dots, y_m) - h\big(g(a_1, \dots, a_n)\big)\right| < \epsilon$. Since g is continuous at $(a_1, \dots, a_n)$, there is $\delta_2 > 0$ such that $|(x_1, \dots, x_n) - (a_1, \dots, a_n)| < \delta_2$ implies $|g(x_1, \dots, x_n) - g(a_1, \dots, a_n)| < \delta_1$. Now, suppose that $|(x_1, \dots, x_n) - (a_1, \dots, a_n)| < \delta_2$. Then $|g(x_1, \dots, x_n) - g(a_1, \dots, a_n)| < \delta_1$. It follows that $\left|h\big(g(x_1, \dots, x_m)\big) - h\big(g(a_1, \dots, a_n)\big)\right| < \epsilon$. Since $\epsilon > 0$ was arbitrary, $h \circ g$ is continuous at $(a_1, \dots, a_n)$. Since $(a_1, \dots, a_n) \in \mathbb{R}^n$ was arbitrary, $h \circ g$ is continuous.

We now proceed by induction on $n \in \mathbb{N}$, with the base case of $n = 2$ having been proved above.

Assume that $f_k: \mathbb{R}^k \to \mathbb{R}$ is continuous, where f_k is defined by $f_k(\mathbf{x}) = x_1 \cdot x_2 \cdots x_k$, and let $f_{k+1}: \mathbb{R}^{k+1} \to \mathbb{R}$ be defined by $f_{k+1}(\mathbf{x}) = x_1 \cdot x_2 \cdots x_k \cdot x_{k+1}$. Let $g_k: \mathbb{R}^{k+1} \to \mathbb{R}^k$ be defined by $g_k(\mathbf{x}) = (x_1, x_2, \dots, x_{k-1}, x_k x_{k+1})$. Then we have

$$(f_k \circ g_k)(\mathbf{x}) = f_k\big(g_k(\mathbf{x})\big) = f_k(x_1, x_2, \dots, x_{k-1}, x_k x_{k+1}) = x_1 \cdot x_2 \cdots x_k \cdot x_{k+1} = f_{k+1}(\mathbf{x})$$

So, $f_{k+1} = f_k \circ g_k$. Therefore, the continuity of f_{k+1} will follow from the continuity of f_k and g_k.

Our inductive assumption was that f_k is continuous.

Now, observe that $g_k(x_1, x_2, \dots, x_{k+1}) = \big(h_1(x_1, \dots, x_{k+1}), h_2(x_1, \dots, x_{k+1}), \dots, h_k(x_1, \dots, x_{k+1})\big)$, where $h_j(x_1, \dots, x_{k+1}) = x_j$ for $j = 1, 2, \dots, k-1$ and $h_k(x_1, \dots, x_{k+1}) = x_k \cdot x_{k+1}$. The functions h_j for $j = 1, 2, \dots, k-1$ are continuous because $|x_j - a_j| \leq \sqrt{(x_1 - a_1)^2 + \cdots (x_{k+1} - a_{k+1})^2}$. Finally, $h_k = b_k \circ c_k$, where $c_k: \mathbb{R}^{k+1} \to \mathbb{R}^2$ is defined by $c_k(x_1, \dots, x_{k+1}) = (x_k, x_{k+1})$ and $b_k: \mathbb{R}^2 \to \mathbb{R}$ is defined by $b_k(x, y) = xy$. The function c_k is continuous because

$$\sqrt{(x_k - a_k)^2 + \dots + (x_{k+1} - a_{k+1})^2} \leq \sqrt{(x_1 - a_1)^2 + \dots + (x_{k+1} - a_{k+1})^2}.$$

The continuity of b_k was the base case for the induction. □

Problem Set 8

LEVEL 1

Let $A = \begin{bmatrix} 1 & 6 & 3 & 4 \\ 2 & -1 & 8 & -3 \\ 5 & -6 & 7 & 20 \end{bmatrix}$. Determine each of the following, if it exists.

1. a_{14}

$a_{14} = \mathbf{4}$

2. a_{42}

a_{42} **does not exist** (the matrix does **not** have four rows)

3. a_{31}

$a_{31} = \mathbf{5}$

4. $a_{11} + a_{22} + a_{33}$

$a_{11} + a_{22} + a_{33} = 1 + (-1) + 7 = \mathbf{7}$

5. $a_{12} - a_{21}$

$a_{12} - a_{21} = 6 - 2 = \mathbf{4}$

6. $2A$

$$2A = \begin{bmatrix} \mathbf{2} & \mathbf{12} & \mathbf{6} & \mathbf{8} \\ \mathbf{4} & \mathbf{-2} & \mathbf{16} & \mathbf{-6} \\ \mathbf{10} & \mathbf{-12} & \mathbf{14} & \mathbf{40} \end{bmatrix}$$

7. $-3A$

$$-3A = \begin{bmatrix} \mathbf{-3} & \mathbf{-18} & \mathbf{-9} & \mathbf{-12} \\ \mathbf{-6} & \mathbf{3} & \mathbf{-24} & \mathbf{9} \\ \mathbf{-15} & \mathbf{18} & \mathbf{-21} & \mathbf{-60} \end{bmatrix}$$

8. What is the size of the matrix A?

$\mathbf{3 \times 4}$

Let $B = \begin{bmatrix} 0 & 2 & 0 \\ 1 & 15 & 1 \\ 8 & 7 & 5 \\ 6 & 9 & 3 \end{bmatrix}$. Determine each of the following, if it exists.

9. b_{14}

b_{14} **does not exist** (the matrix does **not** have four columns)

10. b_{42}

$b_{42} = \mathbf{9}$

11. b_{31}

$b_{31} = \mathbf{8}$

12. $b_{11} + b_{22} + b_{33}$

$b_{11} + b_{22} + b_{33} = 0 + 15 + 5 = \mathbf{20}$

13. $b_{12} - b_{21}$

$b_{12} - b_{21} = 2 - 1 = \mathbf{1}$

14. $4B$

$$4B = \begin{bmatrix} \mathbf{0} & \mathbf{8} & \mathbf{0} \\ \mathbf{4} & \mathbf{60} & \mathbf{4} \\ \mathbf{32} & \mathbf{28} & \mathbf{20} \\ \mathbf{24} & \mathbf{36} & \mathbf{12} \end{bmatrix}$$

15. $-2B$

$$-2B = \begin{bmatrix} \mathbf{0} & \mathbf{-4} & \mathbf{0} \\ \mathbf{-2} & \mathbf{-30} & \mathbf{-2} \\ \mathbf{-16} & \mathbf{-14} & \mathbf{-10} \\ \mathbf{-12} & \mathbf{-18} & \mathbf{-6} \end{bmatrix}$$

16. What is the size of the matrix B?

$\mathbf{4 \times 3}$

Compute each of the following, if possible:

17. $\begin{bmatrix} 4 & 3 \\ 2 & 1 \end{bmatrix} + \begin{bmatrix} 1 & 2 \\ 3 & 4 \end{bmatrix}$

$$\begin{bmatrix} \mathbf{5} & \mathbf{5} \\ \mathbf{5} & \mathbf{5} \end{bmatrix}$$

18. $\begin{bmatrix} 1 & 0 & -1 \\ 0 & 1 & 1 \end{bmatrix} + \begin{bmatrix} 1 & 0 & 2 \\ 0 & 1 & 3 \end{bmatrix}$

$$\begin{bmatrix} \mathbf{2} & \mathbf{0} & \mathbf{1} \\ \mathbf{0} & \mathbf{2} & \mathbf{4} \end{bmatrix}$$

19. $\begin{bmatrix} 1 & 0 & -1 \\ 0 & 1 & 1 \end{bmatrix} + \begin{bmatrix} 1 & 2 \\ 3 & 4 \end{bmatrix}$

Does not exist (the two matrices have different sizes)

20. $-2\begin{bmatrix} 4 & 3 \\ 2 & 1 \end{bmatrix} + 5\begin{bmatrix} 1 & 2 \\ 3 & 4 \end{bmatrix}$

$\begin{bmatrix} \mathbf{-3} & \mathbf{4} \\ \mathbf{11} & \mathbf{18} \end{bmatrix}$

21. $\begin{bmatrix} 1 & 2 & 3 & 4 \\ 1 & -2 & 3 & -4 \\ 1 & -2 & 3 & 4 \end{bmatrix} + \begin{bmatrix} 2 & 3 & 5 & -3 \\ 2 & -1 & 8 & -1 \\ 5 & -2 & 7 & 3 \end{bmatrix}$

$\begin{bmatrix} \mathbf{3} & \mathbf{5} & \mathbf{8} & \mathbf{1} \\ \mathbf{3} & \mathbf{-3} & \mathbf{11} & \mathbf{-5} \\ \mathbf{6} & \mathbf{-4} & \mathbf{10} & \mathbf{7} \end{bmatrix}$

22. $3\begin{bmatrix} 1 & 2 & 3 & 4 \\ 1 & -2 & 3 & -4 \\ 1 & -2 & 3 & 4 \end{bmatrix} + 4\begin{bmatrix} 2 & 3 & 5 & -3 \\ 2 & -1 & 8 & -1 \\ 5 & -2 & 7 & 3 \end{bmatrix}$

$\begin{bmatrix} \mathbf{11} & \mathbf{18} & \mathbf{29} & \mathbf{0} \\ \mathbf{11} & \mathbf{-10} & \mathbf{41} & \mathbf{-16} \\ \mathbf{23} & \mathbf{-14} & \mathbf{37} & \mathbf{24} \end{bmatrix}$

23. $\begin{bmatrix} 1 \\ i \\ -i \\ -1 \end{bmatrix} + [1 \;\; i \;\; i \;\; 1]$

Does not exist (the two matrices have different sizes)

24. $[i \;\; i \;\; i \;\; i] + [1+2i \;\; 1+3i \;\; 1+4i \;\; 1+5i]$

$[\mathbf{1+3}\boldsymbol{i} \;\; \mathbf{1+4}\boldsymbol{i} \;\; \mathbf{1+5}\boldsymbol{i} \;\; \mathbf{1+6}\boldsymbol{i}]$

25. $5[i \;\; i \;\; i \;\; i] + 7[1+2i \;\; 1+3i \;\; 1+4i \;\; 1+5i]$

$[\mathbf{7+19}\boldsymbol{i} \;\; \mathbf{7+26}\boldsymbol{i} \;\; \mathbf{7+33}\boldsymbol{i} \;\; \mathbf{7+40}\boldsymbol{i}]$

For each of the following, determine if the given pair of vectors v and w are linearly independent or linearly dependent in the given vector space V:

26. $V = \mathbb{Q}^4$, $v = (3, 2, 2, -1)$, $w = \left(-1, -\frac{2}{3}, -\frac{2}{3}, -\frac{1}{3}\right)$

$-3w = -3\left(-1, -\frac{2}{3}, -\frac{2}{3}, -\frac{1}{3}\right) = (3, 2, 2, 1)$. Since $-3(-1) = 3$, but $-3\left(-\frac{1}{3}\right) \neq -1$, v and w are **not** scalar multiples of each other. Therefore, v and w are **linearly independent**.

27. $V = \mathbb{R}^3, v = (1, \sqrt{2}, 1), w = (\sqrt{2}, 2, \sqrt{2})$

$\sqrt{2}v = \sqrt{2}(1, \sqrt{2}, 1) = (\sqrt{2}, 2, \sqrt{2}) = w$. So, v and w **are** scalar multiples of each other. Therefore, v and w are **linearly dependent**.

28. $V = \mathbb{C}^5, v = (1, i, 2 - i, 0, 3i), w = (-i, 1, -1 - 2i, 0, 3)$

$-iv = -i(1, i, 2 - i, 0, 3i) = (-i, 1, -1 - 2i, 0, 3) = w$. So, v and w **are** scalar multiples of each other. Therefore, v and w are **linearly dependent**.

29. $V = M_{22}^{\mathbb{Q}}, v = \begin{bmatrix} a & b \\ \frac{a}{2} & 3b \end{bmatrix}, w = \begin{bmatrix} 1 & \frac{b}{a} \\ \frac{1}{2} & 3 \end{bmatrix} (a \neq 0, a \neq b)$

$aw = a\begin{bmatrix} 1 & \frac{b}{a} \\ \frac{1}{2} & 3 \end{bmatrix} = \begin{bmatrix} a & b \\ \frac{a}{2} & 3a \end{bmatrix}$. Since $a \cdot 1 = a$, but $a \cdot 3 = 3a \neq 3b$, v and w are **not** scalar multiples of each other. Therefore, v and w are **linearly independent**.

30. $V = \{ax^2 + bx + c \mid a, b, c \in \mathbb{R}\}, v = x, w = x^2$

If $k \in \mathbb{R}$, then $kx \neq x^2$. So, x and x^2 are **not** scalar multiples of each other. Therefore, v and w are **linearly independent**.

Let V and W be vector spaces over $\mathbb{R}$. Determine if each of the following functions is a linear transformation:

31. $f: \mathbb{R} \to \mathbb{R}$ defined by $f(x) = 2x + 1$

$f(0) = 2 \cdot 0 + 1 = 1 \neq 0$. Since the image of 0 under a linear transformation is 0, f is **not** a linear transformation.

32. $g: \mathbb{R} \to \mathbb{R}^2$ defined by $g(x) = (2x, 3x)$

$$g(ax + by) = \big(2(ax + by), 3(ax + by)\big) = (2ax + 2by, 3ax + 3by)$$
$$= (2ax, 3ax) + (2by, 3by) = a(2x, 3x) + b(2y, 3y) = ag(x) + bg(y).$$

By Theorem 8.18, g is a linear transformation.

33. $h: \mathbb{R}^3 \to \mathbb{R}^3$ defined by $h\big((x, y, z)\big) = (x + y, x + z, z - y)$

$$h\big(a(x, y, z) + b(s, t, w)\big) = h\big((ax + bs, ay + bt, az + bw)\big)$$
$$= (ax + bs + ay + bt, ax + bs + az + bw, az + bw - ay - bt)$$
$$= (ax + ay, ax + az, az - ay) + (bs + bt, bs + bw, bw - bt)$$
$$a(x + y, x + z, z - y) + b(s + t, s + w, w - t) = ah\big((x, y, z)\big) + bh\big((s, t, w)\big).$$

By Theorem 8.18, h is a linear transformation.

LEVEL 2

Determine if each of the following subsets of $\mathbb{R}^2$ is a subspace of $\mathbb{R}^2$:

34. $A = \{(x, y) \mid x + y = 0\}$

Since $0 + 0 = 0$, $(0, 0) \in A$.

Let $(x, y), (z, w) \in A$. Then $x + y = 0$ and $z + w = 0$.Therefore,

$$(x + z) + (y + w) = (x + y) + (z + w) = 0 + 0 = 0.$$

So, $(x, y) + (z, w) = (x + z, y + w) \in A$.

Let $(x, y) \in A$ and $k \in \mathbb{R}$. Then $x + y = 0$. So, $kx + ky = k(x + y) = k \cdot 0 = 0$ (by Problem 63 in Problem Set 3).

So, $k(x, y) = (kx, ky) \in A$.

By Theorem 8.1, A is a subspace of $\mathbb{R}^2$.

35. $B = \{(x, y) \mid xy = 0\}$

Since $0 \cdot 1 = 0$, we have $(0, 1) \in B$. Since $1 \cdot 0 = 0$, we have $(1, 0) \in B$. Adding these two vectors gives us $(1, 0) + (0, 1) = (1, 1)$. However, $1 \cdot 1 = 1 \neq 0$, and so, $(1, 1) \notin B$. So, B is not closed under addition. Therefore, B is **not** a subspace of $\mathbb{R}^2$.

36. $C = \{(x, y) \mid 2x = 3y\}$

Since $2 \cdot 0 = 0$ and $3 \cdot 0 = 0$, $2 \cdot 0 = 3 \cdot 0$. Therefore, $(0, 0) \in C$.

Let $(x, y), (z, w) \in C$. Then $2x = 3y$ and $2z = 3w$. Therefore,

$$2(x + z) = 2x + 2z = 3y + 3w = 3(y + w).$$

So, $(x, y) + (z, w) = (x + z, y + w) \in C$.

Let $(x, y) \in C$ and $k \in \mathbb{R}$. Then $2x = 3y$. So, $2(kx) = k(2x) = k(3y) = 3(ky)$.

So, $k(x, y) = (kx, ky) \in C$.

By Theorem 8.1, A is a subspace of $\mathbb{R}^2$.

37. $D = \{(x, y) \mid x \in \mathbb{Q}\}$

Since $1 \in \mathbb{Q}$, $(1, 0) \in D$. Now, $\sqrt{2}(1, 0) = \left(\sqrt{2}, 0\right) \notin D$ because $\sqrt{2} \notin \mathbb{Q}$. So, D is not closed under scalar multiplication. Therefore, D is **not** a subspace of $\mathbb{R}^2$.

Compute each of the following, if possible:

38. $\begin{bmatrix} 2 & 0 & -3 \\ 0 & 1 & 4 \end{bmatrix} \cdot \begin{bmatrix} 1 & 1 & 3 & 0 \\ 1 & -4 & 2 & 0 \\ 2 & 0 & 1 & -4 \end{bmatrix}$

$$\begin{bmatrix} 2 & 0 & -3 \\ 0 & 1 & 4 \end{bmatrix} \cdot \begin{bmatrix} 1 & 1 & 3 & 0 \\ 1 & -4 & 2 & 0 \\ 2 & 0 & 1 & -4 \end{bmatrix} = \begin{bmatrix} \mathbf{-4} & \mathbf{2} & \mathbf{3} & \mathbf{12} \\ \mathbf{9} & \mathbf{-4} & \mathbf{6} & \mathbf{-16} \end{bmatrix}$$

39. $[3 \ \ -1 \ \ 5] \cdot \begin{bmatrix} -4 \\ -7 \\ 2 \end{bmatrix}$

$$[3 \ \ -1 \ \ 5] \cdot \begin{bmatrix} -4 \\ -7 \\ 2 \end{bmatrix} = -12 + 7 + 10 = \mathbf{5}$$

40. $\begin{bmatrix} -4 \\ -7 \\ 2 \end{bmatrix} \cdot [3 \ \ -1 \ \ 5]$

$$\begin{bmatrix} -4 \\ -7 \\ 2 \end{bmatrix} \cdot [3 \ \ -1 \ \ 5] = \begin{bmatrix} \mathbf{-12} & \mathbf{4} & \mathbf{-20} \\ \mathbf{-21} & \mathbf{7} & \mathbf{-35} \\ \mathbf{6} & \mathbf{-2} & \mathbf{10} \end{bmatrix}$$

41. $\begin{bmatrix} a & b & c \\ d & e & f \\ g & h & i \end{bmatrix} \cdot \begin{bmatrix} 1 & 0 & 1 \\ 0 & 2 & 0 \\ 3 & 1 & 4 \end{bmatrix}$

$$\begin{bmatrix} a & b & c \\ d & e & f \\ g & h & i \end{bmatrix} \cdot \begin{bmatrix} 1 & 0 & 1 \\ 0 & 2 & 0 \\ 3 & 1 & 4 \end{bmatrix} = \begin{bmatrix} \boldsymbol{a + 3c} & \boldsymbol{2b + c} & \boldsymbol{a + 4c} \\ \boldsymbol{d + 3f} & \boldsymbol{2e + f} & \boldsymbol{d + 4f} \\ \boldsymbol{g + 3i} & \boldsymbol{2h + i} & \boldsymbol{g + 4i} \end{bmatrix}$$

42. $\begin{bmatrix} 1 & 0 & -1 \\ 0 & 1 & 1 \end{bmatrix} \cdot \begin{bmatrix} 1 & 0 & 2 \\ 0 & 1 & 3 \end{bmatrix}$

Does not exist (the number of rows of the second matrix is **not** equal to the number of columns of the first matrix)

Consider $\mathbb{C}$ as a vector space over itself.

43. Give an example of a function $f: \mathbb{C} \to \mathbb{C}$ such that f is additive, but **not** a linear transformation.

Define $f: \mathbb{C} \to \mathbb{C}$ by $f(z) = \text{Re } z$. If $z = x + yi$ and $w = u + vi$, then

$$f(z + w) = f\big((x + yi) + (u + vi)\big) = f\big((x + u) + (y + v)i\big)$$
$$= x + u = \text{Re } z + \text{Re } w = f(z) + f(w).$$

So, f is additive.

Now, $f(i \cdot 1) = f(i) = 0$ and $i \cdot f(1) = i \cdot 1 = i$. So, $f(i \cdot 1) \neq i \cdot f(1)$. Therefore, f is **not** homogenous.

So, f is **not** a linear transformation.

44. Give an example of vector spaces V and W and a homogenous function $g: V \to W$ that is **not** a linear transformation.

Let $V = \mathbb{R}^2$ and $W = \mathbb{R}$, considered as vector spaces over the field $\mathbb{R}$. Define $g: \mathbb{R}^2 \to \mathbb{R}$ by $g\big((v,w)\big) = (v^3 + w^3)^{\frac{1}{3}}$. Then for $k \in \mathbb{R}$, we have

$$g\big(k(v,w)\big) = g\big((kv,kw)\big) = ((kv)^3 + (kw)^3)^{\frac{1}{3}} = (k^3v^3 + k^3w^3)^{\frac{1}{3}} = \big(k^3(v^3 + w^3)\big)^{\frac{1}{3}}$$
$$= (k^3)^{\frac{1}{3}}(v^3 + w^3)^{\frac{1}{3}} = k(v^3 + w^3)^{\frac{1}{3}} = k \cdot g\big((v,w)\big).$$

Therefore, g is homogeneous.

Now, we have

$$g\big((1,0)\big) + g\big((0,1)\big) = (1^3 + 0^3)^{\frac{1}{3}} + (0^3 + 1^3)^{\frac{1}{3}} = 1 + 1 = 2.$$
$$g\big((1,0) + (0,1)\big) = g\big((1,1)\big) = (1^3 + 1^3)^{\frac{1}{3}} = 2^{\frac{1}{3}}.$$

So, $g\big((1,0) + (0,1)\big) \neq g\big((1,0)\big) + g\big((0,1)\big)$, and therefore, g is **not** additive.

So, g is **not** a linear transformation.

LEVEL 3

Let $P = \{ax^2 + bx + c \mid a,b,c \in \mathbb{R}\}$ be the vector space of polynomials of degree at most 2 with real coefficients (see part 3 of Example 3.24). Define the linear transformation $D: P \to P$ by $D(ax^2 + bx + c) = 2ax + b$. Find the matrix of T with respect to each of the following bases:

45. The standard basis $B = \{1, x, x^2\}$

$D(1) = 0 = 0x^2 + 0x + 0$, $D(x) = 1 = 0x^2 + 0x + 1$, and $D(x^2) = 2x = 0x^2 + 2x + 0$.

Therefore, $\boldsymbol{\mathcal{M}_T} = \begin{bmatrix} \mathbf{0} & \mathbf{0} & \mathbf{0} \\ \mathbf{0} & \mathbf{0} & \mathbf{2} \\ \mathbf{0} & \mathbf{1} & \mathbf{0} \end{bmatrix}$.

46. $C = \{x + 1, x^2 + 1, x^2 + x\}$

We have the following:

$$D(x+1) = 1 = \frac{1}{2}(x+1) + \frac{1}{2}(x^2+1) - \frac{1}{2}(x^2+x)$$
$$D(x^2+1) = 2x = 1(x+1) - 1(x^2+1) + 1(x^2+x)$$
$$D(x^2+x) = 2x + 1 = \frac{3}{2}(x+1) - \frac{1}{2}(x^2+1) + \frac{1}{2}(x^2+x)$$

Therefore, $\mathcal{M}_T(C) = \begin{bmatrix} \frac{1}{2} & 1 & \frac{3}{2} \\ \frac{1}{2} & -1 & -\frac{1}{2} \\ -\frac{1}{2} & 1 & \frac{1}{2} \end{bmatrix}$.

Notes: (1) It's a little more challenging to express the images of the given basis elements as a linear combination of the given basis elements. For example, how do we express 1 as a linear combination of $x + 1$, $x^2 + 1$, and $x^2 + x$. Well, we need to find weights c_1, c_2, and c_3 such that

$$1 = c_1(x + 1) + c_2(x^2 + 1) + c_3(x^2 + x).$$

We rewrite each side of this equation as a linear combination of x^2, x, and 1 as follows:

$$1 = c_1x + c_1 + c_2x^2 + c_2 + c_3x^2 + c_3x$$

$$0x^2 + 0x + 1 = (c_2 + c_3)x^2 + (c_1 + c_3)x + (c_1 + c_2)$$

Since $x^2, x, 1$ are linearly independent, we must have $c_2 + c_3 = 0$, $c_1 + c_3 = 0$, and $c_1 + c_2 = 1$. Subtracting the first equation from the third equation gives $c_1 - c_3 = 1$. Adding this equation to the second equation gives $2c_1 = 1$. Dividing by 2, we get $c_1 = \frac{1}{2}$. Using this value for c_1 and the second and third equations, we get $c_2 = \frac{1}{2}$ and $c_3 = -\frac{1}{2}$. This is how we wrote $D(x + 1)$ as a linear combination of $x + 1$, $x^2 + 1$, and $x^2 + x$ in the solution above.

The computations for $D(x^2 + 1)$ and $D(x^2 + x)$ can be done similarly.

(2) In a computational linear algebra course, you learn a procedure for solving systems of linear equations called Gauss-Jordan reduction. This procedure will always allow you to express arbitrary vectors as linear combinations of basis vectors. In fact, a single Gauss-Jordan reduction can be used to express the images of as many vectors as we like as a linear combination of basis vectors. Performing tedious computations like this lies outside the scope of this book and so, we leave it for the interested reader to investigate themselves.

Prove each of the following:

47. An arbitrary intersection of subspaces of a vector space V over a field F is a subspace of V.

Proof: Let V be a vector space over a field F and let $\boldsymbol{X}$ be a set of subspaces of V. For each $U \in \boldsymbol{X}$, $0 \in U$ because $U \leq V$. So, $0 \in \bigcap \boldsymbol{X}$. Let $v, w \in \bigcap \boldsymbol{X}$. For each $U \in \boldsymbol{X}$, $v, w \in U$, and so, $v + w \in U$ because $U \leq V$. Therefore, $v + w \in \bigcap \boldsymbol{X}$. Let $v \in \bigcap \boldsymbol{X}$ and $k \in F$. For each $U \in \boldsymbol{X}$, $v \in U$, and so, $kv \in U$ because $U \leq V$. Therefore, $kv \in \bigcap \boldsymbol{X}$. By Theorem 8.1, $\bigcap \boldsymbol{X} \leq V$. □

48. A finite set with at least two vectors is linearly dependent if and only if one of the vectors in the set can be written as a linear combination of the other vectors in the set.

Proof: Suppose that $S = \{v_1, v_2, \dots, v_n\}$ is a linearly dependent set with at least two elements. Then there are weights $c_1, c_2, \dots, c_n$ not all 0 such that $c_1v_1 + c_2v_2 + \cdots + c_nv_n = 0$. Without loss of generality, assume that $c_1 \neq 0$. We have $c_1v_1 = -c_2v_2 - \cdots - c_nv_n$, and so, $v_1 = -\frac{c_2}{c_1}v_2 - \cdots - \frac{c_n}{c_1}v_n$. So, v_1 can be written as a linear combination of the other vectors in S.

Now, suppose that one of the vectors in S can be written as a linear combination of the other vectors in the set. Without loss of generality, assume that $v_1 = c_2v_2 + \cdots + c_nv_n$. Then we have

$$v_1 - c_2v_2 - \cdots - c_nv_n = 0.$$

Since the weight of v_1 is 1, this is a nontrivial dependence relation. This shows that S is a linearly dependent set. □

49. Let V and W be vector spaces with V finite-dimensional, let $U \leq V$, and let $T \in \mathcal{L}(U, W)$. Then there is an $S \in \mathcal{L}(V, W)$ such that $S(v) = T(v)$ for all $v \in U$.

Proof: Suppose that $U \leq V$ with $\dim V = n$. Let $B = \{v_1, v_2, \dots, v_k\}$ be a basis of U. Since $v_1, v_2, \dots, v_k$ are linearly independent, by the note following Theorem 8.31, we can extend B to a basis B' of V, say $B' = \{v_1, v_2, \dots, v_k, v_{k+1}, \dots, v_n\}$. Define S by $S(c_1v_1 + \cdots + c_nv_n) = T(c_1v_1 + \cdots + c_kv_k)$. Since every vector in V can be written as a linear combination of the vectors in B', $\operatorname{dom} S = V$. Also, $\operatorname{ran} S \subseteq \operatorname{ran} T \subseteq W$. So, $S: V \to W$. Since T is linear, so is S. If $v \in U$, then v can be written as a linear combination of the vectors in B, say $v = c_1v_1 + \cdots + c_kv_k$. Then we have

$$\begin{aligned} S(v) = S(c_1v_1 + \cdots + c_kv_k) &= S(c_1v_1 + \cdots + c_kv_k + 0v_{k+1} + \cdots + 0v_n) \\ &= T(c_1v_1 + \cdots + c_kv_k) = T(v). \end{aligned}$$

□

LEVEL 4

Let $A = \begin{bmatrix} a & b \\ c & d \end{bmatrix}$ be a 2×2 matrix. We say that A is **invertible** if there is a 2×2 matrix B such that $AB = I$ and $BA = I$, where $I = \begin{bmatrix} 1 & 0 \\ 0 & 1 \end{bmatrix}$. Verify each of the following:

50. The inverse of $A = \begin{bmatrix} 1 & 1 \\ 0 & 1 \end{bmatrix}$ is $B = \begin{bmatrix} 1 & -1 \\ 0 & 1 \end{bmatrix}$.

$$\begin{bmatrix} 1 & 1 \\ 0 & 1 \end{bmatrix} \cdot \begin{bmatrix} 1 & -1 \\ 0 & 1 \end{bmatrix} = \begin{bmatrix} 1+0 & -1+1 \\ 0+0 & 0+1 \end{bmatrix} = \begin{bmatrix} 1 & 0 \\ 0 & 1 \end{bmatrix} = I$$

$$\begin{bmatrix} 1 & -1 \\ 0 & 1 \end{bmatrix} \cdot \begin{bmatrix} 1 & 1 \\ 0 & 1 \end{bmatrix} = \begin{bmatrix} 1+0 & 1+(-1) \\ 0+0 & 0+1 \end{bmatrix} = \begin{bmatrix} 1 & 0 \\ 0 & 1 \end{bmatrix} = I$$

Therefore, the inverse of A is B.

51. The zero matrix $\mathbf{0} = \begin{bmatrix} 0 & 0 \\ 0 & 0 \end{bmatrix}$ is **not** invertible.

For any 2×2 matrix $\begin{bmatrix} a & b \\ c & d \end{bmatrix}$, we have $\begin{bmatrix} 0 & 0 \\ 0 & 0 \end{bmatrix} \cdot \begin{bmatrix} a & b \\ c & d \end{bmatrix} = \begin{bmatrix} 0 & 0 \\ 0 & 0 \end{bmatrix} \neq \begin{bmatrix} 1 & 0 \\ 0 & 1 \end{bmatrix}$.

52. If $ad - bc \neq 0$, then $A = \begin{bmatrix} a & b \\ c & d \end{bmatrix}$ is invertible. In this case, what is the multiplicative inverse of A?

Let $B = \frac{1}{ad-bc}\begin{bmatrix} d & -b \\ -c & a \end{bmatrix}$. Then we have

$$AB = \begin{bmatrix} a & b \\ c & d \end{bmatrix} \cdot \frac{1}{ad-bc}\begin{bmatrix} d & -b \\ -c & a \end{bmatrix} = \frac{1}{ad-bc}\begin{bmatrix} ad - bc & a(-b) + ba \\ cd - dc & c(-b) + da \end{bmatrix}$$

$$= \frac{1}{ad-bc}\begin{bmatrix} ad - bc & 0 \\ 0 & ad - bc \end{bmatrix} = \begin{bmatrix} 1 & 0 \\ 0 & 1 \end{bmatrix} = I.$$

$$BA = \frac{1}{ad-bc}\begin{bmatrix} d & -b \\ -c & a \end{bmatrix} \cdot \begin{bmatrix} a & b \\ c & d \end{bmatrix} = \frac{1}{ad-bc}\begin{bmatrix} da - bc & db - bd \\ -ca + ac & -cb + ad \end{bmatrix}$$

$$= \frac{1}{ad-bc}\begin{bmatrix} ad - bc & 0 \\ 0 & ad - bc \end{bmatrix} = \begin{bmatrix} 1 & 0 \\ 0 & 1 \end{bmatrix} = I.$$

This shows that B is the multiplicative inverse of A.

Give an example of each of the following:

53. Vector spaces U and V with $U \subseteq V$ such that U is closed under scalar multiplication, but U is not a subspace of V

Let $V = \mathbb{R}^2$ and $U = \{(x, y) \mid x = 0 \text{ or } y = 0 \text{ (or both)}\}$. Let $(x, y) \in U$ and $k \in \mathbb{R}$. Then $k(x, y) = (kx, ky)$. If $x = 0$, then $kx = 0$. If $y = 0$, then $ky = 0$. So, $k(x, y) \in U$. So, U is closed under scalar multiplication. Now, $(0, 1)$ and $(1, 0)$ are in U, but $(1, 1) = (0, 1) + (1, 0) \notin U$. So, $U \not\leq V$. □

54. A vector space V and subspaces U and W, where $U \cup W$ is **not** a subspace of V

Let $V = \mathbb{R}^2$, $U = \{(x, y) \mid x = 0\}$, and $W = \{(x, y) \mid y = 0\}$. By part 1 of Example 8.2, $U \leq V$. By a very similar argument, $W \leq V$. However, $U \cup W$ is exactly the same set as the set U in Problem 53, which we have already shown is **not** a subspace of V. □

55. Let U and W be subspaces of a vector space V. Determine necessary and sufficient conditions for $U \cup W$ to be a subspace of V.

Theorem: Let U and W be subspaces of a vector space V. Then $U \cup W$ is a subspace of V if and only if $U \subseteq W$ or $W \subseteq U$.

Proof: Let U and W be subspaces of a vector space V. If $U \subseteq W$, then $U \cup W = W$, and so, $U \cup W$ is a subspace of V. Similarly, if $W \subseteq U$, then $U \cup W = U$, and so, $U \cup W$ is a subspace of V.

Suppose that $U \not\subseteq W$ and $W \not\subseteq U$. Let $x \in U \setminus W$ and $y \in W \setminus U$. Suppose that $x + y \in U$. We have $-x \in U$ because U is a subspace of V. So, $y = (-x + x) + y = -x + (x + y) \in U$, contradicting $y \in W \setminus U$. So, $x + y \notin U$. A similar argument shows that $x + y \notin W$. So, $x + y \notin U \cup W$. It follows that $U \cup W$ is not closed under addition, and therefore, $U \cup W$ is **not** a subspace of V. □

Note: The conditional statement $p \to q$ can be read "q is necessary for p" or "p is sufficient for q." Furthermore, $p \leftrightarrow q$ can be read "p is necessary and sufficient for q" (as well as "q is necessary and sufficient for p").

So, when we are asked to determine necessary and sufficient conditions for a statement p to be true, we are being asked to find a statement q that is logically equivalent to the statement p.

Usually if we are being asked for necessary and sufficient conditions, the hope is that we will come up with an equivalent statement that is easier to understand and/or visualize than the given statement.

Let $T: V \to W$ be a linear transformation and let $v_1, v_2, \ldots, v_n \in V$. Prove the following:

56. If T is injective and $v_1, v_2, \ldots, v_n$ are linearly independent in V, then $T(v_1), T(v_2), \ldots, T(v_n)$ are linearly independent in W.

Proof: Let T be injective and assume that $v_1, v_2, \ldots, v_n$ are linearly independent. Let $c_1, \ldots, c_n \in F$ be such that $c_1T(v_1) + \cdots + c_nT(v_n) = 0$. Since T is a linear transformation, $T(c_1v_1 + \cdots + c_nv_n) = 0$. Since T is injective and $T(0) = 0$, we have $c_1v_1 + \cdots + c_nv_n = 0$. By the linear independence of $v_1, v_2, \ldots, v_n$, $c_1 = c_2 = \cdots = c_n = 0$. So, $T(v_1), T(v_2), \ldots, T(v_n)$ are linearly independent. □

57. If T is surjective and $\text{span}\{v_1, v_2, \ldots, v_n\} = V$, then $\text{span}\{T(v_1), T(v_2), \ldots, T(v_n)\} = W$.

Proof: Let T be surjective and assume that $\text{span}\{v_1, v_2, \ldots, v_n\} = V$. Let $w \in W$. Since T is surjective, there is $v \in V$ such that $T(v) = w$. Since $\text{span}\{v_1, v_2, \ldots, v_n\} = V$, $v = c_1v_1 + \cdots + c_nv_n$ for some $c_1, \ldots, c_n \in F$. Then $T(v) = T(c_1v_1 + \cdots + c_nv_n) = c_1T(v_1) + \cdots + c_nT(v_n)$. Therefore, $w = T(v)$ is in $\text{span}\{T(v_1), T(v_2), \ldots, T(v_n)\}$. So, $W \subseteq \text{span}\{T(v_1), T(v_2), \ldots, T(v_n)\}$. Since $T: V \to W$ and W is closed under taking linear combinations, $\text{span}\{T(v_1), T(v_2), \ldots, T(v_n)\} \subseteq W$. Thus, we have $\text{span}\{T(v_1), T(v_2), \ldots, T(v_n)\} = W$. □

Determine if each linear transformation is diagonalizable:

58. $T: \mathbb{R}^2 \to \mathbb{R}^2$ defined by $T\big((x, y)\big) = (y, 2x)$

***T* is diagonalizable.**

Proof: We find the eigenvalues of T by solving the equation $T\big((x, y)\big) = \lambda(x, y)$, or equivalently, $(y, 2x) = (\lambda x, \lambda y)$. Equating the first components and second components gives us the two equations $y = \lambda x$ and $2x = \lambda y$. Substituting λx for y in the second equation gives $2x = \lambda^2 x$, or equivalently, $\lambda^2 x - 2x = 0$. Using distributivity on the left-hand side of this equation gives $x(\lambda^2 - 2) = 0$. So, $x = 0$ or $\lambda^2 - 2 = 0$. If $x = 0$, then $y = \lambda(0) = 0$. So, $(x, y) = (0, 0)$. Since an eigenvector must be nonzero, we reject $x = 0$. The equation $\lambda^2 - 2 = 0$ has the two solutions $\lambda = \sqrt{2}$ and $\lambda = -\sqrt{2}$. These are the two eigenvalues of T.

Next, let's find the eigenvectors corresponding to the eigenvalue $\lambda = \sqrt{2}$. In this case, we have $T\big((x,y)\big) = \sqrt{2}(x,y)$, or equivalently, $(y, 2x) = \big(\sqrt{2}x, \sqrt{2}y\big)$. So, $y = \sqrt{2}x$ and $2x = \sqrt{2}y$. These two equations are actually equivalent. Indeed, if we multiply each side of the first equation by $\sqrt{2}$, we get $\sqrt{2}y = \sqrt{2} \cdot \sqrt{2}x$, or equivalently, $\sqrt{2}y = 2x$.

So, we use only one of the equations, say $y = \sqrt{2}x$. So, the eigenvectors of T corresponding to the eigenvalue $\lambda = \sqrt{2}$ are all nonzero vectors of the form $\big(x, \sqrt{2}x\big)$. In particular, if we let $x = 1$, we get the eigenvector $\big(1, \sqrt{2}\big)$.

Let's also find the eigenvectors corresponding to the eigenvalue $\lambda = -\sqrt{2}$. In this case, we have $T\big((x,y)\big) = -\sqrt{2}(x,y)$, or equivalently, $(y, 2x) = \big(-\sqrt{2}x, -\sqrt{2}y\big)$. So, $y = -\sqrt{2}x$ and $2x = -\sqrt{2}y$. These two equations are equivalent. Indeed, if we multiply each side of the first equation by $-\sqrt{2}$, we get $-\sqrt{2}y = -\sqrt{2}\big(-\sqrt{2}x\big)$, or equivalently, $-\sqrt{2}y = 2x$.

So, we use only one of the equations, say $y = -\sqrt{2}x$. So, the eigenvectors of T corresponding to the eigenvalue $\lambda = -\sqrt{2}$ are all nonzero vectors of the form $\big(x, -\sqrt{2}x\big)$. In particular, if we let $x = 1$, we get the eigenvector $\big(1, -\sqrt{2}\big)$.

It follows that $B = \big\{\big(1, \sqrt{2}\big), \big(1, -\sqrt{2}\big)\big\}$ is a basis of eigenvectors of V and we have

$$T\left(\left(1, \sqrt{2}\right)\right) = \sqrt{2}\left(1, \sqrt{2}\right)$$

$$T\left(\left(1, -\sqrt{2}\right)\right) = -\sqrt{2}\left(1, -\sqrt{2}\right)$$

Therefore, the matrix of T with respect to B is $\mathcal{M}_T(B) = \begin{bmatrix} \sqrt{2} & 0 \\ 0 & -\sqrt{2} \end{bmatrix}$.

Since $\mathcal{M}_T(B)$ is a diagonal matrix, T is diagonalizable.

59. $U: \mathbb{C}^2 \to \mathbb{C}^2$ defined by $U\big((z,w)\big) = (z + iw, iz - w)$.

***U* is not diagonalizable.**

Proof: We find the eigenvalues of U by solving the equation $U\big((z,w)\big) = \lambda(z,w)$, or equivalently, $(z + iw, iz - w) = (\lambda z, \lambda w)$. Equating the first components and second components gives us the two equations $z + iw = \lambda z$ and $iz - w = \lambda w$. From the second equation, we must have $\lambda w + w = iz$, or equivalently, $w(\lambda + 1) = iz$. Multiplying the first equation by i gives the equation $iz - w = \lambda iz$. Replacing iz by $w(\lambda + 1)$ gives us $w(\lambda + 1) - w = \lambda w(\lambda + 1)$, or equivalently, $w(\lambda + 1 - 1) = w\lambda(\lambda + 1)$. So, we get $w\lambda - w\lambda(\lambda + 1) = 0$, or equivalently, $w\lambda(1 - \lambda - 1) = 0$ or $-w\lambda^2 = 0$. So, $\lambda = 0$ or $w = 0$. Since $w = 0$ implies $z = 0$ and an eigenvector must be nonzero, we reject $w = 0$. It follows that $\lambda = 0$ is the only eigenvalue of U.

Let's find the eigenvectors corresponding to the eigenvalue $\lambda = 0$. In this case, we have $U((z,w)) = 0(z,w)$, or equivalently, $(z+iw, iz-w) = (0,0)$. So, $z+iw=0$ and $iz-w=0$. These two equations are actually equivalent. Indeed, if we multiply each side of the first equation by i, we get $iz-w=0$.

So, we use only one of the equations, say $iz-w=0$, or equivalently, $w=iz$. So, the eigenvectors of U corresponding to the eigenvalue $\lambda=0$ are all nonzero vectors of the form $(z,iz)=z(1,i)$. Since all these vectors are scalar multiples of each other, a basis of eigenvectors has just one vector. Therefore, U is **not** diagonalizable.

Prove each of the following:

60. Let V and W be vector spaces over a field F. Then $\mathcal{L}(V,W)$ is a vector space over F, where addition and scalar multiplication are defined as in Theorem 8.19.

Proof: We first prove that $(\mathcal{L}(V,W),+)$ is a commutative group.

(Closure) Let $S,T \in \mathcal{L}(V,W)$, let $v,w \in V$, and let $a,b \in F$. Then

$$(S+T)(av+bw) = S(av+bw) + T(av+bw) = aS(v) + bS(w) + aT(v) + bT(w)$$
$$= a\big(S(v)+T(v)\big) + b\big(S(w)+T(w)\big) = a(S+T)(v) + b(S+T)(w).$$

So, $S+T \in \mathcal{L}(V,W)$.

(Associativity) Let $S,T,U \in \mathcal{L}(V,W)$ and let $v \in V$. Since addition is associative in W, we have

$$\big((S+T)+U\big)(v) = (S+T)(v) + U(v) = \big(S(v)+T(v)\big) + U(v)$$
$$= S(v) + \big(T(v)+U(v)\big) = S(v) + (T+U)(v) = \big(S+(T+U)\big)(v).$$

So, $(S+T)+U = S+(T+U)$.

(Commutativity) Let $S,T \in \mathcal{L}(V,W)$ and let $v \in V$. Since addition is commutative in W, we have

$$(S+T)(v) = S(v)+T(v) = T(v)+S(v) = (T+S)(v).$$

So, $S+T = T+S$.

(Identity) Define $0\colon V \to W$ by $0(v)=0$ for all $v \in V$. Then for all $v,w \in V$ and $a,b \in F$,

$$0(av+bw) = 0 = 0+0 = a\cdot 0 + b \cdot 0 = a0(v) + b0(w).$$

So, $0 \in \mathcal{L}(V,W)$.

For any $T \in \mathcal{L}(V,W)$ and $v \in V$, we have $(T+0)(v) = T(v)+0(v) = T(v)+0 = T(v)$ and we have $(0+T)(v) = 0(v)+T(v) = 0+T(v) = T(v)$. So, $T+0 = 0+T = T$.

(Inverse) Let $T \in \mathcal{L}(V,W)$ and define S by $S(v) = -T(v)$ for all $v \in V$. Then for all $v,w \in V$ and $a,b \in F$,

$$S(av+bw) = -T(av+bw) = -\big(aT(v)+bT(w)\big) = a\big(-T(v)\big) + b\big(-T(w)\big) = aS(v) + bS(w).$$

So, $S \in \mathcal{L}(V,W)$.

If $v \in V$, then

$$(T+S)(v) = T(v) + S(v) = T(v) + \big(-T(v)\big) = 0 = 0(v).$$
$$(S+T)(v) = S(v) + T(v) = -T(v) + T(v) = 0 = 0(v).$$

So, $T + S = S + T = 0$. Therefore, $S = -T$.

Now, let's prove that $\mathcal{L}(V, W)$ has the remaining vector space properties.

(Closure under scalar multiplication) Let $k \in F$, let $T \in \mathcal{L}(V, W)$, let $v, w \in V$, and let $a, b \in F$. Then

$$(kT)(av + bw) = kT(av + bw) = k\big(aT(v) + bT(w)\big) = k(aT(v)) + k\big(bT(w)\big)$$
$$= (ka)T(v) + (kb)T(w) = (ak)T(v) + (bk)\big(T(w)\big) = a\big(kT(v)\big) + b\big(kT(w)\big)$$
$$= a(kT)(v) + b(kT)(w).$$

So, $kT \in \mathcal{L}(V, W)$.

(Scalar multiplication identity) Let 1 be the multiplicative identity of F, let $T \in \mathcal{L}(V, W)$, and let $v \in V$. Then $(1T)(v) = 1T(v) = T(v)$. So, $1T = T$.

(Associativity of scalar multiplication) Let $j, k \in F$, let $T \in \mathcal{L}(V, W)$, and let $v \in V$. Then since multiplication is associative in W, we have

$$\big((jk)T\big)(v) = (jk)T(v) = j\big(kT(v)\big) = j(kT)(v) = \big(j(kT)\big)(v).$$

So, $(jk)T = j(kT)$.

(Distributivity of 1 scalar over 2 vectors) Let $k \in F$, let $S, T \in \mathcal{L}(V, W)$, and let $v \in V$. Since multiplication is distributive over addition in W, we have

$$\big(k(S+T)\big)(v) = k(S+T)(v) = k\big(S(v) + T(v)\big) = kS(v) + kT(v)$$
$$= (kS)(v) + (kT)(v) = (kS + kT)(v).$$

So, $k(S+T) = kS + kT$.

(Distributivity of 2 scalars over 1 vector) Let $j, k \in F$, let $S \in \mathcal{L}(V, W)$, and let $v \in V$. Since multiplication is distributive over addition in W, we have

$$\big((j+k)S\big)(v) = (j+k)S(v) = jS(v) + kS(v) = (jS)(v) + (kS)(v) = (jS + kS)(v).$$

So, $(j+k)S = jS + kS$. □

61. Let V be a vector space over a field F. Then $\mathcal{L}(V)$ is a linear algebra over F, where addition and scalar multiplication are defined as in Theorem 8.19 and vector multiplication is given by composition of linear transformations.

Proof: By Problem 60, $(\mathcal{L}(V), +)$ is a vector space over F. We now go through the properties for vector multiplication.

(Closure) We showed right before Example 8.20 that if V, W, and U are vector spaces over F, and $T: V \to W$, $S: W \to U$ are linear transformations, then the composition $S \circ T: V \to W$ is a linear transformation. If $S, T \in \mathcal{L}(V)$, then $V = W = U$, and so, $S \circ T: V \to V$. Therefore, $ST = S \circ T \in \mathcal{L}(V)$.

(Associativity) Let $S, T, U \in \mathcal{L}(V)$ and let $v \in V$. Then

$$((ST)U)(v) = (ST)(U(v)) = S\Big(T\big(U(v)\big)\Big) = S((TU)(v)) = \big(S(TU)\big)(v).$$

Since $v \in V$ was arbitrary, $(ST)U = S(TU)$.

(Identity) Define $I: V \to V$ by $I(v) = v$ for all $v \in V$. Then for all $v, w \in V$ and $a, b \in F$,

$$I(av + bw) = av + bw = aI(v) + bI(w).$$

So, $I \in \mathcal{L}(V)$.

For any $T \in \mathcal{L}(V)$ and $v \in V$, we have $(TI)(v) = T\big(I(v)\big) = T(v)$ and $(IT)(v) = I\big(T(v)\big) = T(v)$. So, $TI = IT = T$.

(Left Distributivity) Let $S, T, U \in \mathcal{L}(V)$ and let $v \in V$. Then we have

$$\begin{aligned}\big(S(T+U)\big)(v) &= S\big((T+U)(v)\big) = S\big(T(v) + U(v)\big) = S\big(T(v)\big) + S\big(U(v)\big)\\ &= (ST)(v) + (SU)(v) = (ST + SU)(v).\end{aligned}$$

Since $v \in V$ was arbitrary, $S(T + U) = ST + SU$.

(Right Distributivity) Let $S, T, U \in \mathcal{L}(V)$ and let $v \in V$. Then we have

$$\big((S+T)U\big)(v) = (S+T)\big(U(v)\big) = S\big(U(v)\big) + T\big(U(v)\big) = (SU)(v) + (TU)(v) = (SU + TU)(v).$$

Since $v \in V$ was arbitrary, $(S + T)U = SU + TU$.

(Compatibility of scalar and vector multiplication) Let $S, T \in \mathcal{L}(V)$ and $k \in F$.

$$\big(k(ST)\big)(v) = k\big((ST)(v)\big) = k\Big(S\big(T(v)\big)\Big) = (kS)\big(T(v)\big) = \big((kS)T\big)(v)$$

So, $k(ST) = (kS)T$.

$$\big(k(ST)\big)(v) = k\Big(\big(ST(v)\big)\Big) = k\Big(S\big(T(v)\big)\Big) = S\Big(k\big(T(v)\big)\Big) = S\big((kT)(v)\big) = \big(S(kT)\big)(v)$$

So, $k(ST) = S(kT)$. □

62. Let $T: V \to W$ and $S: W \to V$ be linear transformations such that $ST = i_V$ and $TS = i_W$. Then S and T are bijections and $S = T^{-1}$.

Proof: By symmetry, it suffices to show that T is a bijection.

Let $v, w \in V$ and suppose that $T(v) = T(w)$. Since $ST = i_V$,

$$v = i_V(v) = (ST)(v) = S\big(T(v)\big) = S\big(T(w)\big) = (ST)(w) = i_V(w) = w.$$

Therefore, T is injective.

Let $w \in W$. Since $TS = i_W$, $w = i_W(w) = (TS)(w) = T\big(S(w)\big)$. So, $w \in \operatorname{ran} T$. Therefore, T is surjective.

Since T is injective and surjective, T is a bijection.

Suppose that $T(v) = w$. Then $v = i_V(v) = (ST)(v) = S(T(v)) = S(w)$. So, $S = T^{-1}$. □

Let V and W be finite-dimensional vector spaces and let $T \in \mathcal{L}(V, W)$. Prove the following:

63. If $\dim V < \dim W$, then T is **not** surjective.

Proof: By Theorem 8.31, $\operatorname{rank} T + \operatorname{nullity} T = \dim V$. So, we have

$$\operatorname{rank} T = \dim V - \operatorname{nullity} T \leq \dim V < \dim W.$$

So, $\dim T[V] = \operatorname{rank} T < \dim W$. Therefore, $T[V] \neq W$, and so, T is **not** surjective. □

64. If $\dim V > \dim W$, then T is **not** injective.

Proof: By Theorem 8.31, $\operatorname{rank} T + \operatorname{nullity} T = \dim V$. So, we have

$$\operatorname{nullity} T = \dim V - \operatorname{rank} T \geq \dim V - \dim W > 0.$$

So, $\dim \ker(T) > 0$. Therefore, $\ker(T) \neq \{0\}$. By Theorem 8.30, T is **not** injective. □

Prove each of the following:

65. Two finite-dimensional vector spaces over a field F are isomorphic if and only if they have the same dimension.

Proof: Let V and W be finite-dimensional vector spaces over a field F. First suppose that $\dim V = \dim W = n$. By part 2 of Example 8.23, both V and W are isomorphic to F^n. Since isomorphism is an equivalence relation, V and W are isomorphic to each other.

Conversely, suppose that V and W are isomorphic. Let $T: V \to W$ be an isomorphism. Since T is an isomorphism, it is surjective. By Problem 63, $\dim V \geq \dim W$. Since T is an isomorphism, it is injective. By Problem 64, $\dim V \leq \dim W$. Therefore, $\dim V = \dim W$.

66. Let $T \in \mathcal{L}(V)$ be invertible and let $\lambda \in F^* = F \setminus \{0\}$. Then λ is an eigenvalue of T if and only if $\frac{1}{\lambda}$ is an eigenvalue of T^{-1}.

Proof: λ is an eigenvalue of T if and only if there is a nonzero $v \in V$ such that $T(v) = \lambda v$. Now,

$$T(v) = \lambda v \Leftrightarrow T^{-1}(T(v)) = T^{-1}(\lambda v) \Leftrightarrow (T^{-1}T)(v) = \lambda T^{-1}(v)$$

$$\Leftrightarrow i_V(v) = \lambda T^{-1}(v) \Leftrightarrow v = \lambda T^{-1}(v) \Leftrightarrow \frac{1}{\lambda} v = T^{-1}(v).$$

So, λ is an eigenvalue of T with corresponding eigenvector v if and only if $T^{-1}(v) = \frac{1}{\lambda} v$ if and only if $\frac{1}{\lambda}$ is an eigenvalue of T^{-1}. □

LEVEL 5

Prove each of the following:

67. Let S be a set of two or more linearly dependent vectors in a vector space V. Then there is a vector v in the set so that $\text{span } S = \text{span } S \setminus \{v\}$.

Proof: Let $S = \{v_1, v_2, \dots, v_n\}$ be a set of two or more linearly dependent vectors in V. By Problem 48, one of the vectors in the set can be written as a linear combination of the other vectors in the set. Without loss of generality, assume that v_n can be written as a linear combination of the other vectors in the set, say $v_n = k_1v_1 + k_2v_2 + \cdots + k_{n-1}v_{n-1}$. We show that $\text{span } S = \text{span } S \setminus \{v_n\}$. Let $v \in \text{span } S$. Then there are weights $c_1, c_2, \dots, c_n$ with $v = c_1v_1 + c_2v_2 + \cdots + c_nv_n$. So, we have

$$v = c_1v_1 + c_2v_2 + \cdots + c_nv_n = c_1v_1 + c_2v_2 + \cdots + c_n(k_1v_1 + k_2v_2 + \cdots + k_{n-1}v_{n-1})$$
$$= (c_1 + c_nk_1)v_1 + (c_2 + c_nk_2)v_2 + \cdots + (c_{n-1} + c_nk_{n-1})v_{n-1} \in \text{span } S \setminus \{v_n\}.$$

So, $\text{span } S \subseteq \text{span } S \setminus \{v_n\}$. Since it is clear that $\text{span } S \setminus \{v_n\} \subseteq \text{span } S$, $\text{span } S = \text{span } S \setminus \{v_n\}$. □

68. A finite set of vectors S in a vector space V is a basis of V if and only if every vector in V can be written uniquely as a linear combination of the vectors in S.

Proof: Suppose that $S = \{v_1, v_2, \dots, v_n\}$ is a basis of V. Then $\text{span}\{v_1, v_2, \dots, v_n\} = V$. So, if $v \in V$, then v can be written as a linear combination of the vectors in S. Suppose there are weights $c_1, c_2, \dots, c_n$ and $d_1, d_2, \dots, d_n$ such that $v = c_1v_1 + c_2v_2 + \cdots + c_nv_n$ and $v = d_1v_1dc_2v_2 + \cdots + d_nv_n$. Then we have $c_1v_1 + c_2v_2 + \cdots + c_nv_n = d_1v_1dc_2v_2 + \cdots + d_nv_n$, and so,

$$(c_1 - d_1)v_1 + (c_2 - d_2)v_2 + \cdots + (c_n - d_n)v_n = 0.$$

Since S is a linearly independent set of vectors, $c_1 - d_1 = 0, c_2 - d_2 = 0, \dots, c_n - d_n = 0$, and therefore, $c_1 = d_1, c_2 = d_2, \dots, c_n = d_n$. So, the expression of v as a linear combination of the vectors in S is unique.

Now, suppose that each vector in V can be written uniquely as a linear combination of the vectors in S. Since each vector in V can be written as a linear combination of the vectors in S, we have that $\text{span}\{v_1, v_2, \dots, v_n\} = V$. Since $0v_1 + 0v_2 + \cdots + 0v_n = 0$, by the uniqueness condition, the only way $c_1v_1 + c_2v_2 + \cdots + c_nv_n = 0$ could be true is if all weights are 0. So, S is linearly independent, and therefore, S is a basis of V. □

69. Let $S = \{v_1, v_2, \dots, v_m\}$ be a set of linearly independent vectors in a vector space V and let $T = \{w_1, w_2, \dots, w_n\}$ be a set of vectors in V such that $\text{span } T = V$. Then $m \leq n$.

Proof: If $V = \{0\}$ or S consists of just one vector, then there is nothing to prove. So, let's assume that $V \neq \{0\}$ and S has at least two vectors. Note that since $V \neq \{0\}$, T has at least one vector.

Let $T_0 = T = \{w_1, w_2, \dots, w_n\}$. Since span $T_0 = V$, v_1 can be written as a linear combination of the vectors in T_0. By Problem 48, $\{w_1, w_2, \dots, w_n, v_1\}$ is linearly dependent. Let $c_1, c_2, \dots, c_n, d$ be weights, not all of which are 0, such that $c_1w_1 + c_2w_2 + \cdots + c_nw_n + dv_1 = 0$. We claim that for some $i = 1, 2, \dots, n$, $c_i \neq 0$. If $d = 0$, then since one of the weights must be nonzero, some c_i must be nonzero. Suppose $d \neq 0$. If every $c_i = 0$, then $dv_1 = 0$. Since $d \neq 0$, $v_1 = 0$, contradicting the linear independence of S. In both cases, we must have $c_i \neq 0$ for some i. Without loss of generality, assume that $c_1 \neq 0$. Then $w_1 = -\frac{c_2}{c_1}w_2 - \cdots - \frac{c_n}{c_1}w_n - \frac{d}{c_1}v_1$. Let $T_1 = \{w_2, \dots, w_n, v_1\}$. By the proof of Problem 67, we have span $\{w_1, w_2, \dots, w_n, v_1\} \subseteq \text{span } T_1$. So, span $T_1 = V$.

At this point, note that if T had just one vector, then $T_1 = \{v_1\}$. Since span $T_1 = V$, v_2 would be a scalar multiple of v_1, contradicting the linear independence of S. So, T has at least two vectors. If S has only two vectors, then we are done. Otherwise, we continue as follows.

Since span $T_1 = V$, $\{w_2, \dots, w_n, v_1, v_2\}$ is linearly dependent. Let $c_2, \dots, c_n, d_1, d_2$ be weights, not all of which are 0, such that $c_2w_2 + \cdots + c_nw_n + d_1v_1 + d_2v_2 = 0$. We claim that for some $i = 2, \dots, n$, $c_i \neq 0$. If $d_1 = 0$ or $d_2 = 0$, we can use the same argument in the last paragraph to show that some c_i must be nonzero. Suppose $d_1 \neq 0$ and $d_2 \neq 0$. If every $c_i = 0$, then $d_1v_1 + d_2v_2 = 0$, contradicting the linear independence of S. In both cases, we must have $c_i \neq 0$ for some i. Without loss of generality, assume that $c_2 \neq 0$. Then $w_2 = -\frac{c_3}{c_2}w_3 - \cdots - \frac{c_n}{c_2}w_n - \frac{d_1}{c_2}v_1 - \frac{d_2}{c_2}v_2$.Let $T_2 = \{w_3, \dots, w_n, v_1, v_2\}$. By the proof of Problem 67, we have span $\{w_2, \dots, w_n, v_1, v_2\} \subseteq \text{span } T_2$. So, span $T_2 = V$.

Observe that if T had just two vectors, then $T_2 = \{v_1, v_2\}$. Since span $T_2 = V$, v_3 could be written as a linear combination of v_1 and v_2, contradicting the linear independence of S. So, T has at least three vectors. If S has only three vectors, then we are done. Otherwise, we continue in the same way.

Assuming $T_{j-1} = \{w_j, \dots, w_n, v_1, v_2, \dots v_{j-1}\}$ and span $T_{j-1} = V$, we have $\{w_j, \dots, w_n, v_1, v_2, \dots v_{j-1}, v_j\}$ linearly dependent. Once again, reindexing the w_i's if necessary, and letting $T_j = \{w_{j+1}, \dots, w_n, v_1, v_2, \dots v_{j-1}, v_j\}$, by an argument just like that given in the first paragraph, we can show that span $\{w_j, \dots, w_n, v_1, v_2, \dots, v_j\} \subseteq \text{span } T_j$. So, span $T_j = V$.

If $j < m$ and T had just j vectors, then $T_j = \{v_1, v_2, \dots, v_j\}$. Since span $T_j = V$, v_{j+1} could be written as a linear combination of the vectors $v_1, v_2, \dots, v_j$, contradicting the linear independence of S. So, T has at least $j + 1$ vectors. If $j = m$, we have shown that $m \leq n$. Otherwise, we continue in the same way. This procedure terminates in m steps. □

70. Let B be a basis of a vector space V with n vectors. Then any other basis of V also has n vectors.

Proof: Let B be a basis of V with n vectors. Let B' be another basis of V with m vectors. Since B' is a basis of V, B' is a linearly independent set of vectors in V. Since B is a basis of V, span $B = V$. By Problem 69, $m \leq n$. Similarly, we have that B is a linearly independent set of vectors in V and span $B' = V$. So, $n \leq m$. Since $m \leq n$ and $n \leq m$, we have $m = n$. So, B' has n vectors. □

71. Let V be a vector space with $\dim V > 1$. Then $\{T \in \mathcal{L}(V) \mid T \text{ is not invertible}\} \nleq \mathcal{L}(V)$.

Proof: Let $X = \{T \in \mathcal{L}(V) | \, T \text{ is not invertible}\}$. First suppose that $\dim V = 2$ and let $\{v_1. v_2\}$ be a basis of V. Let T be the linear transformation such that $T(v_1) = v_1$ and $T(v_2) = v_1$. Since T is not injective, T is not invertible. Let U be the linear transformation such that $U(v_1) = 0$ and $U(v_2) = -v_1 + v_2$. Since $v_1 \neq 0$ and $U(0) = 0$, U is not injective, and therefore, U is not invertible. Now, we have $(T + U)(v_1) = T(v_1) + U(v_1) = v_1 + 0 = v_1$, $(T + U)(v_2) = T(v_2) + U(v_2) = v_1 - v_1 + v_2 = v_2$. So, $T + U = i_V$, which is invertible. Therefore, X is not closed under addition, and so, $X \nleq \mathcal{L}(V)$.

Now, let $\dim V = n > 2$. Let T be the linear transformation such that $T(v_1) = T(v_2) = v_1$ and $T(v_i) = v_i$ for each $i = 3, \dots, n$. Since T is not injective, T is not invertible. Let U be the linear transformation such that $U(v_2) = -v_1 + v_2$ and $U(v_i) = 0$ for each $i \neq 2$. Since U is not injective, U is not invertible. Now, we have $(T + U)(v_2) = T(v_2) + U(v_2) = v_1 - v_1 + v_2 = v_2$ and for all $i \neq 2$, $(T + U)(v_i) = T(v_i) + U(v_i) = v_i + 0 = v_i$. So, $T + U = i_V$, which is invertible. Therefore, X is not closed under addition, and so, $X \nleq \mathcal{L}(V)$. □

72. Let V be an n-dimensional vector space over a field F. Then there is a linear algebra isomorphism $G: \mathcal{L}(V) \to M_{nn}^F$.

Proof: Let $B = \{v_1, \dots, v_n\}$ be a basis of V and define $G: \mathcal{L}(V) \to M_{nn}^F$ by $G(T) = \mathcal{M}_T(B)$.

Suppose that $G(T) = G(U)$, so that $\mathcal{M}_T(B) = \mathcal{M}_U(B)$. Suppose that $\mathcal{M}_T(B) = \begin{bmatrix} a_{11} & \cdots & a_{1n} \\ \vdots & & \vdots \\ a_{n1} & \cdots & a_{nn} \end{bmatrix}$. Then we also have $\mathcal{M}_U(B) = \begin{bmatrix} a_{11} & \cdots & a_{1n} \\ \vdots & & \vdots \\ a_{n1} & \cdots & a_{nn} \end{bmatrix}$. Let $j \in \{1, 2, \dots, n\}$. Then we have

$$T(v_j) = a_{1j}v_1 + a_{2j}v_2 + \cdots + a_{nj}v_n = U(v_j).$$

Since $j \in \{1, 2, \dots, n\}$ was arbitrary, for all $j = 1, 2, \dots, n$, $T(v_j) = U(v_j)$. Since B is a basis of V, for all $v \in V$, $T(v) = U(v)$. Therefore, $T = U$. So, G is injective.

Let $\begin{bmatrix} a_{11} & \cdots & a_{1n} \\ \vdots & & \vdots \\ a_{n1} & \cdots & a_{nn} \end{bmatrix} \in M_{nn}^F$. Define $T \in \mathcal{L}(V)$ on B by $T(v_j) = a_{1j}v_1 + a_{2j}v_2 + \cdots + a_{nj}v_n$ for each $j = 1, 2, \dots, n$. Then clearly $G(T) = \begin{bmatrix} a_{11} & \cdots & a_{1n} \\ \vdots & & \vdots \\ a_{n1} & \cdots & a_{nn} \end{bmatrix}$. So, G is surjective.

Let $T, U \in \mathcal{L}(V)$ and let $a, b \in F$. Suppose that $G(T) = \begin{bmatrix} a_{11} & \cdots & a_{1n} \\ \vdots & & \vdots \\ a_{n1} & \cdots & a_{nn} \end{bmatrix}$ and $G(U) = \begin{bmatrix} b_{11} & \cdots & b_{1n} \\ \vdots & & \vdots \\ b_{n1} & \cdots & b_{nn} \end{bmatrix}$.

Then for each $j = 1, 2, \dots, n$, we have

$$T(v_j) = a_{1j}v_1 + a_{2j}v_2 + \cdots + a_{nj}v_n \text{ and } U(v_j) = b_{1j}v_1 + b_{2j}v_2 + \cdots + b_{nj}v_n.$$

So, $(aT + bU)(v_j) = (aa_{1j} + bb_{1j})v_1 + \cdots + (aa_{nj} + bb_{nj})v_n$.

Therefore,

$$G(aT+bU)=\begin{bmatrix} aa_{11}+bb_{11} & \cdots & aa_{1n}+bb_{1n} \\ \vdots & & \vdots \\ aa_{n1}+bb_{n1} & \cdots & aa_{nn}+bb_{nn} \end{bmatrix}=\begin{bmatrix} aa_{11} & \cdots & aa_{1n} \\ \vdots & & \vdots \\ aa_{n1} & \cdots & aa_{nn} \end{bmatrix}+\begin{bmatrix} bb_{11} & \cdots & bb_{1n} \\ \vdots & & \vdots \\ bb_{n1} & \cdots & bb_{nn} \end{bmatrix}$$
$$=a\begin{bmatrix} a_{11} & \cdots & a_{1n} \\ \vdots & & \vdots \\ a_{n1} & \cdots & a_{nn} \end{bmatrix}+b\begin{bmatrix} b_{11} & \cdots & b_{1n} \\ \vdots & & \vdots \\ b_{n1} & \cdots & b_{nn} \end{bmatrix}=aG(T)+bG(U).$$

So, G is a vector space homomorphism (in other words, G is a linear transformation).

Now, let $T,U\in\mathcal{L}(V)$. Suppose that $G(T)=\begin{bmatrix} a_{11} & \cdots & a_{1n} \\ \vdots & & \vdots \\ a_{n1} & \cdots & a_{nn} \end{bmatrix}$ and $G(U)=\begin{bmatrix} b_{11} & \cdots & b_{1n} \\ \vdots & & \vdots \\ b_{n1} & \cdots & b_{nn} \end{bmatrix}$. Then for each $j=1,2,\dots,n$, $T(v_j)=a_{1j}v_1+a_{2j}v_2+\cdots+a_{nj}v_n$, $U(v_j)=b_{1j}v_1+b_{2j}v_2+\cdots+b_{nj}v_n$.

We have $(TU)(v_j)=T\left(U(v_j)\right)=T(b_{1j}v_1+b_{2j}v_2+\cdots+b_{nj}v_n)=c_{1j}v_1+c_{2j}v_2+\cdots+c_{nj}v_n$, where $c_{ij}=a_{i1}b_{1j}+a_{i2}b_{2j}+\cdots+a_{in}b_{nj}$. Also, we have

$\begin{bmatrix} a_{11} & \cdots & a_{1n} \\ \vdots & & \vdots \\ a_{n1} & \cdots & a_{nn} \end{bmatrix}\cdot\begin{bmatrix} b_{11} & \cdots & b_{1n} \\ \vdots & & \vdots \\ b_{n1} & \cdots & b_{nn} \end{bmatrix}=\begin{bmatrix} c_{11} & \cdots & c_{1n} \\ \vdots & & \vdots \\ c_{n1} & \cdots & c_{nn} \end{bmatrix}$, where $c_{ij}=a_{i1}b_{1j}+a_{i2}b_{2j}+\cdots+a_{in}b_{nj}$.

It follows that $G(TU)=\begin{bmatrix} c_{11} & \cdots & c_{1n} \\ \vdots & & \vdots \\ c_{n1} & \cdots & c_{nn} \end{bmatrix}=\begin{bmatrix} a_{11} & \cdots & a_{1n} \\ \vdots & & \vdots \\ a_{n1} & \cdots & a_{nn} \end{bmatrix}\cdot\begin{bmatrix} b_{11} & \cdots & b_{1n} \\ \vdots & & \vdots \\ b_{n1} & \cdots & b_{nn} \end{bmatrix}=G(T)\cdot G(U)$.

Let $i\in\mathcal{L}(V)$ be the identity function, so that $i(v_j)=v_j$ for each $j=1,2,\dots,n$. Then $G(i)=I$, where $I=\begin{bmatrix} 1 & 0 & \cdots & 0 \\ 0 & 1 & \cdots & 0 \\ \vdots & \vdots & \ddots & \vdots \\ 0 & 0 & \cdots & 1 \end{bmatrix}$, the identity for multiplication of $n\times n$ matrices.

Therefore, G is a ring homomorphism. Since G is both a vector space homomorphism and a ring homomorphism, G is a linear algebra homomorphism. Since G is also bijective, G is a linear algebra isomorphism. □

<u>**Problem Set 9**</u>

LEVEL 1

Let $\mathcal{L} = \{+,\ \cdot, S, <, 0\}$ be the language of number theory. Write each of the following abbreviated formulas in its unabbreviated form:

1. $\big(\forall v_0\big((v_0 \cdot 1 = v_0) \wedge (1 \cdot v_0 = v_0)\big)\big)$

$$\Big(\Big(\forall v_0\,\big(=\big(\cdot\,(v_0\,S(0))\big)\,v_0\big)\Big)\Big) \wedge \big(=\big(\cdot\,(S(0)\,v_0)\,v_0\big)\big)\Big)$$

2. $\big((2 \cdot (v_0 + 3)) + (1 + 0)\big) < \big((0 + 1) \cdot v_1\big) \wedge \forall v_0(v_0 \cdot 0 = 0)$

$$\Bigg(\Bigg(<\Bigg(+\Bigg(\cdot\Bigg(S(S(0))\Big(+\Big(v_0\,S\big(S(S(0))\big)\Big)\Big)\Bigg)\Bigg)(+(S(0)\,0))\Bigg)\cdot(+(0\,S(0))\,v_1)\Bigg)\Bigg)\wedge\Big(\forall v_0\Big(=\big((\cdot\,(v_0\,0))\,0\big)\Big)\Big)\Bigg)$$

Let $\mathcal{L} = \{\in\}$ be the language of set theory. For each of the following formulas, determine whether each occurrence of each variable is free or bound:

3. $v_0 \rightarrow \forall v_1\big(v_1 = v_2 \vee \exists v_2(v_0 = v_2)\big)$

Both occurrences of v_0 are free. The first and only occurrence of v_1 is bound. The first occurrence of v_2 is free and the second occurrence of v_2 is bound.

4. $\forall v_0 \exists v_1(v_0 \in v_1 \rightarrow v_1 = v_2) \wedge \exists v_0 \forall v_2(v_1 \in v_2)$

The first and only occurrence of v_0 is bound. The first two occurrences of v_1 are bound and the third occurrence is free. The first occurrence of v_2 is free and the second occurrence is bound.

LEVEL 2

Give an example of sentences ϕ and ψ and expressions α and β in the language of set theory such that $(\phi \wedge \psi) = (\alpha \wedge \beta)$ but $\phi \neq \alpha$ where

5. ϕ and ψ are abbreviated.

Let $\phi = \forall x(x \in x)$, $\psi = \forall y(y \in y \wedge y \in y)$, $\alpha = \forall x(x \in x) \wedge \forall y(y \in y$, $\beta = y \in y)$.

6. ϕ and ψ are unabbreviated.

Let $\phi = \big(\forall x(\in (x\ x))\big)$, $\psi = \Big(\forall y\big((\in (y\ y)) \wedge (\in (y\ y))\big)\Big)$, $\alpha = \big(\forall x(\in (x\ x))\big) \wedge (\forall y((\in (y\ y))$, $\beta = (\in (y\ y))))$.

Determine if each of the following is a logical axiom:

7. $\forall x \exists y(x < y) \to \exists y(y < y)$

This is **not** a logical axiom. y is **not** substitutable for x in $\exists y(x < y)$. A counterexample is given by $(\mathbb{N}, <)$. In this structure, $\forall x \exists y(x < y)$ is true, but $\exists y(y < y)$ is false.

8. $\forall y\big(C(y) \to C(y)\big) \to \big(C(a) \to C(a)\big)$

This **is** a logical axiom. It is a substitution axiom.

9. $\forall x\big(C(y) \to C(x)\big) \to \big(C(y) \to \forall x C(x)\big)$

This **is** a logical axiom. It is a universal derivation axiom.

10. $\Big[\Big(\forall x\big(S(x)\big) \to \forall y\big(S(y)\big)\Big) \to S(z)\Big] \to \Big[\forall x\big(S(x)\big) \to \Big(\forall y\big(S(y)\big) \to S(z)\Big)\Big]$

This **is** a logical axiom. It is a tautology.

Prove each of the following:

11. Let H and K be the subsets of S_3 (the symmetric group $(S(\{1,2,3\}),\circ)$—see Example 3.13) defined by $H = \{(1), (123), (132)\}$ and $K = \{(1), (12)\}$. Then H and K are subgroups of S_3.

Proof: Here is the multiplication table of S_3.

$(S_3, \circ)$	**(1)**	**(12)**	**(13)**	**(23)**	**(123)**	**(132)**
(1)	(1)	(12)	(13)	(23)	(123)	(132)
(12)	(12)	(1)	(132)	(123)	(23)	(13)
(13)	(13)	(123)	(1)	(132)	(12)	(23)
(23)	(23)	(132)	(123)	(1)	(13)	(12)
(123)	(123)	(13)	(23)	(12)	(132)	(1)
(132)	(132)	(23)	(12)	(13)	(1)	(123)

We now show that H and K are subgroups of S_3. The restricted tables look as follows:

$(H,\circ)$	(1)	(123)	(132)
(1)	(1)	(123)	(132)
(123)	(123)	(132)	(1)
(132)	(132)	(1)	(123)

$(K,\circ)$	(1)	(12)
(1)	(1)	(12)
(12)	(12)	(1)

From the tables, we can see that each of H and K are closed under $\circ$. Also, each of H and K contain the identity (1). In H, (123) and (132) are inverses of each other, and in K, (12) is its own inverse. It follows that both H and K are subgroups of S_3. □

12. Let G be a group and let $a \in G$. The **centralizer** of a in G is $C_a(G) = \{x \in G \mid ax = xa\}$. Then $C_a(G)$ is a subgroup of G.

Proof: Let $x, y \in C_a(G)$. Then $a(xy) = (ax)y = (xa)y = x(ay) = x(ya) = (xy)a$. So, $xy \in C_a(G)$. Since $x, y \in G$ were arbitrary, we see that $C_a(G)$ is closed under the group operation. Next, let $x \in C_a(G)$. Then $ax^{-1} = ax^{-1}a^{-1}a = a(ax)^{-1}a = a(xa)^{-1}a = aa^{-1}x^{-1}a = x^{-1}a$. Therefore, $x^{-1} \in C_a(G)$. Since $x \in C_a(G)$ was arbitrary, $C_a(G)$ is closed under taking inverses. Finally, we have $ae = a = ea$, and so, $e \in C_a(G)$. Therefore, $C_a(G)$ is a subgroup of G. □

13. $C_a(R)$ is a subring of R, where $C_a(R)$ is the centralizer of an element a in a ring R (provide a definition of $C_a(R)$ before proving the result).

Proof: We define $C_a(R) = \{x \in R \mid ax = xa\}$. Let $x, y \in C_a(R)$. Then

$$a(x + y) = ax + ay = xa + ya = (x + y)a.$$

$$a(xy) = (ax)y = (xa)y = x(ay) = x(ya) = (xy)a .$$

It follows that $x + y \in C_a(R)$ and $xy \in C_a(R)$. Since $x, y \in C_a(R)$ were arbitrary, we see that $C_a(R)$ is closed under $+$ and $\cdot$. Next, let $x \in C_a(R)$. Then we have

$$a(-x) = a(-1x) = -1(ax) = -1(xa) = (-1x)a = (-x)a.$$

(Note that we used Problem 79 from Problem Set 3 here.) So, $-x \in C_a(R)$. Since $x \in C_a(R)$ was arbitrary, $C_a(R)$ is closed under taking inverses. Finally, $a \cdot 1 = a = 1a$, and so, $1 \in C_a(R)$. Therefore, $C_a(R)$ is a subring of R. □

14. Let G be a group with H a nonempty subset of G. Then H is a subgroup of G if and only if for all $g, h \in H$, $gh^{-1} \in H$.

Proof: Let G be a group with H a nonempty subset of G.

First, let H be a subgroup of G and let $g, h \in H$. Since H has the inverse property, h^{-1} exists in H. Since H is closed under the group operation, $gh^{-1} \in H$. Since $g, h \in H$ were arbitrary, we have shown that for all $g, h \in H$, $gh^{-1} \in H$.

Conversely, suppose that for all $g, h \in H$, $gh^{-1} \in H$. Since $H \neq \emptyset$, there is $g \in H$. Therefore, $e = gg^{-1} \in H$. Now, let $h \in H$. Since $e \in H$, $h^{-1} = eh^{-1} \in H$. Finally, let $g, h \in H$. We just showed that $h^{-1} \in H$. Therefore, $gh = g(h^{-1})^{-1} \in H$. So, H is closed under the group operation. Thus, H is a subgroup of G. □

15. Let G be a group with H and K subgroups of G. If G is commutative, then $HK = \{hk \in G \mid h \in H \text{ and } k \in K\}$ is a subgroup of G. Is this result still true if G is not commutative?

Proof: Since $e = ee \in HK$, we see that $HK \neq \emptyset$. Let $x, y \in HK$. There are $h_1, h_2 \in H$ and $k_1, k_2 \in K$ such that $x = h_1k_1$ and $y = h_2k_2$. Then we have

$$xy^{-1} = (h_1k_1)(h_2k_2)^{-1} = h_1k_1k_2^{-1}h_2^{-1} = (h_1h_2^{-1})(k_1k_2^{-1}) \in HK.$$

By Problem 14 above, $HK \leq G$.

If G is not commutative, the result can be false. A counterexample is given by $G = S_3$, $H = \{(1), (12)\}$, $K = \{(1), (13)\}$ In this case, $HK = \{(1), (12), (13), (132)\}$ and HK is not closed under the group operation. Indeed, $(132)(132) = (123)$. □

16. Let $X = \{a + bi \mid a, b \in \mathbb{Z}\}$ be the set of Gaussian integers. Then X is a subring of $\mathbb{C}$.

Proof: Let $a + bi, c + di \in X$. Then $(a + bi) + (c + di) = (a + c) + (b + d)i$. Since $\mathbb{Z}$ is closed under addition, $a + c \in \mathbb{Z}$ and $b + d \in \mathbb{Z}$. Therefore, $(a + bi) + (c + di) \in X$, and so, X is closed under addition. Also, $(a + bi)(c + di) = (ac - bd) + (ad + bc)i$. Since $\mathbb{Z}$ is closed under multiplication, $ac, bd, ad, bc \in \mathbb{Z}$. Since $\mathbb{Z}$ has the additive inverse property, $-(bd) \in \mathbb{Z}$. Since $\mathbb{Z}$ is closed under addition, we have $ad + bc \in \mathbb{Z}$ and $ac - bd = ac + (-(bd)) \in \mathbb{Z}$. Therefore, $(a + bi)(c + di) \in X$, and so, X is closed under multiplication. The additive inverse of $a + bi$ is $-a - bi = -a + (-b)i$. Since $\mathbb{Z}$ has the additive inverse property, $-a, -b \in \mathbb{Z}$, and so, $-a - bi \in X$. Finally, $1 = 1 + 0i \in X$ because $1, 0 \in \mathbb{Z}$. It follows that X is a subring of $\mathbb{C}$. □

Let G be a commutative group with identity e. Prove that each of the following subsets of G is a subgroup of G:

17. $H = \{x \in G \mid x^2 = e\}$

Proof: Since $e^2 = e$, $H \neq \emptyset$. Let $x, y \in H$. Then

$$(xy^{-1})^2 = xy^{-1}xy^{-1} = x^2(y^2)^{-1} = e^2(e^2)^{-1} = ee^{-1} = ee = e.$$

So, $xy^{-1} \in H$. By Problem 14, H is a subgroup of G. □

18. $K = \{x^2 \mid x \in G\}$

Proof: Since $e = e^2$, $e \in K$. So, $K \neq \emptyset$. Let $x^2, y^2 \in K$. Then

$$x^2(y^2)^{-1} = x^2(y^{-1})^2 = xxy^{-1}y^{-1} = xy^{-1}xy^{-1} = (xy^{-1})^2$$

So, $x^2(y^2)^{-1} \in K$. By Problem 14, K is a subgroup of G. □

19. $L = \{x \in G \mid x \text{ has finite order}\}$

Proof: Since $e^1 = e$, e has order 1, and so, $e \in L$. So, $L \neq \emptyset$.

Let $x, y \in L$. Then there are $m, n \in \mathbb{N}$ such that $x^m = e$ and $y^n = e$. Then we also have $(y^{-1})^n = (y^n)^{-1} = e^{-1} = e$. So,

$$(xy^{-1})^{mn} = x^{mn}(y^{-1})^{mn} = (x^m)^n((y^{-1})^n)^m = e^n e^m = e.$$

So, xy^{-1} has order less than or equal to mn. In particular, xy^{-1} has finite order. Therefore, $xy^{-1} \in L$. By Problem 14, L is a subgroup of G □

LEVEL 3

Prove each of the following:

20. Recall that a set of statements Σ of sentential logic **tautologically implies** a statement ϕ if and only if every truth assignment for the propositional variables in Σ and ϕ that satisfies every element of Σ also satisfies ϕ. Let Γ be an arbitrary set of first-order formulas, let $\Phi = \{\phi_1, \phi_2, \dots, \phi_n\}$ be a set of n first-order formulas, and let ψ be a first-order formula. Suppose that Φ tautologically implies ψ and for each $k = 1, 2, \dots, n$, $\Gamma \vdash \phi_k$. Then $\Gamma \vdash \psi$.

Proof: The formula $\phi_1 \to \big(\phi_2 \to (\phi_3 \to \cdots \to (\phi_n \to \psi)\cdots)\big)$ is a tautology. Therefore, it is a logical axiom. Since $\Gamma \vdash \phi_1$, by modus ponens, $\Gamma \vdash \phi_2 \to (\phi_3 \to \cdots \to (\phi_n \to \psi)\cdots)$. Since $\Gamma \vdash \phi_2$, again by modus ponens, $\Gamma \vdash \phi_3 \to \cdots \to (\phi_n \to \psi)$. Continuing in this fashion a total of n times, we finally get $\Gamma \vdash \psi$. □

21. Let Λ be the set of logical axioms, let Γ be a set of first-order formulas, and let ϕ be a first-order formula. Suppose that $\Lambda \cup \Gamma$ tautologically implies ϕ. Then $\Gamma \vdash \phi$. (Hint: Use Problem 78 from Problem Set 1.)

Proof: Suppose that $\Gamma \cup \Lambda$ tautologically implies ϕ. Then there is a finite subset $\{\gamma_1, \dots, \gamma_m, \lambda_1, \dots, \lambda_n\}$ that tautologically implies ϕ. $\gamma_1 \to \Big(\gamma_2 \to \Big(\gamma_3 \to \cdots \to \Big(\gamma_m \to \big(\lambda_1 \to (\lambda_2 \to \cdots \to (\lambda_n \to \phi))\big)\Big)\cdots\Big)\Big)$ is a tautology. Therefore, it is a logical axiom. As we saw in Problem 20, by applying modus ponens to this tautology $m + n$ times, we get $\Gamma \vdash \phi$. □

Let Γ be a set of first-order formulas and let ϕ, ψ be first-order formulas. Prove each of the following:

22. $\Gamma \vdash \phi \to \psi$ if and only if $\Gamma \cup \{\phi\} \vdash \psi$.

Proof: First, assume that $\Gamma \vdash \phi \to \psi$. Let $\theta_1, \theta_2, \dots, \theta_n$ be a derivation of $\phi \to \psi$ from Γ. Then the sequence $\phi, \theta_1, \theta_2, \dots, \theta_n, \psi$ is a derivation of ψ from $\Gamma \cup \{\phi\}$.

Conversely, we will prove by induction that if $\Gamma \cup \{\phi\} \vdash \psi$, then $\Gamma \vdash \phi \to \psi$.

If $\psi = \phi$, then clearly $\vdash \phi \to \psi$.

If ψ is a logical axiom or an element of Γ, then $\Gamma \vdash \psi$ and ψ tautologically implies $\phi \to \psi$. By Problem 20, $\Gamma \vdash \phi \to \psi$.

Finally, if ψ is obtained by modus ponens from θ and $\theta \to \psi$, then by the inductive assumption, we have $\Gamma \vdash \phi \to \theta$ and $\Gamma \vdash \phi \to (\theta \to \psi)$. Since $\{\phi \to \theta, \phi \to (\theta \to \psi)\}$ tautologically implies $\phi \to \psi$, by Problem 20, $\Gamma \vdash \phi \to \psi$. □

23. $\Gamma \cup \{\phi\} \vdash \neg\psi$ if and only if $\Gamma \cup \{\psi\} \vdash \neg\phi$.

Proof: Assume that $\Gamma \cup \{\phi\} \vdash \neg\psi$. By Problem 22, $\Gamma \vdash \phi \to \neg\psi$. Now, $\phi \to \neg\psi$ tautologically implies $\psi \to \neg\phi$. So, by Problem 20, $\Gamma \vdash \psi \to \neg\phi$. By Problem 22 again, $\Gamma \cup \{\psi\} \vdash \neg\phi$.

The proof of the converse follows by symmetry, □

24. The following are equivalent: (a) There is a formula ϕ such that $\Gamma \vdash \phi$ and $\Gamma \vdash \neg\phi$. (b) For every formula ψ, $\Gamma \vdash \psi$.

Proof: Suppose there is a formula ϕ such that $\Gamma \vdash \phi$ and $\Gamma \vdash \neg\phi$. Let ψ be an arbitrary formula. Since $\{\phi, \neg\phi\}$ tautologically implies ψ, by Problem 20, $\Gamma \vdash \psi$.

Conversely suppose that for every formula ψ, $\Gamma \vdash \psi$. Let ϕ be any formula. Then $\neg\phi$ is also a formula. By our assumption, $\Gamma \vdash \phi$ and $\Gamma \vdash \neg\phi$. □

Prove each of the following:

25. There are exactly two ring homomorphisms from $\mathbb{Z}$ to itself.

Proof: The identity function $i_{\mathbb{Z}}: \mathbb{Z} \to \mathbb{Z}$ and the zero function $0: \mathbb{Z} \to \mathbb{Z}$ are both ring homomorphisms from $\mathbb{Z}$ to itself. We show that these are the only ones.

Let $f: \mathbb{Z} \to \mathbb{Z}$ be a ring homomorphism and suppose that $f(1) = n$. Then we also have

$$f(1) = f(1 \cdot 1) = f(1) \cdot f(1) = n \cdot n = n^2.$$

So, $n^2 = n$. Therefore, $n^2 - n = 0$, and so, $n(n-1) = 0$. It follows that $n = 0$ or $n = 1$.

Now, it is easy to show that $f(k) = k \cdot f(1)$ for all $k \in \mathbb{Z}$. For $k \in \mathbb{N}$, use the principle of mathematical induction. Then if $k < 0$, we have $f(k) = f(-(-k)) = -f(-k) = -(-k)f(1) = kf(1)$. $f(0) = 0$ follows from Theorem 9.41.

So, if $f(1) = 0$, then for all $k \in \mathbb{Z}$, $f(k) = kf(1) = k \cdot 0 = 0$. In this case, f is the zero function.

If $f(1) = 1$, then for all $k \in \mathbb{Z}$, $f(k) = kf(1) = k \cdot 1 = k$. In this case, f is the identity function. □

26. Ring isomorphism is an equivalence relation.

Proof: $i_R: R \to R$ is a bijection and if $x, y \in R$, then $i_R(x + y) = x + y = i_R(x) + i_R(y)$ and $i_R(xy) = xy = i_R(x)i_R(y)$. Also, $i_R(1_R) = 1_R$. So, i_R is an isomorphism from R to itself. Therefore, $\cong$ is reflexive.

Suppose that $f: R \to S$ is an isomorphism from R to S. We already know that $f^{-1}: S \to R$ is a bijection from S to R. Let $x, y \in S$ and let $z, w \in R$ with $f(z) = x$ and $f(w) = y$. Then $f(z + w) = f(z) + f(w) = x + y$, and so, $f^{-1}(x + y) = z + w = f^{-1}(x) + f^{-1}(y)$. Also, $f(zw) = f(z)f(w) = xy$, so that $f^{-1}(xy) = zw = f^{-1}(x)f^{-1}(y)$. Finally, $f^{-1}(1_S) = 1_R$. Therefore, f^{-1} is an isomorphism from S to R, and so, $\cong$ is symmetric.

Suppose that $f: R \to S, g: S \to T$ are isomorphisms. We already know that $g \circ f: R \to T$ is a bijection. If $x, y \in R$, then we have

$$(g \circ f)(x + y) = g(f(x + y)) = g(f(x) + f(y))$$
$$= g(f(x)) + g(f(y)) = (g \circ f)(x) + (g \circ f)(y)$$

and

$$(g \circ f)(xy) = g(f(xy)) = g(f(x) \cdot f(y))$$
$$= g(f(x)) \cdot g(f(y)) = (g \circ f)(x) \cdot (g \circ f)(y).$$

Also, $(g \circ f)(1_R) = g(f(1_R)) = g(1_S) = 1_T$. So, $g \circ f$ is an isomorphism from R to T, and so, $\cong$ is transitive.

Since $\cong$ is reflexive, symmetric, and transitive, $\cong$ is an equivalence relation. □

27. If we let $\text{Aut}(R)$ be the set of automorphisms of a ring R, then $(\text{Aut}(R), \circ)$ is a group, where $\circ$ is composition. $\text{Aut}(R)$ is called the **automorphism group** of R.

Proof: Let R be a ring and let $f, g \in \text{Aut}(R)$. Then $g \circ f$ is an automorphism by the proof of transitivity from Problem 26. So $\text{Aut}(R)$ is closed under $\circ$. We proved that $\circ$ is associative in $S(R)$ in Theorem 3.12. Since $\text{Aut}(R) \subseteq S(R)$, $\circ$ is associative in $\text{Aut}(R)$ as well. We proved that i_R satisfies $i_R \circ f = f$ and $f \circ i_R = f$ for all $f \in \text{Aut}(R)$ in the same example and we proved i_R is an automorphism in Problem 26. Let $f \in \text{Aut}(R)$. By Theorem 2.60, we have $f^{-1} \circ f = f \circ f^{-1} = i_R$, and we proved that f^{-1} is an automorphism in Problem 26. It follows that $(\text{Aut}(R), \circ)$ is a group. □

28. Let R be a ring and for each $x \in R$, define the function $\phi_x: R \to R$ by $\phi_x(y) = xyx^{-1}$. Let $\text{Inn}(R) = \{\phi_x \mid x \in R\}$. Then $\text{Inn}(R)$ is a subgroup of $\text{Aut}(R)$. $\text{Inn}(R)$ is called the **inner automorphism group** of R.

Proof: We first show that for each $x \in R$, ϕ_x is an automorphism of R. So, let $x, y, z \in R$. Then we have the following:

$$\phi_x(y + z) = x(y + z)x^{-1} = (xy + xz)x^{-1} = xyx^{-1} + xzx^{1} = \phi_x(y) + \phi_x(z)$$
$$\phi_x(yz) = x(yz)x^{-1} = xy \cdot 1 \cdot zx^{-1} = (xyx^{-1})(xzx^{-1}) = \phi_x(y) \cdot \phi_x(z)$$
$$\phi_x(1) = x1x^{-1} = xx^{-1} = 1$$

It follows that $\text{Inn}(R) \subseteq \text{Aut}(R)$. Now, for all $y \in R$, $\phi_e(y) = eye^{-1} = ye = y$. So, ϕ_e is the identity automorphism of R. It follows that $\text{Inn}(R)$ contains the identity of $\text{Aut}(R)$. Let $\phi_x, \phi_y \in \text{Inn}(R)$. Then for all $z \in R$,

$$(\phi_x \circ \phi_y)(z) = \phi_x\left(\phi_y(z)\right) = \phi_x(yzy^{-1}) = xyzy^{-1}x^{-1} = (xy)z(xy)^{-1} = \phi_{xy}(z).$$

So, $\phi_x \circ \phi_y = \phi_{xy} \in \operatorname{Inn}(R)$ and therefore, $\operatorname{Inn}(R)$ is closed under composition. Let $\phi_x \in \operatorname{Inn}(R)$. Then for all $z \in R$,

$$(\phi_x \circ \phi_{x^{-1}})(z) = \phi_x(\phi_{x^{-1}}(z)) = \phi_x(x^{-1}zx) = xx^{-1}zxx^{-1} = z = \phi_e(z).$$

So, $\phi_x \circ \phi_{x^{-1}} = \phi_e$. It follows that $(\phi_x)^{-1} = \phi_{x^{-1}} \in \operatorname{Inn}(R)$. So, $\operatorname{Inn}(R)$ is closed under taking inverses. Therefore, $\operatorname{Inn}(R) \leq \operatorname{Aut}(R)$. □

29. Let G be a group with H and K subgroups of G, and let $G = H \cup K$. Then $H = G$ or $K = G$.

Proof: Let G be a group, let H and K be subgroups of G, and let $G = H \cup K$. Suppose toward contradiction that $H \neq G$ and $K \neq G$. Then there exist $a \in G \setminus H$ and $b \in G \setminus K$. Since $G = H \cup K$, we have $a \in K$ and $b \in H$. Since G is a group, $ab \in G$. So, $ab \in H$ or $ab \in K$. Without loss of generality, let $ab \in H$. Since $b \in H$ and H is a group, $b^{-1} \in H$, and thus, $a = ae = a(bb^{-1}) = (ab)b^{-1} \in H$. This contradicts our assumption that $a \in G \setminus H$. So, we must have $H = G$ or $K = G$. □

30. $\mathbb{Z}[x]_0$ is a subgroup of $\mathbb{Z}[x]_1$ and $\mathbb{Z}[x]_1$ is a subgroup of $\mathbb{Z}[x]_2$.

(Note that $\mathbb{Z}[x]_n = \{a_n x^n + a_{n-1}x^{n-1} + \cdots + a_1 x + a_0 \mid a_0, a_1, \ldots, a_n \in \mathbb{Z}\}$. By Problem 84 in Problem Set 3, $(\mathbb{Z}_n[x], +)$ is a commutative group for $n = 0$, 1, and 2, where addition is defined in the "usual way.")

Proof: $\mathbb{Z}[x]_0 = \{a_0 \mid a_0 \in \mathbb{Z}\} = \mathbb{Z}$.

$\mathbb{Z}[x]_1 = \{a_1 x + a_0 \mid a_0, a_1 \in \mathbb{Z}\}$.

$\mathbb{Z}[x]_2 = \{a_2 x^2 + a_1 x + a_0 \mid a_0, a_1, a_2 \in \mathbb{Z}\}$.

Let $a_0 \in \mathbb{Z}[x]_0 = \mathbb{Z}$. Then $a_0 = 0x + a_0 \in \mathbb{Z}[x]_1$. So, $\mathbb{Z}[x]_0 \subseteq \mathbb{Z}[x]_1$. Since $\mathbb{Z}[x]_0$ is a group under addition, $\mathbb{Z}[x]_0 \leq \mathbb{Z}[x]_1$.

Similarly, if $a_1 x + a_0 \in \mathbb{Z}[x]_1$, then $a_1 x + a_0 = 0x^2 + a_1 x + a_0 \in \mathbb{Z}[x]_2$. Therefore, $\mathbb{Z}[x]_1 \subseteq \mathbb{Z}[x]_2$. Since $\mathbb{Z}[x]_1$ is a group under addition, $\mathbb{Z}[x]_1 \leq \mathbb{Z}[x]_2$. □

LEVEL 4

Prove each of the following:

31. Let Γ be a set of first-order formulas and let ϕ be a first-order formula. Suppose that $\Gamma \cup \{\phi\}$ is inconsistent. Then $\Gamma \vdash \neg\phi$.

Proof: Since $\Gamma \cup \{\phi\}$ is inconsistent, there is a formula ψ such that $\Gamma \cup \{\phi\} \vdash \psi$ and $\Gamma \cup \{\phi\} \vdash \neg\psi$. By Problem 22, $\Gamma \vdash \phi \to \psi$ and $\Gamma \vdash \phi \to \neg\psi$. Since $\{\phi \to \psi, \phi \to \neg\psi\}$ tautologically implies $\neg\phi$, by Problem 20, $\Gamma \vdash \neg\phi$. □

32. For any formula ϕ, $\vdash \exists x \forall y \phi \to \forall y \exists x \phi$.

Proof: By Problem 22, it suffices to show that $\exists x \forall y \phi \vdash \forall y \exists x \phi$. By Universal Generalization, it suffices to show that $\exists x \forall y \phi \vdash \exists x \phi$. By Problem 23, it suffices to show that $\neg \exists x \phi \vdash \neg \exists x \forall y \phi$. By the Generalized De Morgan's Laws, it suffices to show that $\forall x \neg \phi \vdash \forall x \neg \forall y \phi$. Again, by Universal Generalization, it suffices to show that $\forall x \neg \phi \vdash \neg \forall y \phi$. By Problem 31, it suffices to show that the set $\{\forall x \neg \phi, \forall y \phi\}$ is inconsistent. Now simply observe that by Universal Elimination, $\forall x \neg \phi \vdash \neg \phi$ and $\forall x \phi \vdash \phi$. □

Determine whether each of the following logical arguments is valid or invalid. A and B are unary relation symbols and a and b are constant symbols. If the argument is valid, provide a deduction. If the argument is invalid, provide a counterexample.

33

$$\frac{\exists x(A(x))}{\forall x(A(x))}$$

34

$$\frac{\begin{array}{c}\forall x(A(x) \vee B(x)) \\ \exists x \neg A(x)\end{array}}{\exists x(B(x))}$$

35

$$\frac{\begin{array}{c}\forall x(A(x) \vee B(x)) \\ \forall x(A(x) \to C(x)) \\ \forall x(B(x) \to C(x))\end{array}}{C(a) \wedge C(b)}$$

36

$$\frac{\begin{array}{c}\exists x(A(x)) \\ B(a)\end{array}}{\exists x(A(x) \wedge B(x))}$$

33. Consider the structure $\mathfrak{A} = (X, A^{\mathfrak{A}})$, where $X = \{a, b\}$, and $A^{\mathfrak{A}}(a)$. Since $A^{\mathfrak{A}}(a)$, we have $\mathfrak{A} \vDash \exists x(A(x))$. Since $\mathfrak{A} \nvDash A(b)$, $\mathfrak{A} \nvDash \forall x A(x)$. Therefore, the argument is **invalid**.

34. Here is a derivation.

1	$\forall x(A(x) \vee B(x))$	Premise
2	$\exists x \neg A(x)$	Premise
3	$\neg A(a)$	Existential Elimination (2)
4	$A(a) \vee B(a)$	Universal Elimination (1)
5	$B(a)$	Disjunctive syllogism (4, 3)
6	$\exists x(B(x))$	Existential Generalization (5)

Therefore, the argument is **valid**.

35. Here is a derivation.

1	$\forall x(A(x) \vee B(x))$	Premise
2	$\forall x(A(x) \to C(x))$	Premise
3	$\forall x(B(x) \to C(x))$	Premise
4	$A(a) \vee B(a)$	Universal Elimination (1)
5	$A(a) \to C(a)$	Universal Elimination (2)
6	$B(a) \to C(a)$	Universal Elimination (3)
7	$C(a) \vee C(a)$	Constructive Dilemma (5, 6, 4)
8	$C(a)$	Redundancy Law (7)
9	$A(b) \vee B(b)$	Universal Elimination (1)
10	$A(b) \to C(b)$	Universal Elimination (2)
11	$B(b) \to C(b)$	Universal Elimination (3)
12	$C(b) \vee C(b)$	Constructive Dilemma (10, 11, 9)
13	$C(b)$	Redundancy Law (12)
14	$C(a) \wedge C(b)$	Conjunctive Introduction (8, 13)

Therefore, the argument is **valid**.

36. Consider the structure $\mathfrak{A} = (X, A^{\mathfrak{A}}, B^{\mathfrak{A}})$, where $X = \{a, b\}$, $A^{\mathfrak{A}}(b)$ and $B^{\mathfrak{A}}(a)$. Since $A^{\mathfrak{A}}(b)$, $\mathfrak{A} \vDash \exists x(A(x))$. Since $B^{\mathfrak{A}}(a)$, $\mathfrak{A} \vDash B(a)$. However, $\mathfrak{A} \nvDash \exists x(A(x) \wedge B(x))$. Therefore, the argument is **invalid**.

Let G and H be groups and let $f: G \to H$ be a homomorphism. Prove each of the following:

37. If G is commutative, then $f[G] = \{f(x) \mid x \in G\}$ is commutative.

Proof: Suppose that G is commutative and let $x, y \in f[G]$. Then there are $g, h \in G$ such that $f(g) = x$ and $f(h) = y$. So, $xy = f(g)f(h) = f(gh) = f(hg) = f(h)f(g) = yx$. Since $x, y \in f[G]$ were arbitrary, $f[G]$ is commutative. □

Note: $f[G] = \{f(x) \mid x \in G\}$ is the **image** of f. You should check carefully that $f[G]$ is a subgroup of H. See the Note following Theorem 11.2 for details.

38. If $n \in \mathbb{Z}$ and $x \in G$, then $f(x^n) = (f(x))^n$.

Proof: If $n = 0$, then $f(x^0) = f(e) = e = (f(x))^0$. For $n \in \mathbb{Z}^+$, we use induction on n.

Base case ($k = 1$): $f(x^1) = f(x) = (f(x))^1$.

Inductive step: Assume that $k \in \mathbb{Z}^+$ and $f(x^k) = \left(f(x)\right)^k$. Then

$$f(x^{k+1}) = f(x^k x) = f(x^k)f(x) = \left(f(x)\right)^k f(x) = \left(f(x)\right)^{k+1}.$$

By the Principle of Mathematical Induction, $f(x^n) = \left(f(x)\right)^n$ for all $n \in \mathbb{Z}^+$.

Finally, if $n \in \mathbb{Z}^+$, then $f(x^{-n}) = f((x^n)^{-1}) = \left(f(x^n)\right)^{-1} = \left(\left(f(x)\right)^n\right)^{-1} = \left(f(x)\right)^{-n}$.

So, for all $n \in \mathbb{Z}$, $f(x^n) = \left(f(x)\right)^n$. □

39. If G is cyclic, then $f[G] = \{f(x) \mid x \in G\}$ is cyclic.

Proof: Suppose that G is cyclic with generator x. We will show that $f[G]$ is cyclic with generator $f(x)$. Let $y \in f[G]$. Then there is $g \in G$ with $f(g) = y$. Since G is generated by x, there is $n \in \mathbb{Z}$ with $g = x^n$. Then $y = f(g) = f(x^n) = \left(f(x)\right)^n \in \langle f(x) \rangle$ (by Problem 38 above). So, $f[G] \subseteq \langle f(x) \rangle$. Conversely, if $y \in \langle f(x) \rangle$, then there is $n \in \mathbb{Z}$ with $y = \left(f(x)\right)^n$. Since $x \in G$, $f(x) \in f[G]$. Since $f[G]$ is a subgroup of H (see the Note following Theorem 11.2), $y = \left(f(x)\right)^n \in f[G]$. Therefore, $\langle f(x) \rangle \subseteq f[G]$. Thus, $\langle f(x) \rangle = f[G]$. □

40. If f is an isomorphism, then $f[Z(G)] = Z(H)$, where $Z(G) = \{x \in G \mid xz = zx \text{ for all } z \in G\}$ is the **center** of G.

Proof: Let $y \in f[Z(G)]$. Then there is $x \in Z(G)$ with $y = f(x)$. Since $x \in Z(G)$, $xz = zx$ for all $z \in G$. So, $f(x)f(z) = f(xz) = f(zx) = f(z)f(x)$ for all $z \in G$. Since $y = f(x)$, we see that $yf(z) = f(z)y$ for all $z \in G$. So, $y \in Z(f[G])$. Since $y \in f[Z(G)]$ was arbitrary, we have $f[Z(G)] \subseteq Z(f[G]) = Z(H)$ (because $f[G] = H$).

Conversely, let $y \in Z(H)$. Since f is an isomorphism, there is $x \in G$ with $y = f(x)$. Let $w \in G$ and let $z = f(w)$. Then

$$xw = f^{-1}(y)f^{-1}(z) = f^{-1}(yz) = f^{-1}(zy) = f^{-1}(z)f^{-1}(y) = wx.$$

So, $x \in Z(G)$. Therefore, $y = f(x) \in f[Z(G)]$. Since $y \in Z(H)$ was arbitrary, we have $Z(H) \subseteq f[Z(G)]$. So, $f[Z(G)] = Z(H)$. □

Prove each of the following:

41. $(\mathbb{Q}, +)$ and $(\mathbb{Z}, +)$ are **not** isomorphic.

Proof: Suppose toward contradiction that $f\colon \mathbb{Q} \to \mathbb{Z}$ is an isomorphism. Since f is surjective, there is $q \in \mathbb{Q}$ such that $f(q) = 1$. Then $2f\left(\frac{q}{2}\right) = f\left(\frac{q}{2}\right) + f\left(\frac{q}{2}\right) = f\left(\frac{q}{2} + \frac{q}{2}\right) = f(q) = 1$. So, $f\left(\frac{q}{2}\right)$ is a solution of the equation $2x = 1$. But this equation has no solution in $\mathbb{Z}$. Therefore, there is no such isomorphism. □

42. $(\mathbb{Q}, +)$ and $(\mathbb{Q}^+, \cdot)$ are **not** isomorphic.

Proof: Suppose toward contradiction that $f: \mathbb{Q} \to \mathbb{Q}^+$ is an isomorphism. Since f is surjective, there is $q \in \mathbb{Q}$ such that $f(q) = 2$. Then $\left(f\left(\frac{q}{2}\right)\right)^2 = f\left(\frac{q}{2}\right) f\left(\frac{q}{2}\right) = f\left(\frac{q}{2} + \frac{q}{2}\right) = f(q) = 2$. So, $f\left(\frac{q}{2}\right)$ is a solution of the equation $x^2 = 2$. But this equation has no solution in $\mathbb{Q}$. So, there is no such isomorphism. □

43. Let G be a group and let H be a nonempty finite subset of G that is closed under the operation of G. Then H is a subgroup of G.

Proof: Let $x \in H$. We need to prove that $x^{-1} \in H$. If $x = e$, then $x^{-1} = e^{-1} = e \in H$. If $x \neq e$, then by closure and induction on $n \in \mathbb{Z}^+$, $x^n \in H$ for all $n \in \mathbb{Z}^+$. Since H is finite, there are $j, k \in \mathbb{Z}^+$ with $j < k$ and $x^j = x^k$. Then $k - j \in \mathbb{Z}^+$ and $x^{k-j} = x^k x^{-j} = x^k x^{-k} = x^0 = e$. Since $x \neq e$, $k - j > 1$, and so, $k - j - 1 \geq 1$. So, $x \cdot x^{k-j-1} = x^{k-j} = e$. So, $x^{k-j-1} = x^{-1}$ and since $k - j - 1 \geq 1$, $x^{k-j-1} \in H$. □

LEVEL 5

Prove each of the following:

44. Unique Readability for Formulas (Theorem 9.12)

Proof: We need the analogues of Theorems 9.6 and 9.8 for formulas. The proofs are similar to the proofs of those theorems. I will leave the proof of the first to the reader and prove the second. We will prove by the principle of induction on formulas that every proper initial segment of a formula has more left than right parentheses. Let's start with an atomic formula, say $(R(t_1 t_2 \cdots t_n))$. A proper initial segment of this formula has one of the following forms:

$$($$
$$(R$$
$$(R($$
$$(R(s_1, \text{ where } s_1 \text{ is a proper initial segment of } t_1$$
$$(R(t_1$$
$$(R(t_1 t_2 \cdots s_k, \text{ where } s_k \text{ is a proper initial segment of } t_k$$
$$(R(t_1 t_2 \cdots t_k$$
$$(R(t_1 t_2 \cdots t_k)$$

It is easy to check that each of these expressions has more left than right parentheses. For example, (has 1 left parenthesis and 0 right parentheses. As another example, if s_1 has i left parentheses and j right parentheses, then by the inductive assumption, $i > j$. It follows that $(R(s_1$ has $i + 2$ left parentheses and j right parentheses, and $i + 2 > i > j$. The other six cases are similar.

Now assume that the statement "every proper initial segment of a formula has more left than right parentheses" is true for the formula ϕ.

A proper initial segment of $(\neg\phi)$ has one of the following forms:

$$($$

$$(\neg$$

$$(\neg\alpha, \text{ where } \alpha \text{ is a proper initial segment of } \phi$$

$$(\neg\phi$$

It is easy to check that each of these expressions has more left than right parentheses. For the third form, we need to use the inductive assumption. For the last form, we use the fact that a formula has the same number of left and right parentheses (this is the analogue of Theorem 9.6 for formulas).

Similar arguments can be made for $(\phi \wedge \psi)$, $(\phi \vee \psi)$, $(\phi \to \psi)$, $(\phi \leftrightarrow \psi)$, $(\forall x\phi)$, and $(\exists x\phi)$.

By the principle of induction on formulas, every proper initial segment of a formula has more left parentheses than right parentheses.

Now, a formula from group (i) (an atomic formula) has the form $\big(R(t_1t_2 \cdots t_n)\big)$ and a formula from group (ii) has the form $(\neg\psi)$. These two formulas cannot be the same because they disagree in the second symbol. Similar arguments can be made for formulas from different groups among (i), (ii), (iii), and (iv).

Now, suppose ϕ can be written as both $\big(R(t_1t_2 \cdots t_n)\big)$ and $\big(S(u_1u_2 \cdots u_n)\big)$. Since the second symbols must agree, we have $R = S$ (as symbols). Deleting the first three symbols from each expression, we now have $t_1t_2 \cdots t_n) = u_1u_2 \cdots u_n)$. Suppose toward contradiction that there is a positive integer k such that $t_k \neq u_k$. Let j be the least such positive integer. Then, as sequences of symbols, t_j is a proper initial segment of u_j or u_j is a proper initial segment of t_j. Without loss of generality, assume that t_j is a proper initial segment of u_j. Since u_j is a term, t_j has more left parentheses than right parentheses, contradicting the analogue of Theorem 9.6. It follows that for each $k \leq n$, we have $t_k = u_k$. Therefore, $\big(R(t_1t_2 \cdots t_n)\big)$ and $\big(S(u_1u_2 \cdots u_n)\big)$ are the same string of symbols.

Next, suppose that ϕ can be written as both $(\neg\psi)$ and $(\neg\theta)$. Then clearly $\psi = \theta$, so that $(\neg\psi)$ and $(\neg\theta)$ are the same string of symbols.

Next, suppose that ϕ can be written as both $(\psi C\theta)$ and $(\tau D\mu)$. Then we must have $\psi C\theta) = \tau D\mu)$. Suppose toward contradiction that $\psi \neq \tau$. Then, as sequences of symbols, ψ is a proper initial segment of τ or τ is a proper initial segment of ψ. Without loss of generality, assume that ψ is a proper initial segment of τ. Since τ is a formula, ψ has more left parentheses than right parentheses, contradicting that formulas have the same number of left and right parentheses. It follows that $\psi = \tau$. So, we have $C\theta) = D\mu)$. So, C and D must be the same symbol, and therefore, $\theta = \mu$.

The case of quantifiers can be handled similarly. □

45. Let x and y be variables. If y does not occur in ϕ, then x is substitutable for y in ϕ_y^x and $\left(\phi_y^x\right)_x^y = \phi$.

Proof: We first prove that for each term u, if y does not occur in u, then $\left(u_y^x\right)_x^y = u$.

If u is a variable such that $u \neq y$, then $u_y^x = \begin{cases} u & \text{if } u \neq x. \\ y & \text{if } u = x. \end{cases}$ So, $\left(u_y^x\right)_x^y = \begin{cases} u & \text{if } u \neq x \\ x & \text{if } u = x \end{cases} = u.$

If u is a constant symbol, then $u \neq x$ and $u \neq y$. So, $\left(u_y^x\right)_x^y = u_x^y = u$.

Suppose that $\left(u_{i\,y}^{\,x}\right)_x^y = u_i$ for each $i = 1,2, \dots, n$ and that u is $(f(u_1u_2 \cdots u_n))$. Then $\left(u_y^x\right)_x^y$ is $\left(f\left(\left(u_{1\,y}^{\,x}\right)_x^y \left(u_{2\,y}^{\,x}\right)_x^y \cdots \left(u_{n\,y}^{\,x}\right)_x^y\right)\right) = (f(u_1u_2 \cdots u_n))$.

By the principle of induction on terms, for each term u, if y does not occur in u, then $\left(u_y^x\right)_x^y = u$.

We now prove that for each formula ϕ, if y does not occur in ϕ, then $\left(\phi_y^x\right)_x^y = \phi$.

If ϕ is the atomic formula $(R(u_1u_2 \cdots u_n))$, then ϕ_y^x is the atomic formula $\left(R\left(u_{1\,y}^{\,x}u_{2\,y}^{\,x} \cdots u_{n\,y}^{\,x}\right)\right)$, x is substitutable for y in ϕ_y^x and $\left(\phi_y^x\right)_x^y = \left(R\left(\left(u_{1\,y}^{\,x}\right)_x^y \left(u_{2\,y}^{\,x}\right)_x^y \cdots \left(u_{n\,y}^{\,x}\right)_x^y\right)\right) = (R(u_1u_2 \cdots u_n)) = \phi$.

If ϕ is $\neg\psi$, where x is substitutable for y in ψ_y^x and $\left(\psi_y^x\right)_x^y = \psi$, then $\phi_y^x = \neg\psi_y^x$. So, x is substitutable for y in ϕ_y^x and $\left(\phi_y^x\right)_x^y = \neg\left(\psi_y^x\right)_x^y = \neg\psi = \phi$.

If ϕ is $\psi \wedge \theta$, where x is substitutable for y in ψ_y^x and θ_y^x, and $\left(\psi_y^x\right)_x^y = \psi$ and $\left(\theta_y^x\right)_x^y = \theta$, then $\phi_y^x = \psi_y^x \wedge \theta_y^x$. So, x is substitutable for y in ϕ_y^x and $\left(\phi_y^x\right)_x^y = \left(\psi_y^x\right)_x^y \wedge \left(\theta_y^x\right)_x^y = \psi \wedge \theta = \phi$.

The arguments for $\psi \vee \theta$, $\psi \to \theta$, and $\psi \leftrightarrow \theta$ are similar.

Suppose that ϕ is $\forall z\psi$, where x is substitutable for y in ψ_y^x and $\left(\psi_y^x\right)_x^y = \psi$. There are two cases to consider.

<u>Case 1</u> ($z = x$): $\phi_y^x = (\forall z\psi)_y^x = \forall x\psi_y^x = \forall x\psi$. Since y does not occur free in $\forall x\psi$ (in fact, y does not occur in $\forall x\psi$), x is substitutable for y in $\forall x\psi = \phi_y^x$ and $\left(\phi_y^x\right)_x^y = (\forall x\psi)_x^y = \forall x\psi_x^y = \forall x\psi = \phi$.

<u>Case 2</u> ($z \neq x$): $\phi_y^x = (\forall z\psi)_y^x = \forall z\psi_y^x$. Since $x \neq y$ and x is substitutable for y in ψ_y^x, x is substitutable for y in $\forall z\psi_y^x = \phi_y^x$. Also, $\left(\phi_y^x\right)_x^y = \left(\forall z\psi_y^x\right)_x^y = \forall z\left(\psi_y^x\right)_x^y = \forall z\psi = \phi$.

The argument for $\exists z\psi$ is similar.

By the principle of induction on formulas, for each formula ϕ, if y does not occur in ϕ, then x is substitutable for y in ϕ_y^x and $\left(\phi_y^x\right)_x^y = \phi$. □

46. Let x and y be variables. Then $\left(\phi_y^x\right)_x^y$ does **not necessarily** equal ϕ.

Proof: Let ϕ be the formula $\forall x(x < y)$. Then $\phi_y^x = \phi = \forall x(x < y)$ and $\left(\phi_y^x\right)_x^y = \forall x(x < x)$. So, we see that $\left(\phi_y^x\right)_x^y \neq \phi$. □

Problem Set 10

LEVEL 1

Determine if each of the following strict linearly ordered sets is well-ordered. Note that $<$ is always the usual order on the given set.

1. $(\mathbb{Z}^+, <)$

Well-ordered

2. $((0,1), <)$ (Note that $(0,1)$ is an open interval and not an ordered pair.)

Not well-ordered ($(0,1)$ has no least element)

3. $([1,8), <)$

Not well-ordered ($(1,8) \subseteq [1,8)$ and $(1,8)$ has no least element)

4. $([1,8) \cap \mathbb{Q}, <)$

Not well-ordered ($(1,8) \cap \mathbb{Q} \subseteq [1,8) \cap \mathbb{Q}$ and $(1,8) \cap \mathbb{Q}$ has no least element)

5. $(\{0,1,2,3,4\}, <)$

Well-ordered

6. $(2\mathbb{N}, <)$

Well-ordered

7. $(X, <_D)$, where $X = \{x \mid x \text{ is a word in the English language}\}$ and $<_D$ is the dictionary order on X.

Well-ordered

Find an ordinal isomorphic to the given well-ordered set.

8. $(2\mathbb{N}+1, <)$

$\boldsymbol{\omega}$

9. $(\{a,b,c,d,e,f,g\}, <)$, where $<$ orders the given set alphabetically

7

10. $(\mathbb{N} \cup \{\star\}, <^\star)$, where $\star$ is an element that is not equal to any natural number and $<^\star$ is defined on $\mathbb{N} \cup \{\star\}$ by

$$<^\star = \{(m,n) \mid (m,n \in \mathbb{N} \text{ and } m < n) \vee (m \in \mathbb{N} \text{ and } n = \star)\}.$$

$\boldsymbol{\omega + 1}$

11. $(\mathbb{Z}^-, >)$

$\boldsymbol{\omega}$

12. $\text{pred}(\mathbb{N}, 0)$

$\mathbf{0}$

13. $\text{pred}(\mathbb{N}, 30)$

$\mathbf{30}$

14. $\text{pred}(\mathbb{N} \cup \{\star\}, \star)$, where $(\mathbb{N} \cup \{\star\}, <^\star)$ is as defined in Problem 10

$\boldsymbol{\omega}$

Use cardinal arithmetic to compute each of the following:

15. $5 + 11$

$\mathbf{16}$

16. $7 + \omega$

$\boldsymbol{\omega}$

17. $\omega + 7$

$\boldsymbol{\omega}$

18. $\omega + \omega$

$\boldsymbol{\omega}$

19. $\omega \cdot \omega$

$\boldsymbol{\omega}$

20. $10 \cdot \omega$

$\boldsymbol{\omega}$

LEVEL 2

Use ordinal arithmetic to write each of the following ordinals in the form $\omega \cdot a + b$, where $a, b \in \omega$.

21. $7 + 3$

$\mathbf{10 = \omega \cdot 0 + 10}$

22. $7 + \omega$

$\boldsymbol{\omega = \omega \cdot 1 + 0}$

23. $\omega + 7$

$\boldsymbol{\omega + 7 = \omega \cdot 1 + 7}$

24. 7ω

$\boldsymbol{\omega = \omega \cdot 1 + 0}$

25. $\omega \cdot 7$

$\boldsymbol{\omega \cdot 7 = \omega \cdot 7 + 0}$

26. $7\omega + 7$

$\boldsymbol{\omega + 7 = \omega \cdot 1 + 7}$

27. $\omega \cdot 7 + 7$

$\boldsymbol{\omega \cdot 7 + 7}$

28. $7 + 7\omega$

$7 + \omega = \boldsymbol{\omega = \omega \cdot 1 + 0}$

29. $7 + \omega \cdot 7$

$\boldsymbol{\omega \cdot 7 = \omega \cdot 7 + 0}$

30. $7\omega + \omega$

$\omega + \omega = \boldsymbol{\omega \cdot 2 = \omega \cdot 2 + 0}$

31. $7\omega + 7\omega$

$\omega + \omega = \boldsymbol{\omega \cdot 2 = \omega \cdot 2 + 0}$

32. $\omega \cdot 7 + \omega$

$\boldsymbol{\omega \cdot 8 = \omega \cdot 8 + 0}$

33. $\omega + \omega \cdot 7$

$\boldsymbol{\omega \cdot 8 = \omega \cdot 8 + 0}$

34. $\omega \cdot 7 + \omega \cdot 7$

$\boldsymbol{\omega \cdot 14 = \omega \cdot 14 + 0}$

Determine if each of the following statements is true or false. If true, explain why. If false, provide a counterexample.

35. Every well-ordered set is isomorphic to an initial segment of the set of natural numbers.

False: For example, $\omega + 1$ is **not** isomorphic to an initial segment of the natural numbers.

36. If $(A, <)$ is an infinite well-ordered set, then ω is isomorphic to an initial segment of $(A, <)$.

True: This follows immediately from Theorem 10.18, together with the fact that ω is the least infinite ordinal.

37. Every transitive set is an ordinal.

False: For example, $\mathcal{P}(\{0,1\}) = \mathcal{P}(\{\emptyset, \{\emptyset\}\}) = \big\{\emptyset, \{\emptyset\}, \{\{\emptyset\}\}, \{\emptyset, \{\emptyset\}\}\big\}$ is transitive by part 5 of Example 10.9. However, this set is **not** linearly ordered because $\{\{\emptyset\}\}$ and $\{\emptyset, \{\emptyset\}\}$ are incomparable (neither is an element of the other). Therefore, $\mathcal{P}(\{0,1\})$ is a transitive set that is **not** an ordinal.

38. Every set that is well-ordered by $\in$ is an ordinal.

False: For example, $\{0, 2\} = \big\{\emptyset, \{\emptyset, \{\emptyset\}\}\big\}$ is well-ordered by $\in$ because it is a subset of the well-ordered set $3 = \{0, 1, 2\}$. However, this set is **not** transitive because $\{\emptyset\} \in \{\emptyset, \{\emptyset\}\} \in \big\{\emptyset, \{\emptyset, \{\emptyset\}\}\big\}$, but $\{\emptyset\} \notin \big\{\emptyset, \{\emptyset, \{\emptyset\}\}\big\}$ (or equivalently, $1 \in 2 \in \{0, 2\}$, but $1 \notin \{0, 2\}$). Therefore, $\{0, 2\}$ is a set that is well-ordered by $\in$ that is **not** an ordinal.

39. If α, β, and γ are ordinals with $\alpha \in \beta$ and $\beta \in \gamma$, then $\alpha \in \gamma$.

True: Since γ is an ordinal, γ is transitive. Therefore, $\beta \in \gamma$ and $\alpha \in \beta$ implies that $\alpha \in \gamma$.

40. There exist two well-ordered sets such that neither one is isomorphic to an initial segment of the other one.

False: This follows from Theorem 10.18.

Use the axioms of ZF to prove each of the following:

41. For all $n \geq 2$, every n-tuple of n sets is a set.

Proof: We prove by induction that for $n \in \mathbb{N}$ with $n \geq 2$ that every n-tuple of n sets is a set.

Base Case ($k = 2$)**:** The base case was done in part 4 of Example 10.1.

Inductive Step: Assume that every k-tuple of k sets is a set and let $x_1, x_2, \ldots, x_k, x_{k+1}$ be sets. By the inductive hypothesis, $x = (x_1, x_2, \ldots x_k)$ is a set. By part 4 of Example 10.1, (x, x_{k+1}) is a set. Therefore, $(x_1, x_2, \ldots, x_k, x_{k+1}) = \big((x_1, x_2, \ldots, x_k), x_{k+1}\big) = (x, x_{k+1})$ is a set.

By the Principle of Mathematical Induction, for all $n \in \mathbb{N}$ with $n \geq 2$, every n-tuple of n sets is a set. □

42. The Cartesian product of two sets is a set.

Proof: Let A and B be sets. By the pairing axiom, $\{A, B\}$ is a set. So, by the union axiom, $A \cup B = \bigcup\{A, B\}$ is a set. By applying the power set axiom twice, we see that $\mathcal{P}\big(\mathcal{P}(A \cup B)\big)$ is a set. If $a \in A$ and $b \in B$, then $\{a\}$ and $\{a, b\}$ are in $\mathcal{P}(A \cup B)$. So, $(a, b) = \big\{\{a\}, \{a, b\}\big\} \in \mathcal{P}\big(\mathcal{P}(A \cup B)\big)$. By Bounded Comprehension, $A \times B = \big\{x \in \mathcal{P}\big(\mathcal{P}(A \cup B)\big) \,\big|\, x = (a, b) \land a \in A \land b \in B\big\}$ is a set. □

Prove that Zorn's Lemma is equivalent to each of the following statements:

43. Let $(P, \leq)$ be a poset with nonempty domain such that each chain in P with nonempty domain has an upper bound in P. Then P contains at least one maximal element.

Proof: First assume that Zorn's Lemma holds and let $(P, \leq)$ be a poset with nonempty domain such that each chain in P with nonempty domain has an upper bound in P. Consider the empty chain, $(\emptyset, \leq)$. Since $P \neq \emptyset$, there is $a \in P$. The statement $\forall x \in \emptyset(x \leq a)$ is vacuously true. Therefore, a is an upper bound of $(\emptyset, \leq)$. By Zorn's Lemma, P contains at least one maximal element.

Next assume that the given statement holds and let $(P, \leq)$ be a poset such that each chain in P has an upper bound in P. Then P must be nonempty. Since each chain in P has an upper bound in P, it follows that each chain in P with nonempty domain has an upper bound in P. The given statement then implies that P contains at least one maximal element. □

44. If X is a set of sets such that, for each $Y \subseteq X$ such that $(Y, \subseteq)$ is a linearly ordered set, $\bigcup Y \in X$, then $(X, \subseteq)$ has at least one maximal element.

Proof: First assume that Zorn's Lemma holds and let X be a set of sets such that, for each $Y \subseteq X$ that is linearly ordered by $\subseteq$, $\bigcup Y \in X$. Since $X \subseteq \mathcal{P}(\bigcup X)$, by part 2 of Example 10.6, $(X, \subseteq)$ is a partially ordered set (if $A \subseteq B$ and $(B, \leq)$ is a partially ordered set, then $(A, \leq)$ is a partially ordered set). Let $(C, \subseteq)$ be a chain in X. Then $C \subseteq X$ and therefore, $\bigcup C \in X$. So, $\bigcup C$ is an upper bound of $(C, \subseteq)$ in X. By Zorn's Lemma, $(X, \subseteq)$ has at least one maximal element.

Next assume that the given statement holds and let $(P, \leq)$ be a poset such that each chain in P has an upper bound in P. Let $X = \{C \subseteq P \mid (C, \leq) \text{ is a chain in } P\}$. Then X is a set of sets. Let $Y \subseteq X$ be such that $(Y \subseteq)$ is a linearly ordered set. It is straightforward to check that $\bigcup Y$ is a linearly ordered set. So, $\bigcup Y \in X$. The given statement implies that $(X, \subseteq)$ has a maximal element, let's call it D. Since $(D, \leq)$ is a chain, D has an upper bound s. Suppose toward contradiction that s is not a maximal element of $(P, \leq)$. Then there is $y \in P$ with $s < y$. It follows that $(D \cup \{y\}, \leq)$ is a chain in P, contradicting the maximality of D. □

Strong Transfinite Induction is the following statement:

If S is a subclass of $\mathbf{ON}$ such that for all $\alpha \in \mathbf{ON}$, $\forall \beta < \alpha\ (\beta \in S) \rightarrow \alpha \in S$, then $S = \mathbf{ON}$.

45. Use Theorem 10.25 (Ordinal Well Ordering Principle) to prove Strong Transfinite Induction.

Proof: Let S be a subclass of $\mathbf{ON}$ such that for all $\alpha \in \mathbf{ON}$, $\forall \beta < \alpha\ (\beta \in S) \rightarrow \alpha \in S$. Assume towards contradiction that $S \neq \mathbf{ON}$. Let $A = \{\alpha \in \mathbf{ON} \mid \alpha \notin S\}$ (so, A is the class of ordinals **not** in S). Since $S \neq \mathbf{ON}$, A is nonempty. So, by the Ordinal Well Ordering Principle, A has an $\in$-least element, let's call it α. Since α is the $\in$-least element of A, $\forall \beta < \alpha\ (\beta \in S)$ is true. Thus, $\alpha \in S$, contradicting that $\alpha \in A$. So, $S = \mathbf{ON}$. □

Prove each of the following:

46. Each natural number is well-ordered by $\in$, and therefore, each natural number is an ordinal.

Proof: We prove by induction on $n \in \omega$ that each natural number is well-ordered by $\in$. 0 is trivially well-ordered by $\in$ because it has no elements and no non-empty subsets.

Suppose that $k \in \omega$ is well-ordered by $\in$. We will show that $k + 1 = k \cup \{k\}$ is well-ordered by $\in$.

To see that $\in$ is antireflexive on $k + 1$, let $j \in k + 1$. Then $j \in k$ or $j = k$. If $j \in k$, then since $\in$ is antireflexive on k, $j \notin j$. If $j = k$, then $j \notin j$ because if $j \in j$, then j would be an element of k with $j \in j$, contradicting that k is antireflexive.

To see that $\in$ is antisymmetric on $k + 1$, let $j, t \in k + 1$. If $j, t \in k$, then since $\in$ is antireflexive on k, we have $j \in t$ and $t \in j$ implies $t = j$. Suppose that $j = k$ or $t = k$. Without loss of generality, let $j = k$. If $j \in t$ and $t \in j$, then since $j = k$ is a transitive set (by part 6 of Example 10.9), $k \in k$. As in the last paragraph, this is a contradiction.

To see that $\in$ is transitive on $k + 1$, let $j, t, u \in k + 1$, and suppose that $j \in t$ and $t \in u$. If $u = k$, then since k is a transitive set (again by part 6 of Example 10.9), $j \in k = u$. If $u \in k$, then again, since k is a transitive set, $t \in k$ and $j \in k$. Since $\in$ is transitive on k (by the inductive hypothesis), we have $j \in u$.

To see that $(k + 1, \in)$ satisfies trichotomy, let $j, t \in k + 1$ and assume that $j \notin t$ and $t \notin j$. If $t = k$, then since $j \notin t$, $j = k = t$. Similarly, if $j = k$, then since $t \notin j$, $t = k = j$. If $t \neq k$ and $j \neq k$, then $t, j \in k$. Since $(k, \in)$ satisfies trichotomy, (by the inductive hypothesis) $t = j$. So, $(k + 1, \in)$ satisfies trichotomy.

Finally, let A be a nonempty subset of $k+1$. If $A \setminus \{k\} \neq \emptyset$, then $A \setminus \{k\} \subseteq k$. Since k is well-ordered by $\in$ (by the inductive hypothesis), $A \setminus \{k\}$ has an $\in$-least element, say j. Since $j \in k$, j is the $\in$-least element of A. If $A \setminus \{k\} = \emptyset$, then k is the only element of A, and therefore, it is the $\in$-least element of A. Therefore, $k+1$ is well-ordered by $\in$.

Since each natural number is transitive (by part 6 of Example 10.9) and well-ordered by $\in$, each natural number is an ordinal. □

47. Let κ and λ be infinite cardinals. Then $\kappa + \lambda = \max\{\kappa, \lambda\}$.

Proof: Without loss of generality, let $\max\{\kappa, \lambda\} = \kappa$. Then $\max\{\kappa, \lambda\} = \kappa \leq \kappa + \lambda$. Also, since $(\kappa \times \{0\}) \cup (\lambda \times \{1\}) \subseteq \kappa \times \lambda$, we have

$$\kappa + \lambda = |(\kappa \times \{0\}) \cup (\lambda \times \{1\})| \leq |\kappa \times \lambda| = \kappa \cdot \lambda \leq \kappa \cdot \kappa = \kappa \text{ (by Theorem 10.34)} = \max\{\kappa, \lambda\}.$$

Since $\kappa + \lambda \leq \max\{\kappa, \lambda\}$ and $\max\{\kappa, \lambda\} \leq \kappa + \lambda$, we have $\kappa + \lambda = \max\{\kappa, \lambda\}$. □

48. Let κ and λ be infinite cardinals. Then $\kappa \cdot \lambda = \max\{\kappa, \lambda\}$.

Proof: Without loss of generality, let $\max\{\kappa, \lambda\} = \kappa$. Then $\max\{\kappa, \lambda\} = \kappa \leq |\kappa \times \lambda| = \kappa \cdot \lambda$ and

$$\kappa \cdot \lambda = |\kappa \times \lambda| \leq |\kappa \times \kappa| = \kappa \cdot \kappa = \kappa \text{ (by Theorem 10.34)} = \max\{\kappa, \lambda\}.$$

Since $\kappa \cdot \lambda \leq \max\{\kappa, \lambda\}$ and $\max\{\kappa, \lambda\} \leq \kappa \cdot \lambda$, we have $\kappa \cdot \lambda = \max\{\kappa, \lambda\}$. □

49. Let κ be an infinite cardinal. Then $\kappa^3 = \kappa$.

Proof: By Theorem 10.34, $\kappa^2 = \kappa \cdot \kappa = \kappa$. So, $\kappa^3 = \kappa^2 \cdot \kappa = \kappa \cdot \kappa = \kappa$, again by Theorem 10.34.

50. Let κ be an infinite cardinal. For each natural number $n \geq 1$, $\kappa^n = \kappa$.

Proof: We prove by induction that $\kappa^n = \kappa$ for all $n \geq 1$.

Base case $(n = 1)$: $\kappa^1 = \kappa$.

Inductive step: Suppose that $\kappa^j = \kappa$. Then $\kappa^{j+1} = \kappa^j \cdot \kappa^1 = \kappa \cdot \kappa = \kappa$.

By the Principle of Mathematical Induction, $\kappa^n = \kappa$ for all $n \geq 1$. □

LEVEL 3

Determine if each of the following ordinals is a cardinal. If so, explain why. If not, describe a bijection between this ordinal and a smaller ordinal. Use ordinal arithmetic where appropriate.

51. 0

0 is a cardinal because all natural numbers are cardinals.

52. 15

15 is a cardinal because all natural numbers are cardinals.

53. $\omega + 7$

$\boldsymbol{\omega + 7}$ **is not a cardinal**. We can define a bijection $f\colon \omega + 7 \to \omega$ by

$$f(\alpha) = \begin{cases} \alpha + 7 & \text{if } \alpha \in \omega. \\ k & \text{if } \alpha = \omega + k. \end{cases}$$

54. $7 + \omega$

$\boldsymbol{7 + \omega = \omega}$ **is a cardinal** because ω is the least infinite cardinal.

55. $\omega \cdot 7$

$\boldsymbol{\omega \cdot 7}$ **is not a cardinal**. We can define a bijection $f\colon \omega \cdot 7 \to \omega$ by

$$f(\alpha) = 7k + n, \text{ where } \alpha = \omega \cdot n + k.$$

56. 7ω

$\boldsymbol{7\omega = \omega}$ **is a cardinal** because ω is the least infinite cardinal.

57. ω^2

$\boldsymbol{\omega^2}$ **is not a cardinal**. Here is a visual representation of a bijection between ω and ω^2:

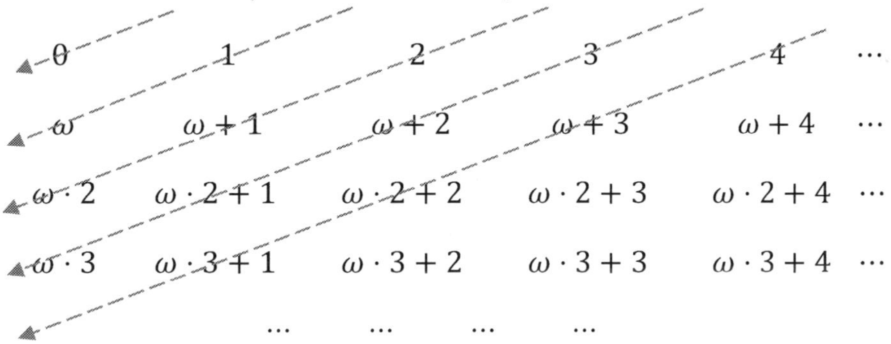

We can now list the elements of ω^2 as $0, 1, \omega, 2, \omega + 1, \omega \cdot 2, 3, \omega + 2, \omega \cdot 2 + 1, \omega \cdot 3, 4, \omega + 3, \ldots$

58. $5\omega^2 + 4$

$\boldsymbol{5\omega^2 + 4 = \omega^2 + 4}$ **is not a cardinal**. We can define a bijection $f\colon \omega^2 + 4 \to \omega^2$ by

$$f(\alpha) = \begin{cases} \alpha + 4 & \text{if } \alpha \in \omega. \\ \alpha & \text{if } \alpha \in \omega^2 \setminus \omega. \\ k & \text{if } \alpha = \omega^2 + k. \end{cases}$$

Use ordinal arithmetic to write each of the following ordinals in the form $\omega^2 \cdot a + \omega \cdot b + c$, where $a, b \in \omega$.

59. $2\omega \cdot \omega$

$\omega \cdot \omega = \boldsymbol{\omega^2} = \boldsymbol{\omega^2 \cdot 1 + \omega \cdot 0 + 0}$

60. $(\omega \cdot \omega) \cdot 2$

$\boldsymbol{\omega^2 \cdot 2 = \omega^2 \cdot 2 + \omega \cdot 0 + 0}$

61. $\omega^2 \cdot 3 + 5\omega$

$\boldsymbol{\omega^2 \cdot 3 + \omega = \omega^2 \cdot 3 + \omega \cdot 1 + 0}$

62. $5\omega + 3\omega^2$

$\omega + \omega^2 = \boldsymbol{\omega^2} = \boldsymbol{\omega^2 \cdot 1 + \omega \cdot 0 + 0}$

63. $5\omega + \omega^2 \cdot 3$

$\omega + \omega^2 \cdot 3 = \boldsymbol{\omega^2 \cdot 3} = \boldsymbol{\omega^2 \cdot 3 + \omega \cdot 0 + 0}$

64. $\omega \cdot 5 + \omega^2 \cdot 3$

$\omega \cdot 5 + \omega^2 \cdot 3 = \boldsymbol{\omega^2 \cdot 3} = \boldsymbol{\omega^2 \cdot 3 + \omega \cdot 0 + 0}$

Prove each of the following:

65. The following two versions of AC are equivalent: (i) If X is a set of nonempty sets, then there is a function $f: X \to \bigcup X$ such that for each $x \in X$, $f(x) \in x$; (ii) If X is a set of nonempty pairwise disjoint sets, then there is a set Y that contains exactly one element from each set in X.

Proof: First assume that (i) holds. Let X be a set of nonempty pairwise disjoint sets and let $f: X \to \bigcup X$ be such that for each $x \in X$, $f(x) \in x$. Since X is a set, by the union axiom, $\bigcup X$ is a set. By bounded comprehension, $Y = \{y \in \bigcup X \mid \exists x(x \in X \wedge y = f(x))\}$ is a set. For each $x \in X$, $f(x) \in x$ and $f(x) \in Y$. So, each set in X has an element in Y. Suppose $x \in X$ and $y \in Y$ with $y \in x$. If $f(z) = y$, then $y \in z$. Since X is pairwise disjoint, $z = x$. So, $f(z) = f(x)$. Therefore, $f(x)$ is the only element of Y that is in x.

Now, assume that (ii) holds and let X be a set of nonempty sets. For each $x \in X$, $\{x\}$ is a set by the pairing axiom. By Problem 42, for each $x \in X$, $x \times \{x\}$ is a set. By replacement, $A = \{x \times \{x\} \mid x \in X\}$ is a set. Also, if $x \times \{x\}, y \times \{y\} \in A$ with $z \in (x \times \{x\}) \cap (y \times \{y\})$, then there are $j \in x$ and $k \in y$ with $z = (j, x)$ and $z = (k, y)$. So, $(j, x) = (k, y)$, and thus, $x = y$. Therefore, $x \times \{x\} = y \times \{y\}$, showing that A is a set of nonempty pairwise disjoint sets. By (ii), there is a set Y containing exactly one element from each set in A. Let $f = \{(a, b) \mid a \in X \wedge (b, a) \in Y\}$. Let's check that $f: X \to \bigcup X$. If $a \in X$, then $a \times \{a\} \in A$. So, there is exactly one element of $a \times \{a\}$ in Y, say (b, a). Since $a \in X$ and $(b, a) \in Y$, $(a, b) \in f$. Also, since $(b, a) \in a \times \{a\}$, $b \in a$. Since $a \in X$ and $b \in a$, we have $b \in \bigcup X$. If $(a, b), (a, b') \in f$, then we have $a \in X$ and $(b, a), (b', a) \in Y$. Since $(b, a), (b', a) \in a \times \{a\}$ and Y contains exactly one element from each set in A, $b = b'$. This shows that f is a function. Finally, let $a \in X$. Then $(f(a), a) \in Y$. So, we have $(f(a), a) \in a \times \{a\}$. Therefore, $f(a) \in a$. □

66. We say that a set of sets X has **finite character** if the following holds: $A \in X$ if and only if every finite subset of A is in X.

Zorn's Lemma is equivalent to the following statement: If X is a set of sets with finite character, then $(X, \subseteq)$ has at least one maximal element (This statement is known as **Tukey's Lemma**).

Proof: First assume that Zorn's Lemma holds and let X be a set of sets with finite character. Let $(C, \subseteq)$ be a chain in X and let $Y = \bigcup C$. If A is a finite subset of Y, then $A \in X$ (because X has finite character). Since X has finite character, $Y \in X$. Therefore, Y is an upper bound of $(C, \subseteq)$ in X. By Zorn's Lemma, $(X, \subseteq)$ has at least one maximal element.

Next assume that Tukey's Lemma holds and let $(P, \leq)$ be a poset such that each chain in P has an upper bound in P. Let $X = \{C \subseteq P \mid (C, \leq) \text{ is a chain in } P\}$. If $A \in X$ and B is a finite subset of X, then $B \in X$ because a subset of a chain is a chain. If $A \notin X$, then A is not a chain. So, there are incomparable elements $a, b \in A$. Then the finite subset $\{a, b\}$ of A is not in X. Therefore, X has finite character. By Tukey's lemma, $(X, \subseteq)$ has a maximal element, let's call it D. Since $(D, \leq)$ is a chain, D has an upper bound s. Suppose toward contradiction that s is not a maximal element of $(P, \leq)$. Then there is $y \in P$ with $s < y$. It follows that $(D \cup \{y\}, \leq)$ is a chain in P, contradicting the maximality of D. □

67. Let $\boldsymbol{P} = \{P_0, P_1, \ldots\}$ be a partition of $\mathbb{N}$ into infinitely many infinite sets (see Problem 251 from Problem Set 2). Define $\lhd$ on $\mathbb{N}$ by

$$\lhd = \{(m, n) \big| \big(\exists k \in \mathbb{N}(m, n \in P_k \wedge m < n) \vee \exists j, k \in \mathbb{N}\big(j < k \wedge m \in P_j \wedge n \in P_k\big)\big\}.$$

$(\mathbb{N}, \lhd)$ is a well-ordered set isomorphic to the ordinal $\omega \cdot \omega$.

Proof: Checking that $(\mathbb{N}, \lhd)$ is a well-ordered set is straightforward.

Define $f \colon \mathbb{N} \to \omega \cdot \omega$ by $f(n) = \omega \cdot k + m$, where $n \in P_k$ and n has exactly m predecessors in P_k under the ordering $\lhd$. Note that for each $n \in \mathbb{N}$, k and m are uniquely determined because $\boldsymbol{P}$ is a partition of $\mathbb{N}$. Also, if $k, m \in \mathbb{N}$, then $\omega \cdot k + m \in \omega \cdot (k+1) \in \bigcup\{\omega \cdot j \mid j \in \omega\} = \omega \cdot \omega$. So, f is in fact a function from $\mathbb{N}$ into $\omega \cdot \omega$. To see that f is injective, suppose that $f(n) = \omega \cdot k + m$, $f(t) = \omega \cdot r + s$, and $f(n) = f(t)$. So, $\omega \cdot k + m = \omega \cdot r + s$. If $k < r$, then $\omega \cdot k + m \in \omega \cdot r \in \omega \cdot r + s$, and therefore, $\omega \cdot k + m \in \omega \cdot r + s$, which is impossible. Similarly, we cannot have $r < k$. So, $r = k$. Thus, we have $\omega \cdot k + m = \omega \cdot k + s$. So, $m = s$. It follows that $n = t$. Since $n, t \in \mathbb{N}$ were arbitrary, f is injective. To see that f is surjective, let $\omega \cdot k + m \in \omega \cdot \omega$. Then $f(n) = \omega \cdot k + m$, where $n \in P_k$ and n has exactly m predecessors in P_k under the ordering $\lhd$.

Now, suppose that $n \lhd t$. If there is $k \in \mathbb{N}$ with $n, t \in P_k$ and $n < t$, then $f(n) = \omega \cdot k + m$ and $f(t) = \omega \cdot k + s$ with $m, s \in \mathbb{N}$ and $m < s$. So, $f(n) < f(t)$. If there are $j, k \in \mathbb{N}$ with $j < k$, $n \in P_j$ and $t \in P_k$, then $f(n) = \omega \cdot j + m$ (for some $m \in \mathbb{N}$) $< \omega \cdot k + t$ (for some $t \in \mathbb{N}$) $= f(t)$. Conversely, if $f(n) < f(t)$ with $f(n) = \omega \cdot k + m$ and $f(t) = \omega \cdot r + s$, then $n \in P_k$ and $t \in P_r$. If $k < r$, then $n \lhd t$ by the definition of $\lhd$. If $k = t$, then $n, t \in P_k$ and $n < t$. So, $n \lhd t$. Therefore, f is an isomorphism. □

68. Let κ be an infinite cardinal and let $X = \bigcup\{\kappa^n \mid n \in \omega\}$. Then $|X| = \kappa$.

Proof: Since $\kappa \subseteq X$, $\kappa \leq |X|$. By Problem 50, for each $n \in \omega$, there is an injection $f_n: \kappa^n \to \kappa$. We define $g: X \to \kappa \times \omega$ by $g((\alpha_1, \alpha_2, \dots, \alpha_n)) = (f_n((\alpha_1, \alpha_2, \dots, \alpha_n)), n)$. It is easy to see that g is an injection. So, $|X| \leq \kappa \cdot \omega = \kappa$. By the Cantor -Schroeder-Bernstein Theorem, $|X| = \kappa$. □

69. Let κ be an infinite cardinal and for each ordinal $\alpha < \kappa$ let X_α be a set such that $|X_\alpha| \leq \kappa$. Then $|\bigcup\{X_\alpha \mid \alpha < \kappa\}| \leq \kappa$.

Proof: For each $\alpha < \kappa$, let $f_\alpha: X_\alpha \to \kappa$ be an injection. Define $g: \bigcup\{X_\alpha \mid \alpha < \kappa\} \to \kappa \times \kappa$ by $g(x) = (f_\alpha(x), \alpha)$, where α is the least ordinal such that $x \in X_\alpha$.

To see that g is injective, let $x, y \in \bigcup\{X_\alpha \mid \alpha < \kappa\}$ with $g(x) = g(y)$. Then $(f_\alpha(x), \alpha) = (f_\beta(y), \beta)$, where α is the least ordinal such that $x \in X_\alpha$ and β is the least ordinal such that $y \in X_\beta$. Since $(f_\alpha(x), \alpha) = (f_\beta(y), \beta)$, we must have $\alpha = \beta$. Therefore, $f_\alpha(x) = f_\beta(y) = f_\alpha(y)$. Since f_α is injective, $x = y$. Since $x, y \in \bigcup\{X_\alpha \mid \alpha < \kappa\}$ were arbitrary, g is injective.

It follows that $|\bigcup\{X_\alpha \mid \alpha < \kappa\}| \leq |\kappa \times \kappa| = \kappa$ (by Theorem 10.34). □

70. Let κ be an infinite cardinal. Define the relation $\lhd$ on $\kappa \times \kappa$ by $(\alpha, \beta) \lhd (\gamma, \delta)$ if and only if

(i) $\max\{\alpha, \beta\} \in \max\{\gamma, \delta\}$; or

(ii) $\max\{\alpha, \beta\} = \max\{\gamma, \delta\}$ and $\alpha \in \gamma$; or

(iii) $\max\{\alpha, \beta\} = \max\{\gamma, \delta\}$ and $\alpha = \gamma$ and $\beta \in \delta$.

Then $(\kappa \times \kappa, \lhd)$ is a well-ordered set.

Proof: To see that $\lhd$ is antireflexive on $\kappa \times \kappa$, let $(\alpha, \beta) \in \kappa \times \kappa$. First assume that $\alpha \leq \beta$ (in the sense that $\alpha \in \beta$ or $\alpha = \beta$). Then $\max\{\alpha, \beta\} = \beta$. Since β is an ordinal, $\beta \notin \beta$, and so, (i) and (iii) fail. Since α is an ordinal, $\alpha \notin \alpha$. So, (ii) fails. The case where $\beta \in \alpha$ is similar. So, $(\alpha, \beta) \ntriangleleft (\alpha, \beta)$. Therefore, $\lhd$ is antireflexive.

To see that $\lhd$ is antisymmetric on $\kappa \times \kappa$, let $(\alpha, \beta), (\gamma, \delta) \in \kappa \times \kappa$ and assume that $(\alpha, \beta) \lhd (\gamma, \delta)$ and $(\gamma, \delta) \lhd (\alpha, \beta)$. First assume that $\alpha \leq \beta$ and $\gamma \leq \delta$, so that $\max\{\alpha, \beta\} = \beta$ and $\max\{\gamma, \delta\} = \delta$. If $\beta \in \delta$, then we cannot have $\delta \in \beta$ or $\delta = \beta$. So, $\beta = \delta$ and $\alpha \in \gamma$. But then we cannot have $\gamma \in \alpha$ or $\gamma = \alpha$. Therefore, $\alpha \leq \beta$ and $\gamma \leq \delta$ is impossible. Similar arguments show that the other three possibilities are impossible as well. So, we cannot have both $(\alpha, \beta) \lhd (\gamma, \delta)$ and $(\gamma, \delta) \lhd (\alpha, \beta)$. Therefore, $\lhd$ is vacuously antisymmetric on $\kappa \times \kappa$.

To see that $\lhd$ is transitive on $\kappa \times \kappa$, let $(\alpha, \beta), (\gamma, \delta), (\eta, \xi) \in \kappa \times \kappa$ and assume that $(\alpha, \beta) \lhd (\gamma, \delta)$ and $(\gamma, \delta) \lhd (\eta, \xi)$. First assume that $\alpha \leq \beta$, $\gamma \leq \delta$, and $\eta \leq \xi$ so that $\max\{\alpha, \beta\} = \beta$, $\max\{\gamma, \delta\} = \delta$, and $\max\{\eta, \xi\} = \xi$. If $\beta \in \delta$ and $\delta \in \xi$, then by the transitivity of ξ, we have $\beta \in \xi$. If $\beta = \delta$ and $\delta \in \xi$, then by a simple substitution, $\beta \in \xi$. If $\beta \in \delta$ and $\delta = \xi$, then $\beta \in \xi$, again by a substitution. If $\beta = \delta$, $\delta = \xi$, $\alpha \in \gamma$, and $\gamma \in \eta$, then by the transitivity of η, $\alpha \in \eta$. This takes care of all possibilities when $\alpha \leq \beta$, $\gamma \leq \delta$, and $\eta \leq \xi$. The other seven cases are similar, and all eight cases together prove that $\lhd$ is transitive on $\kappa \times \kappa$.

To see that $(\kappa \times \kappa, \lhd)$ satisfies trichotomy, let $(\alpha, \beta), (\gamma, \delta) \in \kappa \times \kappa$ and assume that $(\alpha, \beta) \ntriangleleft (\gamma, \delta)$. First assume that $\alpha \leq \beta$ and $\gamma \leq \delta$. Since $(\alpha, \beta) \ntriangleleft (\gamma, \delta)$, $\beta \notin \delta$. So, either $\delta \in \beta$ or $\beta = \delta$. If $\delta \in \beta$, then $(\gamma, \delta) \lhd (\alpha, \beta)$. If $\beta = \delta$, then $\alpha \notin \gamma$. So, either $\gamma \in \alpha$ or $\alpha = \gamma$. If $\gamma \in \alpha$, then $(\gamma, \delta) \lhd (\alpha, \beta)$. If $\alpha = \gamma$, then $(\alpha, \beta) = (\gamma, \delta)$. The other three cases are similar, and all four cases together prove that $(\kappa \times \kappa, \lhd)$ satisfies trichotomy.

Finally, let S be a nonempty subset of $\kappa \times \kappa$. Let $A = \{\max\{\gamma, \delta\} \,|\, (\gamma, \delta) \in S\}$. Since $S \neq \emptyset$, $A \neq \emptyset$. Let η be the $\in$-least element of A. Let $B = \{\delta \mid (\delta, \eta) \in S\}$ and let $C = \{\delta \mid (\eta, \delta) \in S\}$. Let γ be the least element of B if it exists and let ϵ be the least element of C if it exists. Note that at least one of γ or ϵ must exist. Finally, let $(\alpha, \beta) = \min\{(\gamma, \eta), (\eta, \epsilon)\}$ under the ordering $\lhd$. Then (α, β) is the $\lhd$-least element of S. Therefore, $(\kappa \times \kappa, \lhd)$ is a well-ordered set. □

LEVEL 4

Let κ, λ, and μ be infinite cardinals. Prove each of the following:

71. If $\mu \leq \lambda$, then $\kappa^\mu \leq \kappa^\lambda$.

Proof: The function $F\colon \kappa^\mu \to \kappa^\lambda$ defined by $F(f) = g$, where $g(\alpha) = f(\alpha)$ for all $\alpha \in \mu$ and $g(\alpha) = 0$ for all $\alpha \in \lambda \setminus \mu$ is injective. □

72. $\kappa^\omega \leq \kappa^\lambda$.

Proof: Since λ is infinite, $\omega \leq \lambda$. So, the result follows from Problem 71. □

73. $\kappa^\omega \leq \kappa^\mu$.

Proof: Since μ is infinite, $\omega \leq \mu$. So, the result follows from Problem 71. □

74. If $\mu \leq \lambda$, then $\kappa^\lambda \cdot \kappa^\mu = \kappa^\lambda$.

Proof: Since $\omega < 2^\omega \leq \kappa^\omega$, by Problems 72 and 73, κ^λ and κ^μ are both infinite. So, by Problem 48, we have $\kappa^\lambda \cdot \kappa^\mu = \max\{\kappa^\lambda, \kappa^\mu\} = \kappa^\lambda$. The last equality follows from Problem 71. □

75. If $\mu \leq \lambda$, then $\kappa^{\lambda+\mu} = \kappa^\lambda$.

Proof: By Problem 47, we have $\lambda + \mu = \max\{\lambda, \mu\} = \lambda$. So, $\kappa^{\lambda+\mu} = \kappa^\lambda$. □

76. $\kappa^\lambda \cdot \kappa^\mu = \kappa^{\lambda+\mu}$.

Proof: We may assume that $\mu \leq \lambda$ (the argument is identical if we assume $\lambda \leq \mu$).

By Problem 74, $\kappa^\lambda \cdot \kappa^\mu = \kappa^\lambda$. By problem 75, $\kappa^{\lambda+\mu} = \kappa^\lambda$. Therefore, , $\kappa^\lambda . \kappa^\mu = \kappa^{\lambda+\mu}$. □

Determine if each of the following statements is true or false. If true, prove it. If false, provide a counterexample.

77. If α is an ordinal, then $\in$ is a transitive relation on $\alpha + 1 = \alpha \cup \{\alpha\}$.

True

Proof: Let $\beta, \gamma, \delta \in \alpha + 1 = \alpha \cup \{\alpha\}$ with $\beta \in \gamma$ and $\gamma \in \delta$. If $\delta = \alpha$, then since α is a transitive set, $\beta \in \alpha$. So, $\beta \in \delta$. If $\delta \in \alpha$, then since α is a transitive set, $\gamma \in \alpha$, and again, since α is a transitive set, $\beta \in \alpha$. Since $\in$ is transitive on α, we have $\beta \in \delta$. So, $\in$ is a transitive relation on $\alpha + 1 = \alpha \cup \{\alpha\}$. □

78. If α is an ordinal, then $(\alpha + 1, \in)$ satisfies trichotomy.

True

Proof: Let $\beta, \gamma \in \alpha + 1 = \alpha \cup \{\alpha\}$ and assume that $\beta \notin \gamma$ and $\gamma \notin \beta$. If $\gamma = \alpha$, then since $\beta \notin \alpha$, we have $\beta = \alpha = \gamma$. Similarly, if $\beta = \alpha$, then since $\gamma \notin \alpha$, we have $\gamma = \alpha = \beta$. If $\beta \neq \alpha$ and $\gamma \neq \alpha$, then $\beta, \gamma \in \alpha$. Since $(\alpha, \in)$ satisfies trichotomy, $\beta = \gamma$. So, $(\alpha + 1, \in)$ satisfies trichotomy. □

79. If α is an ordinal, then $\alpha + 1 = \alpha \cup \{\alpha\}$ is well-ordered by $\in$.

True

Proof: By Problem 77, $\in$ is transitive on $\alpha + 1 = \alpha \cup \{\alpha\}$.

By Problem 78, $(\alpha + 1, \in)$ satisfies trichotomy.

Now, let $B \subseteq \alpha + 1 = \alpha \cup \{\alpha\}$ be nonempty.

If $B \setminus \{\alpha\} \neq \emptyset$, then $B \setminus \{\alpha\}$ is a nonempty subset of α, and therefore, has a least element β (because α is well-ordered by $\in$). Since $\beta \in \alpha$, β is the least element of B as well.

If $B \setminus \{\alpha\} = \emptyset$, then $B = \{\alpha\}$, and the least element of B is α. It follows that $\alpha + 1 = \alpha \cup \{\alpha\}$ is well- ordered by $\in$. □

80. If α is a transitive set, then $\alpha + 1 = \alpha \cup \{\alpha\}$ is also a transitive set.

True

Proof: Let $\beta \in \alpha + 1 = \alpha \cup \{\alpha\}$ and let $\gamma \in \beta$. Then $\beta \in \alpha$ or $\beta \in \{\alpha\}$. If $\beta \in \alpha$, then γ is in α because α is an ordinal and therefore transitive. If $\beta \in \{\alpha\}$, then $\beta = \alpha$, and so, $\gamma \in \alpha$ because $\gamma \in \beta$. □

81. If α is an ordinal, then $\alpha + 1 = \alpha \cup \{\alpha\}$ is also an ordinal.

True

Proof: This follows immediately from Problems 79 and 80. □

82. If α is an ordinal and $x \in \alpha$, then x is an ordinal.

True

Proof: Let α be an ordinal and let $x \in \alpha$. Since α is transitive, $x \subseteq \alpha$. Since any subset of a well-ordered set is well-ordered, x is well-ordered by $\in$.

Now, suppose that $y \in x$ and $z \in y$. Since $x \subseteq \alpha$, we have $y \in \alpha$. Since α is transitive, $z \in \alpha$. Since $x, y, z \in \alpha$, $z \in y$, $y \in x$, and α is strictly linearly ordered by $\in$, we have $z \in x$. Therefore, x is transitive.

Since x is well-ordered by $\in$ and transitive, x is an ordinal. □

83. If κ a cardinal and $x \in \kappa$, then x is a cardinal.

False

ω_1 is a cardinal and $\omega + 1 \in \omega_1$, but $\omega + 1$ is **not** a cardinal.

84. If α is an ordinal and $x \in \alpha$, then $x = \text{pred}(\alpha, x)$.

True

Proof: Let $z \in x$. Since $x \subseteq \alpha$, we have $z \in \alpha$. Therefore, $z \in \text{pred}(\alpha, x)$. So, $x \subseteq \text{pred}(\alpha, x)$.

Now, let $z \in \text{pred}(\alpha, x)$. Then $z \in \alpha$ and $z \in x$. In particular, $z \in x$. So, $\text{pred}(\alpha, x) \subseteq x$.

Since we have $x \subseteq \text{pred}(\alpha, x)$ and $\text{pred}(\alpha, x) \subseteq x$, it follows that $x = \text{pred}(\alpha, x)$. □

Let α and β be ordinals. Prove each of the following:

85. If α is isomorphic to β, then $\alpha = \beta$.

Proof: Suppose that α and β are isomorphic ordinals and let $f: \alpha \cong \beta$. We will show that $f(\gamma) = \gamma$ for each $\gamma \in \alpha$. Suppose toward contradiction that $S = \{\gamma \in \alpha \mid f(\gamma) \neq \gamma\} \neq \emptyset$. By the Ordinal Well Ordering Principle, S has a least element, say δ. Suppose that $f(\delta) = \eta$. By Problem 84, $\eta = \text{pred}(\beta, \eta)$. Also, $\text{pred}(\beta, \eta) = \{f(\tau) \mid \tau \in \delta\} = \{\tau \mid \tau \in \delta\} = \delta$. So, $f(\delta) = \eta = \delta$, contradicting our assumption that $\delta \in S$. So, $S = \emptyset$ and $f(\gamma) = \gamma$ for each $\gamma \in \alpha$. So, $\gamma \in \alpha$ if and only if $\gamma \in \beta$. Therefore, $\alpha = \beta$. □

86. Exactly one of the following holds: $\alpha = \beta$, $\alpha \in \beta$, or $\beta \in \alpha$.

Proof: Let α and β be ordinals. If $\alpha \in \beta$ and $\beta \in \alpha$, then by the transitivity of α, $\alpha \in \alpha$. But then α is an element of α with $\alpha \in \alpha$, contradicting that α is antireflexive. So, $\alpha \in \beta$ and $\beta \in \alpha$ cannot both hold simultaneously. Also, if $\alpha \in \beta$ and $\alpha = \beta$, then $\alpha \in \alpha$, which once again is impossible. Similarly, we cannot have both $\alpha = \beta$ and $\beta \in \alpha$. So, at most one of $\alpha \in \beta$, $\alpha = \beta$, or $\beta \in \alpha$ holds.

Now, since α and β are well-ordered, by Theorem 10.18, $\alpha \cong \beta$, there is $\gamma \in \beta$ such that $\alpha \cong \text{pred}(\beta,\gamma)$, or there is $\gamma \in \alpha$ such that $\beta \cong \text{pred}(\alpha,\gamma)$. If $\alpha \cong \beta$, then by Problem 85, $\alpha = \beta$. If $\alpha \cong \text{pred}(\beta,\gamma)$, then by Problem 84, $\alpha \cong \gamma$. So, once again by Problem 85, $\alpha = \gamma$. Thus, $\alpha \in \beta$. If $\beta \cong \text{pred}(\alpha,\gamma)$, then again by Problem 84, $\beta \cong \gamma$. So, once again by Problem 85, we have $\beta = \gamma$. Thus, $\beta \in \alpha$. □

Prove each of the following:

87. The bounded comprehension schema follows from the replacement schema.

Proof: Let $b_1, b_2, \dots, b_n, a$ be sets and let ϕ be a first-order formula with free variables $w, y_1, y_2, \dots, y_n, x$. Let $\psi(u, v, y_1, y_2, \dots, y_n, x)$ be the formula

$$\phi(u, y_1, y_2, \dots, y_n, x) \wedge v = u.$$

Suppose that $\psi(u, w, y_1, y_2, \dots, y_n, x) \wedge \psi(u, v, y_1, y_2, \dots, y_n, x)$. Then $w = u$ and $v = u$. So, $w = v$. Therefore, we have

$$\forall u \forall w \forall v \forall y_1 \forall y_2 \cdots \forall y_n \forall x \left(\left(\phi(u, w, y_1, y_2, \dots, y_n, x) \wedge \phi(u, v, y_1, y_2, \dots, y_n, x) \right) \to w = v \right).$$

By the replacement schema, $\exists z \forall v \left(v \in z \leftrightarrow \exists u \left(u \in a \wedge \psi(u, v, b_1, b_2, \dots, b_n, a) \right) \right)$.

Let $z = \left\{ v \mid \exists u \left(u \in a \wedge \psi(u, v, b_1, b_2, \dots, b_n, a) \right) \right\}$.

Now, $\psi(u, v, b_1, b_2, \dots, b_n, a)$ is equivalent to $\phi(u, b_1, b_2, \dots, b_n, a) \wedge v = u$.

Therefore, $u \in a \wedge \psi(u, v, b_1, b_2, \dots, b_n, a)$ is equivalent to $u \in a \wedge \phi(u, b_1, b_2, \dots, b_n, a) \wedge v = u$.

Let $z' = \{ v \mid v \in a \wedge \phi(v, b_1, b_2, \dots, b_n, a) \}$. We will now show that $z = z'$.

First, let $v \in z'$. Then $v \in a$ and $\phi(v, b_1, b_2, \dots, b_n, a)$. Therefore, there is a u such that $u \in a$ and $\phi(u, b_1, b_2, \dots, b_n, a)$ and $v = u$. So, $v \in z$. Therefore, $z' \subseteq z$.

Now, let $v \in z$. Then there is a u such that $u \in a$, $\phi(u, b_1, b_2, \dots, b_n, a)$, and $v = u$. So, $v \in a$ and $\phi(v, b_1, b_2, \dots, b_n, a)$. Thus, $v \in z'$. Therefore, $z \subseteq z'$.

Since $z' \subseteq z$ and $z \subseteq z'$, we have $z = z'$, completing the proof. □

88. A transitive set of ordinals is an ordinal.

Proof: Let A be a transitive set of ordinals. We need to show that A is well-ordered by $\in$.

Let $\alpha \in A$. Since α is an ordinal, $\alpha \notin \alpha$, for otherwise, α is an element of itself that violates the antireflexivity of α. So $\in$ is antireflexive on A.

Let $\alpha, \beta \in A$. If $\alpha \in \beta$ and $\beta \in \alpha$, then by the transitivity of α, $\alpha \in \alpha$, which once again is impossible. So, $\alpha \in \beta$ and $\beta \in \alpha$ cannot hold, and so, $\in$ is antisymmetric on A.

Let $\alpha, \beta, \gamma \in A$ with $\alpha \in \beta$ and $\beta \in \gamma$. Then by the transitivity of γ, $\alpha \in \gamma$. So, $\in$ is transitive on A.

Let $\alpha, \beta \in A$. Since α and β are ordinals, by Problem 86, exactly one of the following holds: $\alpha = \beta$, $\alpha \in \beta$, or $\beta \in \alpha$. So, $(A, \in)$ satisfies trichotomy.

Let B be a nonempty subset of A and let $\alpha \in B$ be arbitrary. If α is not the $\in$-least element of B, let β be the $\in$-least element of $B \cap \alpha$. If $\gamma \in B$ with $\gamma \in \beta$, then by the transitivity of α, $\gamma \in \alpha$. Therefore, $\gamma \in B \cap \alpha$, contradicting that β is the $\in$-least element of $B \cap \alpha$. So, β is the $\in$-least element of B. It follows that A is well-ordered by $\in$.

Since A is transitive and well-ordered by $\in$, A is an ordinal. □

89. For any ordinal α, there exists a strictly increasing cofinal function $f: cf(\alpha) \to \alpha$.

Proof: If $\alpha = \beta + 1$, then $cf(\alpha) = 1$ and the function $f: 1 \to \alpha$ defined by $f(0) = \beta$ is a strictly increasing cofinal function.

Let α be a limit ordinal and let $g: cf(\alpha) \to \alpha$ be any cofinal function. We define $f: cf(\alpha) \to \alpha$. Assuming $f(\eta)$ has been defined for each ordinal $\eta < \xi$, let $f(\xi) = \max\{g(\xi), \bigcup\{f(\eta) + 1 \mid \eta < \xi\}\}$.

First observe that since $\xi < cf(\alpha)$, we must have $\bigcup\{f(\eta) + 1 \mid \eta < \xi\} < \alpha$ (otherwise, the function $h: \xi \to \alpha$ defined by $h(\eta) = f(\eta) + 1$ would be a cofinal function). Therefore, for each $\xi < cf(\alpha)$, $f(\xi) < \alpha$. So, $f: cf(\alpha) \to \alpha$. Since $f(\xi) \geq g(\xi)$ for each $\xi < cf(\alpha)$, f is cofinal.

Finally, if $\xi < \tau$, $f(\xi) < f(\xi) + 1 \leq \bigcup\{f(\eta) + 1 \mid \eta < \tau\} \leq f(\tau)$. So, f is strictly increasing. □

Let $\tau(n)$ be the formula $\phi(n) \wedge \psi(n)$, where

$$\phi(n) \equiv \forall m(m \notin n) \vee \exists k \forall m\big(m \in n \leftrightarrow (m \in k \vee m = k)\big)$$

and

$$\psi(n) \equiv \forall k\Big(k \in n \to \big(\forall m(m \notin k) \vee \exists m\big(m \in n \wedge \forall j(j \in k \leftrightarrow (j \in m \vee j = m)\big)\big)\Big)$$

90. Use the axiom of infinity, the bounded comprehension schema, and the axiom of foundation to prove that $\{n \mid \tau(n)\}$ is the set of natural numbers.

Proof: Let $A = \{n \mid \tau(n)\}$. We first prove by induction on $n \in \omega$ that $\omega \subseteq A$. 0 satisfies $\forall m(m \notin n)$, and therefore $\phi(0)$ is true. Also, $\psi(0)$ is vacuously true. So, $\tau(0)$ is true. Therefore, $0 \in A$. Now assume $k \in A$. Then $k + 1 = k \cup \{k\}$. So, $\forall m\big(m \in k + 1 \leftrightarrow (m \in k \vee m = k)\big)$. So, $\phi(k + 1)$ is true. We need to show that $\psi(k + 1)$ is true. Let $t \in k + 1$. Then $t \in k$ or $t = k$. If $t \in k$, then since $\psi(k)$ is true (by induction), we have $\forall m(m \notin t) \vee \exists m\big(m \in k \wedge \forall j(j \in t \leftrightarrow (j \in m \vee j = m)\big)$. Since $k \subseteq k + 1$, $m \in k$ implies $m \in k + 1$. So, we have $\forall m(m \notin t) \vee \exists m\big(m \in k + 1 \wedge \forall j(j \in t \leftrightarrow (j \in m \vee j = m)\big)$. If $t = k$, then since $\phi(k)$ is true (by induction), $\forall m(m \notin k) \vee \exists j \forall m\big(m \in k \leftrightarrow (m \in j \vee m = j)\big)$. If $\forall m(m \notin k)$ holds, then $\psi(k + 1)$ is true. If $\forall m(m \notin k)$ does not hold, then $k - 1 \in k$ and we have $\forall j\big(j \in k \leftrightarrow (j \in k - 1 \vee j = k - 1)\big)$. Since $k \subseteq k + 1$, $k - 1 \in k + 1$. It follows that we have $\exists m\big(m \in k + 1 \wedge \forall j(j \in k \leftrightarrow (j \in m \vee j = m)\big)$ (namely, $m = k - 1$). Therefore, once again, $\psi(k + 1)$ is true. Since $\phi(k + 1)$ is true and $\psi(k + 1)$ is true, $\tau(k + 1)$ is true, and $k + 1 \in A$. By the Principle of Mathematical Induction, $\omega \subseteq A$.

We now prove that $A \subseteq \omega$. Suppose that there is $x \in A$ such that $x \notin \omega$. We inductively define a sequence $x_0, x_1, \ldots, x_n, \ldots$ such that $x_0 = x$ and for all $n \in \omega$, $x_n \in A \setminus \omega$ and $x_{n+1} \in x_n$. Assuming x_n has been defined, first note that $x_n \neq \emptyset$ because $\emptyset = 0 \in \omega$. Therefore, since $\phi(x_n)$ is true, $\exists k \forall m\big(m \in x_n \leftrightarrow (m \in k \vee m = k)\big)$. In other words, there is a set k such that $x_n = k \cup \{k\}$. Let $x_{n+1} = k$. Then $\tau(x_{n+1})$ is true, and so, $x_{n+1} \in A$. If $x_{n+1} \in \omega$, then $x_n = x_{n+1} \cup \{x_{n+1}\} \in \omega$, which it is not. So, $x_{n+1} \notin \omega$.

The set $\{x_n \mid n \in \omega\}$ has no $\in$-least element, contradicting the axiom of foundation. It follows that there is no $x \in A$ such that $x \notin \omega$. So, $A \subseteq \omega$.

Since $\omega \subseteq A$ and $A \subseteq \omega$, $\{n \mid \tau(n)\} = A = \omega$. □

91. Prove that if there is a set x such that $x = \{x\}$, then $\{n \mid \tau(n)\}$ is **not** the set of natural numbers.

Proof: Let $A = \{n \mid \tau(n)\}$. Suppose that x satisfies $x = \{x\}$. Then $x \cup \{x\} = x \cup x = x$. Since x is the only element of x, we see that $x \in A$. We now show by induction that for all $n \in \omega$, $x \neq n$. Since $x \in x$ and $\emptyset$ has no elements, $x \neq \emptyset = 0$. Assume that $x \neq k$. If $k + 1 = x$, then $k \cup \{k\} = x$. So, $k \in x$. But the only element of x is x. So, $x = k$, contradicting our assumption that $x \neq k$. By the Principle of Mathematical Induction, for all $n \in \omega$, $x \neq n$. Thus, $x \notin \omega$. Therefore, $A \neq \omega$. □

LEVEL 5

Let X be a nonempty set of ordinals. Prove each of the following:

92. $\bigcup X$ is a transitive set.

Proof: If $\alpha \in \bigcup X$, then there is $\beta \in X$ with $\alpha \in \beta$. Since β is an ordinal, β is transitive. Therefore, $\alpha \subseteq \beta$. Since $\beta \subseteq \bigcup X$, $\alpha \subseteq \bigcup X$. Since $\alpha \in \bigcup X$ was arbitrary, $\bigcup X$ is transitive. □

93. $\bigcup X$ is well-ordered by $\in$.

Proof: Let $\beta, \gamma, \delta \in \bigcup X$ with $\beta \in \gamma$ and $\gamma \in \delta$. By the definition of the union, there are $x, y, z \in X$ with $\beta \in x$, $\gamma \in y$, and $\delta \in z$. By the definition of X, x, y, and z are ordinals. By the solution to Problem 82 above, β, γ, and δ are ordinals. By Problem 39, $\beta \in \delta$. So, $\in$ is transitive on $\bigcup X$.

Next, let $\beta, \gamma \in \bigcup X$. As in the last paragraph, there are ordinals $x, y \in X$ with $\beta \in x$ and $\gamma \in y$. We may assume that $x \leq y$ (that is $x \in y$ or $x = y$). Therefore, since y is transitive, $\beta \in y$. Since $\beta, \gamma \in y$ and y is an ordinal, we must have $\beta \in \gamma$, $\beta = \gamma$, or $\gamma \in \beta$. So, $(\bigcup X, \in)$ satisfies trichotomy.

Finally, let Y be a nonempty subset of $\bigcup X$ and let α be any element of Y. By the definition of union, $\alpha \in X$, and so, α is an ordinal. In particular, α is well-ordered by $\in$. If $\alpha \cap Y \neq \emptyset$, then it has a least element, which is also the least element of Y. Otherwise, α is the least element of Y.

It follows that $\bigcup X$ is well-ordered by $\in$. □

94. If $\alpha \in X$, then $\alpha \leq \bigcup X$.

Proof: Suppose $\alpha \in X$. Then $\alpha \subseteq \bigcup X$. Since α and $\bigcup X$ are both ordinals ($\bigcup X$ is an ordinal by Problems 92 and 93), we must have $\alpha \in \bigcup X$, $\alpha = \bigcup X$, or $\bigcup X \in \alpha$. Now, if $\bigcup X \in \alpha$, then since $\alpha \subseteq \bigcup X$, $\bigcup X \in \bigcup X$, which is impossible. So, $\alpha \in \bigcup X$ or $\alpha = \bigcup X$. That is $\alpha \leq \bigcup X$. □

95. $\bigcup X$ is the least ordinal greater than or equal to all elements of X.

Proof: By Problems 92 and 93, $\bigcup X$ is an ordinal.

By Problem 94, $\bigcup X$ is greater than or equal to all elements of X.

Now, suppose that there is an ordinal β such that $\alpha \in \beta$ for all $\alpha \in X$. Let $\gamma \in \bigcup X$. Then there is $\alpha \in X$ with $\gamma \in \alpha$. Since $\alpha \in X$, we have $\alpha \in \beta$. By the transitivity of β, $\gamma \in \beta$. Since $\gamma \in \bigcup X$ was arbitrary, $\bigcup X \subseteq \beta$. We cannot have $\beta \in \bigcup X$, because then we would have $\beta \in \beta$. Therefore, $\bigcup X \leq \beta$. It follows that $\bigcup X$ is the **least** ordinal greater than or equal to all elements of X. □

96. $\bigcap X$ is an ordinal.

Proof: If $\bigcap X = \emptyset$, then $\bigcap X = 0$, which is an ordinal. So, we may assume that $\bigcap X \neq \emptyset$.

We first show that $\bigcap X$ is transitive.

Let $\alpha \in \bigcap X$ and $\beta \in \alpha$. If $\gamma \in X$, then we have $\beta \in \alpha \in \gamma$. So, $\beta \in \gamma$ for all $\gamma \in X$. Therefore, $\beta \in \bigcap X$. Since $\beta \in \alpha$ was arbitrary, $\alpha \subseteq \bigcap X$. Since $\alpha \in \bigcap X$ was arbitrary, $\bigcap X$ is transitive.

Now, let $\beta, \gamma, \delta \in \bigcap X$ with $\beta \in \gamma$ and $\gamma \in \delta$. Let x be any element of X. Then $\beta, \gamma, \delta \in x$. By the definition of X, x is an ordinal. By the solution to Problem 82 above, β, γ, and δ are ordinals. By Problem 39, $\beta \in \delta$. So, $\in$ is transitive on $\bigcap X$.

Next, let $\beta, \gamma \in \bigcap X$. As in the last paragraph, let x be some ordinal in X. Then $\beta, \gamma \in x$. Since x is an ordinal, we must have $\beta \in \gamma$, $\beta = \gamma$, or $\gamma \in \beta$. So, $(\bigcap X, \in)$ satisfies trichotomy.

Finally, let Y be a nonempty subset of $\bigcap X$ and let α be any element of X. Then since $\bigcap X \subseteq \alpha$, we have $Y \subseteq \alpha$. Since α is an ordinal, Y has a least element.

It follows that $\bigcap X$ is well-ordered by $\in$. □

97. $\bigcap X$ is the least ordinal in X.

Proof: Let α be the least ordinal in X. Let $\beta \in \alpha$ and $\gamma \in X$. Since α is the least ordinal in X, $\gamma = \alpha$ or $\alpha \in \gamma$. In either case, $\beta \in \gamma$. So, $\beta \in \gamma$ for every $\gamma \in X$. Therefore, $\beta \in \bigcap X$. Since $\beta \in \alpha$ was arbitrary, $\alpha \subseteq \bigcap X$. Now, let $\beta \in \bigcap X$. Since $\alpha \in X$, $\beta \in \alpha$. Since $\beta \in \bigcap X$ was arbitrary, $\bigcap X \subseteq \alpha$. Since $\alpha \subseteq \bigcap X$ and $\bigcap X \subseteq \alpha$, we have $\bigcap X = \alpha$. Therefore, $\bigcap X$ is the least ordinal in X. □

98. If every ordinal in X is a cardinal, then $\bigcup X$ is a cardinal.

Proof: By Problem 95, $\bigcup X$ is an ordinal. Suppose toward contradiction that $|\bigcup X| \in \bigcup X$. Then there is $\kappa \in X$ with $|\bigcup X| \in \kappa$. Again, by the same problem just mentioned, $\kappa \in \bigcup X$. So, $\kappa = |\kappa| \leq |\bigcup X|$. Therefore, we have $\kappa \in \kappa$, which is impossible. Thus, $|\bigcup X| = \bigcup X$. So, $\bigcup X$ is a cardinal. □

Determine if each of the following statements is true or false. If true, prove it. If false, provide a counterexample.

99. Every infinite cardinal is a limit ordinal.

True

Proof: Every infinite successor ordinal is equinumerous with its predecessor. In other words, if $\alpha + 1$ is infinite, then $|\alpha + 1| = |\alpha|$, and therefore, $\alpha + 1$ is not a cardinal. It follows that every infinite cardinal is a limit ordinal. □

100. Every infinite successor cardinal is regular.

True

Proof: Let κ be an infinite successor cardinal. Then $\kappa = \alpha^+$ for some ordinal α. Suppose toward contradiction that $cf(\kappa) < \kappa$. Then there is an ordinal $\beta < \kappa$ and a cofinal function $f: \beta \to \kappa$. It follows that $\kappa = \bigcup\{f(\gamma) \mid \gamma \in \beta\}$. Now, for each $\gamma \in \beta$, $|f(\gamma)| < \kappa$. So, $|f(\gamma)| \leq |\alpha|$. Also, since $\beta < \kappa$, we have $|\beta| < \kappa$. So, $|\beta| \leq |\alpha|$. We have $\kappa = |\bigcup\{f(\gamma) \mid \gamma \in \beta\}| \leq |\alpha| \leq \alpha < \kappa$, a contradiction. □

101. Every limit cardinal is regular.

False

$cf(\omega_\omega) = \omega$. Therefore, ω_ω is a limit cardinal that is not regular.

Prove each of the following:

102. The set of natural numbers exists. Use only the axiom of infinity and the bounded comprehension schema.

Proof: By the axiom of infinity, there is a set x such that $\emptyset \in x \wedge \forall y(y \in x \to y \cup \{y\} \in x)$. By the bounded comprehension schema, we can define the set $A = \bigcap\{z \in \mathcal{P}(x) \mid \phi(z)\}$, where $\phi(z)$ is the formula $\emptyset \in z \wedge \forall k(k \in z \to k \cup \{k\} \in z)$.

Recall from Lesson 2 that the set of natural numbers, ω is defined to be the unique set satisfying (i) $\emptyset \in \omega$, (ii) $n \in \omega$ implies $n^+ \in \omega$, and (iii) if X is any set satisfying (i) and (ii), then $\omega \subseteq X$.

We now show that A is the unique set satisfying (i), (ii), and (iii) above.

Let $z \in \mathcal{P}(x)$ with $\phi(z)$ true. Then $\emptyset \in z$. Since $z \in \mathcal{P}(x)$ was an arbitrary element satisfying $\phi(z)$, $\emptyset \in A$. So, (i) holds.

Let $n \in A$ and let $z \in \mathcal{P}(x)$ such that $\phi(z)$. Then $n \in z$. Since $\phi(z)$ and $n \in z$, $n^+ \in z$. Since $z \in \mathcal{P}(x)$ with $\phi(z)$ was arbitrary, $n^+ \in A$. So, (ii) holds.

Let X be a set satisfying (i) and (ii). Then $\phi(X)$ holds. So, $\phi(x \cap X)$ holds. Since $x \cap X \subseteq x$, we have $x \cap X \in \mathcal{P}(x)$, and so, $A \subseteq x \cap X$. Since $x \cap X \subseteq X$, we have $A \subseteq X$. Therefore, (iii) holds.

So, A is a set satisfying (i), (ii), and (iii).

If B also satisfies (i), (ii), and (iii), then by (iii), we have $A \subseteq B$ and $B \subseteq A$. Thus, $A = B$. This shows that A is the unique set satisfying (i), (ii), and (iii).

So, A is the set of natural numbers. □

103. For every partially ordered set $(P, \leq_P)$, there is a linearly ordered set $(P, \leq_L)$ such that $\leq_P \subseteq \leq_L$.

Proof: Let $(P, \leq_P)$ be a partially ordered set and define X as follows:

$$X = \{(K, \leq_K) \mid K \subseteq P \wedge (K, \leq_K) \text{ is a linearly ordered set} \wedge \forall x, y \in K(x \leq_P y \rightarrow x \leq_K y\}$$

$(X, \subseteq)$ is a partially ordered set. Let $(C, \subseteq)$ be a chain in X. Then $\bigcup C \in X$. So, $\bigcup C$ is an upper bound of $(C, \subseteq)$ in X. By Zorn's Lemma, $(X, \subseteq)$ has at least one maximal element, say $(L, \leq_L)$. Suppose toward contradiction that $L \neq P$. Let $x \in P \setminus L$, let $M = L \cup \{x\}$, and let

$$\leq_M = \leq_L \cup \{(x, y) \mid y \in L \wedge \exists z \in L(x \leq_P z \wedge z \leq_L y)\}$$
$$\cup \{(y, x) \mid \neg(y \in L \wedge \exists z \in L(x \leq_P z \wedge z \leq_L y))\}.$$

Then $(M, \leq_M)$ is a linearly ordered set, contradicting the maximality of $(L, \leq_L)$. So $L = P$ and $(P, \leq_L)$ is a linearly ordered set such that $\leq_P \subseteq \leq_L$. □

104. Let $(W, <)$ be a well-ordered set. There is a unique ordinal α such that $W \cong \alpha$.

Proof: Let $(W, <)$ be a well-ordered set and let $B = \{a \in W \mid \exists \beta(\beta \text{ is an ordinal} \wedge \text{pred}(W, a) \cong \beta)\}$. Let f be the function with domain B, where $f(a)$ is the unique ordinal β_a such that $\text{pred}(W, a) \cong \beta_a$. The ordinal β_a is unique by Problem 85.

Let C be the range of f. Note that C is a set of ordinals. We first show that C is transitive.

To see this, let $\beta_a \in C$ and $\gamma \in \beta_a$. Let $g\colon \text{pred}(W, a) \cong \beta_a$. Since $\gamma \in \beta_a$, there is $b \in \text{pred}(W, a)$ such that $g(b) = \gamma$. Let h be the restriction of g to $\text{pred}(W, b)$. Let's check that $h\colon \text{pred}(W, b) \cong \gamma$.

First note that if $c \in \text{pred}(W, b)$, then $c < b$, and so, $h(c) = g(c) \in g(b) = \gamma$. So, h does in fact map $\text{pred}(W, b)$ to γ.

To see that h is injective, let $c, d \in \text{pred}(W, b)$ with $c < d$. Then $h(c) = g(c) \in g(d) = h(d)$, and in particular, $h(c) \neq h(d)$.

To see that h is surjective, let $\delta \in \gamma$. Since β_a is an ordinal, by Problem 39, $\delta \in \beta_a$. Since g is surjective, there is $c \in \text{pred}(W, a)$ with $g(c) = \delta$. Since $g(c) = \delta \in \gamma = g(b)$ and g is a homomorphism, $c < b$. Therefore, $c \in \text{pred}(W, b)$, and so, $h(c) = g(c)$.

To see that h is a homomorphism, let $c, d \in \text{pred}(W, b)$. Then $c < d$ if and only if $g(c) \in g(d)$ if and only if $h(c) \in h(d)$.

Also, by Problem 82 , γ is an ordinal. It follows that $\gamma \in C$ and therefore, C is transitive.

Since C is a transitive set of ordinals, by Problem 88, C is an ordinal, say $C = \alpha$.

We now show that B is an initial segment of W. Let $b \in B$ and $c < b$. Let $g: \text{pred}(W, b) \cong \beta_b$. Since $c < b$, $c \in \text{pred}(W, b)$. Let $g(c) = \gamma \in \beta_b$. Then by a reason similar to what was done above, $\text{pred}(W, c) \cong \gamma$.

Next, we show that $f: B \to \alpha$ is an isomorphism.

To see that f is injective, let $c, d \in B$ with $c < d$ and let $g: \text{pred}(W, d) \cong \beta_d$. By Theorem 10.17, g is f restricted to $\text{pred}(W, d)$. Since $c < d$, $f(c) = g(c) \in \beta_d = f(d)$. In particular, $f(c) \neq f(d)$.

f is surjective by the definition of α.

To see that f is a homomorphism, let $c, d \in B$. The argument for injectivity showed that $c < d$ implies $f(c) \in f(d)$. If $f(c) \in f(d)$, let $g: \text{pred}(W, d) \cong \beta_d = f(d)$. By Theorem 10.17, g is f restricted to $\text{pred}(W, d)$. So, $g(c) \in g(d)$. Since g is an isomorphism, $c < d$.

Finally, we show that $B = W$. Assume toward contradiction that B is a proper initial segment of W. Then $W \setminus B \neq \emptyset$. Let $b \in W \setminus B$ with $B = \text{pred}(W, b)$. Then $\text{pred}(W, b) \cong \alpha$. So, $b \in B$, a contradiction. □

105. Let α be a limit ordinal and let $f: \alpha \to \beta$ be a strictly increasing cofinal function. Then $cf(\alpha) = cf(\beta)$.

Proof: Let α be a limit ordinal and let $f: \alpha \to \beta$ be a strictly increasing cofinal function. Also, let $g: cf(\alpha) \to \alpha$ be a cofinal function. Then $f \circ g: cf(\alpha) \to \beta$ is cofinal. To see this, let $\delta \in \beta$. Since f is cofinal, there is $\gamma \in \alpha$ with $\delta \leq f(\gamma)$. Since g is cofinal, there is $\eta \in cf(\alpha)$ with $\gamma \leq g(\eta)$. Since f is increasing, $f(\gamma) \leq f(g(\eta))$. So, we have $\delta \leq f(\gamma) \leq f(g(\eta)) = (f \circ g)(\eta)$.

Since we have found a cofinal function from $cf(\alpha)$ to β, $cf(\beta) \leq cf(\alpha)$.

Now, let $h: cf(\beta) \to \beta$ be cofinal. Define $k: cf(\beta) \to \alpha$ by

$$k(\xi) = \text{"the least } \eta < \alpha \text{ such that } h(\xi) < f(\eta)\text{."}$$

Then k is cofinal. To see this, let $\gamma \in \alpha$. Then $f(\gamma) \in \beta$. Since h is cofinal, there is $\xi \in cf(\beta)$ with $f(\gamma) \leq h(\xi)$. If $k(\xi) = \eta$, then $h(\xi) < f(\eta)$. So, $f(\gamma) < f(\eta)$. Since f is strictly increasing, $\gamma < \eta = k(\xi)$.

Since we have found a cofinal function from $cf(\beta)$ to α, $cf(\alpha) \leq cf(\beta)$.

Since $cf(\beta) \leq cf(\alpha)$ and $cf(\alpha) \leq cf(\beta)$, we have $cf(\alpha) = cf(\beta)$. □

106. For any ordinal α, $cf\big(cf(\alpha)\big) = cf(\alpha)$.

Proof: If $\alpha = \beta + 1$, then $cf(\alpha) = 1$ and so, $cf\big(cf(\alpha)\big) = cf(1) = 1$. So, the result holds.

Let α be a limit ordinal. Then $cf(\alpha)$ is also a limit ordinal. By Problem 89 there is a strictly increasing cofinal function $f: cf(\alpha) \to \alpha$. By Problem 105, $cf\big(cf(\alpha)\big) = cf(\alpha)$. □

Problem Set 11

LEVEL 1

Let $D_3 = \langle r, s \mid r^3 = e, s^2 = e, sr = r^2s \rangle$, let $H = \langle r \rangle = \{e, r, r^2\}$ and let $K = \langle s \rangle = \{e, s\}$ (see part 3 of Example 9.47).

1. Determine if H is a normal subgroup of D_3.

H is a normal subgroup of D_3. We will show this by brute force:

$$srs = r^2ss = r^2s^2 = r^2e = r^2$$
$$(rs)r(rs) = r(sr)rs = r(r^2s)rs = r^3(sr)s = e(r^2s)s = r^2s^2 = r^2$$
$$(r^2s)r(r^2s) = r^2sr^3s = r^2ses = r^2ss = r^2s^2 = r^2$$
$$sr^2s = s(sr) = s^2r = er = r$$
$$(rs)r^2(rs) = rsr^3s = rses = rss = rs^2 = re = r$$
$$(r^2s)r^2(r^2s) = r^2sr^3rs = r^2sers = r^2srs = r^2(r^2s)s = r^3rs^2 = ere = r.$$

We do not need to check conjugation by the elements in H because we already know that H is a subgroup of D_3. We see that for all $g \in D_3$ and $h \in H$, we have $ghg^{-1} \in H$, and therefore, $H \lhd D_3$.

2. Determine if K is a normal subgroup of D_3.

K is **not** normal in S_3 because $(rs)s(rs)^{-1} = (rs)s(rs) = rs^2rs = rers = rrs = r^2s$ and $r^2s \notin K$.

Compute each of the following:

3. $[\mathbb{Z} : n\mathbb{Z}]$, where $n \in \mathbb{Z}$

$\mathbb{Z}/n\mathbb{Z} = \{\mathbb{Z}, 1 + \mathbb{Z}, 2 + \mathbb{Z}, \dots, (n-1) + \mathbb{Z}\}$. So, $[\mathbb{Z} : n\mathbb{Z}] = \boldsymbol{n}$.

4. $|G \cap H|$ where G and H are groups such that $|G| = 14$ and $|H| = 45$.

Since $G \cap H \le G$, $|G \cap H| \mid 14$. Since $G \cap H \le H$, $|G \cap H| \mid 45$. So, $|G \cap H| \mid \gcd(14, 45) = 1$. Therefore, $|G \cap H| = \mathbf{1}$.

LEVEL 2

Prove each of the following:

5. Let $(R, +, \cdot)$ be a ring and define addition and multiplication on $R \times R$ componentwise, as was done in part 4 of Example 9.40. Then $(R, +, \cdot)$ is isomorphic to a subring of $(R \times R, +, \cdot)$.

Proof: Let $S = \{(x, x) \mid x \in R\}$. Then $S \subseteq R \times R$, $(0, 0) \in S$, and $(1, 1) \in S$. If $(x, x), (y, y) \in S$, then we have $(x, x) - (y, y) = (x - y, x - y) \in S$ and $(x, x) \cdot (y, y) = (xy, xy) \in S$. So, S is a subring of $R \times R$ (note that we used Problem 14 from Problem Set 9 here).

Define $f: R \to S$ by $f(x) = (x, x)$. Clearly, f is bijective. If $x, y \in R$, we have

$$f(x + y) = (x + y, x + y) = (x, x) + (y, y) = f(x) + f(y).$$
$$f(xy) = (xy, xy) = (x, x) \cdot (y, y) = f(x) \cdot f(y).$$

So, f is a homomorphism. Therefore, $f: R \cong S$. □

6. $\text{Inn}(R)$ is a normal subgroup of $\text{Aut}(R)$ (see Problems 27 and 28 from Problem Set 9 for the definitions of $\text{Aut}(R)$ and $\text{Inn}(R)$).

Proof: By Problem 28 from Problem Set 9, $\text{Inn}(R) \leq \text{Aut}(R)$.

Let $f \in \text{Aut}(R)$, $\phi_x \in \text{Inn}(R)$, and $y \in R$. Then

$$(f \circ \phi_x \circ f^{-1})(y) = f\left(\phi_x(f^{-1}(y))\right) = f(xf^{-1}(y)x^{-1})$$
$$= f(x)f(f^{-1}(y))f(x^{-1}) = f(x)y(f(x))^{-1} = \phi_{f(x)}(y).$$

So, $f \circ \phi_x \circ f^{-1} = \phi_{f(x)} \in \text{Inn}(R)$. Therefore, $\text{Inn}(R) \lhd \text{Aut}(R)$. □

7. Let G be a group of finite order and let H and K be subgroups of G with H a subgroup of K. Then $[G: H] = [G: K] \cdot [K: H]$.

By Langrange's Theorem, we have

$$[G: H] = \frac{|G|}{|H|} = \frac{|G|}{|K|} \cdot \frac{|K|}{|H|} = [G: K] \cdot [K: H].$$

8. Let V be a vector space over a field F, let W be a subspace of V, and define $\pi: V \to V/W$ by $\pi(v) = v + W$. Then π is a linear transformation.

Proof: Let $v, w \in V$ and $a, b \in F$. Then we have

$$\pi(av + bw) = (av + bw) + W = (av + W) + (bw + W)$$
$$= a(v + W) + b(w + W) = a\pi(v) + b\pi(w).$$

By Theorem 8.18, π is a linear transformation. □

9. Let G and H be groups. Then $(G \times H)/(G \times \{e_H\}) \cong H$.

Proof: Define a function $f: G \times H \to H$ by $f(x, y) = y$. It is clear that f is a surjective homomorphism. Now, $(x, y) \in \ker(f)$ if and only if $f(x, y) = e_H$ if and only if $y = \{e_H\}$ if and only if $(x, y) \in G \times \{e_H\}$. So, $\ker(f) = G \times \{e_H\}$. By the First Homomorphism Theorem, $(G \times H)/(G \times \{e_H\}) \cong H$. □

Use the First Isomorphism Theorem for rings to prove each of the following:

10. $\mathbb{Z}_{15}/\{0, 3, 6, 9, 12\} \cong \mathbb{Z}_3$

Proof: Define a function $f: \mathbb{Z}_{15} \to \mathbb{Z}_3$ by $f([k]_{15}) = [k]_3$. If $[k_1]_{15} = [k_2]_{15}$, then $k_1 \equiv_{15} k_2$. So, $15|k_2 - k_1$. Therefore, there is $n \in \mathbb{Z}$ such that $k_2 - k_1 = 15n = 3(5n)$. So, $3|k_2 - k_1$, and therefore, $k_1 \equiv_3 k_2$. Thus, $[k_1]_3 = [k_2]_3$. So, f is well-defined.

It is clear that f is surjective. To see that f is a homomorphism, observe that

$$f([k_1 + k_2]_{15}) = [k_1 + k_2]_3 = [k_1]_3 + [k_2]_3 = f([k_1]_{15}) + f([k_2]_{15})$$
$$f([k_1 k_2]_{15}) = [k_1 k_2]_3 = [k_1]_3 \cdot [k_2]_3 = f([k_1]_{15}) \cdot f([k_2]_{15})$$
$$f([1]_{15}) = [1]_3$$

Now, $[k]_{15} \in \ker(f)$ if and only if $f([k]_{15}) = [0]_3$ if and only if $[k]_3 = [0]_3$ if and only if $3|k$ if and only if $k \in \{0, 3, 6, 9, 12\}$. So, $\ker(f) = \{0, 3, 6, 9, 12\}$.

By the First Homomorphism Theorem, $\mathbb{Z}_{15}/\{0, 3, 6, 9, 12\} \cong \mathbb{Z}_3$. □

11. $(\mathbb{Z}_2 \times \mathbb{Z}_2)/\{(0,0), (0,1)\} \cong \mathbb{Z}_2$

Proof: Define a function $f: \mathbb{Z}_2 \times \mathbb{Z}_2 \to \mathbb{Z}_2$ by $f(j, k) = j$. It is clear that f is a surjective homomorphism. Now, $(j, k) \in \ker(f)$ if and only if $f(j, k) = 0$ if and only if $j = 0$ if and only if $(j, k) \in \{(0,0), (0,1)\}$. So, $\ker(f) = \{(0,0), (0,1)\}$.

By the First Homomorphism Theorem, $(\mathbb{Z}_2 \times \mathbb{Z}_2)/\{(0,0), (0,1)\} \cong \mathbb{Z}_2$. □

Consider $\mathbb{R}^* = \mathbb{R} \setminus \{0\}$ as a group under multiplication.

12. Find a subgroup H of $\mathbb{R}^*$ such that $[\mathbb{R}^*: H] = 2$.

Let $H = \mathbb{R}^+$ be the set of positive real numbers. Since 1 is a positive real number, the product of two positive real numbers is a positive real number, and the multiplicative inverse (reciprocal) of a positive real number is a positive real number, H is a subgroup of $\mathbb{R}^*$. The two cosets of H in $\mathbb{R}^*$ are H and $(-1)H$. Note that H is the set of positive real numbers and $(-1)H$ is the set of negative real numbers. It follows that $[\mathbb{R}^*: H] = 2$.

LEVEL 3

Determine if each of the following statements is true or false. If true, explain why. If false, provide a counterexample:

13. A commutative ring R is a field if and only if the only ideals of R are $\{0\}$ and R.

True

Proof: Let R be a commutative ring. If R is a field then by part 7 of Example 11.6, the only ideals of R are $\{0\}$ and R.

Conversely, assume that the only ideals of R are $\{0\}$ and R. Let $a \in R$ with $a \neq 0$, and let $I = \{ax \mid x \in R\}$ be the principal ideal generated by a. By part 3 of Example 11.6, I is an ideal of R.

Since $a = ae \in I$ and $a \neq 0$, $I \neq \{0\}$. By assumption, we must have $I = R$. Since $1 \in R$, $1 \in I$. Therefore, there is $b \in R$ such that $ab = 1$. So, $b = a^{-1}$. Since $a \neq 0$ was arbitrary, we have shown that R has the multiplicative inverse property, and therefore, R is a field. □

14. If $\boldsymbol{X}$ is a nonempty set of normal subgroups of a group G then $\bigcap \boldsymbol{X}$ is a normal subgroup of G.

True

Proof: Let $\boldsymbol{X}$ be a nonempty set of normal subgroups of a group G. Since for all $H \in \boldsymbol{X}$, $e \in H$, $e \in \bigcap \boldsymbol{X}$, and so, $\bigcap \boldsymbol{X} \neq \emptyset$. Let $g, h \in \bigcap \boldsymbol{X}$. Then for all $H \in \boldsymbol{X}$, $g, h \in H$. By Problem 14 from Problem Set 9, for all $H \in \boldsymbol{X}$, $gh^{-1} \in H$. So, $gh^{-1} \in \bigcap \boldsymbol{X}$. Again, by Problem 14 from Problem Set 9, $\bigcap \boldsymbol{X}$ is a subgroup of G. Now, let $h \in \bigcap \boldsymbol{X}$, and let $g \in G$. Then for all $H \in \boldsymbol{X}$, $h \in H$, and since each H is a normal subgroup of G, $ghg^{-1} \in H$. So, $ghg^{-1} \in \bigcap \boldsymbol{X}$. Therefore, $\bigcap \boldsymbol{X}$ is a normal subgroup of G. □

15. If $\boldsymbol{X}$ is a nonempty set of normal subgroups of a group G then $\bigcup \boldsymbol{X}$ is a normal subgroup of G.

False

Since $(\mathbb{Z}, +)$ is a commutative group, all subgroups are normal. In particular $(2\mathbb{Z}, +)$ and $(3\mathbb{Z}, +)$ are normal subgroups of $(\mathbb{Z}, +)$. Now, $2, 3 \in (2\mathbb{Z}, +) \cup (3\mathbb{Z}, +)$, but $2 + 3 = 5 \notin (2\mathbb{Z}, +) \cup (3\mathbb{Z}, +)$. So, $(2\mathbb{Z}, +) \cup (3\mathbb{Z}, +)$ is not closed under addition and is therefore not a subgroup of $(\mathbb{Z}, +)$.

16. If $\boldsymbol{X}$ is a nonempty set of ideals of a ring R, then $\bigcap \boldsymbol{X}$ is an ideal of R.

True

Proof: Let $\boldsymbol{X}$ be a nonempty set of ideals of a ring R. Since for all $I \in \boldsymbol{X}$, $0 \in I$, $0 \in \bigcap \boldsymbol{X}$, and so, $\bigcap \boldsymbol{X} \neq \emptyset$. Let $x, y \in \bigcap \boldsymbol{X}$. Then for all $I \in \boldsymbol{X}$, $x, y \in I$. By Problem 14 from Problem Set 9, for all $I \in \boldsymbol{X}$, $x - y \in I$. So, $x - y \in \bigcap \boldsymbol{X}$. Again, by Problem 14 from Problem Set 9, $(\bigcap \boldsymbol{X}, +)$ is a subgroup of $(R, +)$. Now, let $x \in \bigcap \boldsymbol{X}$, and let $y \in R$. Then for all $I \in \boldsymbol{X}$, $x \in I$, and since each I is an ideal of R, for all $I \in \boldsymbol{X}$, $xy \in I$. So, $xy \in \bigcap \boldsymbol{X}$. Therefore, $\bigcap \boldsymbol{X}$ absorbs R, and so, $\bigcap \boldsymbol{X}$ is an ideal of R. □

17. If $\boldsymbol{X}$ is a nonempty set of ideals of a ring R, then $\bigcup \boldsymbol{X}$ is an ideal of R.

False

Since $(2\mathbb{Z}, +, \cdot)$ and $(3\mathbb{Z}, +, \cdot)$ are ideals of $(\mathbb{Z}, +, \cdot)$, the same argument as in Problem 15 shows that the union of ideals is **not** always an ideal.

Consider the following subset of S_4:

$$A_4 = \{(1), (123), (124), (134), (234), (132), (142), (143), (243), (12)(34), (13)(24), (14)(23)\}.$$

18. Prove that A_4 is a subgroup of S_4.

Proof: It can be checked that A_4 is a subgroup of S_4 by brute force. In other words, simply construct the multiplication table of A_4 and check that closure and the inverse property hold. □

19. Prove that A_4 has no subgroup of order 6.

Proof: Suppose toward contradiction that H is a subgroup of A_4 of order 6. Since $|A_4| = 12$ and $|H| = 6$, $[A_4 : H] = \frac{12}{6} = 2$. Let's show that for all $x \in A_4$, $x^2 \in H$. If $x \in H$, then $x^2 \in H$ because H is a group. Suppose toward contradiction that $x \in A_4 \setminus H$ and $x^2 \notin H$. Then $xH \neq H$ and $x^2H \neq H$. Since $[A_4 : H] = 2$, we must have $xH = x^2H$. It follows that $x = x^1 = x^{-1+2} = x^{-1}x^2 \in H$ (by Theorem 11.13), contradicting our assumption that $x \in A_4 \setminus H$. So, for all $x \in A_4$, $x^2 \in H$. For any 3-cycle (abc), we have $(abc) = (abc)^4 = ((abc)^2)^2$. So, every 3-cycle in A_4 is in H. But there are eight 3-cycles, contradicting our assumption that $|H| = 6$. Therefore, A_4 has no subgroup of order 6. □

Prove each of the following:

20. Let G be a group and let $H = \langle xyx^{-1}y^{-1} \mid x, y \in G \rangle$. Then H is a normal subgroup of G such that G/H is commutative. H is known as the commutator subgroup of G.

Proof: We will refer to an element of the form $xyx^{-1}y^{-1}$ as a commutator.

Since $e = eee^{-1}e^{-1}$, we see that e is a commutator, and therefore, $e \in H$.

Next note that the inverse of the commutator $xyx^{-1}y^{-1}$ is $(xyx^{-1}y^{-1})^{-1} = yxy^{-1}x^{-1}$, which is also a commutator. It follows that if h and k are products of commutators, then so is hk^{-1}. Therefore, by Problem 14 from Problem $H \leq G$.

Now, let h be a commutator and let $g \in G$. There are $x, y \in G$ such that $h = xyx^{-1}y^{-1}$. Then we have

$$ghg^{-1} = g(xyx^{-1}y^{-1})g^{-1} = gx(g^{-1}g)y(g^{-1}g)x^{-1}(g^{-1}g)y^{-1}g^{-1}$$
$$= (gxg^{-1})(gyg^{-1})(gx^{-1}g^{-1})(gy^{-1}g^{-1}) = (gxg^{-1})(gyg^{-1})(gxg^{-1})^{-1}(gyg^{-1})^{-1}.$$

Therefore, ghg^{-1} is a commutator.

Now, if $h \in H$, then there are commutators $h_1, h_2, \dots h_n$ such that $h = h_1 \cdot h_2 \cdots h_n$. Then we have

$$ghg^{-1} = gh_1 \cdot h_2 \cdots h_n g^{-1} = (gh_1g^{-1})(gh_2g^{-1}) \cdots (gh_ng^{-1})$$

This is a product of commutators, and therefore, $ghg^{-1} \in H$.

So, $H \lhd G$.

Now, let $xH, yH \in G/H$. Then $x, y \in G$, and so, $(yx)^{-1}xy = x^{-1}y^{-1}xy \in H$. By Theorem 11.13, we have $yxH = xyH$. So, $(yH)(xH) = yxH = xyH = (xH)(yH)$. Since $xH, yH \in G/H$ were arbitrary, we see that G/H is commutative. □

21. Let G be a group and let H be a subgroup of G such that $[G : H] = 2$. Then $H \lhd G$.

Proof: Suppose that $[G : H] = 2$. Then there are two left cosets of H in G, say H and xH and there are two right cosets of H in G, say H and Hy. Since the left cosets and right cosets each form a partition of G, we must have $xH = G \setminus H$ and $Hx = G \setminus H$. So, $xH = Hx$. By Theorem 11.23, $H \lhd G$. □

22. Every element of $\mathbb{Q}/\mathbb{Z}$ has finite order.

Proof: Let $\frac{a}{b}+\mathbb{Z} \in \mathbb{Q}/\mathbb{Z}$, where $a, b \in \mathbb{Z}$ and $b > 0$. Then $b\left(\frac{a}{b}+\mathbb{Z}\right) = a + \mathbb{Z} = \mathbb{Z}$ because $a \in \mathbb{Z}$. Therefore, $\left|\frac{a}{b}+\mathbb{Z}\right| \leq b$. □

Note: Recall from Lesson 9 that if G is a group and $x \in G$, we define the **order of the element x**, written $|x|$, to be the order of the cyclic subgroup $|\langle x\rangle|$. Equivalently, $|x|$ is the least positive integer n such that $x^n = e$ (or $nx = e$ if we are using additive notation), if such an n exists (otherwise x has infinite order). The proof of the equivalence of these two definitions can be found in Lesson 12 (Theorems 12.1 and 12.2).

23. Let G be a commutative group. Then $G/\{x \in G \mid x^2 = e\} \cong \{x^2 \mid x \in G\}$.

Proof: Define a function $f: G \to \{x^2 \mid x \in G\}$ by $f(x) = x^2$. It is clear that f is surjective. To see that f is a homomorphism, observe that

$$f(xy) = (xy)^2 = xyxy = xxyy = x^2y^2 = f(x)f(y).$$

Now, $x \in \ker(f)$ if and only if $f(x) = e$ if and only if $x^2 = e$ if and only if $x \in \{x \in G \mid x^2 = e\}$. So, $\ker(f) = \{x \in G \mid x^2 = e\}$.

By the First Homomorphism Theorem, $G/\{x \in G \mid x^2 = e\} \cong \{x^2 \mid x \in G\}$. □

Let R be a ring.

24. Let I be an ideal of R such that $I \neq R$ and R/I is commutative. Is R necessarily commutative? If so, prove it. If not, provide a counterexample.

Let A and B be rings with A commutative and B not commutative. Then $A \times B$ is a noncommutative ring, where addition and multiplication are defined as follows:

$$(a, b) + (c, d) = (a + c, b + d) \text{ and } (a, b)(c, d) = (ac, bd)$$

Let $I = \{(0, b) \mid b \in B\}$. Then I is an ideal of $A \times B$ and $A \times B/I \cong A$, which is commutative.

25. Describe such a ring R with exactly 2 maximal ideals.

Solution: Let F and K be fields. The only ideals of F are $\{0\}$ and F and the only ideals of K are $\{0\}$ and K. Let $R = F \times K$, let $I = \{(f, 0) \mid f \in F\}$ and let $J = \{(0, k) \mid k \in K\}$. Then I and J are the only two maximal ideals of R. As a specific example, $\mathbb{Q} \times \mathbb{Q}$ has exactly 2 maximal ideals.

LEVEL 4

Prove each of the following:

26. Let G be a group and let $H \lhd G$ such that $|H|$ and $|G/H|$ are finite. Then $|G|$ is finite.

Proof: Suppose that $|H| = m$ and $|G/H| = n$. Let $x_1H, x_2H, \dots, x_nH$ be the n cosets of H in G. By Theorem 11.12, we have $G = x_1H \cup x_2H \cup \cdots \cup x_nH$ and by Theorem 11.14, we have $|x_1H| = |x_2H| = \cdots = |x_nH| = |H|$. Therefore,

$$|G| = |x_1H \cup x_2H \cup \cdots \cup x_nH| = |x_1H| + |x_2H| + \cdots + |x_nH| = n|H| = nm.$$

So, $|G|$ is finite. □

27. Let G be a group, let H be a subgroup of G, and let Y be a selector for the collection of cosets of H in G. For each $x \in Y$, let $f_x: xH \to H$ be a bijection and define $g: G \to H \times \{zH \mid z \in G\}$ by $g(y) = (f_x(y), xH)$, where $x \in Y$ and $y \in xH$. Then g is a bijection.

Proof: Let $y \in G$. By Theorem 11.12, there is exactly one coset xH such that $y \in xH$. So, $f_x(y)$ is an element of H, and so, $g(y) = (f_x(y), xH) \in H \times \{zH \mid z \in G\}$.

To see that g is injective, suppose that $g(y_1) = g(y_2)$. Then there are $x_1, x_2 \in Y$ with $y_1 \in x_1H$, $y_2 \in x_2H$, $f_{x_1}(y_1) = f_{x_2}(y_2)$, and $x_1H = x_2H$. Since $x_1, x_2 \in x_1H$ and Y contains only one element from each coset, we must have $x_1 = x_2$. So, $f_{x_1}(y_1) = f_{x_1}(y_2)$. Since f_{x_1} is injective (because it is bijective), $y_1 = y_2$.

To see that g is surjective, let $(z, xH) \in H \times \{zH \mid z \in G\}$. Choose $w \in Y$ such that $xH = wH$. Let $y = f_w^{-1}(z)$. Then $y \in wH$, and so, $g(y) = (f_w(y), wH) = (z, xH)$. □

28. Let $\mathbb{Z}_n = \{[k]_n \mid k \in \mathbb{Z}\}$, where $[k]_n$ is the equivalence class of k under the equivalence $\equiv_n$ and consider the commutative ring $(\mathbb{Z}_n, +, \cdot)$, where addition and multiplication are defined by $[x]_n + [y]_n = [x + y]_n$ and $[xy]_n = [x]_n \cdot [y]_n$ (see Problem 100 from Problem Set 3). Then $\mathbb{Z}/n\mathbb{Z} \cong \mathbb{Z}_n$.

Proof: Define $f: \mathbb{Z}/n\mathbb{Z} \to \mathbb{Z}_n$ by $f(x + n\mathbb{Z}) = [x]$. Suppose that $f(x + n\mathbb{Z}) = f(y + n\mathbb{Z})$, so that $[x] = [y]$. Then $x \equiv_n y$. So, $n|x - y$, and therefore, there is $k \in \mathbb{Z}$ such that $x - y = nk$. So, $x - y \in n\mathbb{Z}$, and so, $x + n\mathbb{Z} = y + n\mathbb{Z}$. Since $x + n\mathbb{Z}, y + n\mathbb{Z} \in \mathbb{Z}/n\mathbb{Z}$ were arbitrary, f is injective.

f is surjective because if $[x] \in \mathbb{Z}_n$, then $x \in \mathbb{Z}$ and $f(x + n\mathbb{Z}) = [x]$.

Now, we have

$$f\big((x + n\mathbb{Z}) + (y + n\mathbb{Z})\big) = f\big((x + y) + n\mathbb{Z}\big) = [x + y] = [x] + [y] = f(x + n\mathbb{Z}) + f(y + n\mathbb{Z}),$$

$$f\big((x + n\mathbb{Z})(y + n\mathbb{Z})\big) = f\big((xy) + n\mathbb{Z}\big) = [xy] = [x] \cdot [y] = f(x + n\mathbb{Z}) \cdot f(y + n\mathbb{Z}).$$

Therefore, $f: \mathbb{Z}/n\mathbb{Z} \cong \mathbb{Z}_n$. □

29. There is a natural one-to-one correspondence between the ideals of $\mathbb{Z}/15\mathbb{Z}$ and $\mathbb{Z}_{15}$.

Proof: $\mathbb{Z}/15\mathbb{Z} = \{15\mathbb{Z}, 1 + 15\mathbb{Z}, \dots, 14 + 15\mathbb{Z}\}$ and $\mathbb{Z}_{15} = \{[0], [1], \dots, [14]\}$.

The ideals of $\mathbb{Z}/15\mathbb{Z}$ are $\mathbb{Z}/15\mathbb{Z}$, $3\mathbb{Z}/15\mathbb{Z}$, $5\mathbb{Z}/15\mathbb{Z}$, $15\mathbb{Z}/15\mathbb{Z}$ and these correspond to the ideals of $\mathbb{Z}_{15}$, which are $\mathbb{Z}_{15}$, $\{[0], [3], [6], [9], [12]\}$, $\{[0], [5], [10]\}$, and $\{[0]\}$. □

30. Let G be a group such that $G/Z(G)$ is cyclic. Then G is commutative.

Proof: Since $G/Z(G)$ is cyclic, there is $a \in G$ such that $G/Z(G) = \langle aZ(G) \rangle$. Let $x, y \in G$. Then $xZ(G) = a^j Z(G)$ and $yZ(G) = a^k Z(G)$ for some integers j and k. So, there are $b, c, d, f \in Z(G)$ such that $xb = a^j c$ and $yd = a^k f$. Let $g = cb^{-1}$ and $h = fd^{-1}$. Since $Z(G)$ is a group, $g, h \in Z(G)$. Therefore,

$$xy = (a^j cb^{-1})(a^k fd^{-1}) = (a^j g)(a^k h) = (ga^j)(a^k h) = ga^{j+k}h = ga^{k+j}h = (ga^k)(a^j h)$$
$$= (a^k g)(ha^j) = a^k(gh)a^j = a^k hga^j = (a^k h)(a^j g) = (a^k fd^{-1})(a^j cb^{-1}) = yx.$$

Since $x, y \in G$ were arbitrary, G is commutative. □

31. Let H and K be finite subgroups of a group G and let $HK = \{hk \mid h \in H \wedge k \in K\}$. Then $|HK| \cdot |H \cap K| = |H| \cdot |K|$.

Proof: We will prove that $\frac{|H|}{|H \cap K|} = \frac{|HK|}{|K|}$ by showing that the number of cosets of $H \cap K$ in H is the same as the number of cosets of K in HK.

Since H is finite, there are finitely many cosets of $H \cap K$ in H, let's say $a_1(H \cap K), \dots, a_m(H \cap K)$. Let $hk \in HK$. Then $h \in H$, and so, h is in some coset of $H \cap K$ in H, say $h \in a_i(H \cap K)$ for some $i \in \{1, 2, \dots, m\}$. So, there is $a \in H \cap K$ with $h = a_i a$. Therefore, $hk = (a_i a)k = a_i(ak)$. Since $a \in H \cap K$, we have $a \in K$. Since $a \in K$, $k \in K$, and K is a group, $ak \in K$. So, $hk \in a_i K$.

Next, we see that $a_i K \cap a_j K \neq \emptyset \Rightarrow a_i K = a_j K \Rightarrow a_j^{-1} a_i \in K \Rightarrow a_j^{-1} a_i \in H \cap K \Rightarrow a_i = a_j \Rightarrow i = j$. So, if $i \neq j$, then $a_i K \cap a_j K = \emptyset$. That is, the cosets $a_i K$ are pairwise disjoint.

Thus, $\frac{|H|}{|H \cap K|} = \frac{|HK|}{|K|}$, and so, $|HK| \cdot |H \cap K| = |H| \cdot |K|$. □

32. There are exactly two groups of order 6, up to isomorphism.

Proof: $(\mathbb{Z}_6, +)$ is a commutative group of order 6 and $(S_3, \circ)$ is a noncommutative group of order 6. Therefore, there are at least two groups of order 6.

Let G be an arbitrary group of order 6. The order of each element must be 1, 2, 3, or 6. If G has an element of order 6, then G is cyclic and therefore, G is isomorphic to $\mathbb{Z}_6$.

Let G be a noncyclic group of order 6. We first show that G has an element of order 3. Suppose toward contradiction that G has no element of order 3. Then every nonidentity element has order 2. Let $a, b \in G$ be any two distinct nonidentity elements of G. Then $a^2 = e$, and so, $a = a^{-1}$. If $ab = e$, then $b = a^{-1} = a$. Therefore, ab cannot have order 1. So, ab has order 2. It follows that $ab = (ab)^{-1} = b^{-1}a^{-1} = ba$. So, $\{1, a, b, ab\}$ is a subgroup of G of order 4, contradicting Theorem 11.17. So, G must have an element of order 3.

Next, we show that G has an element of order 2. Suppose toward contradiction that G has no element of order 2. Then every nonidentity element has order 3. In other words, there are exactly 5 elements of order 3. If a is an element of order 3, then a is not equal to its own inverse. So, by pairing each nonidentity element with its inverse, we see that there must be an even number of nonidentity elements, contradicting that there are 5 such elements. So, G must have an element of order 2.

Let $a \in G$ have order 3 and let $b \in G$ have order 2. Then it is easy to see that the 6 distinct elements of G are e, a, a^2, b, ba, and ba^2.

It's easy to see that we cannot have $ab = e, a, a^2$, or b. So, $ab = ba$ or $ab = ba^2 = ba^{-1}$.

If $ab = ba$, then $(ba)^2 = baba = abba = ab^2a = aea = a^2$, $(ba)^3 = a^2ba = a^3b = b$, $(ba)^4 = bba = a$, $(ba)^5 = aba = ba^2$, and $(ba)^6 = ba^2ba = b^2a^3 = ee = e$. So, G is a cyclic group generated by ba, contradicting our assumption that G is not cyclic.

If $ab = ba^2$, then $G = \langle a, b \mid a^3 = e, b^2 = e, ab = ba^2 \rangle$. This group is isomorphic to S_3 (or D_3). □

33. Let V be a vector space over a field F, let W be a subspace of V, and define $\pi: V \to V/W$ by $\pi(v) = v + W$. By Problem 8 above, π is a linear transformation. Suppose that U is a vector space over F and that $f: V \to U$ is a linear transformation such that $W \subseteq \ker(f)$. Then there is a unique linear transformation $g: V/W \to U$ such that $f = g \circ \pi$.

Proof: Define $g: V/W \to U$ by $g(v + W) = f(v)$. To see that g is well-defined, suppose that $v + W = w + W$. Then $v - w \in W$. Since $W \subseteq \ker(f)$, $f(v - w) = 0_U$. Since f is a linear transformation, $f(v) - f(w) = 0$, and so, $f(v) = f(w)$.

Now, we have

$$g\big((v + W) + (w + W)\big) = g\big((v + w) + W\big) = f(v + w)$$
$$= f(v) + f(w) = g(v + W) + g(w + W).$$
$$g\big(k(v + W)\big) = g(kv + W) = f(kv) = kf(v) = kg(v + W)$$

Therefore, g is a linear transformation.

Next, if $v \in V$, then $(g \circ \pi)(v) = g\big(\pi(v)\big) = g(v + W) = f(v)$. So, $f = g \circ \pi$.

Finally, we check uniqueness. Suppose that $h: V/W \to U$ satisfies $f = h \circ \pi$. Let $v + W \in V/W$. Since $\pi(v) = v + W$, we have

$$h(v + W) = h\big(\pi(v)\big) = (h \circ \pi)(v) = f(v) = (g \circ \pi)(v) = g\big(\pi(v)\big) = g(v + W).$$

So, $h = g$, as desired. □

34. Let R be a ring, let I be an ideal of R, and let S be a subring of R with $I \subseteq S \subseteq R$. Then S/I is a subring of R/I. Furthermore, if S is an ideal of R, then S/I is an ideal of R/I.

Proof: Assume that I is an ideal of R and S is a subring of R with $I \subseteq S \subseteq R$.

By Note 1 following Theorem 11.41, I is an ideal of S. Therefore, S/I is a ring.

Now, if $x + I \in S/I$, then $x \in S$. Since S is a subset of R, $x \in R$. Therefore, we have $x + I \in R/I$. Since $x + I \in S/I$ was arbitrary, $S/I \subseteq R/I$. It follows that S/I is a subring of R/I.

Now, assume that S is an ideal of R. Let $y + I \in S/I$ and let $x + I \in R/I$. Then $y \in S$ and $x \in R$. So, $xy \in S$ and $yx \in S$. It follows that $(x + I)(y + I) = xy + I \in S/I$ and $(y + I)(x + I) = yx + I \in S/I$. Therefore, S/I is an ideal of R/I. □

35. Let R be a ring and let I be an ideal of R. Then every subring of R/I is of the form S/I for some subring S of R with $I \subseteq S \subseteq R$. Furthermore, every ideal of R/I is of the form S/I for some ideal S of R with $I \subseteq S \subseteq R$.

Proof: Assume that I is an ideal of R and let K be a subring of R/I. We define S as follows: $S = \{x \in R \mid x + I \in K\}$.

Let's show that S is a subring of R. Since K is a subring of R/I, $0 + I, 1 + I \in K$. Therefore, $0, 1 \in S$. Let $x, y \in S$. Then $x + I, y + I \in K$. Since K is a subring of R/I, by Problem 14 in Problem Set 9, we have $(y - x) + I = (y + I) - (x + I) \in K$. So, $y - x \in S$. Also, we have $xy + I = (x + I)(y + I) \in K$. So, $xy \in S$. Therefore, S is a subring of R.

Next, let's show that $I \subseteq S$. To see this, let $x \in I$. Then $x + I = I$. Since K is a subring of R/I and I is the identity of R/I, $x + I = I \in K$. Therefore, $x \in S$.

We now show that $S/I = K$. Let $x + I \in S/I$. Then $x \in S$. By the definition of S, $x + I \in K$. Therefore, $S/I \subseteq K$. Conversely, let $x + I \in K$. By the definition of S, $x \in S$. So, $x + I \in S/I$. Thus, $K \subseteq S/I$. Since $S/I \subseteq K$ and $K \subseteq S/I$, we have $S/I = K$.

Finally, assume that K is an ideal of R/I. Let $y \in S$ and $x \in R$. By the definition of S, $y + I \in K$. Since K is an ideal of R/I, $xy + I = (x + I)(y + I) \in K$ and $yx + I = (y + I)(x + I) \in K$. By the definition of S, $xy \in S$ and $yx \in S$. Thus, S is an ideal of R. □

36. Let V be a vector space over a field F and let U and W be subspaces of V with $W \subseteq U \subseteq V$. Then U/W is a subspace of V/W.

Proof: Assume that U and W are subspaces of V with $W \subseteq U \subseteq V$.

Since W and U are both vector spaces with the same operations and $W \subseteq U$, W is a subspace of U. Therefore, U/W is a vector space.

Now, if $x + W \in U/W$, then $x \in U$. Since U is a subset of V, $x \in V$. Therefore, we have $x + W \in V/W$. Since $x + W \in U/W$ was arbitrary, $U/W \subseteq V/W$. It follows that U/W is a subspace of V/W. □

37. Let V be a vector space over a field F and let W be a subspace of V. Then every subspace of V/W is of the form U/W for some subspace U of V with $W \subseteq U \subseteq V$.

Proof: Assume that W is a subspace of V and let K be a subspace of V/W. We define U as follows: $U = \{x \in V \mid x + W \in K\}$.

Let's show that U is a subspace of V. Since K is a subspace of V/W, $0+W=W \in K$. Therefore, $0 \in U$. Let $x, y \in U$ and let $k \in F$ Then $x+W, y+W \in K$. Since K is a subspace of V/W, we have $(x+y)+W=(x+W)+(y+W) \in K$. So, $x+y \in U$. Again, since K is a subspace of V/W, we have $kx+W=k(x+W) \in K$. So, $kx \in U$. Therefore, U is a subspace of V.

Next, let's show that $W \subseteq U$. To see this, let $x \in W$. Then $x+W=W$. Since K is a subspace of V/W and W is the identity of V/W, $x+W=W \in K$. Therefore, $x \in U$.

We now show that $U/W=K$. Let $x+W \in U/W$. Then $x \in U$. By the definition of U, $x+W \in K$. Therefore, $U/W \subseteq K$. Conversely, let $x+W \in K$. By the definition of U, $x \in U$. So, $x+W \in U/W$. Thus, $K \subseteq U/W$. Since $U/W \subseteq K$ and $K \subseteq U/W$, we have $U/W=K$. □

38. The vector space version of the Fourth Isomorphism Theorem

Fourth Isomorphism Theorem—Vector Space Version: Let V be a vector space over a field F, let W be a subspace of V, let $\mathcal{X}$ be the set of all subspaces of V that contain W, let $\mathcal{Y}$ be the set of subspaces of V/W, and define $f: \mathcal{X} \to \mathcal{Y}$ by $f(U)=U/W$. Then f is a bijection.

Proof: Let's first check that f is a function from $\mathcal{X}$ to $\mathcal{Y}$. Let $U \in \mathcal{X}$. Then U is a subspace of V containing W. By Theorem 11.50, U/W is a subspace of V/W. So, $f(U)=U/W \in \mathcal{Y}$, as desired.

Next, let's check that f is injective. Suppose that $f(U)=f(Z)$. Then $U/W=Z/W$. Let $x \in U$. Then $x+W \in U/W$. Since $U/W=Z/W$, $x+W \in Z/W$. Therefore, $x \in Z$. So, $U \subseteq Z$. A symmetrical argument shows that $Z \subseteq U$. Thus, $U=Z$.

To see that f is surjective, let $K \in \mathcal{Y}$. Then K is a subspace of V/W. By Theorem 11.50, $K=U/W$ for some subspace U of V with $W \subseteq U \subseteq V$. So, $U \in \mathcal{X}$ and $f(U)=U/W=K$. □

39. The ring version of the Fourth Isomorphism Theorem

Fourth Isomorphism Theorem—Ring Version: Let R be a ring, let I be an ideal of R, let $\mathcal{X}$ be the set of all subrings of R that contain I, let $\mathcal{Y}$ be the set of subrings of R/I, and define $f: \mathcal{X} \to \mathcal{Y}$ by $f(S)=S/I$. Then f is a bijection.

Proof: Let's first check that f is a function from $\mathcal{X}$ to $\mathcal{Y}$. Let $S \in \mathcal{X}$. Then S is a subring of R containing I. By Theorem 11.48, S/I is a subring of R/I. So, $f(S)=S/I \in \mathcal{Y}$, as desired.

Next, let's check that f is injective. Suppose that $f(S)=f(T)$. Then $S/I=T/I$. Let $x \in S$. Then $x+I \in S/I$. Since $S/I=T/I$, $x+I \in T/I$. Therefore, $x \in T$. So, $S \subseteq T$. A symmetrical argument shows that $T \subseteq S$. Thus, $S=T$.

To see that f is surjective, let $K \in \mathcal{Y}$. Then K is a subring of R/I. By Theorem 11.49, $K=S/I$ for some subring S of R with $I \subseteq S \subseteq R$. So, $S \in \mathcal{X}$ and $f(S)=S/I=K$. □

LEVEL 5

Consider the structure $({}^{\mathbb{R}}\mathbb{R}, +, \cdot)$, where addition and multiplication are defined pointwise.

40. Prove that $({}^{\mathbb{R}}\mathbb{R}, +, \cdot)$ is a ring.

Proof: For $f, g \in {}^{\mathbb{R}}\mathbb{R}$, we define $f + g$ and fg to be the functions in ${}^{\mathbb{R}}\mathbb{R}$ such that for all $x \in \mathbb{R}$,

$$(f+g)(x) = f(x) + g(x) \text{ and } (fg)(x) = f(x) \cdot g(x).$$

Then $f + g$ and fg are in ${}^{\mathbb{R}}\mathbb{R}$, so that ${}^{\mathbb{R}}\mathbb{R}$ is closed under addition and multiplication. To see that addition and multiplication are associative in ${}^{\mathbb{R}}\mathbb{R}$, we have for each $f, g \in {}^{\mathbb{R}}\mathbb{R}$ and each $x \in \mathbb{R}$,

$$\big(f + (g+h)\big)(x) = f(x) + (g+h)(x) = f(x) + \big(g(x) + h(x)\big) = \big(f(x) + g(x)\big) + h(x)$$
$$= (f+g)(x) + h(x) = \big((f+g)+h\big)(x).$$

$$\big(f(gh)\big)(x) = f(x) \cdot (gh)(x) = f(x) \cdot \big(g(x) \cdot h(x)\big) = \big(f(x) \cdot g(x)\big) \cdot h(x)$$
$$= (fg)(x) \cdot h(x) = \big((fg)h\big)(x).$$

Let $0 \in {}^{\mathbb{R}}\mathbb{R}$ be defined by $0(x) = 0$ for all $x \in \mathbb{R}$ and let $1 \in {}^{\mathbb{R}}\mathbb{R}$ be defined by $1(x) = 1$ for all $x \in \mathbb{R}$. Then for all $x \in \mathbb{R}$, we have

$$(0+f)(x) = 0(x) + f(x) = 0 + f(x) = f(x) \qquad (f+0)(x) = f(x) + 0(x) = f(x) + 0 = f(x)$$

$$(1f)(x) = 1(x) \cdot f(x) = 1f(x) = f(x) \quad (f \cdot 1)(x) = f(x) \cdot 1(x) = f(x) \cdot 1 = f(x).$$

So, 0 is an additive identity in ${}^{\mathbb{R}}\mathbb{R}$ and 1 is a multiplicative identity in ${}^{\mathbb{R}}\mathbb{R}$.

The additive inverse of $f \in {}^{\mathbb{R}}\mathbb{R}$ is $-f$ defined by $(-f)(x) = -f(x)$. Indeed, for each $x \in \mathbb{R}$, we have

$$\big(f + (-f)\big)(x) = f(x) + (-f)(x) = f(x) + \big(-f(x)\big) = 0 = 0(x).$$

To see that addition is commutative in ${}^{\mathbb{R}}\mathbb{R}$, we have for each $f, g \in {}^{\mathbb{R}}\mathbb{R}$ and each $x \in \mathbb{R}$,

$$(f+g)(x) = f(x) + g(x) = g(x) + f(x) = (g+f)(x).$$

It follows that $({}^{\mathbb{R}}\mathbb{R}, +, \cdot)$ is a ring. □

41. Prove that for each $x \in \mathbb{R}$, $I_x = \{f \in {}^{\mathbb{R}}\mathbb{R} \mid f(x) = 0\}$ is an ideal of ${}^{\mathbb{R}}\mathbb{R}$.

Proof: Let $x \in \mathbb{R}$ and let $I_x = \{f \in {}^{\mathbb{R}}\mathbb{R} \mid f(x) = 0\}$. If $f, g \in I_x$, then we have

$$(f+g)(x) = f(x) + g(x) = 0 + 0 = 0.$$

So, I_x is closed under addition.

Since $0(x) = 0$, $0 \in I_x$. So, 0 is an additive identity element for I_x.

If $f \in I_x$, then $(-f)(x) = -f(x) = -0 = 0$, so that $-f \in I_x$. Therefore, I_x has the additive inverse property.

So, $(I_x, +)$ is a subgroup of $({}^{\mathbb{R}}\mathbb{R}, +)$.

Let $f \in I_x$ and $g \in {}^{\mathbb{R}}\mathbb{R}$. Then for each $x \in \mathbb{R}$, we have $(fg)(x) = f(x)g(x) = 0 \cdot g(x) = 0 = 0(x)$ and $(gf)(x) = g(x)f(x) = g(x) \cdot 0 = 0 = 0(x)$. Therefore, $fg \in I_x$ and $gf \in I_x$. So, I_x absorbs ${}^{\mathbb{R}}\mathbb{R}$. It follows that I_x is an ideal of ${}^{\mathbb{R}}\mathbb{R}$. □

42. Prove that the only ideal of ${}^{\mathbb{R}}\mathbb{R}$ containing I_x and not equal to I_x is ${}^{\mathbb{R}}\mathbb{R}$.

Proof: Let K be an ideal containing I_x, such that $K \neq I_x$. Then there is $f \in K \setminus I_x$. Let $g \in {}^{\mathbb{R}}\mathbb{R} \setminus K$ be arbitrary. Define $h, k \in {}^{\mathbb{R}}\mathbb{R}$ by $h(z) = \begin{cases} 0 & \text{if } z = x \\ g(z) & \text{if } z \neq x \end{cases}$ and $k(z) = \begin{cases} \frac{g(z)}{f(z)} & \text{if } z = x \\ 0 & \text{if } z \neq x \end{cases}$. Then, we have $g = h + fk$. Indeed, if $z \neq x$, then $(h + fk)(z) = h(z) + f(z)k(z) = g(z) + f(z) \cdot 0 = g(z)$ and if $z = x$, then $(h + fk)(z) = (h + fk)(x) = h(x) + f(x)k(x) = 0 + f(x) \cdot \frac{g(x)}{f(x)} = g(x)$.

Now, $h \in I_x$ because $h(x) = 0$. Since $I_x \subseteq K$, $h \in K$. Since $f \in K$ and K is an ideal, $fk \in K$. Since $(K, +)$ is a group, $g = h + fk \in K$. Since $g \in {}^{\mathbb{R}}\mathbb{R} \setminus K$ was arbitrary, it follows that $K = {}^{\mathbb{R}}\mathbb{R}$. □

Recall: If I is an ideal of a ring R such that the only ideal containing I not equal to I itself is R, then I is called a **maximal ideal** of R. In the solution above, we proved that for each $x \in \mathbb{R}$, I_x is a maximal ideal of ${}^{\mathbb{R}}\mathbb{R}$.

Prove each of the following:

43. Let G be a group and let N be a normal subgroup of G such that N and G/N are finitely generated. Then G is finitely generated.

Proof: Let $N = \langle n_1, n_2, \dots, n_j \rangle$ and let $G/N = \langle x_1N, x_2N, \dots, x_kN \rangle$. We will show that G is generated by all products of the form $\prod \left(a_i\left(\prod b_j^i\right)\right)$, where each a_i is some x_t or its inverse for some t and each b_j^i is some n_i or its inverse.

Let $x \in G$. Then $xN = \prod a_i N$, where each a_i is some x_t or its inverse. So, $x = xe = \prod a_i y_i$, where each $y_i \in N$. For each i, $y_i = \prod b_j^i$, where each b_j^i is some n_i or its inverse. So, $x = \prod \left(a_i\left(\prod b_j^i\right)\right)$. □

44. Let G be a group and let H be a subgroup of G such that H is the only subgroup of G with index $[G{:}H]$. Then H is normal in G.

Proof: First assume that G is finite. If $|K| = |H|$, by Lagrange's Theorem, $[G{:}K] = \frac{|G|}{|K|} = \frac{|G|}{|H|} = [G{:}H]$. Since H is the only subgroup of G with index $[G{:}H]$, $K = H$. Therefore, H is the only subgroup of G of order $|H|$.

Now, let $x \in G$. It is easily checked that $xHx^{-1} \leq G$. Define $f\colon H \to xHx^{-1}$ by $f(y) = xyx^{-1}$.Then f is a bijection. So, $|H| = |xHx^{-1}|$. Therefore, $xHx^{-1} = H$. By Theorem 11.23, H is normal in G.

Next, assume that G is infinite and let $x \in G$. Define $f: G/H \to G/xHx^{-1}$ by $f(yH) = xyHx^{-1}$. Since $xyHx^{-1} = (xyx^{-1})xHx^{-1}$, $f(yH) = xyHx^{-1} \in G/xHx^{-1}$.

If $f(yH) = f(zH)$, then $xyHx^{-1} = xzHx^{-1}$, and so, $yH = zH$. So, f is injective.

If $yxHx^{-1} \in G/xHx^{-1}$, then $f(x^{-1}yxH) = xx^{-1}yxHx^{-1} = yxHx^{-1}$. So, f is surjective.

It follows that $[G: H] = [G: xHx^{-1}]$. Since H is the only subgroup of G with index $[G: H]$, we have $xHx^{-1} = H$. By Theorem 11.23, H is normal in G. □

45. Let G be a group and for each $x \in G$, define the function $\phi_x: G \to G$ by $\phi_x(y) = xyx^{-1}$. Let $\text{Inn}(G) = \{\phi_x \mid x \in G\}$. Then $G/Z(G) \cong \text{Inn}(G)$, where $Z(G)$ is the center of G.

Proof: By Problem 28 in Problem Set 9, $\text{Inn}(G)$ is a group.

Define a function $f: G \to \text{Inn}(G)$ by $f(x) = \phi_x$. It is clear that f is surjective. To see that f is a homomorphism, observe that by the proof of Problem 28, we have

$$f(xy) = \phi_{xy} = \phi_x\phi_y = f(x)f(y).$$

Now, $x \in \ker(f)$ if and only if $f(x) = \phi_e$ if and only if for all $y \in G$, $\phi_x(y) = \phi_e(y)$ if and only if for all $y \in G$, $xyx^{-1} = eye^{-1} = y$ if and only if for all $y \in G$, $xy = yx$ if and only if $x \in Z(G)$.

By the First Homomorphism Theorem, $G/Z(G) \cong \text{Inn}(G)$. □

46. Let G and H be groups, let $N \lhd G$, and let $K \lhd H$. Then $(G \times H)/(N \times K) \cong G/N \times H/K$.

Proof: Define a function $f: G \times H \to G/N \times H/K$ by $f(x, y) = (xN, yK)$. It is clear that f is surjective. To see that f is a homomorphism, observe that

$$f\big((x, y)(z, w)\big) = f(xz, yw) = (xzN, ywK) = \big((xN)(zN), (yK)(wK)\big)$$
$$= (xN, yK)(zN, wK) = f(x, y)f(z, w).$$

Now, $(x, y) \in \ker(f)$ if and only if $(xN, yK) = (N, K)$ if and only if $xN = N$ and $yK = K$ if and only if $x \in N$ and $y \in K$ if and only if $(x, y) \in N \times K$. So, $\ker(f) = N \times K$.

By the First Homomorphism Theorem, $(G \times H)/(N \times K) \cong G/N \times H/K$. □

47. Let $m, n \in \mathbb{Z}^+$. Then $m\mathbb{Z} \cap n\mathbb{Z} = \text{lcm}(m, n)\mathbb{Z}$ and $m\mathbb{Z} + n\mathbb{Z} = \text{gcd}(m, n)\mathbb{Z}$.

Proof: Let $k \in m\mathbb{Z} \cap n\mathbb{Z}$. Then $k \in m\mathbb{Z}$ and $k \in n\mathbb{Z}$. So, there are integers a and b such that $k = ma$ and $k = nb$. So, $m|k$ and $n|k$. By Problem 102 in Problem Set 4, $\text{lcm}(m, n)|k$. Thus, there is an integer c such that $k = \text{lcm}(m, n) \cdot c$. So, $k \in \text{lcm}(m, n)\mathbb{Z}$. Since $k \in m\mathbb{Z} \cap n\mathbb{Z}$ was arbitrary, we have $m\mathbb{Z} \cap n\mathbb{Z} \subseteq \text{lcm}(m, n)\mathbb{Z}$.

Next, let $k \in \text{lcm}(m, n)\mathbb{Z}$. Then there is an integer c such that $k = \text{lcm}(m, n) \cdot c$. Since $\text{lcm}(m, n)$ is a multiple of m, there is an integer a such that $\text{lcm}(m, n) = ma$. So, $k = (ma)c = m(ac)$. Since $a, c \in \mathbb{Z}$ and $\mathbb{Z}$ is closed under multiplication, $ac \in \mathbb{Z}$. So, $k \in m\mathbb{Z}$. A similar argument shows that $k \in n\mathbb{Z}$. Therefore, $k \in m\mathbb{Z} \cap n\mathbb{Z}$. Since $k \in \text{lcm}(m, n)\mathbb{Z}$ was arbitrary, $\text{lcm}(m, n)\mathbb{Z} \subseteq m\mathbb{Z} \cap n\mathbb{Z}$.

Since $m\mathbb{Z} \cap n\mathbb{Z} \subseteq \text{lcm}(m,n)\mathbb{Z}$ and $\text{lcm}(m,n)\mathbb{Z} \subseteq m\mathbb{Z} \cap n\mathbb{Z}$, we have $m\mathbb{Z} \cap n\mathbb{Z} = \text{lcm}(m,n)\mathbb{Z}$.

Let $k \in m\mathbb{Z} + n\mathbb{Z}$. Then there are integers a and b such that $k = ma + nb$. Since m and n are multiples of $\gcd(m,n)$, there are integers c and d such that $m = \gcd(m,n) \cdot c$ and $n = \gcd(m,n) \cdot d$. So, we have $k = \gcd(m,n) \cdot ac + \gcd(m,n) \cdot bd = \gcd(m,n)\,(ac + bd)$. Since $a, b, c, d \in \mathbb{Z}$ and $\mathbb{Z}$ is closed under addition and multiplication, $ac + bd \in \mathbb{Z}$. So, $k \in \gcd(m,n)\,\mathbb{Z}$. Since $k \in m\mathbb{Z} + n\mathbb{Z}$ was arbitrary, $m\mathbb{Z} + n\mathbb{Z} \subseteq \gcd(m,n)\,\mathbb{Z}$.

Next, let $k \in \gcd(m,n)\,\mathbb{Z}$. Then there is an integer c such that $k = \gcd(m,n) \cdot c$. By Theorem 4.30, there are integers a and b such that $\gcd(m,n) = am + bn$. So, $k = (am + bn)c = m(ac) + n(bc)$. Since $a, b, c \in \mathbb{Z}$ and $\mathbb{Z}$ is closed under multiplication, $ac, bd \in \mathbb{Z}$. So, $k \in m\mathbb{Z} + n\mathbb{Z}$. Since $k \in \gcd(m,n)\,\mathbb{Z}$ was arbitrary, $\gcd(m,n)\,\mathbb{Z} \subseteq m\mathbb{Z} + n\mathbb{Z}$.

Since $m\mathbb{Z} + n\mathbb{Z} \subseteq \gcd(m,n)\,\mathbb{Z}$ and $\gcd(m,n)\,\mathbb{Z} \subseteq m\mathbb{Z} + n\mathbb{Z}$, we have $m\mathbb{Z} + n\mathbb{Z} = \gcd(m,n)\,\mathbb{Z}$. □

48. Let $a, b \in \mathbb{Z}^+$ with $\gcd(a,b) = 1$. The groups $(\mathbb{Z} \times \mathbb{Z})/\langle(a,b)\rangle$ and $\mathbb{Z}$ are isomorphic.

Proof: Define a function $f\colon \mathbb{Z} \times \mathbb{Z} \to \mathbb{Z}$ by $f(j,k) = jb - ka$. To see that f is a homomorphism, observe that

$$f\big((j,k) + (m,n)\big) = f(j+m, k+n) = (j+m)b - (k+n)a$$
$$= (jb - ka) + (mb - na) = f(j,k) + f(m,n).$$

To see that f is surjective, let $x \in \mathbb{Z}$. Since $\gcd(a,b) = 1$, by Theorem 4.30, there are integers m and n such that $1 = ma + nb$. So, $x = x(ma + nb) = (xn)b - (-xm)a$. It follows that $f(xn, -xm) = x$.

Now, $(j,k) \in \ker(f)$ if and only if $f(j,k) = 0$ if and only if $jb - ka = 0$ if and only if $jb = ka$ if and only if there is an integer n such that $j = na$ and $k = nb$ if and only if there is an integer n such that $(j,k) = (na, nb)$ if and only if $(j,k) \in \langle(a,b)\rangle$. So, $\ker(f) = \langle(a,b)\rangle$.

By the First Homomorphism Theorem, $(\mathbb{Z} \times \mathbb{Z})/\langle(a,b)\rangle \cong \mathbb{Z}$. □

Let $f\colon \mathcal{X} \to \mathcal{Y}$ be the bijection defined in the Fourth Isomorphism Theorem and let $A, B \in \mathcal{X}$. Prove each of the following:

49. $A \leq B$ if and only if $f(A) \leq f(B)$.

Proof: I will prove the group version. In what follows $\mathcal{X}$ is the set of all subgroups of G that contain N, $\mathcal{Y}$ is the set of subgroups of G/N, and $f\colon \mathcal{X} \to \mathcal{Y}$ is the bijection defined by $f(H) = H/N$.

Suppose that $A \leq B$. Then $f(A) = A/N$ and $f(B) = B/N$. Let $x + N \in A/N$. Then $x \in A$. Since $A \leq B$, $x \in B$. So, $x + N \in B/N$. Since $x + N \in A/N$ was arbitrary, $A/N \leq B/N$. In other words, $f(A) \leq f(B)$.

Now, suppose that $f(A) \leq f(B)$, so that $A/N \leq B/N$. Let $x \in A$. Then $x + N \in A/N$. Since $A/N \leq B/N$, $x + N \in B/N$. So, $x \in B$. Since $x \in A$ was arbitrary, $A \leq B$. □

50. If $A \subseteq B$, then $[B: A] = [f(B): f(A)]$.

Proof: Define a function $f: (B/N)/(A/N) \to B/A$ by $f\big((x+N) + A/N\big) = x + A$.

If $(x+N) + A/N = (y+N) + A/N$, then $(x-y) + N = (x+N) - (y+N) \in A/N$. So, $x - y \in A$, and therefore, $x + A = y + A$. So, f is well-defined.

Now, if by $f\big((x+N) + A/N\big) =$ by $f\big((y+N) + A/N\big)$, then $x + A = y + A$. So, $x - y \in A$. It follows that $(x+N) - (y+N) = (x-y) + N \in A/N$. Thus, $(x+N) + A/N = (y+N) + A/N$. So, f is injective.

It is obvious that f is surjective, and therefore, f is a bijection. It follows that$[f(B): f(A)] = [B: A]$. □

51. Let $H \in \mathcal{X}$. Then $H \lhd G$ if and only if $f(H) \lhd G/N$ (prove this for the group version). State and prove the analogous result for the ring version of the Fourth Isomorphism Theorem.

Proof: Suppose that $H \lhd G$. By Theorem 11.46, $f(H) = H/N \lhd G/N$. Conversely, suppose that $f(H) \lhd G/N$. So, $H/N \lhd G/N$. By Theorem 11.47, $H \lhd G$. □

Ring version: Let $f: \mathcal{X} \to \mathcal{Y}$ be the bijection defined in the ring version of the Fourth Isomorphism Theorem. Let $S \in \mathcal{X}$. Prove that $S \lhd R$ if and only if $f(S) \lhd R/I$.

Proof of ring version: Recall that $\mathcal{X}$ is the set of all subrings of R that contain I, $\mathcal{Y}$ is the set of subrings of R/I, and $f: \mathcal{X} \to \mathcal{Y}$ is the bijection defined by $f(S) = S/I$.

Suppose that $S \lhd R$. By Theorem 11.48, $f(S) = S/I \lhd R/I$. Conversely, suppose that $f(S) \lhd R/I$. So, $S/I \lhd R/I$. By Theorem 11.49, $S \lhd R$. □

Find each of the following:

52. All groups of order 8, up to isomorphism

$\mathbb{Z}_8$, $\mathbb{Z}_4 \times \mathbb{Z}_2$, and $\mathbb{Z}_2 \times \mathbb{Z}_2 \times \mathbb{Z}_2$ are commutative groups of order 8. $\mathbb{Z}_2 \times \mathbb{Z}_2 \times \mathbb{Z}_2$ does not have any elements of order 4 or 8, and so, it is not isomorphic to the other two groups. $\mathbb{Z}_4 \times \mathbb{Z}_2$ does not have an element of order 8, and so, it is not isomorphic to $\mathbb{Z}_8$.

The Dihedral group $D_4 = \langle r, s \mid r^4 = e, s^2 = e, sr = r^3 s\rangle = \{e, r, r^2, r^3, s, rs, r^2 s, r^3 s\}$ is a group of order 8 that is **not** commutative, and so, it is not isomorphic to the previous three groups.

Let $Q = \{1, i, j, k, -1, -i, -j, -k\}$. As in part 1 of Example 11.54, we make Q into a group by letting 1 be the identity, and defining multiplication so that $(-1)^2 = 1$, $i^2 = j^2 = k^2 = -1$, $(-1)i = -i$, $(-1)j = -j$, $(-1)k = -k$, $ij = k$, $jk = i$, $ki = j$, $ji = -k$, $kj = -i$, and $ik = -j$. Q is also a group of order 8 that is **not** commutative, and so, it is not isomorphic to any of the three commutative groups above. To see that Q is not isomorphic D_4, observe that the only element of order 2 in Q is -1, whereas in D_4, s and r^2 are two distinct elements of order 2.

Now, let G be any group of order 8. The order of each element must be 1, 2, 4, or 8.

If G has an element of order 8, then G is cyclic and therefore, G is isomorphic to $\mathbb{Z}_8$.

Next, assume that every nonidentity element of G has order 2. Then every element is equal to its own inverse ($a \cdot a = a^2 = e$ implies that $a = a^{-1}$). So, if $a, b \in G$. Then $ab = (ab)^{-1} = b^{-1}a^{-1} = ba$. Therefore, G is commutative. So, we have $G = \{e, a, b, c, ab, ac, bc, abc\}$. We show that G is isomorphic to $\mathbb{Z}_2 \times \mathbb{Z}_2 \times \mathbb{Z}_2$. To see this, define $f: G \to \mathbb{Z}_2 \times \mathbb{Z}_2 \times \mathbb{Z}_2$ by $f(e) = (0,0,0)$, $f(a) = (1,0,0)$, $f(b) = (0,1,0)$, $f(c) = (0,0,1)$, $f(ab) = (1,1,0)$, $f(ac) = (1,0,1)$, $f(bc) = (0,1,1)$, and $f(a,b,c) = (1,1,1)$. Clearly f is bijective and it is easy to verify that f is a homomorphism. For example, $f(ab) = (1,1,0) = (1,0,0) + (0,1,0) = f(a) + f(b)$.

Assume that G has at least one element a of order 4. Let b be any element not in $\langle a \rangle$ (so, b is not equal to e, a, a^2, or a^3). Then $G = \{e, a, a^2, a^3, b, ba, ba^2, ba^3\}$. Since the left cancellation law holds in groups, we cannot have b^2 equal to b, ba, ba^2, or ba^3. Also, b^2 cannot be equal to a or a^3 because then b would have order 8, and so, G would be isomorphic to $\mathbb{Z}_8$. So, $b^2 = e$ or $b^2 = a^2$. Similarly, $ab = ba$ or $ab = ba^3$ ($ab = ba^2 \Rightarrow b^{-1}ab = a^2 \Rightarrow (b^{-1}ab)^2 = (a^2)^2 \Rightarrow b^{-1}a^2b = e \Rightarrow a^2 = e$, which is not true).

Let's look at each of these four cases separately.

$b^2 = e, ab = ba$: In this case, it is easy to check that G is commutative and isomorphic to $\mathbb{Z}_4 \times \mathbb{Z}_2$.

$b^2 = e, ab = ba^3$: In this case, we have $ba = ba(bb^{-1}) = b(ab)b^{-1} = b(ba^3)b^{-1} = b^2a^3b = a^3b$. So, we get $G = \langle a, b \mid a^4 = e, b^2 = e, ba = a^3b \rangle$, which is isomorphic to D_4.

$b^2 = a^2, ab = ba$: In this case, once again, G is commutative, b has order 4, and a^2 has order 2. So, it is easy to check that G is isomorphic to $\mathbb{Z}_4 \times \mathbb{Z}_2$.

$b^2 = a^2, ab = ba^3$: In this case, we get a group isomorphic to Q. Define $f: G \to Q$ by $f(a) = i$, $f(a^2) = -1, f(a^3) = -i, f(b) = j, f(ba) = -k, f(ba^2) = -j$, and $f(ba^3) = k$. Clearly f is a bijection and it is easy to verify that f is also a homomorphism. So, G is isomorphic to Q. □

53. All groups of order 10, up to isomorphism

$\mathbb{Z}_{10}$ is a commutative group of order 10.

$D_5 = \langle r, s \mid r^5 = e, s^2 = e, sr = r^4s \rangle = \{e, r, r^2, r^3, r^4, r^4s, rs, r^2s, r^3s, r^4s\}$ is a group of order 10 that is **not** commutative, and so, it is not isomorphic to $\mathbb{Z}_{10}$.

We will now prove that these are the only two groups of order 10, up to isomorphism.

Let G be an arbitrary group of order 10. The order of each element must be 1, 2, 5, or 10. If G has an element of order 10, then G is cyclic and therefore, G is isomorphic to $\mathbb{Z}_{10}$.

Let G be a noncyclic group of order 10. G has an element of order 5 and an element of order 2 for the same reason that a group of order 6 has an element of order 3 and an element of order 2 (see Problem 32).

Let $a \in G$ have order 5 and let $b \in G$ have order 2. Then the 10 distinct elements of G are $e, a, a^2, a^3, a^4, b, ba, ba^2, ba^3$, and ba^4.

It is not too hard to verify that we cannot have $ab = e, a, a^2, a^3, a^4$ or b. So, $ab = ba, ba^2, ba^3$, or ba^4.

If $ab = ba$, then it can be checked that ba generates G, contradicting our assumption that G is not cyclic.

If $ab = ba^2$, then $a = ba^2b$ and so, $a^2 = (ba^2b)(ba^2b) = ba^4b$. Also, we have $a^2 = bab$. Therefore, $ba^4b = bab$, and so, $a^4 = a$, which is not true. So, ab cannot be equal to ba^2.

If $ab = ba^3$, then $a = ba^3b$ and so, $a^2 = (ba^3b)(ba^3b) = ba^6b = bab = bba^3 = a^3$, which is not true. So, ab cannot be equal to ba^3.

If $ab = ba^4$, then $G = \langle a, b \mid a^5 = e, b^2 = e, ab = ba^4 \rangle$. This group is isomorphic to D_5. □

$G = \{1, 4, 11, 14, 16, 19, 26, 29, 31, 34, 41, 44\}$ is a group under multiplication modulo 45. Express G as each of the following:

54. An external direct product of cyclic groups

Since $|G| = 12 = 2^2 \cdot 3$, G is isomorphic to $\mathbb{Z}_2 \times \mathbb{Z}_2 \times \mathbb{Z}_3 \cong \mathbb{Z}_6 \times \mathbb{Z}_2$ or $\mathbb{Z}_4 \times \mathbb{Z}_3 \cong \mathbb{Z}_{12}$.

We have the following:

$$\langle 1 \rangle = \{1\}$$
$$\langle 4 \rangle = \{1, 4, 16, 19, 31, 34\}$$
$$\langle 11 \rangle = \{1, 11, 31, 26, 16, 41\}$$
$$\langle 14 \rangle = \{1, 14, 16, 44, 31, 29\}$$
$$\langle 16 \rangle = \{1, 16, 31\}$$
$$\langle 19 \rangle = \{1, 19\}$$
$$\langle 26 \rangle = \{1, 26\}$$
$$\langle 29 \rangle = \{1, 29, 31, 44, 16, 14\}$$
$$\langle 31 \rangle = \{1, 31, 16\}$$
$$\langle 34 \rangle = \{1, 34, 31, 19, 16, 4\}$$
$$\langle 41 \rangle = \{1, 41, 16, 26, 31, 11\}$$
$$\langle 44 \rangle = \{1, 44\}$$

So, the order diagram of G is $\{(1,1), (2,3), (3,2), (6,6)\}$. Since this matches the order diagram of $\mathbb{Z}_6 \times \mathbb{Z}_2$, we see that as an external direct product of cyclic groups, we have $G \cong \mathbb{Z}_6 \times \mathbb{Z}_2$, or equivalently, $G \cong \mathbb{Z}_2 \times \mathbb{Z}_2 \times \mathbb{Z}_3$.

55. An internal direct product of cyclic groups

$G \cong \langle 4 \rangle \times \langle 26 \rangle$. This works because 4 is an element of maximal order (the order of 4 is 6), and 26 is an element of order $\frac{|G|}{|\langle 4 \rangle|} = \frac{12}{6} = 2$ and $26 \notin \langle 4 \rangle$.

Prove the following:

56. Isomorphic finite groups have the same order diagram, where the order diagram of G is $d(G) = \{(n, x_n) \mid n \in \mathbb{Z}^+ \wedge x_n = |\{x \in G \mid |x| = n\}| > 0\}$.

Proof: Let G and H be isomorphic finite groups and let $f: G \to H$ be an isomorphism. Let $x \in G$. We show that $|x| = |f(x)|$. To see this, suppose that $x^n = e_G$. By Problem 38 in Problem Set 9, $(f(x))^n = f(x^n) = f(e_G) = e_H$. So, $|f(x)| \leq |x|$. Similarly, if $(f(x))^n = e_H$, then we have $x^n = f^{-1}(f(x^n)) = f^{-1}((f(x))^n) = f^{-1}(e_H) = e_G$ .So, $|x| \leq |f(x)|$.

Since $|f(x)| \leq |x|$ and $|x| \leq |f(x)|$, we have $|x| = |f(x)|$.

Now, suppose that $(n, x_n) \in d(G)$. Then there are n distinct elements in G of order n. For each element $x \in G$ of order n, the element $f(x) \in H$ also has order n. Since f is a bijection, there is a one-to-one correspondence between the elements of order n in G and the elements of order n in H. So, $(n, x_n) \in d(H)$. Since $(n, x_n) \in d(G)$ was arbitrary, $d(G) \subseteq d(H)$. A symmetrical argument shows that $d(H) \subseteq d(G)$. Therefore $d(G) = d(H)$, as desired. □

List all the commutative groups of the given order, up to isomorphism:

57. 31

Since 31 is prime, the only commutative group of order 31, up to isomorphism, is $\mathbb{Z}_{31}$.

58. 91

Since $91 = 7 \cdot 13$, the only commutative group of order 91, up to isomorphism, is $\mathbb{Z}_7 \times \mathbb{Z}_{13}$ or $\mathbb{Z}_{91}$.

59. 210

Since $210 = 2 \cdot 3 \cdot 5 \cdot 7$, the only commutative group of order 210, up to isomorphism, is

$$\mathbb{Z}_2 \times \mathbb{Z}_3 \times \mathbb{Z}_5 \times \mathbb{Z}_7 \text{ or } \mathbb{Z}_{210}.$$

60. 625

Since $625 = 5^4$, the commutative groups of order 625, up to isomorphism, are

$$\mathbb{Z}_5 \times \mathbb{Z}_5 \times \mathbb{Z}_5 \times \mathbb{Z}_5, \mathbb{Z}_{25} \times \mathbb{Z}_5 \times \mathbb{Z}_5, \mathbb{Z}_{25} \times \mathbb{Z}_{25}, \mathbb{Z}_{125} \times \mathbb{Z}_5, \text{ and } \mathbb{Z}_{625}.$$

61. 8575

Since $8575 = 5^2 \cdot 7^3$, the commutative groups of order 8575, up to isomorphism, are

$$\mathbb{Z}_5 \times \mathbb{Z}_5 \times \mathbb{Z}_7 \times \mathbb{Z}_7 \times \mathbb{Z}_7,\ \mathbb{Z}_{25} \times \mathbb{Z}_7 \times \mathbb{Z}_7 \times \mathbb{Z}_7,\ \mathbb{Z}_5 \times \mathbb{Z}_5 \times \mathbb{Z}_{49} \times \mathbb{Z}_7,\ \mathbb{Z}_{25} \times \mathbb{Z}_{49} \times \mathbb{Z}_7$$

$$\mathbb{Z}_5 \times \mathbb{Z}_5 \times \mathbb{Z}_{343}, \text{ and } \mathbb{Z}_{25} \times \mathbb{Z}_{343}.$$

Note: We can also write these groups as follows:

$$\mathbb{Z}_{35} \times \mathbb{Z}_{35} \times \mathbb{Z}_7,\ \mathbb{Z}_{175} \times \mathbb{Z}_7 \times \mathbb{Z}_7,\ \mathbb{Z}_{245} \times \mathbb{Z}_{35},\ \mathbb{Z}_{1225} \times \mathbb{Z}_7$$

$$\mathbb{Z}_{1715} \times \mathbb{Z}_5, \text{ and } \mathbb{Z}_{8575}.$$

62. 183,495,637

Since $183{,}495{,}637 = 13^3 17^4$, the commutative groups of order 183,495,637, up to isomorphism, are

$$\mathbb{Z}_{13} \times \mathbb{Z}_{13} \times \mathbb{Z}_{13} \times \mathbb{Z}_{17} \times \mathbb{Z}_{17} \times \mathbb{Z}_{17} \times \mathbb{Z}_{17}$$

$$\mathbb{Z}_{169} \times \mathbb{Z}_{13} \times \mathbb{Z}_{17} \times \mathbb{Z}_{17} \times \mathbb{Z}_{17} \times \mathbb{Z}_{17}$$

$$\mathbb{Z}_{2197} \times \mathbb{Z}_{17} \times \mathbb{Z}_{17} \times \mathbb{Z}_{17} \times \mathbb{Z}_{17}$$

$$\mathbb{Z}_{13} \times \mathbb{Z}_{13} \times \mathbb{Z}_{13} \times \mathbb{Z}_{289} \times \mathbb{Z}_{17} \times \mathbb{Z}_{17}$$

$$\mathbb{Z}_{169} \times \mathbb{Z}_{13} \times \mathbb{Z}_{289} \times \mathbb{Z}_{17} \times \mathbb{Z}_{17}$$

$$\mathbb{Z}_{2197} \times \mathbb{Z}_{289} \times \mathbb{Z}_{17} \times \mathbb{Z}_{17}$$

$$\mathbb{Z}_{13} \times \mathbb{Z}_{13} \times \mathbb{Z}_{13} \times \mathbb{Z}_{289} \times \mathbb{Z}_{289}$$

$$\mathbb{Z}_{169} \times \mathbb{Z}_{13} \times \mathbb{Z}_{289} \times \mathbb{Z}_{289}$$

$$\mathbb{Z}_{2197} \times \mathbb{Z}_{289} \times \mathbb{Z}_{289}$$

$$\mathbb{Z}_{13} \times \mathbb{Z}_{13} \times \mathbb{Z}_{13} \times \mathbb{Z}_{4913} \times \mathbb{Z}_{17}$$

$$\mathbb{Z}_{169} \times \mathbb{Z}_{13} \times \mathbb{Z}_{4913} \times \mathbb{Z}_{17}$$

$$\mathbb{Z}_{2197} \times \mathbb{Z}_{4913} \times \mathbb{Z}_{17}$$

$$\mathbb{Z}_{13} \times \mathbb{Z}_{13} \times \mathbb{Z}_{13} \times \mathbb{Z}_{83{,}521}$$

$$\mathbb{Z}_{169} \times \mathbb{Z}_{13} \times \mathbb{Z}_{83{,}521}$$

$$\mathbb{Z}_{2197} \times \mathbb{Z}_{83{,}521}$$

Problem Set 12

LEVEL 1

Find the order of each of the following elements in $(\mathbb{Z}_4, +)$.

1. 0

$1 \cdot 0 \equiv_4 0$. So, $|0| = \mathbf{1}$.

2. 1

$2 \cdot 1 \equiv_4 2$, $3 \cdot 1 \equiv_4 3$, $4 \cdot 1 = 4 \equiv_4 0$. So, $|1| = \mathbf{4}$.

3. 2

$2 \cdot 2 \equiv_4 4 \equiv_4 0$. So, $|2| = \mathbf{2}$.

4. 3

$2 \cdot 3 \equiv_4 3 + 3 \equiv_4 6 \equiv_4 2$, $3 \cdot 3 \equiv_4 2 + 3 \equiv_4 5 \equiv_4 1$, $4 \cdot 3 = 1 + 3 \equiv_4 4 \equiv_4 0$. So, $|3| = \mathbf{4}$.

List the elements of each of the following groups:

5. U_{11}

$U_{11} = \{1, 2, 3, 4, 5, 6, 7, 8, 9, 10\}$

6. U_{15}

$U_{15} = \{1, 2, 4, 7, 8, 11, 13, 14\}$

7. U_{20}

$U_{20} = \{1, 3, 7, 9, 11, 13, 17, 19\}$

Find the inverse of each element in each of the following groups:

8. U_5

$U_5 = \{1, 2, 3, 4\}$

Since $1 \cdot 1 = 1$, $\mathbf{1^{-1} = 1}$.

Since $2 \cdot 3 = 6 \equiv_5 1$, $\mathbf{2^{-1} = 3}$ and $\mathbf{3^{-1} = 2}$.

Since $4 \cdot 4 = 16 \equiv_5 1$, $\mathbf{4^{-1} = 4}$.

9. U_9

$U_9 = \{1, 2, 4, 5, 7, 8\}$

Since $1 \cdot 1 = 1$, $\mathbf{1^{-1} = 1}$.

Since $2 \cdot 5 = 10 \equiv_9 1$, $\mathbf{2^{-1} = 5}$ and $\mathbf{5^{-1} = 2}$.

Since $4 \cdot 7 = 28 \equiv_9 1$, $\mathbf{4^{-1} = 7}$ and $\mathbf{7^{-1} = 4}$.

Since $8 \cdot 8 = 64 \equiv_9 1$, $\mathbf{8^{-1} = 8}$.

10. U_{28}

$U_{28} = \{1, 3, 5, 9, 11, 13, 15, 17, 19, 23, 25, 27\}$

Since $1 \cdot 1 = 1$, $\mathbf{1^{-1} = 1}$.

Since $3 \cdot 19 = 57 \equiv_{28} 1$, $\mathbf{3^{-1} = 19}$ and $\mathbf{19^{-1} = 3}$.

Since $5 \cdot 17 = 85 \equiv_{28} 1$, $\mathbf{5^{-1} = 17}$ and $\mathbf{17^{-1} = 5}$.

Since $9 \cdot 25 = 225 \equiv_{28} 1$, $\mathbf{9^{-1} = 25}$ and $\mathbf{25^{-1} = 9}$.

Since $11 \cdot 23 = 253 \equiv_{28} 1$, $\mathbf{11^{-1} = 23}$ and $\mathbf{23^{-1} = 11}$.

Since $13 \cdot 13 = 169 \equiv_{28} 1$, $\mathbf{13^{-1} = 13}$.

Since $15 \cdot 15 = 225 \equiv_{28} 1$, $\mathbf{15^{-1} = 15}$.

Since $27 \cdot 27 = 729 \equiv_{28} 1$, $\mathbf{27^{-1} = 27}$.

11. Draw the multiplication table for U_{28}.

$U_{28} = \{1, 3, 5, 9, 11, 13, 15, 17, 19, 23, 25, 27\}$

$\cdot$	1	3	5	9	11	13	15	17	19	23	25	27
1	1	3	5	9	11	13	15	17	19	23	25	27
3	3	9	15	27	5	11	17	23	1	13	19	25
5	5	15	25	17	27	9	19	1	11	3	13	23
9	9	27	17	25	15	5	23	13	3	11	1	19
11	11	5	27	15	9	3	25	19	13	1	23	17
13	13	11	9	5	3	1	27	25	23	19	17	15
15	15	17	19	23	25	27	1	3	5	9	11	13
17	17	23	1	13	19	25	3	9	15	27	5	11
19	19	1	11	3	13	23	5	15	25	17	27	9
23	23	13	3	11	1	19	9	27	17	25	15	5
25	25	19	13	1	23	17	11	5	27	15	9	3
27	27	25	23	19	17	15	13	11	9	5	3	1

LEVEL 2

Find the order of each of the following elements in $(\mathbb{Z}_{40}, +)$.

12. 32

$2 \cdot 32 \equiv_{40} 64 \equiv_{40} 24$, $3 \cdot 32 \equiv_{40} 24 + 32 \equiv_{40} 56 \equiv_{40} 16$,

$4 \cdot 32 \equiv_{40} 16 + 32 \equiv_{40} 48 \equiv_{40} 8$, $5 \cdot 32 \equiv_{40} 8 + 32 \equiv_{40} 40 \equiv_{40} 0$.

So, $|32| = \mathbf{5}$.

13. 6

$2 \cdot 6 \equiv_{40} 12$, $3 \cdot 6 \equiv_{40} 18$, $4 \cdot 6 \equiv_{40} 24$, $5 \cdot 6 \equiv_{40} 30$, $6 \cdot 6 \equiv_{40} 36$, $7 \cdot 6 \equiv_{40} 42 \equiv_{40} 2$

$8 \cdot 6 \equiv_{40} 48 \equiv_{40} 8$, $9 \cdot 6 \equiv_{40} 54 \equiv_{40} 14$, $10 \cdot 6 \equiv_{40} 60 \equiv_{40} 20$, $11 \cdot 6 \equiv_{40} 66 \equiv_{40} 26$

$12 \cdot 6 \equiv_{40} 72 \equiv_{40} 32$, $13 \cdot 6 \equiv_{40} 78 \equiv_{40} 38$, $14 \cdot 6 \equiv_{40} 84 \equiv_{40} 4$, $15 \cdot 6 \equiv_{40} 90 \equiv_{40} 10$

$16 \cdot 6 \equiv_{40} 96 \equiv_{40} 16$, $17 \cdot 6 \equiv_{40} 102 \equiv_{40} 22$, $18 \cdot 6 \equiv_{40} 108 \equiv_{40} 28$, $19 \cdot 6 \equiv_{40} 114 \equiv_{40} 34$

$20 \cdot 6 \equiv_{40} 120 \equiv_{40} 0$

So, $|6| = \mathbf{20}$.

Determine if each of the following statements is true or false. If true, provide a proof. If false, provide a counterexample.

14. Let $(H, +)$ be a subgroup of $(\mathbb{Z}, +)$. There is $n \in H$ such that $H = n\mathbb{Z}$.

True

Proof: If $H = \{0\}$, then $H = 0\mathbb{Z}$. If $H \neq \{0\}$, then since H is closed under additive inverses, H must contain a positive integer. Let n be the least positive integer in H. We will show that $H = n\mathbb{Z}$. If $x \in n\mathbb{Z}$, then there is a $k \in \mathbb{Z}$ such that $x = nk$. Since H is closed under addition , $x = nk \in H$. So, $n\mathbb{Z} \subseteq H$. Conversely, let $x \in H$ and let $d = \gcd(x, n)$. By Theorem 4.30, there are integers a and b such that $ax + bn = d$. Since $n\mathbb{Z}$ is closed under addition, $d \in n\mathbb{Z}$. Since $d \leq n$, we must have $d = n$. It follows that x is a multiple of n, and so, $x \in n\mathbb{Z}$. So, $H \subseteq n\mathbb{Z}$. Since $n\mathbb{Z} \subseteq H$ and $H \subseteq n\mathbb{Z}$, we have $H = n\mathbb{Z}$, as desired. □

15. For each $n \in \mathbb{Z}^+$, U_n is a cyclic group.

False: $U_{15} = \{1, 2, 4, 7, 8, 11, 13, 14\}$ is not cyclic. Observe that $11^2 \equiv_{15} 121 \equiv_{15} 1$. So, $\langle 11 \rangle$ is a subgroup of U_{15} of order 2. Similarly, $14^2 \equiv_{15} 196 \equiv_{15} 1$. So, $\langle 14 \rangle$ is also a subgroup of U_{15} of order 2. However, by Theorem 12.6, a cyclic group of order 8 has **exactly one** subgroup of order 2. Therefore, U_{15} **cannot** be cyclic.

Note: An alternative proof that U_{15} is not cyclic would be to just compute the order of each element and observe that no element has order 8.

Compute each of the following:

16. 3^{31} modulo 7

By Fermat's Little Theorem, $3^7 \equiv_7 3$, or equivalently, $3^6 \equiv_7 1$.

Therefore, $3^{31} = 3^{30} \cdot 3 = (3^6)^5 \cdot 3 \equiv_7 1^5 \cdot 3 \equiv_7 \mathbf{3}$.

17. 128^{129} modulo 17.

By Fermat's Little Theorem, $128^{17} \equiv_{17} 128$, or equivalently, $128^{16} \equiv_{17} 1$. Therefore, we have $128^{129} = 128^{128} \cdot 128 = (128^{16})^8 \cdot 128 \equiv_{17} 1^8 \cdot 128 \equiv_{17} 128 \equiv_{17} \mathbf{9}$.

LEVEL 3

Find each of the following:

18. The order of $(123)(45)(6789)$ in S_{10}

$(123)^2 = (123)(123) = (132)$ and $(123)^3 = (132)(123) = (1)$. Therefore, $|123| = 3$. Similarly, $|(45)| = 2$ and $|(6789)| = 4$. So, $|(123)(45)(6789)| = \text{lcm}(3,2,4) = \mathbf{12}$.

Note: If you don't understand this solution, you can just compute the order directly by multiplying the given permutation by itself 12 times. While performing the computations, try to understand why this solution works.

19. All subgroups of $(\mathbb{Z}_{30}, +)$.

Since $(\mathbb{Z}_{30}, +)$ is cyclic, by Theorems 12.6 and 11.17, for each divisor d of 30, there is exactly one subgroup of order d and there are no other subgroups. The divisors of 30 are $1, 2, 3, 5, 6, 10, 15$, and 30. The corresponding subgroups are as follows:

$$\langle 0 \rangle = \{0\}$$
$$\langle 15 \rangle = \{0, 15\}$$
$$\langle 10 \rangle = \{0, 10, 20\}$$
$$\langle 6 \rangle = \{0, 6, 12, 18, 24\}$$
$$\langle 5 \rangle = \{0, 5, 10, 15, 20, 25\}$$
$$\langle 3 \rangle = \{0, 3, 6, 9, 12, 15, 18, 21, 24, 27\}$$
$$\langle 2 \rangle = \{0, 2, 4, 6, 8, 10, 12, 14, 16, 18, 20, 22, 24, 26, 28\}$$
$$\langle 1 \rangle = \mathbb{Z}_{30}$$

20. All generators of $(\mathbb{Z}_{75},+)$.

The generators of $\mathbb{Z}_{75}$ are the positive integers less than 75 that are relatively prime with 75. Here is a list:

$$1, 2, 4, 7, 8, 11, 13, 14, 16, 17, 19, 22, 23, 26, 28, 29, 31, 32, 34, 37, 38, 41,$$
$$43, 44, 46, 47, 49, 52, 53, 56, 58, 59, 61, 62, 64, 67, 68, 71, 73, 74$$

Let G be a group and let $a \in G$. Prove each of the following:

21. $|a^{-1}| = |a|$.

Proof: If $a^n = e$, then $(a^{-1})^n = (a^n)^{-1} = e^{-1} = e$.

Conversely, if $(a^{-1})^n = e$, then $a^n = ((a^{-1})^{-1})^n = ((a^{-1})^n)^{-1} = e^{-1} = e$.

It follows that $|a^{-1}| = |a|$. □

22. $|a| = 1$ if and only if $a = e$.

Proof: If $a = e$, then $a^1 = e$. So, $|a| = 1$.

Conversely, suppose that $|a| = 1$. Then $a = a^1 = e$. □

23. The order of any conjugate of a is equal to the order of a.

Proof: We first prove by induction on $n \in \mathbb{N}$ that $(xax^{-1})^n = xa^nx^{-1}$. For the base case $(k = 0)$, we have $(xax^{-1})^0 = e$ by definition and $xa^0x^{-1} = xex^{-1} = xx^{-1} = e$. Assuming that $k \in \mathbb{N}$ and $(xax^{-1})^k = xa^kx^{-1}$, we get

$$(xax^{-1})^{k+1} = (xax^{-1})^k(xax^{-1}) = (xa^kx^{-1})(xax^{-1}) = xa^k(x^{-1}x)ax^{-1}$$
$$= xa^keax^{-1} = xa^kax^{-1} = xa^{k+1}x^{-1}$$

By the Principle of Mathematical Induction, for all $n \in \mathbb{N}$, $(xax^{-1})^n = xa^nx^{-1}$.

It follows that $(xax^{-1})^n = e$ if and only if $xa^nx^{-1} = e$ if and only if

$$a^n = ea^ne = (x^{-1}x)a^n(x^{-1}x) = x^{-1}(xa^nx^{-1})x = x^{-1}ex = x^{-1}x = e.$$

So, $|xax^{-1}| = |a|$, as desired. □

24. If p is prime, $a \neq e$, and $a^p = e$, then $|a| = p$.

Proof: Since $a^p = e$, $|a| \leq p$. If $n \leq p$ and $a^n = e$, then $n|p$. Since p is prime, $n = p$. It follows that p is the least positive integer such that $a^p = e$. So, $|a| = p$. □

LEVEL 4

Prove each of the following:

25. Let G be a group and let $a, b \in G$. Then $|ab| = |ba|$.

Proof: Let $m, n \in \mathbb{Z}^+$ and assume that $|ab| = m$ and $|ba| = n$. Then $(ab)^m = e$ and $(ba)^n = e$. Since $(ab)^m = e$, we have

$$(a^{-1}a)(ba)^m = (a^{-1}a)b(ab)^{m-1}a = a^{-1}(ab)(ab)^{m-1}a = a^{-1}(ab)^m a = a^{-1}ea = a^{-1}a = e.$$

It follows that $n = |ba| \leq m$.

A symmetrical argument shows that $m = |ab| \leq n$. So, $|ab| = m = n = |ba|$.

Now, assume that $|ba|$ is infinite. If $(ab)^m = e$ for some $m \in \mathbb{Z}^+$, then the previous argument shows that $|ba| \leq m$, contradicting our assumption that $|ba|$ is infinite. So, $|ab|$ is infinite.

A symmetrical argument shows that if $|ab|$ is infinite, then $|ba|$ is infinite. □

26. $\Phi(p^n) = p^n - p^{n-1}$ for any prime number p and $n \in \mathbb{Z}^+$ (Φ is Euler's totient function).

Proof: The nonnegative integers less than p^n that have a common divisor with p^n are all the multiples of p. Since every pth integer is a multiple of p, we see that there are $\frac{p^n}{p} = p^{n-1}$ nonnegative multiples of p less than p^n. So, there are p^{n-1} nonnegative integers less than p^n that are **not** relatively prime with p^n. It follows that the number of positive integers less than p^n that are relatively prime with p^n is $p^n - p^{n-1}$. So, $\Phi(p^n) = p^n - p^{n-1}$. □

27. Let $n \in \mathbb{Z}^+$ and $[a]_n \in U_n$. Then $[a]_n^{\Phi(n)} \equiv_n [1]_n$, where Φ is Euler's totient function.

Proof: Since U_n is a group of order $\Phi(n)$, we can write $U_n = \left\{[a_1]_n, [a_2]_n, \dots, \left[a_{\Phi(n)}\right]_n\right\}$. If we multiply each element in U_n by $[a]_n$, we get the same group: $U_n = \left\{[a_1 \cdot a]_n, [a_2 \cdot a]_n, \dots, \left[a_{\Phi(n)} \cdot a\right]_n\right\}$. So, we have $[a_1]_n \cdot [a_2]_n \cdots \left[a_{\Phi(n)}\right]_n = [a_1 \cdot a]_n, [a_2 \cdot a]_n \cdots \left[a_{\Phi(n)} \cdot a\right]_n$, or equivalently,

$$\left[a_1 \cdot a_2 \cdots a_{\Phi(n)}\right]_n = \left[a_1 \cdot a \cdot a_2 \cdot a \cdots a_{\Phi(n)} \cdot a\right]_n$$
$$= \left[(a_1 \cdot a_2 \cdots a_{\Phi(n)})a^{\Phi(n)}\right]_n = \left[a_1 \cdot a_2 \cdots a_{\Phi(n)}\right]_n \cdot \left[a^{\Phi(n)}\right]_n$$

So, we have $\left[a_1 \cdot a_2 \cdots a_{\Phi(n)}\right]_n \cdot [1]_n = \left[a_1 \cdot a_2 \cdots a_{\Phi(n)}\right]_n = \left[a_1 \cdot a_2 \cdots a_{\Phi(n)}\right]_n \cdot \left[a^{\Phi(n)}\right]_n$. Since U_n is a group, the left cancellation law holds, and so, $[1]_n = \left[a^{\Phi(n)}\right]_n = [a]_n^{\Phi(n)}$. □

28. The Chinese Remainder Theorem

Proof: We will prove the theorem by induction on $n \geq 1$. The base case of $k = 1$ is trivial. Assuming that the theorem holds for any system of k linear congruences satisfying the given conditions, assume that $n_1, n_2, \dots, n_k, n_{k+1}$ are pairwise relatively prime integers, let $c_1, c_2, \dots, c_k, c_{k+1} \in \mathbb{Z}$, and consider the following system of linear congruences:

$$\begin{gathered} x \equiv_{n_1} c_1, \\ x \equiv_{n_2} c_2, \\ \vdots \quad \vdots \quad \vdots \\ x \equiv_{n_k} c_k, \\ x \equiv_{n_{k+1}} c_{k+1}, \end{gathered}$$

By induction, we can let $x \equiv_{n_1 \cdot n_2 \cdots n_k} d$ be a unique solution of the first n linear congruences. Consider the following system of linear congruences:

$$\begin{gathered} x \equiv_{n_1 \cdot n_2 \cdots n_k} d, \\ x \equiv_{n_{k+1}} c_{k+1}. \end{gathered}$$

Since $\gcd(n_1 \cdot n_2 \cdots n_k, n_{k+1}) = 1$, by Theorem 12.32, this system has a unique solution, up to congruence modulo $n_1 \cdot n_2 \cdots n_k \cdot n_{k+1}$. This is a unique solution of the original system, up to congruence modulo $n_1 \cdot n_2 \cdots n_k \cdot n_{k+1}$.

By the Principle of Mathematical Induction, the Chinese Remainder Theorem holds. □

Find each of the following:

29. All subgroups of $\mathbb{Z}_5 \times \mathbb{Z}_3$

$\mathbb{Z}_5 \times \mathbb{Z}_3$ is cyclic with generator $(1,1)$. Since $|\mathbb{Z}_5 \times \mathbb{Z}_3| = |\mathbb{Z}_5| \cdot |\mathbb{Z}_3| = 5 \cdot 3 = 15$, by Theorems 12.6 and 11.17, for each divisor d of 15, there is exactly one subgroup of order d and there are no other subgroups. The divisors of 15 are $1, 3, 5$, and 15. The corresponding subgroups are as follows:

$$\begin{gathered} \langle 0 \rangle = \{0\} \\ \langle (0,1) \rangle = \{(0,0), (0,1), (0,2)\} \\ \langle (1,0) \rangle = \{(0,0), (1,0), (2,0), (3,0), (4,0)\} \\ \langle (1,1) \rangle = \mathbb{Z}_5 \times \mathbb{Z}_3 \end{gathered}$$

30. $\Phi(1024)$, where Φ is Euler's totient function

$\Phi(1024) = \Phi(2^{10}) = 2^{10} - 2^9 = 1024 - 512 = \mathbf{512}$.

31. $\Phi(625)$, where Φ is Euler's totient function

$\Phi(625) = \Phi(5^4) = 5^4 - 5^3 = 625 - 125 = \mathbf{500}$.

LEVEL 5

Let $a, b \in \mathbb{Z}^+$ with $\gcd(a, b) = 1$. Prove each of the following:

32. $(\mathbb{Z}_{ab}, +, \cdot) \cong (\mathbb{Z}_a, +, \cdot) \times (\mathbb{Z}_b, +, \cdot)$.

Proof: We define a function $f: \mathbb{Z}_{ab} \to \mathbb{Z}_a \times \mathbb{Z}_b$ by $f([k]_{ab}) = ([k]_a, [k]_b)$. We first show that f is well-defined. Suppose that $[k_1]_{ab} = [k_2]_{ab}$. Then $ab | k_2 - k_1$. So, there is an integer j such that $k_2 - k_1 = abj$. This equation shows that $a | k_2 - k_1$ and $b | k_2 - k_1$. So, $[k_1]_a = [k_2]_a$ and $[k_1]_b = [k_2]_b$. Therefore, $([k_1]_a, [k_1]_b) = ([k_2]_a, [k_2]_b)$.

Next, we show that f is a homomorphism. We have

$$f([k_1]_{ab} + [k_2]_{ab}) = f([k_1 + k_2]_{ab}) = ([k_1 + k_2]_a, [k_1 + k_2]_b)$$
$$= ([k_1]_a + [k_2]_a, [k_1]_b + [k_2]_b) = ([k_1]_a, [k_1]_b) + ([k_2]_a, [k_2]_b)$$
$$= f([k_1]_{ab}) + f([k_2]_{ab}).$$

$$f([k_1]_{ab} \cdot [k_2]_{ab}) = f([k_1 \cdot k_2]_{ab}) = ([k_1 \cdot k_2]_a, [k_1 \cdot k_2]_b)$$
$$= ([k_1]_a \cdot [k_2]_a, [k_1]_b \cdot [k_2]_b) = ([k_1]_a, [k_1]_b) \cdot ([k_2]_a, [k_2]_b) = f([k_1]_{ab}) \cdot f([k_2]_{ab}).$$

To see that f is injective, let $f([k_1]_{ab}) = f([k_2]_{ab})$. So, $([k_1]_a, [k_1]_b) = ([k_2]_a, [k_2]_b)$. It follows that $[k_1]_a = [k_2]_a$ and $[k_1]_b = [k_2]_b$. So, $a | k_2 - k_1$ and $b | k_2 - k_1$. Since $\gcd(a, b) = 1$, by Theorem 4.38, $\operatorname{lcm}(a, b) = ab$. So, by Problem 102 from Problem Set 4, $ab | k_2 - k_1$. Therefore, $[k_1]_{ab} = [k_2]_{ab}$.

Since f is an injection between two finite sets with the same number of elements, f is surjective. □

33. $U_{ab} \cong U_a \times U_b$.

Proof: We define $g: U_{ab} \to U_a \times U_b$ by $g(x) = f(x)$, where f is the function given in Problem 32 above. We need only show that $x \in U_{ab}$ if and only if $(x, x) \in U_a \times U_b$. Well, $x \in U_{ab}$ if and only if $\gcd(x, ab) = 1$ if and only if $\gcd(x, a) = 1$ and $\gcd(x, b) = 1$ if and only if $x \in U_a$ and $x \in U_b$. □

34. If $n \in \mathbb{Z}^+$ and $n = p_1^{m_1} \cdot p_2^{m_2} \cdots p_k^{m_k}$ is the prime factorization of n and Φ is Euler's totient function, then $\Phi(n) = (p_1^{m_1} - p_1^{m_1 - 1})(p_1^{m_2} - p_1^{m_2 - 1}) \cdots (p_1^{m_k} - p_1^{m_k - 1})$.

Proof: Using Problem 33 above, it is easy to show by induction that $U_n \cong U_{p_1^{m_1}} \times U_{p_2^{m_2}} \times \cdots \times U_{p_k^{m_k}}$. Also, by Problem 26 above, we have $\left|U_{p_i^{m_i}}\right| = \Phi(p_i^{m_i}) = p_i^{m_i} - p_i^{m_i - 1}$. It follows that

$$\Phi(n) = |U_n| = \left|U_{p_1^{m_1}} \times U_{p_2^{m_2}} \times \cdots \times U_{p_k^{m_k}}\right|$$
$$= \left|U_{p_1^{m_1}}\right| \cdot \left|U_{p_2^{m_2}}\right| \cdots \left|U_{p_k^{m_k}}\right| = (p_1^{m_1} - p_1^{m_1 - 1})(p_1^{m_2} - p_1^{m_2 - 1}) \cdots (p_1^{m_k} - p_1^{m_k - 1})$$ □

35. $\Phi(1{,}047{,}816) = 299{,}376$.

Proof: The prime factorization of 1,047,816 is $2^3 \cdot 3^5 \cdot 7^2 \cdot 11$. It follows from Problem 34 above that

$$\Phi(1{,}047{,}816) = \Phi(2^3 \cdot 3^5 \cdot 7^2 \cdot 11) = (2^3 - 2^2)(3^5 - 3^4)(7^2 - 7)(11)$$
$$= 4 \cdot 162 \cdot 42 \cdot 11 = \mathbf{299, 376}.$$
□

Prove each of the following:

36. Let $a, b \in \mathbb{Z}$, let $n \in \mathbb{Z}^+$, let $d = \gcd(a, n)$ and suppose that $d|b$. The linear congruence $ax \equiv_n b$ has exactly d solutions, up to congruence modulo n.

Proof: We know by Theorem 12.22 that the linear congruence $ax \equiv_n b$ has a solution. Let's call this solution c. Since $\gcd(a, n)$ is a divisor of both a and n, there are integers s and t such that $a = \gcd(a, n) \cdot s$ and $n = \gcd(a, n) \cdot t$. We will show that for each $i = 1, 2, \dots, d - 1$, $c + it$ is a solution of $ax \equiv_n b$.

Since $a = \gcd(a, n) \cdot s$ and $n = \gcd(a, n) \cdot t$, we have $at = \gcd(a, n) \cdot st = sn$. Therefore, we have $a(c + it) = ac + ait = ac + sni \equiv_n ac \equiv_n b$, as desired.

Next, let's show that if $0 \le i < j \le \gcd(a, n) - 1$, then $c + it \not\equiv_n c + jt$. Suppose toward contradiction that $c + it \equiv_n c + jt$. Then $it \equiv_n jt$, and so, $n|(j - i)t$. Therefore, $(j - i)t = nr$ for some $r \in \mathbb{Z}$. So, we have $(j - i)n = (j - i)\gcd(a, n) \cdot t = \gcd(a, n) \cdot nr$. So, $j - i = \gcd(a, n) \cdot r$. Therefore, $\gcd(a, n) \le j - i \le \gcd(a, n) - 1 - i < \gcd(a, n)$, a contradiction. □

37. Let G be a group, let $m, n \in \mathbb{Z}^+$, and let $a \in G$ with $|a| = n$. Then $\gcd(m, n) = 1$ if and only if $|a^m| = n$.

Proof: Assume that $\gcd(m, n) = 1$. By Theorem 4.30, there are $s, t \in \mathbb{Z}$ such that $sm + tn = 1$. Then $a = a^1 = a^{sm+tn} = a^{sm}a^{tn} = (a^m)^s(a^n)^t = (a^m)^s e^t = (a^m)^s e = (a^m)^s$. Let $x \in \langle a \rangle$ be arbitrary. Then there is $k \in \mathbb{Z}$ such that $x = a^k$. So, $x = a^k = ((a^m)^s)^k = (a^m)^{sk} \in \langle a^m \rangle$. It follows that a^m is a generator for $\langle a \rangle$, and so $|a^m| = |\langle a^m \rangle| = |\langle a \rangle| = |a| = n$.

Conversely, assume that $|a^m| = n$. Then $G = \langle a^m \rangle$, and so, $a \in \langle a^m \rangle$. Therefore, there is an integer k such that $a^1 = a = (a^m)^k = a^{mk}$. By Theorem 12.4, $mk \equiv_n 1$. So, m has a multiplicative inverse in $\mathbb{Z}_n$. Therefore, $m \in U_n$, and so, $\gcd(m, n) = 1$. □

38. Let G be a group, let $m, n \in \mathbb{Z}^+$, and let $a \in G$ with $|a| = n$. Then $m \cdot |a^m| = \operatorname{lcm}(m, n)$.

Proof: Since $n|\operatorname{lcm}(m, n)$, by the Note following Theorem 12.4, $a^{\operatorname{lcm}(m,n)} = e$. It follows that

$$(a^m)^{\frac{\operatorname{lcm}(m,n)}{m}} = a^{\operatorname{lcm}(m,n)} = e.$$

So, $|a^m| \le \frac{\operatorname{lcm}(m,n)}{m}$, or equivalently, $m \cdot |a^m| \le \operatorname{lcm}(m, n)$.

Let $t < n$ and assume that $(a^m)^t = e$. Then $a^{mt} = e$. By the Note following Theorem 12.4, $n|mt$. Since $m|mt$ as well, mt is a common multiple of m and n. Therefore, $\text{lcm}(m,n) \leq mt$, or equivalently, $\frac{\text{lcm}(m,n)}{m} \leq t$. So, If $t < \frac{\text{lcm}(m,n)}{m}$, then $(a^m)^t \neq e$. Thus, $|a^m| \geq \frac{\text{lcm}(m,n)}{m}$, or equivalently, we have $m \cdot |a^m| \geq \text{lcm}(m,n)$.

Since $m \cdot |a^m| \leq \text{lcm}(m,n)$ and $m \cdot |a^m| \geq \text{lcm}(m,n)$, we have $m \cdot |a^m| = \text{lcm}(m,n)$. □

39. Let G be a finite commutative group and let $k \in \mathbb{Z}^+$ such that k divides $|G|$. Then G has a subgroup of order k.

Proof: We prove this by induction on $m = |G|$.

Base case ($m = 0$): If $|G| = 1$, then $G = \{e\}$. The only positive divisor of 1 is $k = 1$ and $G = \{e\}$ is a subgroup of G of order 1.

Inductive step: Assume that for any group H of order $j < m$ and $k \in \mathbb{Z}^+$ dividing j, there is a subgroup of H of order k. Let G be a group of order m, suppose that $k > 1$ divides m, and let p be a prime factor of k. By the Fundamental Theorem of Commutative Groups, G has a cyclic subgroup K of order p^t for some $t \in \mathbb{Z}^+$. By Theorem 12.6, K has a subgroup L of order p, and L is also a subgroup of G. Now, G/L is a commutative group of order $\frac{m}{p}$. Since $\frac{k}{p}$ divides $\frac{m}{p}$, by the inductive hypothesis, there is a subgroup M/L of H/L, where M is a subgroup of G and $|M/L| = \frac{k}{p}$. We have $|M| = \left(\frac{|M|}{|L|}\right) \cdot |L| = \frac{k}{p} \cdot p = k$.

By the Principle of Mathematical Induction, for all $k \in \mathbb{Z}^+$ with k dividing the order of G, there is a subgroup of G of order k. □

Problem Set 13

LEVEL 1

Determine if each of the following statements is true or false. If true, provide a proof. If false, provide a counterexample.

1. Let $r \in \mathbb{R}$ and define $f\colon \mathbb{R} \to \mathbb{R}$ by $f(x) = r$. Then for any $a \in \mathbb{R}$, $f'(a) = 0$.

True

Proof:

$$f'(a) = \lim_{x\to a}\frac{f(x)-f(a)}{x-a} = \lim_{x\to a}\frac{r-r}{x-a} = \lim_{x\to a}\frac{0}{x-a} = 0 \qquad \square$$

2. If $f\colon \mathbb{R} \to \mathbb{R}$ is continuous, then f is differentiable on $\mathbb{R}$.

False

Let $f(x) = |x|$. Then f is continuous on $\mathbb{R}$, but f is not differentiable at 0. To see that f is not differentiable at 0, observe that

$$f'(0) = \lim_{x\to 0}\frac{f(x)-f(0)}{x-0} = \lim_{x\to 0}\frac{|x|-0}{x} = \lim_{x\to 0}\frac{|x|}{x}.$$

Since $\lim_{x\to 0^+}\frac{|x|}{x} = \lim_{x\to 0^+}\frac{x}{x} = \lim_{x\to 0^+} 1 = 1$ and $\lim_{x\to 0^-}\frac{|x|}{x} = \lim_{x\to 0^-}\frac{-x}{x} = \lim_{x\to 0^-}(-1) = -1$, it follows that $f'(0)$ does not exist. $\square$

3. If $f\colon \mathbb{R} \to \mathbb{R}$ is differentiable, then f' is continuous on $\mathbb{R}$.

False

Let $f(x) = \begin{cases} x^2\sin\frac{1}{x}, & x \neq 0 \\ 0 & , x = 0 \end{cases}$

Then $f'(x) = \begin{cases} 2x\sin\frac{1}{x} - \cos\frac{1}{x}, & x \neq 0 \\ 0 & , x = 0 \end{cases}$

Observe that f is differentiable at all nonzero values of x and we have

$$f'(0) = \lim_{h\to 0}\frac{f(0+h)-f(0)}{h} = \lim_{h\to 0}\frac{h^2\sin\frac{1}{h} - 0}{h} = \lim_{h\to 0} h\sin\frac{1}{h}.$$

We now use the Squeeze Theorem (see Problem 74 from Problem Set 5). Since $-1 \leq \sin\frac{1}{h} \leq 1$ for all $h \neq 0$, we have $-h \leq h\sin\frac{1}{h} \leq h$ for all $h \neq 0$. Since $\lim_{h\to 0}(-h) = 0$ and $\lim_{h\to 0} h = 0$, it follows from the Squeeze Theorem that $\lim_{h\to 0} h\sin\frac{1}{h} = 0$. So, $f'(0)$ exists. Therefore, f is differentiable on $\mathbb{R}$.

Next, observe that $\lim_{x \to 0} f'(x)$ does not exist, because as x approaches 0, $\frac{1}{x}$ tends toward $+$ or $-\infty$, and so $\cos\frac{1}{x}$ oscillates between – 1 and 1. So, f' is not continuous at $x = 0$.

Compute the derivative of each of the following:

4. $\cos x$

$$\frac{d}{dx}[\cos x] = \lim_{h \to 0} \frac{\cos(x+h) - \cos x}{h} = \lim_{h \to 0} \frac{\cos x \cos h - \sin x \sin h - \cos x}{h}$$
$$= \lim_{h \to 0} \frac{(\cos x \cos h - \cos x) - \sin x \sin h}{h} = \lim_{h \to 0} \frac{\cos x\,(\cos h - 1) - \sin x \sin h}{h}$$
$$= \lim_{h \to 0} \frac{\cos x\,(\cos h - 1)}{h} - \lim_{h \to 0} \frac{\sin x \sin h}{h} = (\cos x) \lim_{h \to 0} \frac{\cos h - 1}{h} - (\sin x) \lim_{h \to 0} \frac{\sin h}{h}$$
$$= (\cos x)(0) - (\sin x)(1) = -\sin x$$ (by Problems 100 and 101 in Problem Set 5).

5. $\tan x$

$$\frac{d}{dx}[\tan x] = \frac{d}{dx}\left[\frac{\sin x}{\cos x}\right] = \frac{(\cos x)(\cos x) - (\sin x)(-\sin x)}{\cos^2 x} = \frac{\cos^2 x + \sin^2 x}{\cos^2 x} = \frac{1}{\cos^2 x} = \sec^2 x.$$

6. $\sec x$

$$\frac{d}{dx}[\sec x] = \frac{d}{dx}\left[\frac{1}{\cos x}\right] = \frac{(\cos x)(0) - (1)(-\sin x)}{\cos^2 x} = \frac{\sin x}{\cos^2 x} = \frac{1}{\cos x} \cdot \frac{\sin x}{\cos x} = \sec x \tan x.$$

7. $\cot x$

$$\frac{d}{dx}[\cot x] = \frac{d}{dx}[(\tan x)^{-1}] = -(\tan x)^{-2}\sec^2 x = -\frac{\cos^2 x}{\sin^2 x} \cdot \frac{1}{\cos^2 x} = -\frac{1}{\sin^2 x} = -\csc^2 x\,.$$

Prove each of the following:

8. For each $n \in \mathbb{Z}^+$,

$$\sum_{i=1}^{n} 1 = n.$$

Proof:

Base Case ($k = 1$):

$$\sum_{i=1}^{1} 1 = 1.$$

Inductive Step: Let $k \in \mathbb{N}$ with $k \geq 1$ and assume that

$$\sum_{i=1}^{k} 1 = k.$$

Then we have

$$\sum_{i=1}^{k+1} 1 = \left(\sum_{i=1}^{k} 1\right) + 1 = k + 1.$$

By the Principle of Mathematical Induction, the result holds for all natural numbers $n \geq 1$. □

9. Let $n \in \mathbb{Z}^+$ and let $r, a_1, a_2, \dots, a_n \in \mathbb{R}$. Then

$$\sum_{i=1}^{n} ra_i = r\sum_{i=1}^{n} a_i.$$

Proof: Let $r \in \mathbb{R}$.

Base Case $(k = 1)$**:**

$$\sum_{i=1}^{1} ra_i = ra_i = r\sum_{i=1}^{1} a_i.$$

Inductive Step: Let $k \in \mathbb{N}$ with $k \geq 1$ and assume that

$$\sum_{i=1}^{k} ra_i = r\sum_{i=1}^{k} a_i.$$

Then we have

$$\sum_{i=1}^{k+1} ra_i = \left(\sum_{i=1}^{k} ra_i\right) + ra_{k+1} = \left(r\sum_{i=1}^{k} a_i\right) + ra_{k+1} = r\left(\left(\sum_{i=1}^{k} a_i\right) + a_{k+1}\right) = r\sum_{i=1}^{k+1} a_i.$$

By the Principle of Mathematical Induction, the result holds for all natural numbers $n \geq 1$. □

10. Let $n \in \mathbb{Z}^+$. For each $i = 1, 2, \dots, n$, let $a_i, b_i \in \mathbb{R}$. Then

$$\sum_{i=1}^{n} (a_i + b_i) = \sum_{i=1}^{n} a_i + \sum_{i=1}^{n} b_i.$$

Proof:

Base Case ($k = 1$):

$$\sum_{i=1}^{1}(a_i + b_i) = a_1 + b_1 = \sum_{i=1}^{1} a_i + \sum_{i=1}^{1} b_i.$$

Inductive Step: Let $k \in \mathbb{N}$ with $k \geq 1$ and assume that

$$\sum_{i=1}^{k}(a_i + b_i) = \sum_{i=1}^{k} a_i + \sum_{i=1}^{k} b_i.$$

Then we have

$$\sum_{i=1}^{k+1}(a_i + b_i) = \left(\sum_{i=1}^{k}(a_i + b_i)\right) + (a_{k+1} + b_{k+1}) = \left(\sum_{i=1}^{k} a_i + \sum_{i=1}^{k} b_i\right) + (a_{k+1} + b_{k+1})$$

$$= \left[\left(\sum_{i=1}^{k} a_i\right) + a_{k+1}\right] + \left[\left(\sum_{i=1}^{k} b_i\right) + b_{k+1}\right] = \sum_{i=1}^{k+1} a_i + \sum_{i=1}^{k+1} b_i.$$

By the Principle of Mathematical Induction, the result holds for all natural numbers $n \geq 1$. □

Let $f: [0, 3] \to \mathbb{R}$ be defined by $f(x) = x$.

11. Let $P = \{[0, 1], [1, 2], [2, 3]\}$. Compute $L_f(P)$ and $U_f(P)$.

$L_f(P) = 0 \cdot 1 + 1 \cdot 1 + 2 \cdot 1 = 0 + 1 + 2 = \mathbf{3}$; $U_f(P) = 1 \cdot 1 + 2 \cdot 1 + 3 \cdot 1 = 1 + 2 + 3 = \mathbf{6}$

12. Let $n \in \mathbb{Z}^+$ and let P_n be the partition of $[0, 3]$ into n equal subintervals. Compute $L_f(P_n)$ and $U_f(P_n)$.

$\Delta x = \frac{3-0}{n} = \frac{3}{n}$ and $P_n = \left\{\left[\frac{3(i-1)}{n}, \frac{3i}{n}\right]\right\}_{i=1}^{n}$. So, we have

$$L_f(P) = \sum_{i=1}^{n} m_i \Delta x = \sum_{i=1}^{n}\left(\frac{3(i-1)}{n}\right)\frac{3}{n} = \frac{9}{n^2}\sum_{i=1}^{n}(i-1) = \frac{9}{n^2}\left(\sum_{i=1}^{n} i - \sum_{i=1}^{n} 1\right)$$

$$= \frac{9}{n^2}\left(\frac{n(n+1)}{2} - n\right) = \frac{9(n+1)}{2n} - \frac{9}{n} = \frac{9n+9}{2n} - \frac{18}{2n} = \frac{9n-9}{2n}$$

$$U_f(P) = \sum_{i=1}^{n} M_i \Delta x = \sum_{i=1}^{n} \frac{3i}{n} \cdot \frac{3}{n} = \frac{9}{n^2}\sum_{i=1}^{n} i = \frac{9}{n^2} \cdot \frac{n(n+1)}{2} = \frac{9(n+1)}{2n} = \frac{9n+9}{2n}$$

13. Compute $\underline{\int_0^3} f(x)\,dx$ and $\overline{\int_0^3} f(x)\,dx$. (You may use Problem 61 below.)

Using Problem 61 below, we have

$$\underline{\int_0^3} f(x) = \sup\left\{\frac{9n-9}{2n}\,\middle|\, n \in \mathbb{Z}^+\right\} = \lim_{n\to\infty} \frac{9n-9}{2n} = \frac{\mathbf{9}}{\mathbf{2}}$$

$$\overline{\int_0^3} f(x)\,dx = \inf\left\{\frac{9n+9}{2n}\,\middle|\, n \in \mathbb{Z}^+\right\} = \lim_{n\to\infty} \frac{9n+9}{2n} = \frac{\mathbf{9}}{\mathbf{2}}$$

14. Is f Riemann integrable on $[0,3]$? If so, compute $\int_0^3 f(x)\,dx$. If not, explain why.

By Problem 13, $\overline{\int_0^3} f(x)\,dx = \underline{\int_0^3} f(x)$, and so, f is Riemann integrable on $[0,3]$ and

$$\int_0^3 f(x)\,dx = \frac{\mathbf{9}}{\mathbf{2}}.$$

LEVEL 2

Prove each of the following:

15. Let $A \subseteq \mathbb{R}$, let $k \in \mathbb{R}$, and let $f: A \to \mathbb{R}$ be differentiable at $a \in A$. Then the function $kf: A \to \mathbb{R}$ defined by $(kf)(x) = k \cdot f(x)$ is differentiable at a and $(kf)'(a) = k \cdot f'(a)$.

Proof:

$$(kf)'(a) = \lim_{x\to a} \frac{(kf)(x) - (kf)(a)}{x-a} = \lim_{x\to a} \frac{k \cdot f(x) - k \cdot f(a)}{x-a} = \lim_{x\to a} \frac{k\big(f(x) - f(a)\big)}{x-a}$$

$$= k \cdot \lim_{x\to a} \frac{f(x) - f(a)}{x-a} = k \cdot f'(a).$$ □

16. Suppose that $(a,b) \subseteq A \subseteq \mathbb{R}$, $f, g: A \to \mathbb{R}$ are differentiable on (a,b), and $f'(x) = g'(x)$ for all $x \in (a,b)$. Then there is a constant k such that for all $x \in (a,b)$, $g(x) = f(x) + k$.

Proof: Define $h: A \to \mathbb{R}$ by $h(x) = g(x) - f(x)$. Then, for all $x \in (a,b)$, $h'(x) = g'(x) - f'(x) = 0$. Let $c, d \in (a,b)$ with $c < d$. Then h is continuous on $[c,d]$ and differentiable on (c,d). So, by the Mean Value Theorem (Theorem 13.25), there is $e \in (c,d)$ with $\frac{h(d)-h(c)}{d-c} = h'(e) = 0$. It follows that $h(d) = h(c)$. Since $c, d \in (a,b)$ were arbitrary, h is constant on (a,b). So, there is a constant k such that for all $x \in (a,b)$, $h(x) = k$. Therefore, $g(x) - f(x) = k$, or equivalently, $g(x) = f(x) + k$. □

17. Suppose that $(a, b) \subseteq A \subseteq \mathbb{R}$, $f, g: A \to \mathbb{R}$ are continuous on A and differentiable on (a, b), $\lim_{x \to b^-} f(x) = \lim_{x \to b^-} g(x) = 0$, $g'(x) \neq 0$ for all $x \in (a, b)$, and $\lim_{x \to b^-} \frac{f'(x)}{g'(x)}$ exists. Then

$$\lim_{x \to b^-} \frac{f(x)}{g(x)} = \lim_{x \to b^-} \frac{f'(x)}{g'(x)}$$

This is Standard Version B of L'Hôpital's Rule (Theorem 13.30).

Proof: If $b \in A$, then since f and g are continuous on A and $\lim_{x \to b^-} f(x) = \lim_{x \to b^-} g(x) = 0$, we must have $f(b) = 0$ and $g(b) = 0$. In general, let $B = A \cup \{b\}$ and define $F, G: B \to \mathbb{R}$ as follows:

$$F(x) = \begin{cases} f(x) & \text{if } x \in A \\ 0 & \text{if } x = b \end{cases} \qquad G(x) = \begin{cases} g(x) & \text{if } x \in A \\ 0 & \text{if } x = b \end{cases}$$

Note that if $b \in A$, then $B = A$, $F = f$, and $G = g$. So, there is no harm in simply assuming that $b \in A$ and $f(b) = g(b) = 0$. It follows that f and g are continuous on $(a, b]$.

For any $x \in (a, b)$, we see that f and g are continuous on $[x, b]$ and differentiable on (x, b). By the Mean Value Theorem (Theorem 13.25), there is $c \in (x, b)$ such that $g'(c) = \frac{g(b) - g(x)}{b - x} = \frac{-g(x)}{b - x}$. Since $g'(c) \neq 0$, we must also have $g(x) \neq 0$. By the Generalized Mean value Theorem (Theorem 13.27), there is $c \in (x, b)$ such that $f(x)g'(c) = g(x)f'(c)$, or equivalently, $\frac{f(x)}{g(x)} = \frac{f'(c)}{g'(c)}$. Since $x < c < b$, as x approaches b from the left, so does c. So, we have

$$\lim_{x \to b^-} \frac{f(x)}{g(x)} = \lim_{x \to b^-} \frac{f'(c)}{g'(c)} = \lim_{c \to b^-} \frac{f'(c)}{g'(c)} = \lim_{x \to b^-} \frac{f'(x)}{g'(x)}.$$

Note that the last equality above is true because we simply changed the name of the variable. □

18. Let A and B be nonempty bounded sets of real numbers with $A \subseteq B$. Then

$$\inf B \leq \inf A \leq \sup A \leq \sup B.$$

Proof: Let $k = \inf B$. Then k is a lower bound of B. So, for all $b \in B$, $k \leq b$. Since $A \subseteq B$, we have that for all $b \in A$, $k \leq b$. So, k is a lower bound of A. Since $\inf A$ is the **greatest** lower bound of A, it follows that $k \leq \inf A$. Since $k = \inf B$, we have $\inf B \leq \inf A$.

Now, let $k = \inf A$. Since $A \neq \emptyset$, there is $a \in A$ with $k \leq a$. Since $\sup A$ is an upper bound of A, we have $a \leq \sup A$. So, $\inf A = k \leq a \leq \sup A$.

Finally, let $k = \sup B$. Then k is an upper bound of B. So, for all $b \in B$, $b \leq k$. Since $A \subseteq B$, we have that for all $b \in A$, $b \leq k$. So, k is an upper bound of A. Since $\sup A$ is the **least** upper bound of A, it follows that $\sup A \leq k$. Since $k = \sup B$, we have $\sup A \leq \sup B$. □

19. Let A be a nonempty bounded set of real numbers, let $c \geq 0$, and let $B = \{cx \mid x \in A\}$. Then $\sup B = c \sup A$ and $\inf B = c \inf A$.

Proof: If $c = 0$, then $B = \{0\}$, and so, $\sup B = 0$, $\inf B = 0$, $c \sup A = 0$, and $c \inf A = 0$. So, we have $\sup B = c \sup A$ and $\inf B = c \inf A$.

Now, assume $c > 0$.

Let $cx \in B$. Then $x \in A$. Since $\sup A$ is an upper bound of A, $x \leq \sup A$. Since $c > 0$, we have $cx \leq c \sup A$. Since $cx \in B$ was arbitrary, $c \sup A$ is an upper bound of B. Since $\sup B$ is the **least** upper bound of B, $\sup B \leq c \sup A$.

Let $x \in A$. Then $cx \in B$. Since $\sup B$ is an upper bound of B, we have $cx \leq \sup B$. Since $c > 0$, it follows that $x \leq \frac{1}{c} \sup B$. Since $x \in A$ was arbitrary, $\frac{1}{c} \sup B$ is an upper bound of A. Since $\sup A$ is the **least** upper bound of A, $\sup A \leq \frac{1}{c} \sup B$, and therefore, $c \sup A \leq \sup B$.

Since $\sup B \leq c \sup A$ and $c \sup A \leq \sup B$, it follows that $\sup B = c \sup A$.

Now, let $cx \in B$. Then $x \in A$. Since $\inf A$ is a lower bound of A, $x \geq \inf A$. Since $c > 0$, we have $cx \geq c \inf A$. Since $cx \in B$ was arbitrary, $c \inf A$ is a lower bound of B. Since $\inf B$ is the **greatest** lower bound of B, $\inf B \geq c \inf A$.

Let $x \in A$. Then $cx \in B$. Since $\inf B$ is a lower bound of B, we have $cx \geq \inf B$. Since $c > 0$, it follows that $x \geq \frac{1}{c} \inf B$. Since $x \in A$ was arbitrary, $\frac{1}{c} \inf B$ is a lower bound of A. Since $\inf A$ is the **greatest** lower bound of A, $\inf A \geq \frac{1}{c} \inf B$, and therefore, $c \inf A \geq \inf B$.

Since $\inf B \geq c \inf A$ and $c \inf A \geq \inf B$, it follows that $\inf B = c \inf A$. □

20. Let $f, g: [a, b] \to \mathbb{R}$ be bounded functions, let $A = \{f(x) \mid x \in [a, b]\}$, $B = \{g(x) \mid x \in [a, b]\}$, and $C = \{f(x) + g(x) \mid x \in [a, b]\}$. Then

$$\sup C \leq \sup A + \sup B \quad \text{and} \quad \inf C \geq \inf A + \inf B.$$

Proof: Let $x \in [a, b]$. Then $f(x) \in A$, and therefore, $\inf A \leq f(x) \leq \sup A$. Also, $g(x) \in B$, and therefore, $\inf B \leq g(x) \leq \sup B$. So, $\inf A + \inf B \leq f(x) + g(x) \leq \sup A + \sup B$. Since $x \in [a, b]$ was arbitrary, $\sup A + \sup B$ is an upper bound of C and $\inf A + \inf B$ is a lower bound of C. Since $\sup C$ is the **least** upper bound of C, $\sup C \leq \sup A + \sup B$. Similarly, since $\inf C$ is the **greatest** lower bound of C, $\inf C \geq \inf A + \inf B$. □

21. Let $f, g: [a, b] \to \mathbb{R}$ be bounded functions that are Riemann integrable on $[a, b]$ and let $f(x) \leq g(x)$ for all $x \in [a, b]$. Then

$$\int_a^b f(x)\, dx \leq \int_a^b g(x)\, dx.$$

Proof: Let $P = \{[x_{i-1}, x_i]\}_{i=1}^n$ be a partition of $[a, b]$ and let $m_i = \inf\{f(x) \mid x_{i-1} \leq x \leq x_i\}$ and $k_i = \inf\{g(x) \mid x_{i-1} \leq x \leq x_i\}$ for $i = 1, 2, \dots, n$. Then $m_i \leq k_i$ for each $i = 1, 2, \dots, n$. So, we have

$$L_f(P) = \sum_{i=1}^n m_i \Delta x_i \leq \sum_{i=1}^n k_i \Delta x_i = L_g(P)$$

Since P was an arbitrary partition of $[a, b]$, we have

$$\underline{\int_a^b} f(x)\,dx = \sup\{L_f(P) \mid P \text{ a partition of } [a, b]\}$$

$$\leq \sup\{L_g(P) \mid P \text{ a partition of } [a, b]\} = \underline{\int_a^b} g(x)\,dx.$$

Since f and g are Riemann integrable on $[a, b]$, we have

$$\int_a^b f(x)\,dx = \underline{\int_a^b} f(x)\,dx \leq \underline{\int_a^b} g(x)\,dx = \int_a^b g(x)\,dx\,. \qquad \square$$

Let $a \in (0,1) \cup (1, \infty)$. Prove each of the following:

22. For all $x \in (0, \infty)$, $\log_a x = \frac{\ln x}{\ln a}$.

Proof: Let $y = \log_a x$. Then $x = a^y$. So, $\ln x = \ln a^y = y \ln a$. Therefore, we have

$$\log_a x = y = \frac{\ln x}{\ln a}.$$

$\square$

23. Let $b \in (0,1) \cup (1, \infty)$. For all $x \in (0, \infty)$, $\log_a x = \frac{\log_b x}{\log_b a}$.

Proof: By Problem 22 above, we have $\log_b x = \frac{\ln x}{\ln b}$, or equivalently, $\ln x = (\ln b)(\log_b x)$.

Therefore, by Problem 22 again and the previous paragraph, we have

$$\log_a x = \frac{\ln x}{\ln a} = \frac{(\ln b)(\log_b x)}{\ln a} = \frac{\log_b x}{\frac{\ln a}{\ln b}} = \frac{\log_b x}{\log_b a}.$$

The first and last equalities follow from Problem 22 above. $\square$

24. If $b, c \in (0, \infty)$, then $\log_a bc = \log_a b + \log_a c$.

Proof: By Problem 22 above and Theorem 13.56, we have

$$\log_a bc = \frac{1}{\ln a} \ln bc = \frac{1}{\ln a}(\ln b + \ln c) = \frac{\ln b}{\ln a} + \frac{\ln c}{\ln a} = \log_a b + \log_a c.$$

$\square$

25. Let $b \in (0, \infty)$ and let $r \in \mathbb{R}$. Then $\log_a(b^r) = r \log_a b$.

Proof: By Problem 22 above and Theorem 13.70, we have

$$\log_a(b^r) = \frac{\ln(b^r)}{\ln a} = \frac{r \ln b}{\ln a} = r \cdot \frac{\ln b}{\ln a} = r \log_a b.$$

$\square$

Approximate $\ln 1.5$ in each of the following ways:

26. Use Problem 47 below to prove that $\frac{1}{3} \leq \ln 1.5 \leq \frac{1}{2}$.

Proof: We have

$$\ln 1.5 = \int_1^{1.5} \frac{1}{t}\, dt.$$

If $t \in [1, 1.5]$, then $\frac{1}{1.5} \leq \frac{1}{t} \leq 1$, or equivalently, $\frac{2}{3} \leq \frac{1}{t} \leq 1$. So, by Problem 47 below together with the fact that $f(x) = \frac{1}{x}$ is integrable on $[1, 1.5]$, we have $\frac{2}{3}(0.5) \leq \ln 1.5 \leq 1(0.5)$, or equivalently, $\frac{1}{3} \leq \ln 1.5 \leq \frac{1}{2}$. □

27. Use Riemann sums to prove that $\frac{25}{66} < \ln 1.5 < \frac{47}{110}$.

Proof: We have

$$\ln 1.5 = \int_1^{1.5} \frac{1}{t}\, dt.$$

Let $f(x) = \frac{1}{x}$ and let $P = \{[1, 1.125], [1.125, 1.25], [1.25, 1.375], [1.375, 1.5]\}$. Then we have

$$L_f(P) = \frac{1}{1.125} \cdot \frac{1}{8} + \frac{1}{1.25} \cdot \frac{1}{8} + \frac{1}{1.375} \cdot \frac{1}{8} + \frac{1}{1.5} \cdot \frac{1}{8} = \frac{763}{1980}$$

$$U_f(P) = \frac{1}{1} \cdot \frac{1}{8} + \frac{1}{1.125} \cdot \frac{1}{8} + \frac{1}{1.25} \cdot \frac{1}{8} + \frac{1}{1.375} \cdot \frac{1}{8} = \frac{1691}{3960}.$$

So, we see that $\frac{763}{1980} \leq \ln 2 \leq \frac{1691}{3960}$. Since $\frac{25}{66} = \frac{750}{1980} < \frac{763}{1980}$ and $\frac{1691}{3960} < \frac{1692}{3960} = \frac{47}{110}$, we see that $\frac{25}{66} < \ln 1.5 < \frac{47}{110}$. □

28. Use a geometric argument to prove that $\ln 1.5 < \frac{5}{12}$.

Proof: The area under the curve $f(x) = \frac{1}{x}$ between $x = 1$ and $x = 1.5$ is less than the area of the trapezoid formed with bases of length $f(1) = 1$ and $f(1.5) = \frac{1}{1.5} = \frac{2}{3}$ and height $1.5 - 1 = \frac{1}{2}$. This trapezoid has area $A = \frac{1}{2}(\text{base1} + \text{base2}) \cdot \text{height} = \frac{1}{2}\left(1 + \frac{2}{3}\right)\left(\frac{1}{2}\right) = \frac{5}{12}$. □

LEVEL 3

Prove each of the following:

29. Let $A, B \subseteq \mathbb{R}$, suppose that $f: A \to \mathbb{R}$, $g: B \to \mathbb{R}$ are differentiable at $a \in A \cap B$, and assume that $g(a) \neq 0$. Let $C = (A \cap B) \setminus \{x \in B \mid g(x) = 0\}$. Then the function $f/g: C \to \mathbb{R}$ defined by $(f/g)(x) = \frac{f(x)}{g(x)}$ is differentiable at a and $(f/g)'(a) = \frac{g(a)f'(a)-f(a)g'(a)}{(g(a))^2}$. This is the **quotient rule** (Theorem 13.18).

Proof: Using the product rule and chain rule, we have

$$\left(\frac{f}{g}\right)'(x) = \frac{d}{dx}\left[\frac{f(x)}{g(x)}\right] = \frac{d}{dx}\left[f(x)\left(g(x)\right)^{-1}\right] = f(x)(-1)\left(g(x)\right)^{-2}g'(x) + \left(g(x)\right)^{-1}f'(x)$$

$$= -\frac{f(x)g'(x)}{\left(g(x)\right)^2} + \frac{f'(x)}{g(x)} = -\frac{f(x)g'(x)}{\left(g(x)\right)^2} + \frac{f'(x)g(x)}{\left(g(x)\right)^2} = \frac{g(x)f'(x) - f(x)g'(x)}{\left(g(x)\right)^2}.$$

Substituting a for x gives the desired result. □

30. Suppose that $f(c)$ is a relative minimum for a function $f: A \to \mathbb{R}$. Then either $f'(c) = 0$ or f is not differentiable at c.

Proof: Suppose that $f(c)$ is a relative minimum for $f: A \to \mathbb{R}$ and f is differentiable at c. First, let's suppose toward contradiction that $f'(c) > 0$. In other words, we have

$$\lim_{x \to c} \frac{f(x) - f(c)}{x - c} > 0.$$

By Problem 72 from Problem Set 5, there is a deleted neighborhood N of c such that $\frac{f(x)-f(c)}{x-c} > 0$ for all $x \in N$. Now, if $x \in N$ and $x < c$, then $x - c < 0$, and therefore,

$$f(x) - f(c) = \frac{f(x) - f(c)}{x - c}(x - c) < 0.$$

So, for $x \in N$ with $x < c$, we have $f(x) < f(c)$, contradicting our assumption that $f(c)$ is a relative minimum for f. So, it is impossible to have $f'(c) > 0$.

Next, let's suppose toward contradiction that $f'(c) < 0$. Then we have

$$\lim_{x \to c} \frac{f(x) - f(c)}{x - c} < 0.$$

Therefore, it follows that

$$\lim_{x \to c} \frac{f(c) - f(x)}{x - c} = -\left[\lim_{x \to c} \frac{f(x) - f(c)}{x - c}\right] > 0.$$

Again, by Problem 72 from Problem Set 5, there is a deleted neighborhood N of c such that $\frac{f(c)-f(x)}{x-c} > 0$ for all $x \in N$. Now, if $x \in N$ and $x > c$, then $x - c > 0$, and therefore,

$$f(c) - f(x) = \frac{f(c) - f(x)}{x - c}(x - c) > 0.$$

So, for $x \in N$ with $x > c$, we have $f(c) > f(x)$, once again contradicting our assumption that $f(c)$ is a relative minimum for f. So, it is also impossible to have $f'(c) < 0$.

Since f is differentiable at c, we must have $f'(c) = 0$. □

31. $-x \leq \sin x \leq x$ for all $x \in [0, \infty)$. Hint: Use the Mean Value Theorem.

Proof: If $x \geq 1$, then $-x \leq -1$. Since $-1 \leq \sin x \leq 1$, we have $-x \leq -1 \leq \sin x \leq 1 \leq x$.

If $x = 0$, we have $\sin 0 = 0$, and so, the result holds.

Assume $0 < x < 1$. Since $f(t) = \sin t$ is differentiable for all real numbers, in particular, it is continuous on $[0, x]$ and differentiable on $(0, x)$. So, by the Mean Value Theorem (Theorem 13.25), there is $c \in (0, x)$ such that $\frac{f(x)-f(0)}{x-0} = f'(c)$, or equivalently, $\frac{\sin x}{x} = \cos c$. Since $-1 \leq \cos x \leq 1$, it follows that $-1 \leq \frac{\sin x}{x} \leq 1$, or equivalently, $-x \leq \sin x \leq x$, as desired. □

32. Suppose that $(a, b) \subseteq A \subseteq \mathbb{R}$ and $f: A \to \mathbb{R}$ is differentiable on (a, b). If $f'(x) \geq 0$ for all $x \in (a, b)$, then f is increasing on (a, b).

Proof: Let $x, y \in (a, b)$ with $x < y$. Then f is differentiable on $[x, y]$. By Theorem 13.11, f is continuous on $[x, y]$. By the Mean Value Theorem (Theorem 13.25), there is $c \in (x, y)$ such that $f'(c) = \frac{f(y)-f(x)}{y-x}$. So, $f(y) - f(x) = f'(c)(y - x)$. Since $x < y$, $y - x > 0$.

Since $f'(c) \geq 0$, we have $f(y) - f(x) \geq 0$, or equivalently, $f(x) \leq f(y)$. Since $x, y \in (a, b)$ with $x < y$ were arbitrary, f is increasing on (a, b).

33. Suppose that $(a, b) \subseteq A \subseteq \mathbb{R}$ and $f: A \to \mathbb{R}$ is differentiable on (a, b). If $f'(x) \leq 0$ for all $x \in (a, b)$, then f is decreasing on (a, b).

Proof: Let $x, y \in (a, b)$ with $x < y$. Then f is differentiable on $[x, y]$. By Theorem 13.11, f is continuous on $[x, y]$. By the Mean Value Theorem (Theorem 13.25), there is $c \in (x, y)$ such that $f'(c) = \frac{f(y)-f(x)}{y-x}$. So, $f(y) - f(x) = f'(c)(y - x)$. Since $x < y$, $y - x > 0$.

Since $f'(c) \leq 0$, we have $f(y) - f(x) \leq 0$, or equivalently, $f(x) \geq f(y)$. Since $x, y \in (a, b)$ with $x < y$ were arbitrary, f is decreasing on (a, b). □

34. Suppose that $(a,c) \cup (c,b) \subseteq A \subseteq \mathbb{R}$, $f,g\colon A \to \mathbb{R}$ are continuous on A and differentiable on $(a,c) \cup (c,b)$, $\frac{\lim_{x\to c} f(x)}{g(x)} = L$, where L is a finite nonzero real number. Also, suppose that $\lim_{x\to c} f(x) = \lim_{x\to c} g(x) = \infty$ (either or both can be $-\infty$) and $g'(x) \neq 0$ for all $x \in (a,c) \cup (c,b)$. If$\lim_{x\to c} \frac{f'(x)}{g'(x)}$ exists, then

$$\lim_{x\to b^-} \frac{f(x)}{g(x)} = \lim_{x\to b^-} \frac{f'(x)}{g'(x)}$$

This is a special case of Infinite Version A of L'Hôpital's Rule.

Proof: Since $\lim_{x\to c} f(x) = \lim_{x\to c} g(x) = \infty$, we have $\lim_{x\to c} \frac{1}{f(x)} = \lim_{x\to c} \frac{1}{g(x)} = 0$ (Check this!). By standard version C of L'Hôpital's rule (Theorem 13.31), we have

$$L = \lim_{x\to c} \frac{f(x)}{g(x)} = \lim_{x\to c} \frac{\frac{1}{g(x)}}{\frac{1}{f(x)}} = \lim_{x\to c} \frac{\frac{d}{dx}\left[\frac{1}{g(x)}\right]}{\frac{d}{dx}\left[\frac{1}{f(x)}\right]} = \lim_{x\to c} \frac{\frac{-g'(x)}{[g(x)]^2}}{\frac{-f'(x)}{[f(x)]^2}} = \lim_{x\to c} \frac{[f(x)]^2 g'(x)}{[g(x)]^2 f'(x)}$$

$$= \lim_{x\to c} \frac{[f(x)]^2}{[g(x)]^2} \cdot \lim_{x\to c} \frac{g'(x)}{f'(x)} = \left[\lim_{x\to c} \frac{f(x)}{g(x)}\right]^2 \cdot \frac{1}{\lim_{x\to c} \frac{f'(x)}{g'(x)}} = \frac{L^2}{\lim_{x\to c} \frac{f'(x)}{g'(x)}}.$$

It follows that $L \cdot \lim_{x\to c} \frac{f'(x)}{g'(x)} = L^2$, or equivalently, $L = \lim_{x\to c} \frac{f'(x)}{g'(x)}$, as desired. □

35. Let A and B be nonempty bounded sets of real numbers and let $A + B = \{a + b \mid a \in A, b \in B\}$. Then $\sup(A + B) = \sup A + \sup B$ and $\inf(A + B) = \inf A + \inf B$.

Proof: Let $a + b \in A + B$, where $a \in A$ and $b \in B$. Since $\sup A$ is an upper bound of A, $a \leq \sup A$. Similarly, $b \leq \sup B$. So, $a + b \leq \sup A + \sup B$. Since $a + b \in A + B$ was arbitrary, it follows that $\sup A + \sup B$ is an upper bound of $A + B$. Since $\sup(A + B)$ is the **least** upper bound of $A + B$, we have $\sup(A + B) \leq \sup A + \sup B$.

Let $a \in A$. Then for each $b \in B$, we have $a + b \in A + B$. Since $\sup(A + B)$ is an upper bound of $A + B$, for each $b \in B$, $a + b \leq \sup(A + B)$. So, for each $b \in B$, $b \leq \sup(A + B) - a$. Therefore, $\sup(A + B) - a$ is an upper bound of B. Since $\sup B$ is the **least** upper bound of B, it follows that $\sup B \leq \sup(A + B) - a$. Therefore, $a \leq \sup(A + B) - \sup B$. Since $a \in A$ was arbitrary, we see that $\sup(A + B) - \sup B$ is an upper bound of A. Since $\sup A$ is the **least** upper bound of A, we have $\sup A \leq \sup(A + B) - \sup B$. Therefore, $\sup A + \sup B \leq \sup(A + B)$.

Since $\sup(A + B) \leq \sup A + \sup B$ and $\sup A + \sup B \leq \sup(A + B)$, $\sup(A + B) = \sup A + \sup B$.

Let $a + b \in A + B$, where $a \in A$ and $b \in B$. Since $\inf A$ is a lower bound of A, $a \geq \inf A$. Similarly, $b \geq \inf B$. So, $a + b \geq \inf A + \inf B$. Since $a + b \in A + B$ was arbitrary, it follows that $\inf A + \inf B$ is a lower bound of $A + B$. Since $\inf(A + B)$ is the **greatest** lower bound of $A + B$, it follows that $\inf(A + B) \geq \inf A + \inf B$.

Let $a \in A$. Then for each $b \in B$, we have $a + b \in A + B$. Since $\inf(A + B)$ is a lower bound of $A + B$, for each $b \in B$, $a + b \geq \inf(A + B)$. So, for each $b \in B$, $b \geq \inf(A + B) - a$. So, $\inf(A + B) - a$ is a lower bound of B. Since $\inf B$ is the **greatest** lower bound of B, it follows that $\inf B \geq \inf(A + B) - a$. Therefore, $a \geq \inf(A + B) - \inf B$. Since $a \in A$ was arbitrary, we see that $\inf(A + B) - \inf B$ is a lower bound of A. Since $\inf A$ is the **greatest** lower bound of A, we have $\inf A \geq \inf(A + B) - \inf B$. Therefore, $\inf A + \inf B \geq \inf(A + B)$.

Since $\inf(A + B) \geq \inf A + \inf B$ and $\inf A + \inf B \geq \inf(A + B)$, $\inf(A + B) = \inf A + \inf B$. □

36. Let $f: [a, b] \to \mathbb{R}$ be continuous on $[a, b]$. Then there is $c \in [a, b]$ such that

$$\int_a^b f(x)\, dx = f(c)(b - a).$$

This result is known as the **Mean Value Theorem for Integrals**.

Proof: Define $F: [a, b] \to \mathbb{R}$ by $F(x) = \int_a^x f(t)\, dt$. By the first form of the Fundamental Theorem of Calculus (Theorem 13.48), F is continuous on $[a, b]$ and differentiable on (a, b). Therefore, by the Mean Value Theorem (Theorem 13.25), there is $c \in (a, b)$ such that

$$F'(c) = \frac{F(b) - F(a)}{b - a} = \frac{\int_a^b f(t)\, dt - \int_a^a f(t)\, dt}{b - a} = \frac{\int_a^b f(t)\, dt - 0}{b - a} = \frac{1}{b - a}\int_a^b f(t)\, dt.$$

Again, by the first form of the Fundamental Theorem of Calculus, we have $F'(c) = f(c)$. Therefore,

$$\frac{1}{b - a}\int_a^b f(t)\, dt = f(c).$$

Equivalently, we have

$$\int_a^b f(x)\, dx = f(c)(b - a).$$

□

37. Let $f: [a, b] \to \mathbb{R}$ be a bounded function that is Riemann integrable on $[a, b]$ and let $c \in \mathbb{R}$. Then cf is Riemann integrable on $[a, b]$ and

$$\int_a^b cf(x)\, dx = c\int_a^b f(x)\, dx.$$

Proof: First assume that $c \geq 0$. Let $P = \{[x_{i-1}, x_i]\}_{i=1}^n$ be a partition of $[a, b]$ and define m_i and M_i for $i = 1, 2, \dots, n$ by $m_i = \inf\{f(x) \mid x_{i-1} \leq x \leq x_i\}$ and $M_i = \sup\{f(x) \mid x_{i-1} \leq x \leq x_i\}$. By Problem 19 above, $cm_i = \inf\{cf(x) \mid x_{i-1} \leq x \leq x_i\}$ and $cM_i = \sup\{cf(x) \mid x_{i-1} \leq x \leq x_i\}$. So, we have the following:

$$L_{cf}(P) = \sum_{i=1}^{n} cm_i \Delta x_i = c\sum_{i=1}^{n} m_i \Delta x_i = cL_f(P)$$

$$U_{cf}(P) = \sum_{i=1}^{n} cM_i \Delta x_i = c\sum_{i=1}^{n} M_i \Delta x_i = cU_f(P)$$

By Problem 19 again, we have the following:

$$\underline{\int_a^b} cf(x)\,dx = \sup\{L_{cf}(P) \mid P \text{ a partition of } [a,b]\} = \sup\{cL_f(P) \mid P \text{ a partition of } [a,b]\}$$

$$= c\sup\{L_f(P) \mid P \text{ a partition of } [a,b]\} = c\underline{\int_a^b} f(x)\,dx$$

$$\overline{\int_a^b} cf(x)\,dx = \inf\{U_{cf}(P) \mid P \text{ a partition of } [a,b]\} = \inf\{cU_f(P) \mid P \text{ a partition of } [a,b]\}$$

$$= c\inf\{U_f(P) \mid P \text{ a partition of } [a,b]\} = c\overline{\int_a^b} f(x)\,dx$$

Since f is Riemann integrable on $[a,b]$, we have

$$\underline{\int_a^b} cf(x)\,dx = c\underline{\int_a^b} f(x)\,dx = c\overline{\int_a^b} f(x)\,dx = \overline{\int_a^b} cf(x)\,dx.$$

It follows that cf is Riemann integrable on $[a,b]$ and $\int_a^b cf(x)\,dx = c\int_a^b f(x)\,dx$.

Next, assume that $c = -1$ and P, m_i, and M_i are defined the same way as above. By the method used in the proof of Problem 62 in Problem Set 5, we see that $-m_i = \sup\{-f(x) \mid x_{i-1} \leq x \leq x_i\} = K_i$ and $-M_i = \inf\{-f(x) \mid x_{i-1} \leq x \leq x_i\} = k_i$ (k_i and K_i are defined as shown). So, we have the following:

$$L_{-f}(P) = \sum_{i=1}^{n} k_i \Delta x_i = \sum_{i=1}^{n} -M_i \Delta x_i = -\sum_{i=1}^{n} M_i \Delta x_i = -U_f(P)$$

$$U_{-f}(P) = \sum_{i=1}^{n} K_i \Delta x_i = \sum_{i=1}^{n} -m_i \Delta x_i = -\sum_{i=1}^{n} m_i \Delta x_i = -L_f(P)$$

By the same argument as above, we have the following:

$$\underline{\int_a^b} -f(x)\,dx = \sup\{L_{-f}(P) \mid P \text{ a partition of } [a,b]\} = \sup\{-U_f(P) \mid P \text{ a partition of } [a,b]\}$$

$$= -\inf\{U_f(P) \mid P \text{ a partition of } [a,b]\} = -\overline{\int_a^b} f(x)\,dx$$

$$\overline{\int_a^b} -f(x)\,dx = \inf\{U_{-f}(P) \mid P \text{ a partition of } [a,b]\} = \inf\{-L_f(P) \mid P \text{ a partition of } [a,b]\}$$

$$= -\sup\{L_f(P) \mid P \text{ a partition of } [a,b]\} = -\underline{\int_a^b} f(x)\,dx$$

Since f is Riemann integrable on $[a, b]$, we have

$$\underline{\int_a^b} -f(x)\,dx = -\overline{\int_a^b} f(x)\,dx = -\underline{\int_a^b} f(x)\,dx = \overline{\int_a^b} -f(x)\,dx.$$

So, $-f$ is Riemann integrable on $[a, b]$ and $\int_a^b -f(x)\,dx = -\int_a^b f(x)\,dx$.

Finally, if $c < 0$, then we have the following:

$$\underline{\int_a^b} cf(x)\,dx = \underline{\int_a^b} -(-cf(x))\,dx = -\overline{\int_a^b} -cf(x)\,dx = -(-c)\overline{\int_a^b} f(x)\,dx = c\overline{\int_a^b} f(x)\,dx$$

$$\overline{\int_a^b} cf(x)\,dx = \overline{\int_a^b} -(-cf(x))\,dx = -\underline{\int_a^b} -cf(x)\,dx = -(-c)\underline{\int_a^b} f(x)\,dx = c\underline{\int_a^b} f(x)\,dx$$

Since f is Riemann integrable on $[a, b]$, we have

$$\underline{\int_a^b} cf(x)\,dx = c\overline{\int_a^b} f(x)\,dx = c\underline{\int_a^b} f(x)\,dx = \overline{\int_a^b} cf(x)\,dx.$$

It follows that cf is Riemann integrable and $\int_a^b cf(x)\,dx = c\int_a^b f(x)\,dx$. □

38. For each $x \in \mathbb{R}$, $e^x = \lim_{h\to 0}(1 + hx)^{\frac{1}{h}}$.

Proof: If $x = 0$, then $\lim_{h\to 0}(1 + hx)^{\frac{1}{h}}. = \lim_{h\to 0}(1 + h \cdot 0)^{\frac{1}{h}} = \lim_{h\to 0} 1^{\frac{1}{h}} = \lim_{h\to 0} 1 = 1 = e^0$.

Now, assume that $x \neq 0$ and let $F(x) = \ln|x|$. Then $F'(x) = \frac{1}{x}$.

Using the definition of the derivative, the Note following Theorem 13.57, and Theorem 13.70, for $t \neq 0$, we have

$$\frac{1}{t} = F'(t) = \lim_{h\to 0}\frac{F(t+h) - F(t)}{h} = \lim_{h\to 0}\frac{\ln|t+h| - \ln|t|}{h} = \lim_{h\to 0}\frac{1}{h}\ln\frac{|t+h|}{|t|}$$

$$= \lim_{h\to 0}\frac{1}{h}\ln\left|\frac{t+h}{t}\right| = \lim_{h\to 0}\ln\left|\frac{t+h}{t}\right|^{\frac{1}{h}} = \lim_{h\to 0}\ln\left|1 + \frac{h}{t}\right|^{\frac{1}{h}} = \lim_{h\to 0}\ln\left(1 + \frac{h}{t}\right)^{\frac{1}{h}}$$

We can drop the absolute value at the end because $1 + \frac{h}{t} > 0$ for h sufficiently close to 0. Specifically, we have $1 + \frac{h}{t} > 0$ whenever $\frac{h}{t} > -1$. If $t > 0$, this inequality is true for all $h > -t$ and if $t < 0$, this inequality is true for all $h < -t$.

Substituting $\frac{1}{x}$ for t gives us $x = \lim_{h\to 0}\ln(1 + hx)^{\frac{1}{h}}$.

Therefore, since e^x is continuous, $e^x = e^{\lim_{h\to 0} \ln(1+hx)^{\frac{1}{h}}} = \lim_{h\to 0} e^{\ln(1+hx)^{\frac{1}{h}}} = \lim_{h\to 0}(1+hx)^{\frac{1}{h}}$. □

39. For each $x \in \mathbb{R}$, $e^x = \lim_{h\to\infty}\left(1+\frac{x}{h}\right)^h$.

Proof: By Problem 38 above, we have $e^x = \lim_{k\to 0}(1+kx)^{\frac{1}{k}}$. It follows that $\lim_{k\to 0^+}(1+kx)^{\frac{1}{k}} = e^x$.

If we let $h = \frac{1}{k}$, then $k = \frac{1}{h}$ and $h \to \infty$ if and only if $k \to 0^+$, and so, we have

$$\lim_{h\to\infty}\left(1+\frac{x}{h}\right)^h = \lim_{k\to 0^+}(1+kx)^{\frac{1}{k}} = e^x.$$

□

40. For each $n \in \mathbb{Z}^+$ with $n \geq 2$,

$$\frac{1}{2}+\frac{1}{3}+\cdots+\frac{1}{n} < \ln n < 1+\frac{1}{2}+\frac{1}{3}+\cdots+\frac{1}{n-1}.$$

Proof: Let $f(x) = \frac{1}{x}$, let $n \geq 2$ and let P be the partition of $[1,n]$ consisting of $n-1$ equal subintervals of $[1,n]$. That is $P = \{[1,2],[2,3],\dots,[n-1,n]\} = \{[i,i+1]\}_{i=1}^{n-1}$. Then, we have the following:

$$L_f(P) = \frac{1}{2}\cdot 1+\frac{1}{3}\cdot 1+\cdots+\frac{1}{n}\cdot 1 = \frac{1}{2}+\frac{1}{3}+\cdots+\frac{1}{n}$$

$$U_f(P) = 1\cdot 1+\frac{1}{2}\cdot 1+\cdots+\frac{1}{n-1}\cdot 1 = 1+\frac{1}{2}+\frac{1}{3}+\cdots+\frac{1}{n-1}$$

Since $\ln n = \int_1^n \frac{1}{t}\,dt$, we have $L_f(P) \leq \ln n \leq U_f(P)$. However, this isn't quite good enough. We need the inequalities to be strict.

Now, consider the partition Q of $[1,2]$ consisting of two equal subintervals of $[1,2]$. That is $Q = \{[1,1.5],[1.5,2]\}$. Then we have seen in part 2 of Example 13.54 that $L_f(Q) = \frac{7}{12}$ and $U_f(Q) = \frac{5}{6}$. Let $R = Q \cup \{[2,3],[3,4]\dots,[n-1,n]\}$. Then we have

$$L_f(R) = \frac{7}{12}+\frac{1}{3}+\cdots+\frac{1}{n} \quad \text{and} \quad U_f(R) = \frac{5}{6}+\frac{1}{2}+\cdots+\frac{1}{n-1}.$$

Notice how we can easily compute $L_f(R)$ and $U_f(R)$ by simply replacing the first terms of $L_f(P)$ and $U_f(P)$ by $\frac{7}{12}$ and $\frac{5}{6}$, respectively.

As before, we have $L_f(R) \leq \ln n \leq U_f(R)$.

Since $\frac{1}{2} = \frac{6}{12} < \frac{7}{12}$, we have $L_f(P) < L_f(R)$. Since $\frac{5}{6} < 1$, we have $U_f(R) < U_f(P)$.

Therefore, we have $L_f(P) < L_f(R) \leq \ln n \leq U_f(R) < U_f(P)$. □

41. For each $x \in \mathbb{R}$, $1 + x \leq e^x$.

Proof: Let $f(x) = e^x - x - 1$. Then $f'(x) = e^x - 1$. Since e^x is strictly increasing on $\mathbb{R}$ it follows that if $x > 0$, then $e^x > e^0 = 1$, and consequently, for $x > 0$, $f'(x) > 1 - 1 = 0$. By part 1 of Corollary 13.26, f is strictly increasing on $[0, \infty)$. Since $f(0) = e^0 - 0 - 1 = 1 - 0 - 1 = 0$. It follows that for all $x > 0$, $f(x) > 0$, and therefore, $e^x > 1 + x$.

Once again, since e^x is strictly increasing on $\mathbb{R}$ it follows that if $x < 0$, then $e^x < e^0 = 1$, and consequently, for $x < 0$, $f'(x) < 1 - 1 = 0$. By part 2 of Corollary 13.26, f is strictly decreasing on $(-\infty, 0)$. Since $f(0) = 0$. It follows that for all $x < 0$, $f(x) > 0$, and therefore, $e^x > 1 + x$.

Finally, if $x = 0$, we have $1 + 0 = 1 = e^0$. □

42. The following inequality holds:

$$\frac{4}{3} \leq \int_0^1 e^{x^2}\, dx \leq e.$$

Proof: By Problem 41 above, for all $x \in \mathbb{R}$, $1 + x^2 \leq e^{x^2}$. So, by Problem 21 above,

$$\int_0^1 e^{x^2}\, dx \geq \int_0^1 (1 + x^2)\, dx = \left[x + \frac{x^3}{3}\right]_0^1 = 1 + \frac{1}{3} = \frac{4}{3}.$$

Now, since x^2 is increasing on $[0, 1]$ (because $\frac{d}{dx}[x^2] = 2x \geq 0$ for all $x \in [0, 1]$) and e^x is increasing on $[0, 1]$, we have e^{x^2} increasing on $[0, 1]$ (see the Note below for a detailed explanation). It follows that for all $x \in [0, 1]$, $e^{x^2} \leq e^{1^2} = e^1 = e$. So, by Problem 21 above,

$$\int_0^1 e^{x^2}\, dx \leq \int_0^1 e\, dx = e \int_0^1 dx = e(1 - 0) = e.$$ □

Note: Let $f: [a, b] \to \mathbb{R}$ and $g: [c, d] \to \mathbb{R}$ be increasing functions with $f([a, b]) \subseteq [c, d]$. Let's verify carefully that $g \circ f: [a, b] \to \mathbb{R}$ is increasing. To this end, let $x, y \in [a, b]$ with $x < y$. Since f is increasing on $[a, b]$, $f(x) < f(y)$. Since $f([a, b]) \subseteq [c, d]$, we have $f(x), f(y) \in [c, d]$ with $f(x) < f(y)$. Since g is increasing on $[c, d]$, $(g \circ f)(x) = g(f(x)) < g(f(y)) = (g \circ f)(y)$, as desired.

LEVEL 4

Prove each of the following:

43. Suppose that $[a, b] \subseteq A \subseteq \mathbb{R}$ and $f, g: A \to \mathbb{R}$ are functions that are continuous on $[a, b]$ and differentiable on (a, b). Then there is $c \in (a, b)$ such that

$$[f(b) - f(a)]g'(c) = [g(b) - g(a)]f'(c).$$

This is the **Generalized Mean Value Theorem** (Theorem 13.27).

Proof: Define $h: A \to \mathbb{R}$ by $h(x) = [f(b) - f(a)]g(x) - [g(b) - g(a)]f(x)$. Then h is continuous on $[a, b]$ and differentiable on (a, b) (Check this!).

Also, note that

$$h(a) = [f(b) - f(a)]g(a) - [g(b) - g(a)]f(a) = f(b)g(a) - f(a)g(a) - g(b)f(a) + g(a)f(a)$$
$$= f(b)g(a) - g(b)f(a),$$

$$h(b) = [f(b) - f(a)]g(b) - [g(b) - g(a)]f(b) = f(b)g(b) - f(a)g(b) - g(b)f(b) + g(a)f(b)$$
$$= g(a)f(b) - f(a)g(b) = f(b)g(a) - g(b)f(a).$$

So, $h(a) = h(b)$.

By Rolles's Theorem (Theorem 13.24), there is $c \in (a, b)$ such that $h'(c) = 0$. This is equivalent to $[f(b) - f(a)]g'(c) - [g(b) - g(a)]f'(c) = 0$ or $[f(b) - f(a)]g'(c) = [g(b) - g(a)]f'(c)$. □

44. Suppose that $f: (a, b) \to \mathbb{R}$ is differentiable on (a, b) with $f'(x) > 0$ for all $x \in (a, b)$. Then f is invertible. Furthermore, if $g: B \to \mathbb{R}$ is the inverse of f, then g is differentiable on B and $g'(f(x)) = \frac{1}{f'(x)}$ for all $x \in (a, b)$.

Proof: Since $f'(x) > 0$ for all $x \in (a, b)$, by Corollary 13.26, f is strictly increasing on (a, b). Suppose $x, y \in (a, b)$ with $x \neq y$. Without loss of generality, we may assume that $x < y$. Since f is strictly increasing, $f(x) < f(y)$. In particular, $f(x) \neq f(y)$. This shows that f is injective. By the Intermediate Value Theorem (Corollary 6.44) and the fact that f is strictly increasing, the range of f is the open interval $(f(a), f(b))$. Therefore, $f: (a, b) \to (f(a), f(b))$ is a bijection, and so, f is invertible.

Let $g: (f(a), f(b)) \to (a, b)$ be the inverse of f. It follows from Problem 56 from Problem Set 6 that g is continuous (Check this carefully!). Let $x, x + h \in (a, b)$, let $y = f(x)$ and $y + k = f(x + h)$. Then we have

$$\frac{g(y + k) - g(y)}{k} = \frac{(x + h) - x}{f(x + h) - y} = \frac{h}{f(x + h) - f(x)} = \frac{1}{\frac{f(x + h) - f(x)}{h}}.$$

Since $f'(x) \neq 0$ for all $x \in (a, b)$, it follows that for each $x \in (a, b)$ with $f(x) = y$,

$$\frac{1}{f'(x)} = \frac{1}{\lim_{h \to 0} \frac{f(x + h) - f(x)}{h}} = \lim_{h \to 0} \frac{1}{\frac{f(x + h) - f(x)}{h}} = \lim_{k \to 0} \frac{g(y + k) - g(y)}{k} = g'(y) = g'(f(x)).$$

□

Note: In the last sequence of equalities above, we have

$$\lim_{h \to 0} \frac{1}{\frac{f(x + h) - f(x)}{h}} = \lim_{k \to 0} \frac{g(y + k) - g(y)}{k}.$$

In particular, we used the fact that for fixed $x \in (a, b)$, $h \to 0$ if and only if $k \to 0$. To see that this is true, first observe that $k = f(x + h) - f(x)$. Since f is continuous, as $h \to 0$, $k \to f(x) - f(x) = 0$. Next, observe that $h = g(y + k) - g(y)$. Since g is continuous, as $k \to 0$, $h \to g(x) - g(x) = 0$.

45. A function $f: A \to \mathbb{R}$ that is differentiable on A is said to be **uniformly differentiable** on A if

$$\forall \epsilon > 0\ \exists \delta > 0\ \forall x, y \in A\left(0 < |y - x| < \delta \to \left|\frac{f(y) - f(x)}{y - x} - f'(x)\right| < \epsilon\right).$$

Let $g: [a, b] \to \mathbb{R}$ be differentiable on $[a, b]$ with g' continuous on $[a, b]$. Then g is uniformly differentiable on $[a, b]$.

Proof: Since g' is continuous on $[a, b]$, by Problem 59 from Problem Set 6, g' is uniformly continuous on $[a, b]$. Let $\epsilon > 0$ be given. Since g' is uniformly continuous on $[a, b]$, there is $\delta > 0$ such that for all $u, v \in [a, b]$, $|v - u| < \delta \to |g'(v) - g'(u)| < \epsilon$.

Suppose that $x, y \in [a, b]$ and $0 < |y - x| < \delta$. By the Mean Value Theorem, there is $c \in (a, b)$ such that $\frac{g(y)-g(x)}{y-x} = g'(c)$. It follows that

$$\left|\frac{g(y) - g(x)}{y - x} - g'(x)\right| = |g'(c) - g'(x)| < \epsilon.$$

Since $\epsilon > 0$ was arbitrary, g is uniformly differentiable on $[a, b]$. □

46. Let $f: [a, b] \to \mathbb{R}$ be a bounded function and let P and Q be partitions of $[a, b]$ with Q a refinement of P. Then $L_f(P) \leq L_f(Q)$ and $U_f(Q) \leq U_f(P)$.

Proof: Let $P = \{[x_{i-1}, x_i]\}_{i=1}^{n}$ and $Q = \{[y_{i-1}, y_i]\}_{i=1}^{t}$. Since Q is a refinement of P, it follows that $\{x_0, x_1, \dots, x_n\} \subseteq \{y_0, y_1, \dots, y_t\}$. So, there are integers $k_0 < k_1 < \cdots < k_n$ with $k_0 = 0$ and $k_n = t$ and $x_i = y_{k_i}$ for each $i = 1, 2, \dots, n$. We then have for each $i = 1, 2, \dots, n$,

$$\Delta x_i = x_i - x_{i-1} = y_{k_i} - y_{k_{i-1}} = \sum_{j=k_{i-1}+1}^{k_i} y_j - y_{j-1} = \sum_{j=k_{i-1}+1}^{k_i} \Delta y_j.$$

For each $i = 1, 2, .., n$, let $m_i = \inf\{f(x) \mid x_{i-1} \leq x \leq x_i\}$ and for each $j = 1, 2, \dots, t$, let $n_j = \inf\{f(x) \mid y_{j-1} \leq x \leq y_j\}$. Then $m_i \leq n_j$ for $k_{i-1} < j \leq k_i$. So,

$$m_i \Delta x_i = m_i \sum_{j=k_{i-1}+1}^{k_i} \Delta y_j = \sum_{j=k_{i-1}+1}^{k_i} m_i \Delta y_j \leq \sum_{j=k_{i-1}+1}^{k_i} n_j \Delta y_j.$$

Therefore, we have

$$L_f(P) = \sum_{i=1}^{n} m_i \Delta x_i \leq \sum_{i=1}^{n} \sum_{j=k_{i-1}+1}^{k_i} n_j \Delta y_j = \sum_{i=1}^{t} n_i \Delta y_i = L_f(Q).$$

Similarly, for each $i = 1, 2, .., n$, let $M_i = \sup\{f(x) \mid x_{i-1} \leq x \leq x_i\}$ and for each $j = 1, 2, \dots, t$, let $N_j = \sup\{f(x) \mid y_{j-1} \leq x \leq y_j\}$. Then $M_i \geq N_j$ for $k_{i-1} < j \leq k_i$. So,

$$M_i \Delta x_i = M_i \sum_{j=k_{i-1}+1}^{k_i} \Delta y_j = \sum_{j=k_{i-1}+1}^{k_i} M_i \Delta y_j \geq \sum_{j=k_{i-1}+1}^{k_i} N_j \Delta y_j.$$

Therefore, we have

$$U_f(P) = \sum_{i=1}^{n} M_i \Delta x_i \geq \sum_{i=1}^{n} \sum_{j=k_{i-1}+1}^{k_i} N_j \Delta y_j = \sum_{i=1}^{t} N_i \Delta y_i = U_f(Q).$$

□

47. Let $f: [a, b] \to \mathbb{R}$ be a bounded function, let $m = \inf\{f(x) \mid x \in [a, b]\}$, and let $M = \sup\{f(x) \mid x \in [a, b]\}$. Then

$$m(b - a) \leq \underline{\int_a^b} f(x)\, dx \leq \overline{\int_a^b} f(x)\, dx \leq M(b - a).$$

Proof: Let P be any partition of $[a, b]$. By Theorem 13.41, we have

$$m(b - a) \leq L_f(P) \leq U_f(P) \leq M(b - a).$$

Since $L_f(P) \leq \underline{\int_a^b} f(x)\, dx$, we have $m(b - a) \leq \underline{\int_a^b} f(x)\, dx$. Similarly, since $\overline{\int_a^b} f(x)\, dx \leq U_f(P)$, we have $\overline{\int_a^b} f(x)\, dx \leq M(b - a)$.

Now, let Q and R be partitions of $[a, b]$ and let $T = Q * R$ (see Note 2 before Example 13.42). By Problem 46, we have $L_f(Q) \leq L_f(T)$ and $U_f(T) \leq U_f(R)$. By Theorem 13.41, we have

$$L_f(Q) \leq L_f(T) \leq U_f(T) \leq U_f(R).$$

So, for any partitions Q and R of $[a, b]$, we have $L_f(Q) \leq U_f(R)$.

It follows that if Q is a fixed partition of $[a, b]$, then $L_f(Q)$ is a lower bound of $\{U_f(P) \mid P \text{ is a partition of } [a, b]\}$. Therefore, $L_f(Q) \leq \inf\{U_f(P) \mid P \text{ is a partition of } [a, b]\}$ (this is true because $\inf\{U_f(P) \mid P \text{ is a partition of } [a, b]\}$ is the **greatest** lower bound of the set $\{U_f(P) \mid P \text{ is a partition of } [a, b]\}$).

Since Q was an arbitrary partition of $[a, b]$, it follows that $\inf\{U_f(P) \mid P \text{ is a partition of } [a, b]\}$ is an upper bound of $\{L_f(P) \mid P \text{ is a partition of } [a, b]\}$. Since $\sup\{L_f(P) \mid P \text{ is a partition of } [a, b]\}$ is the **least** upper bound of the set $\{L_f(P) \mid P \text{ is a partition of } [a, b]\}$, it follows that

$$\sup\{L_f(P) \mid P \text{ is a partition of } [a, b]\} \leq \inf\{U_f(P) \mid P \text{ is a partition of } [a, b]\}.$$

Therefore,

$$\underline{\int_a^b} f(x)\, dx = \sup\{L_f(P) \mid P \text{ is a partition of } [a,b]\}$$

$$\leq \inf\{U_f(P) \mid P \text{ is a partition of } [a,b]\} = \overline{\int_a^b} f(x)\, dx. \quad \square$$

48. Let $f, g: [a,b] \to \mathbb{R}$ be bounded functions, let $A = \{L_f(P) \mid P \text{ is a partition of } [a,b]\}$, $B = \{L_g(P) \mid P \text{ is a partition of } [a,b]\}$, and $C = \{L_f(P) + L_g(P) \mid P \text{ is a partition of } [a,b]\}$. Then $\sup C = \sup A + \sup B$. State and prove an analogous result involving upper sums.

Proof: Let $L_f(P) + L_g(P) \in C$. Then $L_f(P) \in A$ and $L_g(P) \in B$. Since $\sup A$ is an upper bound of A, we have $L_f(P) \leq \sup A$. Similarly, $L_g(P) \leq \sup B$. So, $L_f(P) + L_g(P) \leq \sup A + \sup B$. Since $L_f(P) + L_g(P) \in C$ was arbitrary, it follows that $\sup A + \sup B$ is an upper bound of C. Since $\sup C$ is the **least** upper bound of C, we have $\sup C \leq \sup A + \sup B$.

Fix $L_f(P) \in A$. Now, let $L_g(Q) \in B$ and let $R = P * Q$ (see Note 2 before Example 13.42). By Problem 46 above, we have $L_f(P) \leq L_f(R)$ and $L_g(Q) \leq L_g(R)$. So, $L_f(P) + L_g(Q) \leq L_f(R) + L_g(R)$. Since $L_f(R) + L_g(R) \in C$, we see that $L_f(P) + L_g(Q) \leq L_f(R) + L_g(R) \leq \sup C$. Therefore, $L_g(Q) \leq \sup C - L_f(P)$. Since $L_g(Q) \in B$ was arbitrary, $\sup C - L_f(P)$ is an upper bound of B. Since $\sup B$ is the **least** upper bound of B, it follows that $\sup B \leq \sup C - L_f(P)$. Therefore, $L_f(P) \leq \sup C - \sup B$. Since $L_f(P) \in A$ was arbitrary, we see that $\sup C - \sup B$ is an upper bound of A. Since $\sup A$ is the **least** upper bound of A, we have $\sup A \leq \sup C - \sup B$. Therefore, $\sup A + \sup B \leq \sup C$.

Since $\sup C \leq \sup A + \sup B$ and $\sup A + \sup B \leq \sup C$, we have $\sup C = \sup A + \sup B$.

Next, we let $A = \{U_f(P) \mid P \text{ is a partition of } [a,b]\}$, $B = \{U_g(P) \mid P \text{ is a partition of } [a,b]\}$, and $C = \{U_f(P) + U_g(P) \mid P \text{ is a partition of } [a,b]\}$. We will prove that $\inf C = \inf A + \inf B$.

Let $U_f(P) + U_g(P) \in C$. Then $U_f(P) \in A$ and $U_g(P) \in B$. Since $\inf A$ is a lower bound of A, we have $U_f(P) \geq \inf A$. Similarly, $U_g(P) \geq \inf B$. So, $U_f(P) + U_g(P) \geq \inf A + \inf B$. Since $U_f(P) + U_g(P) \in C$ was arbitrary, it follows that $\inf A + \inf B$ is a lower bound of C. Since $\inf C$ is the **greatest** lower bound of C, we have $\inf C \geq \inf A + \inf B$.

Fix $U_f(P) \in A$. Now, let $U_g(Q) \in B$ and let $R = P * Q$ (see Note 2 before Example 13.42). By Problem 46 above, we have $U_f(R) \leq U_f(P)$ and $U_g(R) \leq U_g(Q)$. So, $U_f(P) + U_g(Q) \geq U_f(R) + U_g(R)$. Since $U_f(R) + U_g(R) \in C$, we see that $U_f(P) + U_g(Q) \geq U_f(R) + U_g(R) \geq \inf C$. Therefore, $U_g(Q) \geq \inf C - U_f(P)$. Since $U_g(Q) \in B$ was arbitrary, $\inf C - U_f(P)$ is a lower bound of B. Since $\inf B$ is the **greatest** lower bound of B, it follows that $\inf B \geq \inf C - U_f(P)$. Therefore, $U_f(P) \geq \inf C - \inf B$. Since $U_f(P) \in A$ was arbitrary, we see that $\inf C - \inf B$ is a lower bound of A. Since $\inf A$ is the **greatest** lower bound of A, we have $\inf A \geq \inf C - \inf B$. So, $\inf A + \inf B \geq \inf C$.

Since $\inf C \geq \inf A + \inf B$ and $\inf A + \inf B \geq \inf C$, we have $\inf C = \inf A + \inf B$. $\square$

49. For all $n \in \mathbb{N}$ and all $x \in [0, \infty)$,

$$e^x \geq \sum_{k=0}^{n} \frac{x^k}{k!}$$

(where $0! = 1$ and $k! = 1 \cdot 2 \cdots k$ for all natural numbers $k \geq 1$).

Proof: For each $n \in \mathbb{Z}^+$, let

$$u_n(x) = \sum_{k=0}^{n} \frac{x^k}{k!}$$

We first prove by induction on n that for all $n \in \mathbb{N}$, $u'_{n+1} = u_n$.

Base case ($t = 0$): $u'_{t+1}(x) = u'_1(x) = \frac{d}{dx}\left[\frac{x^0}{0!} + \frac{x^1}{1!}\right] = \frac{d}{dx}[1 + x] = 1 = \frac{x^0}{0!} = u_0(x) = u_t(x)$.

Inductive step: Assume that $t \geq 0$ and $u'_{t+1}(x) = u_t(x)$. Then we have

$$u'_{t+2}(x) = \frac{d}{dx}\left[u_{t+1}(x) + \frac{x^{t+2}}{(t+2)!}\right] = \frac{d}{dx}[u_{t+1}(x)] + \frac{d}{dx}\left[\frac{x^{t+2}}{(t+2)!}\right] = u_t(x) + \frac{x^{t+1}}{(t+1)!} = u_{t+1}(x).$$

If $x = 0$, we have

$$\sum_{k=0}^{t} \frac{x^k}{k!} = 1 + x + \frac{x^2}{2!} + \cdots + \frac{x^t}{t!} = 1 + 0 + \frac{0^2}{2!} + \cdots + \frac{0^t}{t!} = 1 = e^0.$$

We now prove by induction on $n \in \mathbb{Z}^+$ that for all $x \in (0, \infty)$,

$$e^x \geq \sum_{k=0}^{n} \frac{x^k}{k!}$$

For $t = 1$, we have $e^x \geq 1 + x = \frac{x^0}{0!} + \frac{x^1}{1!}$ by Problem 41 above. For the inductive step, assume $t \geq 1$ and that we have

$$e^x \geq \sum_{k=0}^{t} \frac{x^k}{k!} = u_t(x).$$

Let $f(x) = e^x - u_{t+1}(x)$. Then for all $x \in \mathbb{R}$, we have $f'(x) = e^x - u'_{t+1}(x) = e^x - u_t(x) \geq 0$. By Problem 32, f is increasing on $\mathbb{R}$. Since $f(0) = e^0 - 1 = 1 - 1 = 0$. It follows that for all $x > 0$, $f(x) \geq 0$, and therefore, $e^x \geq u_{t+1}(x)$. □

LEVEL 5

Determine if each of the following statements is true or false. If true, provide a proof. If false, provide a counterexample.

50. If $f: \mathbb{R} \to \mathbb{R}$ satisfies $f(0) = 0$ and $f'(0) = 1$, then there is a neighborhood N of 0 such that f is strictly monotonic on N.

False

Solution: Define $f: \mathbb{R} \to \mathbb{R}$ by $f(x) = \begin{cases} 2x^2 \sin\frac{1}{x} + x \text{ if } x \neq 0 \\ 0 \quad\quad \text{if } x = 0 \end{cases}$. By definition, $f(0) = 0$. Also, we have

$$f'(0) = \lim_{x \to 0} \frac{f(x) - f(0)}{x - 0} = \lim_{x \to 0} \left(2x \sin\frac{1}{x} + 1\right) = 0 + 1 = 1.$$

(Use the Squeeze Theorem (Problem 74 from Problem Set 5) to show that $\lim_{x \to 0} 2x \sin\frac{1}{x} = 0$.)

Now, if $x \neq 0$, then $f'(x) = 2x^2\left(\cos\frac{1}{x}\right)\left(-\frac{1}{x^2}\right) + 4x \sin\frac{1}{x} + 1 = -2\cos\frac{1}{x} + 4x \sin\frac{1}{x} + 1$.

$$f'\left(\frac{1}{2n\pi}\right) = -2\cos 2n\pi + \frac{2}{n\pi}\sin 2n\pi + 1 = -2 + 0 + 1 = -1 < 0.$$

$$f'\left(\frac{2}{(4n+1)\pi}\right) = -2\cos\frac{(4n+1)\pi}{2} + \frac{8}{(4n+1)\pi}\sin\frac{(4n+1)\pi}{2} + 1 = 0 + \frac{8}{(4n+1)\pi} + 1 > 0.$$

Since the sequences $\left(\frac{1}{2n\pi}\right)$ and $\left(\frac{2}{(4n+1)\pi}\right)$ both converge to 0, we see that f' takes on both positive and negative values in every neighborhood of 0. Therefore, f is **not** strictly monotonic on any neighborhood of 0.

51. Suppose that $f: (0, \infty) \to \mathbb{R}$ is differentiable on $(0, \infty)$, $f'(x)$ is uniformly continuous on $(0, \infty)$, and $\lim_{x \to \infty} f(x)$ is a finite real number. Then $\lim_{x \to \infty} f'(x) = 0$.

True

Proof: Assume toward contradiction that $\lim_{x \to \infty} f'(x) \neq 0$. Without loss of generality, we may assume that $\lim_{x \to \infty} f'(x) \nless 0$ (if $\lim_{x \to \infty} f'(x) < 0$, simply replace f by $-f$). Then there is $\epsilon > 0$ and a sequence (x_n) such that $x_n \to \infty$ and $f'(x_n) \geq \epsilon > 0$. Since $f'(x)$ is uniformly continuous on $(0, \infty)$, we can choose $\delta > 0$ so that

$$\forall x, y \in (0, \infty)\left(|y - x| < \delta \to |f'(y) - f'(x)| < \frac{\epsilon}{2}\right).$$

Then we have

$$|y - x_n| < \delta \to |f'(y) - f'(x_n)| < \frac{\epsilon}{2} \to -\frac{\epsilon}{2} < f'(y) - f'(x_n) < \frac{\epsilon}{2}$$
$$\to f'(y) > f'(x_n) - \frac{\epsilon}{2} \to f'(y) > \epsilon - \frac{\epsilon}{2} \to f'(y) > \frac{\epsilon}{2}.$$

Now, $|y - x_n| < \delta$ if and only if $-\delta < y - x_n < \delta$ if and only if $x_n - \delta < y < x_n + \delta$ if and only if $y \in (x_n - \delta, x_n + \delta)$.

By the Mean Value Theorem, there is a c between $x_n - \delta$ and $x_n + \delta$ such that

$$\frac{f(x_n + \delta) - f(x_n - \delta)}{(x_n + \delta) - (x_n - \delta)} = f'(c)$$

So,

$$\frac{|f(x_n + \delta) - f(x_n - \delta)|}{2\delta} = \left|\frac{f(x_n + \delta) - f(x_n - \delta)}{2\delta}\right| = \left|\frac{f(x_n + \delta) - f(x_n - \delta)}{(x_n + \delta) - (x_n - \delta)}\right| = |f'(c)| > \frac{\epsilon}{2}.$$

It follows that $|f(x_n + \delta) - f(x_n - \delta)| > \frac{\epsilon}{2} \cdot 2\delta = \epsilon\delta$.

Since $\epsilon\delta > 0$, there is $K \in \mathbb{N}$ such that $x > K \to |f(x) - L| < \frac{\epsilon\delta}{2}$, where $\lim_{x \to \infty} f(x) = L$.

Therefore, by the Triangle Inequality and the fact that $x_n \to \infty$, for n sufficiently large, we have

$$|f(x_n + \delta) - f(x_n - \delta)| \leq |f(x_n + \delta) - L| + |L - f(x_n - \delta)| < \frac{\epsilon\delta}{2} + \frac{\epsilon\delta}{2} = \epsilon\delta.$$

We have now reached a contradiction. □

Prove each of the following:

52. Let $f: [a, b] \to \mathbb{R}$ be a monotonic function. Then f is Riemann integrable on $[a, b]$.

Proof: First let's assume that f is increasing on $[a, b]$. Then for all $x \in [a, b]$, $f(a) \leq f(x) \leq f(b)$. Therefore, f is bounded on $[a, b]$. We will use Lemma 13.46 to show that f is Riemann integrable on $[a, b]$.

So, let $\epsilon > 0$. By the Archimedean Property of $\mathbb{R}$ (Theorem 5.11), there is $n \in \mathbb{N}$ such that $n > \frac{(b-a)(f(b)-f(a))}{\epsilon}$. Let P be the partition consisting of n equal subintervals of $[a, b]$. In other words, if we let $x_i = a + \frac{i(b-a)}{n}$, then $P = \{[x_{i-1}, x_i]\}_{i=1}^{n}$. Then we have the following:

$$L_f(P) = \sum_{i=1}^{n} f(x_{i-1})\Delta x_i = \frac{b-a}{n}\sum_{i=1}^{n} f(x_{i-1})$$

$$U_f(P) = \sum_{i=1}^{n} f(x_i)\Delta x_i = \frac{b-a}{n}\sum_{i=1}^{n} f(x_i)$$

Therefore,

$$U_f(P) - L_f(P) = \frac{b-a}{n}\sum_{i=1}^{n} f(x_i) - \frac{b-a}{n}\sum_{i=1}^{n} f(x_{i-1})$$

$$= \frac{b-a}{n}\left(\sum_{i=1}^{n}\big(f(x_i) - f(x_{i-1})\big)\right) = \frac{(b-a)\big(f(b) - f(a)\big)}{n} < \epsilon.$$

By Lemma 13.46, f is Riemann integrable on $[a, b]$.

Next, let's assume that f is decreasing on $[a, b]$. Then for all $x \in [a, b]$, $f(b) \leq f(x) \leq f(a)$. Therefore, once again, f is bounded on $[a, b]$. We will again use Lemma 13.46 to show that f is Riemann integrable on $[a, b]$.

So, let $\epsilon > 0$. By the Archimedean Property of $\mathbb{R}$ (Theorem 5.11), there is $n \in \mathbb{N}$ such that $n > \frac{(b-a)(f(a)-f(b))}{\epsilon}$. Let P be the partition consisting of n equal subintervals of $[a, b]$. In other words, if we let $x_i = a + \frac{i(b-a)}{n}$, then $P = \{[x_{i-1}, x_i]\}_{i=1}^{n}$. Then we have the following:

$$L_f(P) = \sum_{i=1}^{n} f(x_i)\Delta x_i = \frac{b-a}{n}\sum_{i=1}^{n} f(x_i)$$

$$U_f(P) = \sum_{i=1}^{n} f(x_{i-1})\Delta x_i = \frac{b-a}{n}\sum_{i=1}^{n} f(x_{i-1})$$

Therefore,

$$U_f(P) - L_f(P) = \frac{b-a}{n}\sum_{i=1}^{n} f(x_{i-1}) - \frac{b-a}{n}\sum_{i=1}^{n} f(x_i)$$

$$= \frac{b-a}{n}\left(\sum_{i=1}^{n}\big(f(x_{i-1}) - f(x_i)\big)\right) = \frac{(b-a)\big(f(a)-f(b)\big)}{n} < \epsilon.$$

By Lemma 13.46, f is Riemann integrable on $[a, b]$. □

53. If P is a partition of $[a, b]$, we define the **norm** of P, written $\|P\|$ to be the maximum length of a subinterval in P. A bounded function $f: [a, b] \to \mathbb{R}$ is Riemann integrable if and only if there is $L \in \mathbb{R}$ such that the following holds: For every $\epsilon > 0$ there is $\delta > 0$ such that whenever P is a partition of $[a, b]$ with $\|P\| < \delta$ and S is any Riemann sum over P, then $|S - L| < \epsilon$. You may use Problem 62 below.

Proof: Assume that f is Riemann integrable on $[a, b]$, let $L = \int_a^b f(x)\,dx$, and let $\epsilon > 0$. By Problem 62, there is $\delta > 0$ such that whenever P is a partition with $\|P\| < \delta$, we have both $L_f(P) > L - \epsilon$ and $U_f(P) < L + \epsilon$. So, if $\|P\| < \delta$ and S is any Riemann sum over P, then we have

$$L - \epsilon < L_f(P) \leq S \leq U_f(P) < L + \epsilon.$$

So, $|S - L| < \epsilon$, as desired.

Now, assume that there is $L \in \mathbb{R}$ such that the following holds: For every $\epsilon > 0$ there is $\delta > 0$ such that whenever P is a partition with $\|P\| < \delta$ and S is any Riemann sum over P, then $|S - L| < \epsilon$ (or equivalently, $L - \epsilon < S < L + \epsilon$).

Let $\epsilon > 0$ be given, let $\delta > 0$ satisfy the given condition with ϵ replaced by $\frac{\epsilon}{2}$, let $P = \{[x_{i-1}, x_i]\}_{i=1}^n$ be a partition with $\|P\| < \delta$, and let $m_i = \inf\{f(x) \mid x_{i-1} \leq x \leq x_i\}$ and $M_i = \sup\{f(x) \mid x_{i-1} \leq x \leq x_i\}$ for each $i = 1, 2, \dots, n$.

Choose tags $c_1, c_2, \dots, c_n$ with $c_i \in [x_{i-1}, x_i]$ such that $f(c_i) > M_i - \frac{\epsilon}{2(b-a)}$ (we can do this because $M_i - \frac{\epsilon}{2(b-a)}$ is **not** an upper bound of $\{f(x) \mid x_{i-1} \leq x \leq x_i\}$) and tags $d_1, d_2, \dots, d_n$ with $d_i \in [x_{i-1}, x_i]$ such that $f(d_i) < m_i + \frac{\epsilon}{2(b-a)}$ (we can do this because $m_i + \frac{\epsilon}{2(b-a)}$ is **not** a lower bound of $\{f(x) \mid x_{i-1} \leq x \leq x_i\}$). Let S_U and S_L be the corresponding Riemann sums. Then we have:

$$S_U = \sum_{i=1}^n f(c_i)\Delta x_i > \sum_{i=1}^n \left(M_i - \frac{\epsilon}{2(b-a)}\right)\Delta x_i = \sum_{i=1}^n M_i \Delta x_i - \frac{\epsilon}{2(b-a)} \sum_{i=1}^n \Delta x_i$$

$$= U_f(P) - \frac{\epsilon}{2(b-a)}(b-a) = U_f(P) - \frac{\epsilon}{2} \geq \overline{\int_a^b} f(x)\, dx - \frac{\epsilon}{2}$$

$$S_L = \sum_{i=1}^n f(d_i)\Delta x_i < \sum_{i=1}^n \left(m_i + \frac{\epsilon}{2(b-a)}\right)\Delta x_i = \sum_{i=1}^n m_i \Delta x_i + \frac{\epsilon}{2(b-a)} \sum_{i=1}^n \Delta x_i$$

$$= L_f(P) + \frac{\epsilon}{2(b-a)}(b-a) = L_f(P) + \frac{\epsilon}{2} \leq \underline{\int_a^b} f(x)\, dx + \frac{\epsilon}{2}$$

The given condition tells us that $S_U < L + \frac{\epsilon}{2}$ and $S_L > L - \frac{\epsilon}{2}$. So, we have

$$\overline{\int_a^b} f(x)\, dx - \frac{\epsilon}{2} < S_U < L + \frac{\epsilon}{2} \quad \text{and} \quad \underline{\int_a^b} f(x)\, dx + \frac{\epsilon}{2} > S_L > L - \frac{\epsilon}{2}.$$

So, we have

$$\overline{\int_a^b} f(x)\, dx < L + \epsilon \quad \text{and} \quad \underline{\int_a^b} f(x)\, dx > L - \epsilon.$$

Also, we have

$$\overline{\int_a^b} f(x)\, dx \geq L \quad \text{and} \quad \underline{\int_a^b} f(x)\, dx \leq L \text{ (Check this!)}.$$

Since $\epsilon > 0$ was arbitrary, by Problem 29 from Problem Set 5, we have

$$\overline{\int_a^b} f(x)\, dx \leq L \quad \text{and} \quad \underline{\int_a^b} f(x)\, dx \geq L.$$

Therefore,

$$L \leq \underline{\int_a^b} f(x)\,dx \leq \overline{\int_a^b} f(x)\,dx \leq L.$$

So, $\underline{\int_a^b} f(x)\,dx = \overline{\int_a^b} f(x)\,dx$, and therefore, f is Riemann integrable on $[a, b]$. □

54. Let $f: [a, b] \to \mathbb{R}$ be a bounded function that is Riemann integrable on $[a, b]$. Let $c, d, e \in [a, b]$. Then

$$\int_c^e f(x)\,dx = \int_c^d f(x)\,dx + \int_d^e f(x)\,dx.$$

Proof: We will first prove several lemmas:

Lemma 1: Let $a < c < b$ and let $f: [a, b] \to \mathbb{R}$ be bounded on $[a, b]$. Then

$$\overline{\int_a^b} f(x)\,dx = \overline{\int_a^c} f(x)\,dx + \overline{\int_c^b} f(x)\,dx \quad \text{and} \quad \underline{\int_a^b} f(x)\,dx = \underline{\int_a^c} f(x)\,dx + \underline{\int_c^b} f(x)\,dx.$$

Proof of Lemma 1: Let $Q = \{[x_{i-1}, x_i]\}_{i=1}^{k}$ be an arbitrary partition of $[a, b]$ and let $R = \{[y_{i-1}, y_i]\}_{i=1}^{j} \cup \{[y_{i-1}, y_i]\}_{i=j+1}^{t}$ be a refinement of Q with $y_j = c$. By Problem 46 above,

$$L_f(Q) \leq L_f(R)$$
$$\leq \sup\{L_f(P) \mid P = \{[x_{i-1}, x_i]\}_{i=1}^{n} \text{ is a partition of } [a, b] \text{ with } c = x_i \text{ for some } i = 1, 2, \dots, n\}.$$

Since Q was an arbitrary partition of $[a, b]$, we see that

$$\sup\{L_f(P) \mid P \text{ is a partition of } [a, b]\}$$
$$\leq \sup\{L_f(P) \mid P = \{[x_{i-1}, x_i]\}_{i=1}^{n} \text{ is a partition of } [a, b] \text{ with } c = x_i \text{ for some } i = 1, 2, \dots, n\}.$$

Now, if $Q = \{[x_{i-1}, x_i]\}_{i=1}^{k}$ is a partition of $[a, b]$ such that $c = x_i$ for some $i = 1, 2, \dots, k$, then Q is a partition of $[a, b]$. It follows that

$$L_f(Q) \leq \sup\{L_f(P) \mid P \text{ is a partition of } [a, b]\}.$$

Since Q was an arbitrary partition of $[a, b]$ such that $c = x_i$ for some $i = 1, 2, \dots, n$, we have

$$\sup\{L_f(P) \mid P = \{[x_{i-1}, x_i]\}_{i=1}^{n} \text{ is a partition of } [a, b] \text{ with } c = x_i \text{ for some } i = 1, 2, \dots, n\}$$
$$\leq \sup\{L_f(P) \mid P \text{ is a partition of } [a, b]\}.$$

It follows that

$$\sup\{L_f(P) \mid P = \{[x_{i-1}, x_i]\}_{i=1}^{n} \text{ is a partition of } [a, b] \text{ with } c = x_i \text{ for some } i = 1, 2, \dots, n\}$$
$$= \sup\{L_f(P) \mid P \text{ is a partition of } [a, b]\}.$$

Therefore, we have

$$\underline{\int_a^c} f(x)\,dx + \underline{\int_c^b} f(x)\,dx$$

$$= \sup\{L_f(P) \mid P \text{ a partition of } [a,c]\} + \sup\{L_f(Q) \mid Q \text{ a partition of } [c,d]\}$$

$$= \sup\{L_f(P) + L_f(Q) \mid P \text{ a partition of } [a,c] \wedge Q \text{ a partition of } [c,d]\} \text{ (by Problem 35 above)}$$

$$= \sup\{L_f(P) \mid P = \{[x_{i-1}, x_i]\}_{i=1}^{n} \text{ is a partition of } [a,b] \text{ with } c = x_i \text{ for some } i = 1,2,\dots,n\}$$

$$= \sup\{L_f(P) \mid P \text{ is a partition of } [a,b]\} = \underline{\int_a^b} f(x)\,dx.$$

An analogous argument shows that

$$\overline{\int_a^c} f(x)\,dx + \overline{\int_c^b} f(x)\,dx = \overline{\int_a^b} f(x)\,dx.$$

□

Lemma 2: Let $a < c < b$ and let $f: [a,b] \to \mathbb{R}$ be bounded on $[a,b]$. Then f is Riemann integrable on $[a,b]$ if and only if f is Riemann integrable on $[a,c]$ and f is Riemann integrable on $[c,b]$.

Proof of Lemma 2: First assume that f is Riemann integrable on $[a,b]$. By Lemma 1 above, we have

$$\int_a^b f(x)\,dx = \underline{\int_a^b} f(x)\,dx = \underline{\int_a^c} f(x)\,dx + \underline{\int_c^b} f(x)\,dx$$

$$\le \overline{\int_a^c} f(x)\,dx + \overline{\int_c^b} f(x)\,dx = \overline{\int_a^b} f(x)\,dx = \int_a^b f(x)\,dx.$$

So,

$$\underline{\int_a^c} f(x)\,dx + \underline{\int_c^b} f(x)\,dx = \overline{\int_a^c} f(x)\,dx + \overline{\int_c^b} f(x)\,dx.$$

We know that $\underline{\int_a^c} f(x)\,dx \le \overline{\int_a^c} f(x)\,dx$. If we have $\underline{\int_a^c} f(x)\,dx < \overline{\int_a^c} f(x)\,dx$, then we must have

$$\underline{\int_c^b} f(x)\,dx = \left(\overline{\int_a^c} f(x)\,dx - \underline{\int_a^c} f(x)\,dx\right) + \overline{\int_c^b} f(x)\,dx > \overline{\int_c^b} f(x)\,dx.$$

This is impossible, and so, we must have $\underline{\int_a^c} f(x)\,dx = \overline{\int_a^c} f(x)\,dx$. It then follows that $\underline{\int_c^b} f(x)\,dx = \overline{\int_c^b} f(x)\,dx$. Therefore, f is Riemann integrable on $[a,c]$ and $[c,b]$.

Conversely, if f is Riemann integrable on $[a, c]$ and f is Riemann integrable on $[c, b]$, then we have

$$\underline{\int_a^b} f(x)\,dx = \underline{\int_a^c} f(x)\,dx + \underline{\int_c^b} f(x)\,dx = \int_a^c f(x)\,dx + \int_c^b f(x)\,dx$$

$$= \overline{\int_a^c} f(x)\,dx + \overline{\int_c^b} f(x)\,dx = \overline{\int_a^b} f(x)\,dx.$$

Therefore, f is Riemann integrable on $[a, b]$, as desired. □

Lemma 3: Let $a < c < b$ and let $f: [a, b] \to \mathbb{R}$ be Riemann integrable on $[a, b]$, $[a, c]$, and $[c, b]$. Then

$$\int_a^b f(x)\,dx = \int_a^c f(x)\,dx + \int_c^b f(x)\,dx.$$

Proof of Lemma 3: By Lemma 1,

$$\int_a^b f(x)\,dx = \underline{\int_a^b} f(x)\,dx = \underline{\int_a^c} f(x)\,dx + \underline{\int_c^b} f(x)\,dx = \int_a^c f(x)\,dx + \int_c^b f(x)\,dx.$$

□

Lemma 4: Let $f: [a, b] \to \mathbb{R}$ be Riemann integrable on $[a, b]$ and let $[c, d] \subseteq [a, b]$. Then f is Riemann integrable on $[c, d]$.

Proof of Lemma 4: By Lemma 2, f is Riemann integrable on $[a, c]$ and $[c, b]$. By Lemma 2 again, f is Riemann integrable on $[c, d]$ and $[d, b]$. In particular, f is Riemann integrable on $[c, d]$. □

Proof of main result: Let $c, d, e \in [a, b]$. We will prove the result by cases.

Case 1 ($c < d < e$): By Lemma 4, f is Riemann integrable on each of $[c, e]$, $[c, d]$, and $[d, e]$. So, by Lemma 3, we have

$$\int_c^e f(x)\,dx = \int_c^d f(x)\,dx + \int_d^e f(x)\,dx.$$

Case 2 ($c < e < d$): By Lemma 4, f is Riemann integrable on each of $[c, d]$, $[c, e]$, and $[e, d]$. So, by Lemma 3, we have

$$\int_c^d f(x)\,dx = \int_c^e f(x)\,dx + \int_e^d f(x)\,dx.$$

It follows that

$$\int_c^e f(x)\,dx = \int_c^d f(x)\,dx - \int_e^d f(x)\,dx = \int_c^d f(x)\,dx + \int_d^e f(x)\,dx.$$

Case 3 ($d < c < e$): By Lemma 4, f is Riemann integrable on each of $[d, e]$, $[d, c]$, and $[c, e]$. So, by Lemma 3, we have

$$\int_d^e f(x)\,dx = \int_d^c f(x)\,dx + \int_c^e f(x)\,dx.$$

It follows that

$$\int_c^e f(x)\,dx = \int_d^e f(x)\,dx - \int_d^c f(x)\,dx = \int_c^d f(x)\,dx + \int_d^e f(x)\,dx.$$

Case 4 ($d < e < c$): By Lemma 4, f is Riemann integrable on each of $[d, c]$, $[d, e]$, and $[e, c]$. So, by Lemma 3, we have

$$\int_d^c f(x)\,dx = \int_d^e f(x)\,dx + \int_e^c f(x)\,dx.$$

It follows that

$$\int_c^e f(x)\,dx = -\int_e^c f(x)\,dx = \int_d^e f(x)\,dx - \int_d^c f(x)\,dx = \int_c^d f(x)\,dx + \int_d^e f(x)\,dx.$$

Case 5 ($e < c < d$): By Lemma 4, f is Riemann integrable on each of $[e, d]$, $[e, c]$, and $[c, d]$. So, by Lemma 3, we have

$$\int_e^d f(x)\,dx = \int_e^c f(x)\,dx + \int_c^d f(x)\,dx.$$

It follows that

$$\int_c^e f(x)\,dx = -\int_e^c f(x)\,dx = \int_c^d f(x)\,dx - \int_e^d f(x)\,dx = \int_c^d f(x)\,dx + \int_d^e f(x)\,dx.$$

Case 6 ($e < d < c$): By Lemma 4, f is Riemann integrable on each of $[e, c]$, $[e, d]$, and $[d, c]$. So, by Lemma 3, we have

$$\int_e^c f(x)\,dx = \int_e^d f(x)\,dx + \int_d^c f(x)\,dx.$$

It follows that

$$\int_c^e f(x)\,dx = -\int_e^c f(x)\,dx = -\int_e^d f(x)\,dx - \int_d^c f(x)\,dx = \int_c^d f(x)\,dx + \int_d^e f(x)\,dx.$$

□

55. Let $f: [a, b] \to \mathbb{R}$ be a bounded function with finitely many discontinuities. Then f is Riemann integrable on $[a, b]$.

Proof: We first prove a lemma:

Lemma: Let $f: [a, b] \to \mathbb{R}$ be a bounded function and for each $n \in \mathbb{N}$ let $[a_n, b_n]$ be a closed interval such that $a < a_n < b_n < b$ and f is Riemann integrable on $[a_n, b_n]$. If $\lim_{n\to\infty} a_n = a$ and $\lim_{n\to\infty} b_n = b$, then f is Riemann integrable on $[a, b]$.

Proof of Lemma: Since f is bounded on $[a, b]$, there is $M \in \mathbb{R}^+$ such that $-M \leq f(x) \leq M$ for all $x \in [a, b]$. Since $b - a > b_n - a_n$ for all $n \in \mathbb{N}$, by problem 21 above, we have

$$-M(b - a) < -M(b_n - a_n) = \int_{a_n}^{b_n} -M\, dx \leq \int_{a_n}^{b_n} f(x)\, dx \leq \int_{a_n}^{b_n} M\, dx = M(b_n - a_n) < M(b - a).$$

So, $\left(\int_{a_n}^{b_n} f(x)\, dx\right)$ is a bounded sequence of real numbers. By the Bolzano-Weierstrass Theorem (Theorem 5.52), $\left(\int_{a_n}^{b_n} f(x)\, dx\right)$ has a convergent subsequence $\left(\int_{a_{n_k}}^{b_{n_k}} f(x)\, dx\right)$. Let L be the limit of this subsequence.

By Lemma 1 in the solution to Problem 54 above and the fact that for each $n \in \mathbb{N}$, f is Riemann integrable on $[a_n, b_n]$, for each $k \in \mathbb{N}$, we have the following:

$$\overline{\int_a^b} f(x)\, dx = \overline{\int_a^{a_{n_k}}} f(x)\, dx + \overline{\int_{a_{n_k}}^{b_{n_k}}} f(x)\, dx + \overline{\int_{b_{n_k}}^{b}} f(x)\, dx$$

$$\leq M(a_{n_k} - a) + \int_{a_{n_k}}^{b_{n_k}} f(x)\, dx + M(b - b_{n_k})$$

$$\underline{\int_a^b} f(x)\, dx = \underline{\int_a^{a_{n_k}}} f(x)\, dx + \underline{\int_{a_{n_k}}^{b_{n_k}}} f(x)\, dx + \underline{\int_{b_{n_k}}^{b}} f(x)\, dx$$

$$\geq -M(a_{n_k} - a) + \int_{a_{n_k}}^{b_{n_k}} f(x)\, dx - M(b - b_{n_k}).$$

Taking the limit as k goes to ∞, we get the following:

$$\overline{\int_a^b} f(x)\, dx \leq M(a - a) + L + M(b - b) = M \cdot 0 + L + M \cdot 0 = L$$

$$\underline{\int_a^b} f(x)\, dx \geq -M(a - a) + L - M(b - b) = -M \cdot 0 + L - M \cdot 0 = L$$

So, we have

$$L \leq \underline{\int_a^b} f(x)\,dx \leq \overline{\int_a^b} f(x)\,dx \leq L.$$

Therefore, $\overline{\int_a^b} f(x)\,dx = \underline{\int_a^b} f(x)\,dx = L$. So, f is Riemann integrable on $[a,b]$. □

Proof of main result: Let $f:[a,b] \to \mathbb{R}$ be bounded on $[a,b]$ with finitely many discontinuities. Let $P = \{[x_{i-1}, x_i]\}_{i=1}^n$ be a partition of $[a,b]$ such that f is continuous on (x_{i-1}, x_i) for each $i = 1, 2, \dots, n$. If $x_{i-1} < y_{i-1} < y_i < x_i$, then f is continuous on $[y_{i-1}, y_i]$ and therefore, f is Riemann integrable on $[y_{i-1}, y_i]$. By the Lemma above, f is Riemann integrable on $[x_{i-1}, x_i]$. By Lemma 2 from Problem 54 above and a straightforward induction, f is Riemann integrable on $[a,b]$. □

56. For each $k = 1, 2, \dots, n$, let $c_k \in (0, \infty)$. Then

$$(c_1 c_2 \cdots c_n)^{\frac{1}{n}} \leq \frac{c_1 + c_2 + \cdots c_n}{n}$$

Proof: Let $M = \frac{c_1+c_2+\cdots+c_n}{n}$ and for each $k = 1, 2, \dots, n$, let $x_k = \frac{c_k}{M} - 1$. By Problem 41 above, for each $k = 1, 2, , , , . n$, we have

$$\frac{c_k}{M} = 1 + x_k \leq e^{x_k}.$$

Since the left- and right-hand sides of each of these n inequalities is positive, we can multiply to get

$$\frac{c_1}{M} \cdot \frac{c_2}{M} \cdots \frac{c_n}{M} \leq e^{x_1} e^{x_2} \cdots e^{x_n} = e^{x_1+x_2+\cdots+x_n} = e^{\frac{c_1+c_2+\cdots+c_n}{M} - n} = e^{n-n} = e^0 = 1.$$

Therefore, we have

$$c_1 c_2 \cdots c_n \leq M^n = \left(\frac{c_1 + c_2 + \cdots + c_n}{n}\right)^n.$$

Thus,

$$(c_1 c_2 \cdots c_n)^{\frac{1}{n}} \leq \frac{c_1 + c_2 + \cdots + c_n}{n}.$$

□

57. For all $n \in \mathbb{Z}^+$ and all $x \in [0, \infty)$,

$$\ln(x+1) = \sum_{k=1}^{n} \frac{(-1)^{k+1} x^k}{k} + \int_0^x \frac{(-1)^n t^n}{t+1}\,dt.$$

Proof: If $n \in \mathbb{Z}^+$ and $x = 0$, then $\ln(0+1) = \ln 1 = 0$ and

$$\sum_{k=1}^{n} \frac{(-1)^{k+1} 0^k}{k} + \int_0^0 \frac{(-1)^n t^n}{t+1}\,dt = 0 + 0 = 0.$$

Now, let $x \in (0, \infty)$. We prove by induction on $n \in \mathbb{Z}^+$ that

$$\frac{1 - (-1)^n x^n}{x+1} = 1 - x + x^2 - x^3 + \cdots + (-1)^{n-1} x^{n-1}.$$

Base case ($k = 1$)**:** We have

$$\frac{1 - (-1)^1 x^1}{x+1} = \frac{1 - (-x)}{x+1} = \frac{1+x}{x+1} = 1.$$

Inductive step: Assume that $k \in \mathbb{Z}^+$ and

$$\frac{1 - (-1)^k x^k}{x+1} = 1 - x + x^2 - x^3 + \cdots + (-1)^{k-1} x^{k-1}$$

Then we have

$$1 - x + x^2 - x^3 + \cdots + (-1)^{k-1} x^{k-1} + (-1)^k x^k = \frac{1 - (-1)^k x^k}{x+1} + (-1)^k x^k$$
$$= \frac{1 - (-1)^k x^k}{x+1} + \frac{(-1)^k x^k (x+1)}{x+1} = \frac{1 - (-1)^k x^k + (-1)^k x^{k+1} + (-1)^k x^k}{x+1}$$
$$= \frac{1 + (-1)^k x^{k+1}}{x+1} = \frac{1 - (-1)^{k+1} x^{k+1}}{x+1}.$$

So, by the Principle of Mathematical Induction, for all $n \in \mathbb{Z}^+$,

$$\frac{1}{x+1} - \frac{(-1)^n x^n}{x+1} = \frac{1 - (-1)^n x^n}{x+1} = 1 - x + x^2 - x^3 + \cdots + (-1)^{n-1} x^{n-1},$$

or equivalently,

$$\frac{1}{x+1} = 1 - x + x^2 - x^3 + \cdots + (-1)^{n-1} x^{n-1} + \frac{(-1)^n x^n}{x+1}.$$

By the second form of the Fundamental Theorem of Calculus (Theorem 13.50), we have

$$\int_0^x \frac{1}{t+1}\, dt = \int_0^x \left(1 - t + t^2 - t^3 + \cdots + (-1)^{n-1} t^{n-1} + \frac{(-1)^n t^n}{t+1}\right) dt$$
$$= \left[t - \frac{t^2}{2} + \frac{t^3}{3} - \frac{t^4}{4} + \cdots + \frac{(-1)^{n-1} t^n}{n}\right]_0^x + \int_0^x \frac{(-1)^n t^n}{t+1}\, dt$$
$$= x - \frac{x^2}{2} + \frac{x^3}{3} - \frac{x^4}{4} + \cdots + \frac{(-1)^{n+1} x^n}{n} + \int_0^x \frac{(-1)^n t^n}{t+1}\, dt$$
$$= \sum_{k=1}^{n} \frac{(-1)^{k+1} x^k}{k} + \int_0^x \frac{(-1)^n t^n}{t+1}\, dt.$$

Now, by definition, we have

$$\ln x = \int_1^x \frac{1}{t}\,dt.$$

We make the change of variables $u = t - 1$. Then $du = dt$, $u = 0$ when $t = 1$ and $u = x - 1$ when $t = x$. So, we have

$$\ln x = \int_0^{x-1} \frac{1}{u+1}\,du.$$

Therefore, we have

$$\ln(x+1) = \int_0^x \frac{1}{u+1}\,du = \int_0^x \frac{1}{t+1}\,dt = \sum_{k=1}^{n} \frac{(-1)^{k+1}x^k}{k} + \int_0^x \frac{(-1)^n t^n}{t+1}\,dt.$$

□

Problem Set 14

LEVEL 1

Determine if each of the following statements is true or false. If true, provide a proof. If false, provide a counterexample.

1. Let $\mathcal{T}$ and $\mathcal{U}$ be topologies on a set S. Then $\mathcal{T} \cap \mathcal{U}$ is a topology on S.

True

Proof: Since $\mathcal{T}$ is a topology on S, $\emptyset \in \mathcal{T}$. Since $\mathcal{U}$ is a topology on S, $\emptyset \in \mathcal{U}$. Since $\emptyset \in \mathcal{T}$ and $\emptyset \in \mathcal{U}$, we have $\emptyset \in \mathcal{T} \cap \mathcal{U}$.

Since $\mathcal{T}$ is a topology on S, $S \in \mathcal{T}$. Since $\mathcal{U}$ is a topology on S, $S \in \mathcal{U}$. Since $S \in \mathcal{T}$ and $S \in \mathcal{U}$, we have $S \in \mathcal{T} \cap \mathcal{U}$.

Let $\boldsymbol{X} \subseteq \mathcal{T} \cap \mathcal{U}$. Then $\boldsymbol{X} \subseteq \mathcal{T}$ and $\boldsymbol{X} \subseteq \mathcal{U}$. Since $\boldsymbol{X} \subseteq \mathcal{T}$ and $\mathcal{T}$ is a topology on S, $\bigcup \boldsymbol{X} \in \mathcal{T}$. Since $\boldsymbol{X} \subseteq \mathcal{U}$ and $\mathcal{U}$ is a topology on S, $\bigcup \boldsymbol{X} \in \mathcal{U}$. Since $\bigcup \boldsymbol{X} \in \mathcal{T}$ and $\bigcup \boldsymbol{X} \in \mathcal{U}$, we have $\bigcup \boldsymbol{X} \in \mathcal{T} \cap \mathcal{U}$.

Let $\boldsymbol{Y} \subseteq \mathcal{T} \cap \mathcal{U}$ with $\boldsymbol{Y}$ finite. Then $\boldsymbol{Y} \subseteq \mathcal{T}$ and $\boldsymbol{Y} \subseteq \mathcal{U}$ Since $\boldsymbol{Y} \subseteq \mathcal{T}$ with $\boldsymbol{Y}$ finite and $\mathcal{T}$ is a topology on S, $\bigcap \boldsymbol{Y} \in \mathcal{T}$. Since $\boldsymbol{Y} \subseteq \mathcal{U}$ with $\boldsymbol{Y}$ finite and $\mathcal{U}$ is a topology on S, $\bigcap \boldsymbol{Y} \in \mathcal{U}$. Since $\bigcap \boldsymbol{Y} \in \mathcal{T}$ and $\bigcap \boldsymbol{Y} \in \mathcal{U}$, we have $\bigcap \boldsymbol{Y} \in \mathcal{T} \cap \mathcal{U}$.

Therefore, $\mathcal{T} \cap \mathcal{U}$ is a topology on S. □

2. Let $(S, \mathcal{T})$ be a topological space and let A be a subspace of S. If B is open in A in the subspace topology, then B is open in S.

False

Let $S = \mathbb{R}$ and let $\mathcal{T}$ be the standard topology on $\mathbb{R}$. Let $A = [0, 1]$ and let $B = \left[0, \frac{1}{2}\right)$. Then we have $B = \left(-1, \frac{1}{2}\right) \cap [0, 1]$. Therefore, B is open in the subspace topology of A. However, B is **not** open in the standard topology of $\mathbb{R}$.

3. Let $(S, \mathcal{T})$ be a topological space and let A be a subspace of S. If B is closed in A in the subspace topology, then B is closed in S.

False

Let $S = \mathbb{R}$ and let $\mathcal{T}$ be the standard topology on $\mathbb{R}$. Let $A = [0, 1)$ and let $B = \left[\frac{1}{2}, 1\right)$. Then we have $A \setminus B = \left[0, \frac{1}{2}\right) = \left(-1, \frac{1}{2}\right) \cap [0, 1)$. So, $A \setminus B$ is open in the subspace topology of A. Therefore, B is closed in the subspace topology of A. However, B is **not** closed in the standard topology of $\mathbb{R}$.

Prove each of the following:

4. Let $(S, \mathcal{T})$ be a T_1-space with $|S| \geq 2$ and let $\mathcal{B}$ be a basis for $(S, \mathcal{T})$. Then $\mathcal{B} \setminus \{S\}$ is a basis for $(S, \mathcal{T})$.

Proof: Let $(S, \mathcal{T})$ be a T_1-space with $|S| \geq 2$ and let $\mathcal{B}$ be a basis for $(S, \mathcal{T})$. We need to show that $\mathcal{B} \setminus \{S\}$ covers S. Let $x \in S$. Since $|S| \geq 2$, there is $y \in S$ with $x \neq y$. Since $(S, \mathcal{T})$ is a T_1-space, there is an open set U with $x \in U$ and $y \notin U$. Since U is a union of basis elements, there is $V \in \mathcal{B}$ with $x \in V$ and $V \subseteq U$. Since $y \notin U$, it follows that $y \notin V$. So, $V \neq S$. Since $x \in S$ was arbitrary, $\mathcal{B} \setminus \{S\}$ covers S. Therefore, $\mathcal{B} \setminus \{S\}$ is a basis for $(S, \mathcal{T})$. □

5. Let (S, d) be a metric space. For all $x, y \in S$, $d(x, y) \geq 0$.

Proof: Let (S, d) be a metric space and let $x, y \in S$. Then

$$2d(x,y) = d(x,y) + d(x,y) = d(x,y) + d(y,x) \geq d(x,x) = 0.$$

So, $d(x, y) \geq 0$, as desired. □

6. A sequence (s_n) is bounded in $\mathbb{R}$ (with the Euclidean metric) if and only if there is $M \in \mathbb{R}^+$ such that $|s_n| \leq M$ for all $n \in \mathbb{N}$.

Proof: Suppose that (s_n) is bounded in $\mathbb{R}$. Then there is $L \in \mathbb{R}^+$ such that for all $n, m \in \mathbb{N}$, $d(s_n, s_m) = |s_n - s_m| \leq L$. By the Triangle Inequality (and SACT), we have

$$|s_n| = |(s_n - s_0) + s_0| \leq |s_n - s_0| + |s_0| \leq L + |s_0|.$$

So, if we let $M = L + |s_0|$, then for all $m \in \mathbb{N}$, $|s_m| \leq M$.

Conversely, suppose that there is $M \in \mathbb{R}^+$ such that $|s_n| \leq M$ for all $n \in \mathbb{N}$. Then for all $n, m \in \mathbb{N}$,

$$d(s_n, s_m) = |s_n - s_m| \leq |s_n| + |s_m| \leq 2M.$$

So (s_n) is bounded in $\mathbb{R}$ with the Euclidean metric. □

7. Let $(S, \mathcal{T})$ be a topological space. A finite union of compact subspaces of S is compact.

Proof: Let $(S, \mathcal{T})$ be a topological space, and for each $k = 0, 1, \dots, n$, let $(A_k, \mathcal{T}_{A_k})$ be a compact subspace of $(S, \mathcal{T})$. Let $A = \bigcup\{A_k \mid k = 0, 1, \dots, n\}$. We will show that $(A, \mathcal{T}_A)$ is compact. To this end, let $\mathcal{C}$ be an open covering of A. Then for each $k = 0, 1, \dots, n$, $\mathcal{C}$ is an open covering of A_k by open sets in S. By Lemma 14.54, for each $k = 0, 1, \dots, n$, there is a finite subcollection $\mathcal{C}_k$ such that $\mathcal{C}_k$ still covers A_k. Then $\bigcup\{\mathcal{C}_k \mid k = 0, 1, \dots, n\}$ is a finite subcollection of $\mathcal{C}$ that covers A. By Lemma 14.54 again, $(A, \mathcal{T}_A)$ is compact. □

8. The Cantor set is nowhere dense in $\mathbb{R}$.

Proof: Since the Cantor set is closed, $\overline{C} = C$, and so we need only show that C has empty interior. Suppose toward contradiction that $U \subseteq C$ is an open interval. Let diam $U = d$. By the Archimedean Property of $\mathbb{R}$, we can find $n \in \mathbb{N}$ with $n > \frac{1}{d}$, or equivalently, $\frac{1}{n} < d$. Now, $C = \bigcap C_k$, where C_k is the union of 2^k closed intervals, each of diameter $\frac{1}{3^k}$. In particular, C_n is the union of 2^n closed intervals, each of diameter $\frac{1}{3^n}$. Since $3^n > n$, we have $\frac{1}{3^n} < \frac{1}{n} < d$. Therefore, U cannot be contained inside of one of these intervals. It follows that there is $x \in U$ with $x \notin C_n$. Since $C \subseteq C_n$, $x \notin C$. Therefore, U is **not** a subset of C, contrary to our assumption. It follows that C has empty interior. □

9. Let $(A, \mathcal{T})$ and $(B, \mathcal{U})$ be topological spaces and let $f: A \to B$. Then f is continuous if and only if for each C that is closed in B, the set $f^{-1}[C]$ is closed in A.

Proof: Let $(A, \mathcal{T})$ and $(B, \mathcal{U})$ be topological spaces and let $f: A \to B$. First assume that f is continuous. Let C be closed in B. Then $B \setminus C$ is open in B. Since f is continuous, $f^{-1}[B \setminus C]$ is open in A. We will show that $f^{-1}[B \setminus C] = A \setminus f^{-1}[C]$. To see this, note that $x \in f^{-1}[B \setminus C]$ if and only if $f(x) \in B \setminus C$ if and only if $f(x) \in B$ and $f(x) \notin C$ if and only if $x \in f^{-1}[B] = A$ and $x \notin f^{-1}[C]$ if and only if $x \in A \setminus f^{-1}[C]$. It follows that $A \setminus f^{-1}[C]$ is open in A, and therefore, $f^{-1}[C]$ is closed in A.

Conversely, assume that for each C that is closed in B, the set $f^{-1}[C]$ is closed in A. Let U be open in B. Then $B \setminus U$ is closed in B. By our assumption, $f^{-1}[B \setminus U]$ is closed in A. By the same argument given in the first paragraph above, we have $f^{-1}[B \setminus U] = A \setminus f^{-1}[U]$. So, $A \setminus f^{-1}[U]$ is closed in A. Therefore, $f^{-1}[U]$ is open in A. Since U was an arbitrary open set in B, f is continuous. □

LEVEL 2

Determine if each of the following statements is true or false. If true, provide a proof. If false, provide a counterexample.

10. Let $K = \left\{\frac{1}{n} \mid n \in \mathbb{Z}^+\right\}$ and $\mathcal{B} = \{(a, b) \mid a, b \in \mathbb{R} \land a < b\} \cup \{(a, b) \setminus K \mid a, b \in \mathbb{R} \land a < b\}$. Then $\mathcal{B}$ is a basis for a topology $\mathcal{T}_K$ on $\mathbb{R}$.

True

Proof: We already know that the bounded open intervals alone cover $\mathbb{R}$. So, $\mathcal{B}$ covers $\mathbb{R}$. By Problem 37 from Problem Set 6, the intersection of two open intervals with nonempty intersection is an open interval. Furthermore, we have $(a, b) \cap [(c, d) \setminus K] = [(a, b) \cap (c, d)] \setminus K$ and we have $[(a, b) \setminus K] \cap [(c, d) \setminus K] = [(a, b) \cap (c, d)] \setminus K$. This shows that $\mathcal{B}$ has the intersection containment property. Therefore, $\mathcal{B}$ is a basis for a topology $\mathcal{T}_K$ on $\mathbb{R}$. □

11. $\mathcal{B} = \{X \subseteq \mathbb{R} \mid \mathbb{R} \setminus X \text{ is finite}\}$ generates a topology $\mathcal{T}$ on $\mathbb{R}$ that is strictly coarser than the standard topology. $\mathcal{T}$ is called the **cofinite topology** on $\mathbb{R}$.

True

Proof: Let $x \in \mathbb{R}$. Then $X = \mathbb{R} \setminus \{x+1\} \in \mathcal{B}$ because $\mathbb{R} \setminus X = \mathbb{R} \setminus (\mathbb{R} \setminus \{x+1\}) = \{x+1\}$, which is finite and $x \in \mathbb{R} \setminus \{x+1\}$ because $x \neq x+1$. So, $\mathcal{B}$ covers $\mathbb{R}$. Let $x \in \mathbb{R}$, and let $X, Y \in \mathcal{B}$ with $x \in X \cap Y$. Then $\mathbb{R} \setminus X$ and $\mathbb{R} \setminus Y$ are both finite and $\mathbb{R} \setminus (X \cap Y) = (\mathbb{R} \setminus X) \cup (\mathbb{R} \setminus Y)$ (by De Morgan's Law) is the union of two finite sets, thus finite. It follows that $X \cap Y \in \mathcal{B}$. Therefore, $\mathcal{B}$ has the intersection containment property. Since $\mathcal{B}$ covers $\mathbb{R}$ and has the intersection containment property, $\mathcal{B}$ is a basis for a topology $\mathcal{T}$ on $\mathbb{R}$.

Note that the topology generated by $\mathcal{B}$ is simply $\mathcal{B} \cup \{\emptyset\}$. $\mathbb{R}$ is in the basis $\mathcal{B}$ because $\mathbb{R} \setminus \mathbb{R} = \emptyset$, which of course is finite. If $\boldsymbol{X} \subseteq \mathcal{B}$, then $\mathbb{R} \setminus \bigcup \boldsymbol{X} = \bigcap\{\mathbb{R} \setminus A \mid A \in \boldsymbol{X}\}$. This is an intersection of finite sets, which is finite. So, $\bigcup \boldsymbol{X} \in \mathcal{B}$. Finally, if $\boldsymbol{Y} \subseteq \mathcal{T}$ and $\boldsymbol{Y}$ is finite, then $\mathbb{R} \setminus \bigcap \boldsymbol{Y} = \bigcup\{\mathbb{R} \setminus A \mid A \in \boldsymbol{Y}\}$. This is a finite union of finite sets, which is finite. So, $\bigcap \boldsymbol{Y} \in \mathcal{B}$.

It follows that $\mathcal{T}$, the topology generated by $\mathcal{B}$, consists of only the cofinite sets together with the empty set.

Since $(0, 1)$ is open in the standard topology of $\mathbb{R}$ and is **not** cofinite, $(0, 1)$ is **not** open in the cofinite topology. Therefore, the cofinite topology on $\mathbb{R}$ is strictly coarser than the standard topology on $\mathbb{R}$. □

12. Let S be a nonempty set and let $\mathcal{T} = \{X \subseteq S \mid S \setminus X \text{ is countable}\} \cup \{\emptyset\}$. Then $\mathcal{T}$ is a topology on S.

True

Proof: $\emptyset \in \mathcal{T}$ by definition. $S \setminus S = \emptyset$, which is countable, and so, $S \in \mathcal{T}$.

Let $\boldsymbol{X} \subseteq \mathcal{T}$. Then $S \setminus A$ is countable for each $A \in \boldsymbol{X}$. Now, $S \setminus \bigcup \boldsymbol{X} = \bigcap\{S \setminus A \mid A \in \boldsymbol{X}\}$. Since this is countable, we have $\bigcup \boldsymbol{X} \in \mathcal{T}$.

Let $\boldsymbol{Y} \subseteq \mathcal{T}$ with $\boldsymbol{Y}$ finite. Then $S \setminus A$ is countable for each $A \in \boldsymbol{Y}$. Now, $S \setminus \bigcap \boldsymbol{Y} = \bigcup\{S \setminus A \mid A \in \boldsymbol{Y}\}$. This is a finite union of countable sets, which is countable. So, $\bigcap \boldsymbol{Y} \in \mathcal{T}$.

It follows that $\mathcal{T}$ is a topology on S. □

13. Let $\mathcal{X}$ be a set of topologies on a set S. Then $\bigcap \mathcal{X}$ is a topology on S.

True

Proof: Let $\mathcal{T} \in \mathcal{X}$. Since $\mathcal{T}$ is a topology on S, $\emptyset \in \mathcal{T}$ and $S \in \mathcal{T}$. Since $\mathcal{T} \in \mathcal{X}$ was arbitrary, we have $\emptyset \in \bigcap \mathcal{X}$ and $S \in \bigcap \mathcal{X}$.

Let $\boldsymbol{X} \subseteq \bigcap \mathcal{X}$. Then $\boldsymbol{X} \subseteq \mathcal{T}$ for each $\mathcal{T} \in \mathcal{X}$. So, $\bigcup \boldsymbol{X} \in \mathcal{T}$ for each $\mathcal{T} \in \mathcal{X}$. Therefore, $\bigcup \boldsymbol{X} \in \bigcap \mathcal{X}$.

Let $\boldsymbol{Y} \subseteq \bigcap \mathcal{X}$ with $\boldsymbol{Y}$ finite. Then $\boldsymbol{Y} \subseteq \mathcal{T}$ for each $\mathcal{T} \in \mathcal{X}$. So, $\bigcap \boldsymbol{Y} \in \mathcal{T}$ for each $\mathcal{T} \in \mathcal{X}$. It follows that $\bigcap \boldsymbol{Y} \in \bigcap \mathcal{X}$.

Therefore, $\bigcap \mathcal{X}$ is a topology on S. □

14. Let $(A, \mathcal{T}_A)$ be a compact subspace of a T_1-space $(S, \mathcal{T})$. Then A is closed in S. (Compare this problem with Theorem 14.55.)

False

Consider $(\mathbb{R}, \mathcal{T})$, where $\mathcal{T}$ is the cofinite topology on $\mathbb{R}$. By part 3 of Example 14.23, this is a T_1-space. If $A \subseteq \mathbb{R}$ is any subset, then $(A, \mathcal{T}_A)$ is compact. To see this, let $\mathcal{C}$ be an open covering of A by open sets in $\mathbb{R}$ and let A_0 be any set in $\mathcal{C}$. Then $S \setminus A_0$ is finite, say $S \setminus A_0 = \{a_1, a_2, \dots, a_n\}$. For each $i = 1, 2, \dots, n$, let $A_i \in \mathcal{C}$ with $a_i \in A_i$. Then the collection $\{A_0, A_1, A_2, \dots, A_n\}$ is a finite subcollection from $\mathcal{C}$ that covers S. By Lemma 14.54, $(A, \mathcal{T}_A)$ is compact.

Not every subset of $\mathbb{R}$ is closed in this topology. In fact, aside from $\mathbb{R}$ itself, the only closed sets in $\mathbb{R}$ are the finite sets. As a specific example, $(\mathbb{N}, \mathcal{T}_{\mathbb{N}})$ is compact by the previous paragraph, but it is not closed because $\mathbb{N}$ is infinite.

Prove each of the following:

15. Let $(S, \mathcal{T})$ be a T_1-space. $(S, \mathcal{T})$ is a T_3-space if and only if for every point $x \in S$ and every open set U containing x, there is an open set V containing x such that $\overline{V} \subseteq U$.

Proof: Suppose that $(S, \mathcal{T})$ is a T_3-space, let $x \in S$ and let U be an open set containing x. Then the set $A = S \setminus U$ is closed in S. Since $(S, \mathcal{T})$ is a T_3-space, there are disjoint open sets V and W such that $x \in V$ and $A \subseteq W$. If $y \in A$, then W is an open set disjoint from V with $y \in W$. So, $y \notin \overline{V}$. Therefore, $y \in \overline{V}$ implies that $y \notin A$, or equivalently, $y \in U$. Since $y \in \overline{V}$ was arbitrary, $\overline{V} \subseteq U$.

Conversely, let $(S, \mathcal{T})$ be a T_1-space such that for every point $x \in S$ and every open set U containing x, there is an open set V containing x such that $\overline{V} \subseteq U$. Let $x \in S$ and let A be a closed set in S not containing x. Then $U = S \setminus A$ is an open set containing x. So, by our assumption, there is an open set W containing x such that $\overline{W} \subseteq U$. Then W and $S \setminus \overline{W}$ are open sets, $x \in W$, $A \subseteq S \setminus \overline{W}$ (if $x \in A$, then $x \notin S \setminus A = U$, and so, $x \notin \overline{W}$). Also, $W \cap (S \setminus \overline{W}) \subseteq W \cap (S \setminus W) = \emptyset$. □

16. Let $(S, \mathcal{T})$ be a T_1-space. $(S, \mathcal{T})$ is a T_4-space if and only if for any closed set A in S and every open set U containing A, there is an open set V containing A such that $\overline{V} \subseteq U$.

Proof: Suppose that $(S, \mathcal{T})$ is a T_4-space, let A be closed in S, and let U be an open set containing A. Then the set $B = S \setminus U$ is closed in S. Since $(S, \mathcal{T})$ is a T_4-space, there are disjoint open sets V and W such that $A \subseteq V$ and $B \subseteq W$. If $y \in B$, then W is an open set disjoint from V with $y \in W$. So, $y \notin \overline{V}$. Therefore, $y \in \overline{V}$ implies that $y \notin B$, or equivalently, $y \in U$. Since $y \in \overline{V}$ was arbitrary, $\overline{V} \subseteq U$.

Conversely, let $(S, \mathcal{T})$ be a T_1-space such that for any closed $A \subseteq S$ and every open set U containing A, there is an open set V containing A such that $\overline{V} \subseteq U$. Let A and B be disjoint closed sets in S. Then $U = S \setminus B$ is an open set containing A. So, by our assumption, there is an open set W containing A such that $\overline{W} \subseteq U$. Then W and $S \setminus \overline{W}$ are open sets, $A \subseteq W$, $B \subseteq S \setminus \overline{W}$ (if $x \in B$, then $x \notin S \setminus B = U$, and so, $x \notin \overline{W}$). Also, $W \cap (S \setminus \overline{W}) \subseteq W \cap (S \setminus W) = \emptyset$. □

17. Let (S, d) be a metric space and let $\overline{d}$ be the standard bounded metric corresponding to d. Then $(S, \overline{d})$ is a metric space and $\overline{d}$ induces the same topology as d.

Proof: Let's check that the 3 properties of a metric space are satisfied for $(S, \overline{d})$. We have $\overline{d}(x, y) = 0$ if and only if $\min\{1, d(x, y)\} = 0$ if and only if $d(x, y) = 0$ if and only if $x = y$**.** This shows that Property 1 holds.

$d(x, y) = d(y, x)$ implies that $\overline{d}(x, y) = \min\{1, d(x, y)\} = \min\{1, d(y, x)\} = \overline{d}(y, x)$, and so, Property 2 holds.

For Property 3, let $x, y, z \in S$. If $d(x, z) < 1$, then

$$\overline{d}(x, z) = d(x, z) \leq d(x, y) + d(y, z),$$

$$\overline{d}(x, z) = d(x, z) \leq 1 \leq d(x, y) + 1,$$

$$\overline{d}(x, z) = d(x, z) \leq 1 \leq 1 + d(y, z),$$

$$\overline{d}(x, z) = d(x, z) \leq 1 \leq 1 + 1,$$

Therefore, $\overline{d}(x, z) \leq \min\{1, d(x, y)\} + \min\{1, d(y, z)\} = \overline{d}(x, y) + \overline{d}(y, z)$.

Otherwise $d(x, z) \geq 1$. In this case, $\overline{d}(x, z) = 1$. If $d(x, y) > 1$ or $d(y, z) > 1$, then $\overline{d}(x, y) = 1$ or $\overline{d}(y, x) = 1$, and so, $\overline{d}(x, y) + \overline{d}(y, z) \geq 1$, and so, $\overline{d}(x, z) = 1 \leq \overline{d}(x, y) + \overline{d}(y, z)$.

Otherwise, we have $\overline{d}(x, z) = 1$, $d(x, y) \leq 1$, and $d(y, z) \leq 1$. So, $\overline{d}(x, y) = d(x, y)$ and $\overline{d}(y, z) = d(y, z)$. Suppose toward contradiction that $\overline{d}(x, z) > \overline{d}(x, y) + \overline{d}(y, z)$. Then we have $d(x, z) \geq 1 > d(x, y) + d(y, z)$, contradicting that (S, d) is a metric space.

Let $B_r(a; \overline{d})$ be an arbitrary open ball in $(S, \overline{d})$. We will show that $B_r(a; d) \subseteq B_r(a; \overline{d})$. To see this, let $x \in B_r(a; d)$. Then $d(x, a) < r$. So, $\overline{d}(x, a) \leq d(x, a) < r$. Thus, $x \in B_r(a; \overline{d})$. Since $x \in B_r(a; d)$ was arbitrary, it follows that $B_r(a; d) \subseteq B_r(a; \overline{d})$. So, the topology induced by d is finer than the topology induced by $\overline{d}$.

Now, let $B_r(a; d)$ be an arbitrary open ball in (S, d) and let $s = \min\{r, 1\}$. We will show that $B_s(a; \overline{d}) \subseteq B_r(a; d)$. To see this, let $x \in B_s(a; \overline{d})$. Then $\overline{d}(x, a) < s$. Since $s \leq 1$, $\overline{d}(x, a) < 1$. It follows that $\overline{d}(x, a) = d(x, a)$. Therefore, $d(x, a) < s \leq r$. So, $x \in B_r(a; d)$. Since $x \in B_s(a; \overline{d})$ was arbitrary, it follows that $B_s(a; \overline{d}) \subseteq B_r(a; d)$. So, the topology induced by $\overline{d}$ is finer than the topology induced by d.

Since the topology induced by d is finer than the topology induced by $\overline{d}$ and the topology induced by $\overline{d}$ is finer than the topology induced by d, d and $\overline{d}$ induce the same topology. □

18. Prove that the following are equivalent for a topological space $(S, \mathcal{T})$:

(i) $(S, \mathcal{T})$ is compact.

(ii) Every covering of S by basic open sets contains a finite subcollection that still covers S.

(iii) For every collection $\mathcal{C}$ of basic closed sets in S with the finite intersection property, we have $\bigcap \mathcal{C} \neq \emptyset$.

Proof: (i → ii) Suppose that $(S, \mathcal{T})$ is compact and let $\mathcal{C}$ be a covering of S by basic open sets. Then $\mathcal{C}$ is a covering of S by open sets. Since $(S, \mathcal{T})$ is compact, there is a finite subcollection of $\mathcal{C}$ that still covers S.

(ii → iii) Let $(S, \mathcal{T})$ be a topological space and suppose that there is a collection $\mathcal{C}$ of basic closed sets in S with the finite intersection property such that $\bigcap \mathcal{C} = \emptyset$. Let $\mathcal{D} = \{S \setminus B \mid B \in \mathcal{C}\}$. Then $\mathcal{D}$ is a collection of basic open sets in S. Since $\bigcap \mathcal{C} = \emptyset$, we have $\bigcup \mathcal{D} = S \setminus \bigcap \mathcal{C} = S \setminus \emptyset = S$ (by De Morgan's law). Let $\mathcal{E}$ be an arbitrary finite subcollection of elements in $\mathcal{D}$, say $\mathcal{E} = \{S \setminus B_1, S \setminus B_2, \dots, S \setminus B_k\}$. Then $\{B_1, B_2, \dots, B_k\}$ is a finite subcollection of $\mathcal{C}$. Since $\mathcal{C}$ has the finite intersection property, there is $x \in B_1 \cap B_2 \cap \cdots \cap B_k$. So, $x \notin S \setminus (B_1 \cap B_2 \cap \cdots \cap B_k) = (S \setminus B_1) \cup (S \setminus B_2) \cup \cdots \cup (S \setminus B_k)$. It follows that $\mathcal{E}$ does not cover S. Since $\mathcal{E}$ was an arbitrary finite subcollection of $\mathcal{D}$, we have found a basic open covering of S without a finite subcollection that covers S.

(iii → ii) Suppose there is a basic open covering $\mathcal{C}$ of S such that no finite subcollection of $\mathcal{C}$ covers S. Let $\mathcal{D} = \{S \setminus U \mid U \in \mathcal{C}\}$. Then $\mathcal{D}$ is a collection of basic closed sets. Since $\bigcup \mathcal{C} = S$, by De Morgan's law, we have $\bigcap \mathcal{D} = S \setminus \bigcup \mathcal{C} = S \setminus S = \emptyset$. Let $\mathcal{E}$ be a finite subcollection of elements in $\mathcal{D}$, say $\mathcal{E} = \{S \setminus U_1, S \setminus U_2, \dots, S \setminus U_k\}$. Then $\{U_1, U_2, \dots, U_k\}$ is a finite subcollection of $\mathcal{C}$. Since $\mathcal{C}$ does not have a finite subcollection that covers S, there is $x \in S$ such that $x \notin U_1 \cup U_2 \cup \cdots \cup U_k$. So, $x \in S \setminus (U_1 \cup U_2 \cup \cdots \cup U_k) = (S \setminus U_1) \cap (S \setminus U_2) \cap \cdots \cap (S \setminus U_k)$. Since $\mathcal{E}$ was an arbitrary finite subcollection of $\mathcal{D}$, we see that $\mathcal{D}$ has the finite intersection property. And yet, we saw that $\bigcap \mathcal{D} = \emptyset$.

(ii → i) Let $(S, \mathcal{T})$ be a topological space such that every covering of S by basic open sets contains a finite subcollection that still covers S. Let $\mathcal{C}$ be an open covering of S. For each $x \in S$ and $U \in C$, let U_x be a basic open set containing x with $U_x \subseteq U$. Then $\mathcal{D} = \{U_x \mid x \in S \wedge U \in \mathcal{C}\}$ is a covering of S by basic open sets. By (ii), there is a finite subcollection $\mathcal{E}$ of $\mathcal{D}$ that still covers S. Let $\mathcal{F} = \{U \mid U_x \in \mathcal{E}\}$. Then $\mathcal{F}$ is a finite subcollection of $\mathcal{C}$ that covers S. Since $\mathcal{C}$ was an arbitrary open covering of S, $(S, \mathcal{T})$ is compact. □

19. $\mathbb{C}^n$ and $\mathbb{R}^{2n}$ with their standard topologies are topologically equivalent.

Proof: Define $f: \mathbb{C}^n \to \mathbb{R}^{2n}$ by $f(a_1 + b_1 i, a_2 + b_2 i, \dots, a_n + b_n i) = (a_1, b_1, a_2, b_2, \dots, a_n, b_n)$.

If

$$f(a_1 + b_1 i, a_2 + b_2 i, \dots, a_n + b_n i) = f(c_1 + d_1 i, c_2 + d_2 i, \dots, c_n + d_n i),$$

then we have $(a_1, b_1, a_2, b_2, \dots, a_n, b_n) = (c_1, d_1, c_2, d_2, \dots, c_n, d_n)$. So, $a_1 = c_1$, $b_1 = d_1$,...,$b_n = d_n$. Therefore, $(a_1 + b_1 i, a_2 + b_2 i, \dots, a_n + b_n i) = (c_1 + d_1 i, c_2 + d_2 i, \dots c_n + d_n i)$. It follows that f is injective.

If $(x_1, x_2, \ldots, x_{2n}) \in \mathbb{R}^{2n}$, then $f(x_1 + x_2 i, x_3 + x_4 i, \ldots, x_{2n-1} + x_{2n} i) = (x_1, x_2, \ldots, x_{2n})$. This shows that f is surjective.

f is continuous because $f^{-1}[B_r(x_1, x_2, \ldots, x_{2n})] = B_r(x_1 + x_2 i, x_3 + x_4 i, \ldots, x_{2n-1} + x_{2n} i)$ for any open ball $B_r(x_1, x_2, \ldots, x_{2n})$ in $\mathbb{R}^{2n}$.

Similarly, f^{-1} is continuous because $f[B_r(x_1 + y_1 i, x_2 + y_2 i, \ldots, x_n + y_n i)] = B_r(x_1, y_1, \ldots, x_n, y_n)$. □

20. Let $(A, \mathcal{T})$ and $(B, \mathcal{U})$ be topological spaces, let $x \in A$, and let $f: A \to B$. If f is continuous at x, then then for every sequence x_n converging to x, the sequence $f(x_n)$ converges to $f(x)$. Furthermore, if $(A, \mathcal{T})$ is a first-countable space, then the converse of this statement holds.

Proof: Let $(A, \mathcal{T})$ and $(B, \mathcal{U})$ be topological spaces, let $x \in A$, and let $f: A \to B$. First assume that f is continuous at x. Let (x_n) be a sequence in A such that $x_n \to x$. Let V be an open set containing $f(x)$. Since f is continuous at x there is an open set U containing x such that $f[U] \subseteq V$. Since $x_n \to x$, there is $K \in \mathbb{N}$ such that $n > K$ implies $x_n \in U$. So, $n > K$ implies $f(x_n) \in f[U]$. Since $f[U] \subseteq V$, it follows that $n > K$ implies $f(x_n) \in V$. Therefore, $f(x_n) \to f(x)$.

Conversely, suppose that $(A, \mathcal{T})$ is first countable and f is **not** continuous at x. Then there is an open set V with $f(x) \in V$ such that for any open set U containing x, $f[U] \nsubseteq V$. Let $\mathcal{B} = \{U_n \mid n \in \mathbb{N}\}$ be a countable basis at x. For each $n \in \mathbb{N}$, let $W_n = \bigcap\{U_k \mid k \leq n\}$. Let $n \in \mathbb{N}$. Since $x \in U_k$ for each $k = 0, 1, \ldots, n$, $x \in W_n$. Since W_n is a finite intersection of open sets, it is open. For each $n \in \mathbb{N}$, let $x_n \in W_n$ with $f(x_n) \notin V$ (we can find x_n because $f[W_n] \nsubseteq V$). As in the proof of Theorem 14.35, $x_n \to x$. Since $f(x_n) \notin V$ for all $n \in \mathbb{N}$, $f(x_n) \nrightarrow f(x)$. □

21. Let $(A, \mathcal{T})$ and $(B, \mathcal{U})$ be metrizable topological spaces. Let $f, g: A \to B$ be continuous and let $C \subseteq A$. Suppose that $f(x) = g(x)$ for all $x \in C$. Then $f(x) = g(x)$ for all $x \in \overline{C}$.

Proof: Let $x \in \overline{C}$. Then there is a sequence (x_n) in C such that $x_n \to x$ (simply choose $x_n \in B_{\frac{1}{n}}(x)$ for each $n \in \mathbb{N}$). Since f is continuous, $f(x_n) \to f(x)$. Since g is continuous, $f(x_n) = g(x_n) \to g(x)$. So, $f(x_n)$ converges to both $f(x)$ and $g(x)$. Since metrizable spaces are T_2-spaces, by Theorem 14.30, $f(x) = g(x)$.

22. A topological space $(S, \mathcal{T})$ is disconnected if and only if there is a surjective continuous function $f: S \to \{0, 1\}$, where $\{0, 1\}$ is given the discrete topology.

Proof: First assume that $(S, \mathcal{T})$ is disconnected and let (U, V) be a disconnection of S. Define $f: S \to \{0, 1\}$ by $f(x) = \begin{cases} 0 & \text{if } x \in U. \\ 1 & \text{if } x \in V. \end{cases}$ Since U and V are nonempty, f is surjective. $f^{-1}[\{0\}] = U$, which is open in S and $f^{-1}[\{1\}] = V$, which is also open in S. So, f is continuous.

Conversely, suppose that $f: S \to \{0, 1\}$ is surjective and continuous. Let $U = f^{-1}[\{0\}]$ and $V = f^{-1}[\{1\}]$. Since f is continuous and $\{0\}$ and $\{1\}$ are open in the discrete topology of $\{0, 1\}$, U and V are both open in S. Since f is surjective, U and V are both nonempty. If $x \in U \cap V$, then $f(x) = 0$ and $f(x) = 1$, which is a contradiction. So, $U \cap V = \emptyset$. Finally, if $x \in S$, then $f(x) \in \{0, 1\}$. If $f(x) = 0$, then $x \in U$. If $f(x) = 1$, then $x \in V$. So, $x \in U$ or $x \in V$. Therefore, $x \in U \cup V$. Since $x \in S$ was arbitrary, $S = U \cup V$. It follows that (U, V) is a disconnection of S. So, $(S, \mathcal{T})$ is disconnected. □

23. Let $(S, \mathcal{T})$ be a topological space and define the relation $\sim$ on S by $x \sim y$ if and only if $x, y \in A$ for some connected subspace $(A, \mathcal{T}_A)$ of $(S, \mathcal{T})$. Then $\sim$ is an equivalence relation on S.

Proof: Let $x \in S$. Then $(\{x\}, \mathcal{T}_{\{x\}})$ is a connected subspace of $(S, \mathcal{T})$ with $x \in A$. So, $x \sim x$, and therefore, $\sim$ is reflexive.

Let $x, y \in S$ with $x \sim y$. Then $x, y \in A$ for some connected subspace $(A, \mathcal{T}_A)$ of $(S, \mathcal{T})$. So, $y, x \in A$ for some connected subspace $(A, \mathcal{T}_A)$ of $(S, \mathcal{T})$. Thus, $y \sim x$, and so, $\sim$ is symmetric.

Let $x, y, z \in S$ with $x \sim y$ and $y \sim z$. Then there are connected subspaces $(A, \mathcal{T}_A)$ and $(B, \mathcal{T}_B)$ of $(S, \mathcal{T})$ such that $x, y \in A$ and $y, z \in B$. Since $y \in A \cap B$, by Theorem 14.85, $(A \cup B, \mathcal{T}_{A \cup B})$ is a connected subspace of $(S, \mathcal{T})$ with $x, y, z \in A \cup B$. So, $x \sim z$, and therefore, $\sim$ is transitive.

Since $\sim$ is reflexive, symmetric, and transitive, $\sim$ is an equivalence relation. □

Let $d: \mathbb{R} \times \mathbb{R} \to \mathbb{R}$ be the Euclidean metric on $\mathbb{R}$, let $\overline{d}: \mathbb{R} \times \mathbb{R} \to \mathbb{R}$ be the standard bounded metric corresponding to d, and let $d^*: {}^{\mathbb{N}}\mathbb{R} \times {}^{\mathbb{N}}\mathbb{R} \to \mathbb{R}$ be the metric on $\mathbb{R}^n$ defined by

$$d^*(\mathbf{x}, \mathbf{y}) = \sup\left\{\frac{\overline{d}(x_k, y_k)}{k+1} \,\middle|\, k \in \mathbb{N}\right\} \text{ (see Theorem 14.48).}$$

Prove each of the following:

24. (s_n) is a Cauchy sequence in $(\mathbb{R}, d)$ if and only if (s_n) is a Cauchy sequence in $(\mathbb{R}, \overline{d})$.

Proof: Suppose that (s_n) is a Cauchy sequence in $(\mathbb{R}, d)$ and let $r \in \mathbb{R}^+$. Then there is $K \in \mathbb{N}$ such that $m \geq n > K$ implies $d(s_m, s_n) < r$. Since $\overline{d}(s_m, s_n) = \min\{1, d(s_m, s_n)\} \leq d(s_m, s_n)$, we have that $m \geq n > K$ implies $\overline{d}(s_m, s_n) < r$. It follows that (s_n) is a Cauchy sequence in $(\mathbb{R}, \overline{d})$.

Conversely, suppose that (s_n) is a Cauchy sequence in $(\mathbb{R}, \overline{d})$ and let $r \in \mathbb{R}^+$. Let $t = \min\{r, 1\}$. Then there is $K \in \mathbb{N}$ such that $m \geq n > K$ implies $\overline{d}(s_m, s_n) < t$. Since $t \leq 1$, it follows that for $m \geq n > K$, $\overline{d}(s_m, s_n) = d(s_m, s_n)$. So, $m \geq n > K$ implies $d(s_m, s_n) < t \leq r$. It follows that (s_n) is a Cauchy sequence in $(\mathbb{R}, d)$. □

25. If $(\mathbf{x}^n)$ is a Cauchy sequence in $({}^{\mathbb{N}}\mathbb{R}, d^*)$, then for each $k \in \mathbb{N}$, the sequence (x_k^n) is a Cauchy sequence in $(\mathbb{R}, d)$.

Proof: Suppose that $(\mathbf{x}^n)$ is a Cauchy sequence in $({}^{\mathbb{N}}\mathbb{R}, d^*)$, let $k \in \mathbb{N}$, let $r \in \mathbb{R}^+$, and let $t = \min\{r, 1\}$. Then there is $K \in \mathbb{N}$ such that $m \geq n > K$ implies $d^*(\mathbf{x^m}, \mathbf{x^n}) < \frac{t}{k}$. So, $m \geq n > K$ implies $\sup\left\{\frac{\overline{d}(x_j^m, x_j^n)}{j} \,\middle|\, j \in \mathbb{N}\right\} < \frac{t}{k}$. So, $\frac{\overline{d}(x_k^m, x_k^n)}{k} \leq \sup\left\{\frac{\overline{d}(x_j^m, x_j^n)}{j} \,\middle|\, j \in \mathbb{N}\right\} < \frac{t}{k}$. Thus, $\overline{d}(x_k^m, y_k^n) < t$. Since $t \leq 1$, it follows that for $m \geq n > K$, $\overline{d}(x_k^m, x_k^n) = d(x_k^m, x_k^n)$. So, $m \geq n > K$ implies $d(x_k^m, x_k^n) < t \leq r$. It follows that (x_k^n) is a Cauchy sequence in $(\mathbb{R}, d)$. □

LEVEL 3

Let $(S, \mathcal{T})$ be a topological space and let A be a subspace of S. Prove each of the following:

26. If B is open in A in the subspace topology and A is open in S, then B is open in S.

Proof: Suppose that B is open in A in the subspace topology and A is open in S. Since B is open in A, there is $U \in \mathcal{T}$ such that $B = U \cap A$. Since A is open in S, we have $A \in \mathcal{T}$. Since $U \in \mathcal{T}$, $A \in \mathcal{T}$, and $\mathcal{T}$ is a topology on S, $B = U \cap A \in \mathcal{T}$. So, B is open in S. □

27. B is closed in A if and only if $B = C \cap A$ for some closed set C of S.

Proof: First suppose that B is closed in A. Then $A \setminus B$ is open in A. Therefore, there is $U \in \mathcal{T}$ such that $A \setminus B = U \cap A$. Let $C = S \setminus U$. Then C is closed in S. We now show that $B = C \cap A$. To see this, first let $x \in B$. Then $x \notin A \setminus B$. So, $x \notin U \cap A$. Since $x \in A$ (because $B \subseteq A$), $x \notin U$. So, $x \in S \setminus U = C$. Since $x \in C$ and $x \in A$, we have $x \in C \cap A$. Next, let $x \in C \cap A$. Then $x \in C$ and $x \in A$. Since $x \in C$ and $C = S \setminus U$, $x \notin U$. So, we have $x \in A$ and $x \notin U$. It follows that $x \in A \setminus U$. So, $x \notin U \cap A = A \setminus B$. Since $x \in A$ and $x \notin A \setminus B$, we have $x \in B$. Therefore, $B = C \cap A$, as desired.

Conversely, suppose that $B = C \cap A$ for some closed set C of S. Let $U = S \setminus C$. Then U is open in S. We now show that $A \setminus B = U \cap A$. To see this, first let $x \in A \setminus B$. Then $x \in A$ and $x \notin B$. Since, $x \in A$ and $x \notin B = C \cap A$, we have $x \notin C$. So, $x \in S \setminus C = U$. Since $x \in U$ and $x \in A$, we have $x \in U \cap A$. Next, let $x \in U \cap A$. Then $x \in U = S \setminus C$ and $x \in A$. So, $x \notin C$. Since $x \notin C$ and $C \cap A \subseteq C$, we have $x \notin C \cap A = B$. Since $x \in A$ and $x \notin B$, we have $x \in A \setminus B$. Since $A \setminus B = U \cap A$ with U open in S, $A \setminus B$ is open in A. Therefore, B is closed in A. □

28. If B is closed in A in the subspace topology and A is closed in S, then B is closed in S.

Proof: Suppose that B is closed in A in the subspace topology and A is closed in S. Since B is closed in A, by Problem 27 there is a closed set C of S such that $B = C \cap A$. Since C and A are closed in S, $S \setminus C$ and $S \setminus A$ are open in S. Since $S \setminus C \in \mathcal{T}$, $S \setminus A \in \mathcal{T}$, and $\mathcal{T}$ is a topology on S, we have that $S \setminus B = S \setminus (C \cap A) = (S \setminus C) \cup (S \setminus A) \in \mathcal{T}$ (by De Morgan's Law). So, $S \setminus B$ is open in S, and therefore, B is closed in S. □

29. If $(S, \mathcal{T})$ is a T_4-space and $(C, \mathcal{T}_C)$ is a closed subspace of $(S, \mathcal{T})$, then $(C, \mathcal{T}_C)$ is a T_4-space.

Proof: Let $(S, \mathcal{T})$ be a T_4-space, let $(C, \mathcal{T}_C)$ be a closed subspace of $(S, \mathcal{T})$, and let A, B be disjoint closed subsets of C. Since A is closed in C and C is closed in S, by Problem 28, A is closed in S. Similarly, B is closed in S. Since $(S, \mathcal{T})$ be a T_4-space, there are open sets U and V with $A \subseteq U$, $B \subseteq V$, and $U \cap V = \emptyset$. $U \cap C$ and $V \cap C$ are open in the subspace topology, $A \subseteq U \cap C$, $B \subseteq V \cap C$, and $(U \cap C) \cap (V \cap C) = (U \cap V) \cap C = \emptyset \cap C = \emptyset$. □

Prove each of the following:

30. Let $(S, \mathcal{T})$ be a topological space and let X and Y be subsets of S. Then $\overline{X \cup Y} = \overline{X} \cup \overline{Y}$.

Proof: Let X and Y be subsets of S and let $x \in \overline{X \cup Y}$. By part 3 of Theorem 14.4, $x \in X \cup Y$ or x is an accumulation point of $X \cup Y$. If $x \in X \cup Y$, then $x \in X$ or $x \in Y$. By part 1 of Theorem 14.4, $X \subseteq \overline{X}$ and $Y \subseteq \overline{Y}$. So, $x \in \overline{X}$ or $x \in \overline{Y}$. Thus, $x \in \overline{X} \cup \overline{Y}$. Now, suppose that x is an accumulation point of $X \cup Y$, but x is **not** an accumulation point of Y. Then there is an open set U with $x \in U$ and such that if $y \in Y$ with $y \neq x$, then $y \notin U$. Let V be any open set containing x and let $W = U \cap V$. Since $x \in W$, W is a nonempty open set. Since x is an accumulation point of $X \cup Y$, there is $y \in X \cup Y$ with $y \in W$ and $y \neq x$. Since $W \subseteq U$, $y \notin Y$. Therefore, $y \in X$. Since $W \subseteq V$, we have $y \in V$. It follows that x is an accumulation point of X. So, $x \in \overline{X}$. Therefore, $x \in \overline{X}$ or $x \in \overline{Y}$, and so, $x \in \overline{X} \cup \overline{Y}$. Since $x \in \overline{X \cup Y}$ was arbitrary, we have $\overline{X \cup Y} \subseteq \overline{X} \cup \overline{Y}$.

Conversely, let $x \in \overline{X} \cup \overline{Y}$. Then $x \in \overline{X}$ or $x \in \overline{Y}$. Without loss of generality, assume that $x \in \overline{X}$. By part 3 of Theorem 14.4, $x \in X$ or x is an accumulation point of X. If $x \in X$, then $x \in X$ or $x \in Y$. So, $x \in X \cup Y$. By part 1 of Theorem 14.4, $X \cup Y \subseteq \overline{X \cup Y}$. So, $x \in \overline{X \cup Y}$. Now, suppose that x is an accumulation point of X and let U be any open set containing x. Then there is $y \in X$ with $y \in U$ and $y \neq x$. So, $y \in X$ or $y \in Y$, and thus, $y \in X \cup Y$. It follows that x is an accumulation point of $X \cup Y$. Therefore, $x \in \overline{X \cup Y}$. Since $x \in \overline{X} \cup \overline{Y}$ was arbitrary, $\overline{X} \cup \overline{Y} \subseteq \overline{X \cup Y}$.

Since $\overline{X \cup Y} \subseteq \overline{X} \cup \overline{Y}$ and $\overline{X} \cup \overline{Y} \subseteq \overline{X \cup Y}$, we have $\overline{X \cup Y} = \overline{X} \cup \overline{Y}$. □

31. Let $(S, \mathcal{T})$ be a topological space and let $A \subseteq B$ with A and B be subspaces of S. The subspace topology of A with respect to S is the same as the subspace topology of A with respect to B.

Proof: Let U be open in the subspace topology of A with respect to S. Then $U = V \cap A$ for some set V that is open in S. Since V is open in S, $W = V \cap B$ is open in B. We have

$$W \cap A = (V \cap B) \cap A = V \cap (B \cap A) = V \cap A = U.$$

Therefore, U is open in the subspace topology of A with respect to B.

Conversely, let U be open in the subspace topology of A with respect to B. Then $U = V \cap A$ for some set V that is open in B. Then $V = W \cap B$ for some set W that is open in S. We have

$$U = V \cap A = (W \cap B) \cap A = W \cap (B \cap A) = W \cap A.$$

Therefore, U is open in the subspace topology of A with respect to S. □

32. Let S be a nonempty set and let $\mathcal{B}$ be a collection of subsets of S. Then the set generated by $\mathcal{B}$, $\{\bigcup\boldsymbol{X} \mid \boldsymbol{X} \subseteq \mathcal{B}\}$, is equal to $\{A \subseteq S \mid \forall x \in A\, \exists B \in \mathcal{B}(x \in B \wedge B \subseteq A)\}$.

Proof: Let $\mathcal{C} = \{\bigcup\boldsymbol{X} \mid \boldsymbol{X} \subseteq \mathcal{B}\}$ and let $\mathcal{D} = \{A \subseteq S \mid \forall x \in A\, \exists B \in \mathcal{B}(x \in B \wedge B \subseteq A)\}$. First, let $A \in \mathcal{C}$. Then there is $\boldsymbol{X} \subseteq \mathcal{B}$ such that $A = \bigcup\boldsymbol{X}$. Let $x \in A$. Then there is a $B \in \boldsymbol{X}$ with $x \in B$. Since $\boldsymbol{X} \subseteq \mathcal{B}$, $B \in \mathcal{B}$. Also, since $B \in \boldsymbol{X}$, $B \subseteq \bigcup\boldsymbol{X} = A$. Therefore, $A \in \mathcal{D}$. Since $A \in \mathcal{C}$ was arbitrary, $\mathcal{C} \subseteq \mathcal{D}$.

Now, let $A \in \mathcal{D}$. For each $x \in A$, there is $B_x \in \mathcal{B}$ such that $x \in B_x$ and $B_x \subseteq A$. If $y \in A$, then $y \in B_y$. So, $y \in \bigcup\{B_x \mid x \in A\}$. So, $A \subseteq \bigcup\{B_x \mid x \in A\}$. If $y \in \bigcup\{B_x \mid x \in A\}$, then $y \in B_x$ for some $x \in A$. Since $B_x \subseteq A$, $y \in A$. Therefore, $\bigcup\{B_x \mid x \in A\} \subseteq A$. It follows that $A = \bigcup\{B_x \mid x \in A\}$. In other words, we have $A = \bigcup\boldsymbol{X}$, where $\boldsymbol{X} = \{B_x \mid x \in A\}$. Since $A \in \mathcal{D}$ was arbitrary, $\mathcal{D} \subseteq \mathcal{C}$.

Since $\mathcal{C} \subseteq \mathcal{D}$ and $\mathcal{D} \subseteq \mathcal{C}$, we have $\mathcal{C} = \mathcal{D}$. □

33. Let $(S, \mathcal{T})$ be a T_2-space and $A \subseteq S$. Then $(A, \mathcal{T}_A)$ is a $\mathcal{T}_2$-space (Recall that $\mathcal{T}_A$ is the subspace topology on A). Determine if the analogous statement is true for T_3-spaces.

Proof: Let $x, y \in A$ with $x \neq y$. Since $(S, \mathcal{T})$ is a T_2-space, there are $U, V \in \mathcal{T}$ with $x \in U$, $y \in V$, and $U \cap V = \emptyset$. Since $x \in A$ and $x \in U$, $x \in A \cap U$. Since $y \in A$ and $y \in V$, $y \in A \cap V$. By the definition of $\mathcal{T}_A$, $A \cap U$ and $A \cap V$ are in $\mathcal{T}_A$. Finally, $(A \cap U) \cap (A \cap V) = A \cap (U \cap V) = A \cap \emptyset = \emptyset$. Since $x, y \in A$ were arbitrary, $(A, \mathcal{T}_A)$ is a T_2-space.

Let $(S, \mathcal{T})$ be a T_3-space and $A \subseteq S$. We will show that $(A, \mathcal{T}_A)$ is a $\mathcal{T}_3$-space.

Since $(S, \mathcal{T})$ is a T_3-space, $(S, \mathcal{T})$ is also a T_2-space. By the first paragraph, $(A, \mathcal{T}_A)$ is a T_2-space. Since every T_2-space is a T_1-space, $(A, \mathcal{T}_A)$ is a T_1-space.

Let $x \in A$ and $B \subseteq A \setminus \{x\}$ with B closed in $\mathcal{T}_A$. By definition, $A \setminus B$ is open in $\mathcal{T}_A$. So, there is $C \in \mathcal{T}$ with $A \setminus B = A \cap C$. Since $C \in \mathcal{T}$, $S \setminus C$ is closed in $(S, \mathcal{T})$. Since $x \in A \setminus B$, $x \in A \cap C$. So, $x \notin S \setminus C$. Since $(S, \mathcal{T})$ is a T_3-space, there are open sets $U, V \in \mathcal{T}$ with $x \in U, S \setminus C \subseteq V$, and $U \cap V = \emptyset$. Since $x \in A$ and $x \in U$, $x \in A \cap U$. Let $b \in B$. Then since $B \subseteq A \setminus \{x\}$, $b \in A$. Since $b \notin A \setminus B$, $b \notin A \cap C$. Since $b \in A$, $b \notin C$. So, $b \in S \setminus C$. Since $b \in B$ was arbitrary, $B \subseteq S \setminus C$. Since $S \setminus C \subseteq V$, $B \subseteq V$. Since $B \subseteq A$ and $B \subseteq V$, $B \subseteq A \cap V$. Finally, $(A \cap U) \cap (A \cap V) = A \cap (U \cap V) = A \cap \emptyset = \emptyset$. Since $x \in A$ was arbitrary and $B \subseteq A \setminus \{x\}$ was an arbitrary closed set, $(A, \mathcal{T}_A)$ is a T_3-space. □

34. Let $(S, \mathcal{T})$ be a topological space and let $D = \{(s, s) \mid s \in S\}$. Then $(S, \mathcal{T})$ is a T_2-space if and only if D is closed in $S \times S$ in the product topology.

Proof: First assume that $(S, \mathcal{T})$ is a T_2-space. We will prove that $(S \times S) \setminus D$ is open in $S \times S$. Let $(x, y) \in (S \times S) \setminus D$. Then $x \neq y$. Since $(S, \mathcal{T})$ is a T_2-space, there are open sets U and V with $x \in U$, $y \in V$, and $U \cap V = \emptyset$. So, $(x, y) \in U \times V$ and $(U \times V) \cap D = \emptyset$. Therefore, $U \times V \subseteq (S \times S) \setminus D$.

Conversely, assume that D is closed in $S \times S$ in the product topology. Let $x, y \in S$ with $x \neq y$. Then $(x, y) \notin D$, and so, (x, y) is in the open set $(S \times S) \setminus D$. Therefore, there is a basic open set $U \times V$ in the product topology such that $(x, y) \in U \times V$ and $U \times V \subseteq (S \times S) \setminus D$. Note that U and V are open in S, $x \in U$, and $y \in V$. If $z \in U \cap V$, then $(z, z) \in U \times V$. Since $U \times V \subseteq (S \times S) \setminus D$, $(z, z) \notin D$, which is clearly not true. So, $U \cap V = \emptyset$. □

35. Define the functions d_1 and d_2 from $\mathbb{C} \times \mathbb{C}$ to $\mathbb{R}$ by $d_1(z, w) = |\text{Re } z - \text{Re } w| + |\text{Im } z - \text{Im } w|$ and $d_2(z, w) = \max\{|\text{Re } z - \text{Re } w|, |\text{Im } z - \text{Im } w|\}$. Then $(\mathbb{C}, d_1)$ and $(\mathbb{C}, d_2)$ are metric spaces such that d_1 and d_2 induce the standard topology on $\mathbb{C}$.

Proof: $d_1(z, w) = 0$ if and only if $|\text{Re } z - \text{Re } w| + |\text{Im } z - \text{Im } w| = 0$ if and only if $|\text{Re } z - \text{Re } w| = 0$ and $|\text{Im } z - \text{Im } w| = 0$ if and only if $\text{Re } z - \text{Re } w = 0$ and $\text{Im } z - \text{Im } w = 0$ if and only if $\text{Re } z = \text{Re } w$ and $\text{Im } z = \text{Im } w$ if and only if $z = w$. So, property 1 holds for d_1. Property 2 follows immediately from the fact that $|x - y| = |y - x|$ for all $x, y \in \mathbb{R}$. Let's verify property 3. Let $z, w, v \in \mathbb{C}$. Then, we have

$$d_1(z, v) = |\text{Re } z - \text{Re } v| + |\text{Im } z - \text{Im } v|$$

$$= |\text{Re } z - \text{Re } w + \text{Re } w - \text{Re } v| + |\text{Im } z - \text{Im } w + \text{Im } w - \text{Im } v| \text{ (by SACT)}$$

$$\leq |\text{Re } z - \text{Re } w| + |\text{Re } w - \text{Re } v| + |\text{Im } z - \text{Im } w| + |\text{Im } w - \text{Im } v| \text{ (by the Triangle Inequality)}$$

$$= |\text{Re } z - \text{Re } w| + |\text{Im } z - \text{Im } w| + |\text{Re } w - \text{Re } v| + |\text{Im } w - \text{Im } v| = d_1(z, w) + d_1(w, v).$$

This shows that $(\mathbb{C}, d_1)$ is a metric space.

$d_2(z, w) = 0$ if and only if $\max\{|\text{Re } z - \text{Re } w|, |\text{Im } z - \text{Im } w|\} = 0$ if and only if $|\text{Re } z - \text{Re } w| = 0$ and $|\text{Im } z - \text{Im } w| = 0$ if and only if $\text{Re } z - \text{Re } w = 0$ and $\text{Im } z - \text{Im } w = 0$ if and only if $\text{Re } z = \text{Re } w$ and $\text{Im } z = \text{Im } w$ if and only if $z = w$. So, property 1 holds for d_2. Property 2 follows immediately from the fact that $|x - y| = |y - x|$ for all $x, y \in \mathbb{R}$. Let's verify property 3. Let $z, w, v \in \mathbb{C}$. Then, we have

$$d_2(z, v) = \max\{|\text{Re } z - \text{Re } v|, |\text{Im } z - \text{Im } v|\}$$

$$= \max\{|\text{Re } z - \text{Re } w + \text{Re } w - \text{Re } v|, |\text{Im } z - \text{Im } w + \text{Im } w - \text{Im } v|\} \text{ (by SACT)}$$

$$\leq \max\{|\text{Re } z - \text{Re } w| + |\text{Re } w - \text{Re } v|, |\text{Im } z - \text{Im } w| + |\text{Im } w - \text{Im } v|\} \text{ (by the Triangle Inequality)}$$

$$\leq \max\{|\text{Re } z - \text{Re } w|, |\text{Im } z - \text{Im } w|\} + \max\{|\text{Re } w - \text{Re } v|, |\text{Im } w - \text{Im } v|\}$$

(In fact, it's not hard to show that for all $a, b, c, d \in \mathbb{R}$, $\max\{a + b, c + d\} \leq \max\{a, c\} + \max\{b, d\}$)

$$= d_2(z, w) + d_2(w, v).$$

This shows that $(\mathbb{C}, d_2)$ is a metric space.

Let $d: \mathbb{C} \times \mathbb{C} \to \mathbb{R}$ be defined by $d(z, w) = |z - w|$. We have already seen in part 1 of Example 14.38 that d induces the standard topology on $\mathbb{C}$. Let's let $\mathcal{T}$ be the standard topology on $\mathbb{C}$.

Now, if $z, w \in \mathbb{C}$, then we have

$$\max\{|\text{Re } z - \text{Re } w|, |\text{Im } z - \text{Im } w|\} \leq \sqrt{(\text{Re } z - \text{Re } w)^2 + (\text{Im } z - \text{Im } w)^2} = |z - w|$$

$$\leq |\text{Re } z - \text{Re } w| + |\text{Im } z - \text{Im } w| \leq 2\max\{|\text{Re } z - \text{Re } w|, |\text{Im } z - \text{Im } w|\}.$$

Therefore, $d_2(z, w) \leq d(z, w) \leq d_1(z, w) \leq 2d_2(z, w)$.

So, if $z \in \mathbb{C}$ and $r \in \mathbb{R}^+$, then $B_{\frac{r}{2}}(z; d_2) \subseteq B_r(z; d_1) \subseteq B_r(z; d) \subseteq B_r(z; d_2)$. For example, to see that $B_{\frac{r}{2}}(z; d_2) \subseteq B_r(z; d_1)$, if $w \in B_{\frac{r}{2}}(z; d_2)$, then $d_2(z, w) < \frac{r}{2}$, so that $2d_2(z, w) < r$. Then since $d_1(z, w) \leq 2d_2(z, w)$, $d_1(z, w) < r$, so that $w \in B_r(z; d_1)$. The other two arguments are similar.

Let U be an element of the topology induced by d_1. For each $z \in U$, let $B_{r_z}(z; d_1) \subseteq U$. Then for each $z \in U$, $B_{\frac{r_z}{2}}(z; d_2) \subseteq U$. Therefore, $\bigcup \left\{B_{\frac{r_z}{2}}(z; d_2) \middle| z \in U\right\} \subseteq U$. Also, if $w \in U$, then $w \in B_{\frac{r_w}{2}}(w; d_2)$, so that $w \in \bigcup \left\{B_{\frac{r_z}{2}}(z; d_2) \middle| z \in U\right\}$. Therefore, $U \subseteq \bigcup \left\{B_{\frac{r_z}{2}}(z; d_2) \middle| z \in U\right\}$. It follows that we have $U = \bigcup \left\{B_{\frac{r_z}{2}}(z; d_2) \middle| z \in U\right\}$. This shows that the topology $\mathcal{T}_2$ induced by d_2 is finer than the topology $\mathcal{T}_1$ induced by d_1. That is $\mathcal{T}_1 \subseteq \mathcal{T}_2$. Similarly, we have $\mathcal{T} \subseteq \mathcal{T}_1$ and $\mathcal{T}_2 \subseteq \mathcal{T}$. These inclusions together show us that $\mathcal{T} = \mathcal{T}_1 = \mathcal{T}_2$. So, d_1 and d_2 induce the standard topology on $\mathbb{C}$. □

36. The square metric on $\mathbb{R}^n$ induces the same topology as the Euclidean metric on $\mathbb{R}^n$. Furthermore, the topology induced by these metrics is the same as the product topology on $\mathbb{R}^n$.

Proof: Let $B_r(\mathbf{x}; \rho)$ be an arbitrary open ball in $(\mathbb{R}^n, \rho)$. We will show that $B_{\frac{r}{\sqrt{n}}}(\mathbf{x}; \rho) \subseteq B_r(\mathbf{x}; d)$. To see this, let $\mathbf{y} \in B_{\frac{r}{\sqrt{n}}}(\mathbf{x}; \rho)$. Then $\max\{|y_1 - x_1|, \dots, |y_n - x_n|\} = \rho(\mathbf{y}, \mathbf{x}) < \frac{r}{\sqrt{n}}$. So, for each $k = 1, 2, \dots, n$, $|y_k - x_k| < \frac{r}{\sqrt{n}}$. Thus, $d(\mathbf{y}, \mathbf{x}) = \sqrt{(y_1 - x_1)^2 + (y_2 - x_2)^2 + (y_n - x_n)^2} \leq \sqrt{n(\frac{r}{\sqrt{n}})^2} = \sqrt{r^2} = r$. Therefore, $\mathbf{y} \in B_r(\mathbf{x}; d)$. So, the topology induced by ρ is finer than the topology induced by d.

Now, let $B_r(\mathbf{x}; d)$ be an arbitrary open ball in $(\mathbb{R}^n, d)$. We will show that $B_r(\mathbf{x}; d) \subseteq B_r(\mathbf{x}; \rho)$. To see this, let $\mathbf{y} \in B_r(\mathbf{x}; d)$. Then $\sqrt{(y_1 - x_1)^2 + (y_2 - x_2)^2 + (y_n - x_n)^2} = |\mathbf{y} - \mathbf{x}| = d(\mathbf{y}, \mathbf{x}) < r$. For each $k = 1, 2, \dots, n$, we have $|y_k - x_k| \leq \sqrt{(y_1 - x_1)^2 + (y_2 - x_2)^2 + (y_n - x_n)^2} = d(\mathbf{y}, \mathbf{x})$. Therefore, $\rho(\mathbf{y}, \mathbf{x}) = \max\{|y_1 - x_1|, \dots, |y_n - x_n|\} \leq d(\mathbf{y}, \mathbf{x}) < r$. So, $\mathbf{y} \in B_r(\mathbf{x}; \rho)$. Thus, the topology induced by d is finer than the topology induced by ρ.

Since the topology induced by ρ is finer than the topology induced by d and the topology induced by d is finer than the topology induced by ρ, ρ and d induce the same topology.

Now, let U be a basic open set in the product topology on $\mathbb{R}^n$, say $U = B_{r_1}(x_1) \times \cdots \times B_{r_n}(x_n)$, where for each $k = 1, 2, \dots, n$, $B_{r_k}(x_k)$ is an open ball in the Euclidean metric on $\mathbb{R}$. Let $r = \min\{r_1, \dots, r_n\}$. We will show that $B_r(\mathbf{x}; \rho) \subseteq U$. To see this, let $\mathbf{y} \in B_r(\mathbf{x}; \rho)$. Then

$$\max\{|y_1 - x_1|, \dots, |y_n - x_n|\} = \rho(\mathbf{y}, \mathbf{x}) < r.$$

So, for each $k = 1, 2, \dots, n$, we have $|y_k - x_k| < r \leq r_k$. So, for each $k = 1, 2, \dots, n$, $y_k \in B_{r_k}(x_k)$. It follows that $\mathbf{y} \in U$. Thus, the topology induced by ρ is finer that the product topology on $\mathbb{R}^n$.

Finally, let $B_r(\mathbf{x}; \rho)$ be an arbitrary open ball in $(\mathbb{R}^n, \rho)$. We will show $B_r(x_1) \times \cdots \times B_r(x_n) \subseteq B_r(\mathbf{x}; \rho)$. To see this, let $\mathbf{y} \in B_r(x_1) \times \cdots \times B_r(x_n)$. Then for each $k = 1, 2, \dots, n$, we have $y_k \in B_r(x_k)$. So, for each $k = 1, 2, \dots, n$, we have $|y_k - x_k| < r$. Therefore, $\rho(\mathbf{y}, \mathbf{x}) = \max\{|y_1 - x_1|, \dots, |y_n - x_n|\} < r$. So, $\mathbf{y} \in B_r(\mathbf{x}; \rho)$. Thus, the product topology on $\mathbb{R}^n$ is finer than the topology induced by ρ.

Since the topology induced by ρ is finer than the product topology on $\mathbb{R}^n$ and the product topology on $\mathbb{R}^n$ is finer than the topology induced by ρ, ρ induces the product topology on $\mathbb{R}^n$. □

37. A totally bounded metric space is bounded.

Proof: Let (S, d) be a totally bounded metric space and let A be a 1-net. Since A is finite, $m = \max\{d(a, b) \mid a, b \in A\}$ is finite. Let $M = m + 2$ and let $x, y \in S$ be arbitrary. Since A is a 1-net, there are $a, b \in A$ with $d(x, a) < 1$ and $d(y, b) < 1$. Then

$$d(x, y) \leq d(x, a) + d(a, b) + d(b, y) < 1 + m + 1 = m + 2 = M.$$

Therefore, (S, d) is bounded by M. □

38. If $(S, \mathcal{T})$ is a sequentially compact metrizable space, then every infinite subset of S has an accumulation point. Give a direct proof.

Proof: Assume that $(S, \mathcal{T})$ is sequentially compact and let A be an infinite subset of S. Then there is a sequence (s_n) in A such that $n \neq m \to s_n \neq s_m$. By sequential compactness, (s_n) has a convergent subsequence (s_{n_k}), let's say $s_{n_k} \to x$. Assume toward contradiction that x is not an accumulation point of A. Then there is an open ball $B_r(x)$ with center x such that $B_r(x) \cap A = \{x\}$. Since $s_{n_k} \to x$, there is $K \in \mathbb{Z}^+$ such that $s_{n_k} \in B_r(x) \cap A = \{x\}$ foir all $k > K$. So, for $k > K$, $s_{n_k} = x$, contradicting our assumption that all members of the sequence (and thus the subsequence) are distinct. □

39. Every sequentially compact metric space is totally bounded.

Proof: Assume that (S, d) is a sequentially compact metric space and let $r \in \mathbb{R}^+$. Let $s_0 \in S$ be arbitrary. If $B_r(s_0)$ is not an r-net, let $s_1 \in S \setminus B_r(s_0)$. We continue inductively. If $B_r(s_0) \cup \cdots \cup B_r(s_n)$ is not an r-net, let $s_n \in S \setminus \left(B_r(s_0) \cup \cdots \cup B_r(s_n)\right)$. This process must stop in finitely many steps, or else the sequence (s_k) would have no convergent subsequence. □

40. A metric space is compact if and only if it is complete and totally bounded.

Proof: First assume that (S, d) is a compact metric space. By Theorem 14.63, (S, d) is sequentially compact. If (s_n) is a Cauchy sequence in S, then by sequential compactness, (s_n) has a convergent subsequence. By Corollary 14.44, (S, d) is complete. By Problem 39 above, (S, d) is totally bounded.

Conversely, assume that (S, d) is a complete, totally bounded metric space. Let (s_n) be a sequence in S. We use the fact that (S, d) is totally bounded to show that (s_n) has a Cauchy subsequence. For each $n \in \mathbb{N}$, let A_n be a $\frac{1}{n+1}$-net. Since A_0 is finite, there is $a_0 \in A_0$ such that infinitely many terms of (s_n) are in $B_1(a_0)$. Let $C_0 = \{n \in \mathbb{N} \mid s_n \in B_1(a_0)\}$. Since A_1 is finite, there is $a_1 \in A_1$ such that infinitely many of the elements of C_0 are in $B_{\frac{1}{2}}(a_1)$. Let $C_1 = \left\{n \in C_0 \,\middle|\, s_n \in B_{\frac{1}{2}}(a_1)\right\}$. Continuing in this way, assuming that C_k is infinite, since A_{k+1} is finite, there is $a_{k+1} \in A_{k+1}$ such that infinitely many of the elements of C_k are in $B_{\frac{1}{k+2}}(a_{k+1})$. Let $C_{k+1} = \left\{n \in C_k \,\middle|\, s_n \in B_{\frac{1}{k+1}}(a_{k+1})\right\}$. Let (n_k) be a strictly increasing sequence of natural numbers such that $n_k \in C_k$ for each $k \in \mathbb{N}$. It is easy to check that (s_{n_k}) is a Cauchy sequence. Now, since (S, d) is complete, this Cauchy sequence converges, giving us a convergent subsequence of (s_n). It follows that (S, d) is sequentially compact. By Theorem 14.63, (S, d) is compact. □

41. Let $(K,\mathcal{T})$ and $(L,\mathcal{U})$ be topological spaces with $(K,\mathcal{T})$ compact and let $f: K \to L$ be a homeomorphism. Then $(L,\mathcal{U})$ is compact.

Proof: Let $\mathcal{C}$ be an open covering of L. Since f is continuous and bijective, $\mathcal{D} = \{f^{-1}[B] \mid B \in \mathcal{C}\}$ is an open covering of K. Since $(K,\mathcal{T})$ is compact, there is a finite subcollection $\mathcal{E} \subseteq \mathcal{D}$ that covers K. Since f^{-1} is continuous and bijective, $\mathcal{H} = \{f[f^{-1}[B]] \mid f^{-1}[B] \in \mathcal{E}\}$ covers L. By part 3 of Example 2.61, $f[f^{-1}[B]] \subseteq B$. Since $\mathcal{H}$ covers L, so does $\mathcal{J} = \{B \mid f^{-1}[B] \in \mathcal{E}\}$. Finally, $\mathcal{J} \subseteq \mathcal{C}$ because if $B \in \mathcal{J}$, then $f^{-1}[B] \in \mathcal{E}$. So, $f^{-1}[B] \in \mathcal{D}$. Therefore, $B \in \mathcal{C}$. □

Note: Since f is surjective, we actually have $f[f^{-1}[B]] = B$. To see this, first note that by part 3 of Example 2.61,, we have $f[f^{-1}[B]] \subseteq B$. For the other inclusion, let $y \in B$. Since f is surjective, there is $x \in f^{-1}[B]$ with $f(x) = y$. Then $y = f(x) \in f[f^{-1}[B]]$. Since $y \in B$ was arbitrary, we have $B \subseteq f[f^{-1}[B]]$. Since $f[f^{-1}[B]] \subseteq B$ and $B \subseteq f[f^{-1}[B]]$, we have $f[f^{-1}[B]] = B$.

42. Let $(S,\mathcal{T})$ be a T_4-space, let A and B be disjoint closed subsets of S, and let a and b be real numbers with $a < b$. Then there is a continuous function $f: S \to [a,b]$ such that $f[A] = \{a\}$ and $f[B] = \{b\}$.

Proof: Let $(S,\mathcal{T})$ be a T_4-space, let A and B be disjoint closed subsets of S, and let a and b be real numbers with $a < b$. By Urysohn's Lemma (Theorem 14.80), there is a continuous function $g: S \to [0,1]$ such that $g[A] = \{0\}$ and $g[B] = \{1\}$. Define $f: S \to [a,b]$ by $f(x) = (b-a)g(x) + a$. Then we have

$$f[A] = (b-a)g[A] + a = (b-a)(0) + a = a,$$
$$f[B] = (b-a)g[B] + a = (b-a)(1) + a = b,$$

and f is continuous by a standard continuity argument. □

43. If $(S_1,\mathcal{T}_1)$ and $(S_2,\mathcal{T}_2)$ are connected topological spaces, then $(S_1 \times S_2,\mathcal{T})$ is connected, where $\mathcal{T}$ is the product topology on $S_1 \times S_2$.

Proof: Let $(S_1,\mathcal{T}_1)$ and $(S_2,\mathcal{T}_2)$ be connected topological spaces and let $f: S_1 \times S_2 \to \{0,1\}$ be continuous (here $\{0,1\}$ is given the discrete topology). By Problem 22 above, it suffices to show that f is constant. For each $a \in S_1$, define $g_a: S_2 \to \{0,1\}$ by $g_a(y) = f(a,y)$. Then g_a is continuous. Since $(S_2,\mathcal{T}_2)$ is connected, by Problem 22 again, g_a is constant. Similarly, for each $b \in S_2$, the function $h_b: S_1 \to \{0,1\}$ defined by $h_b(x) = f(x,b)$ is constant. Let $(a,b),(c,d) \in S_1 \times S_2$. Then we have $f(a,b) = g_a(b) = g_a(d) = f(a,d) = h_d(a) = h_d(c) = f(c,d)$. So, f is constant. □

44. The topological space $(S,\mathcal{T})$ is locally connected if and only if each point of S has a basis $\mathcal{B}$ consisting of connected subspaces of S ($\mathcal{B}$ is a basis at x if (i) $V \in \mathcal{B} \to x \in V$; and (ii) for any open set U containing x, there is a $V \in \mathcal{B}$ such that $V \subseteq U$).

Proof: First assume that $(S,\mathcal{T})$ is locally connected and let $x \in S$. For each $U \in \mathcal{T}$ with $x \in U$, let V_U be an open set containing x such that $(V,\mathcal{T}_V)$ is connected and $V_U \subseteq U$. Let $\mathcal{B} = \{V_U \mid U \in \mathcal{T} \wedge x \in U\}$. Then clearly $\mathcal{B}$ is a basis at x consisting of connected subspaces of S.

Conversely, suppose that each point of S has a basis consisting of connected subspaces of S. Let $x \in S$ and let U be an open set containing x. Let $\mathcal{B}$ be a basis of x consisting of connected subspaces of S. Then by (ii), there is $V \in \mathcal{B}$ such that $V \subseteq U$. □

45. Local connectedness is a topological property.

Proof: Let $(A, \mathcal{T})$ and $(B, \mathcal{U})$ be topological spaces with $(A, \mathcal{T})$ locally connected, let $f: A \to B$ be a homeomorphism, let $y \in B$ and let W be an open set containing y. Then $y = f(x)$ for some $x \in A$ and $U = f^{-1}[W]$ is an open set containing x. Since $(A, \mathcal{T})$ is locally connected, there is an open set V containing x such that $(V, \mathcal{T}_V)$ is connected and $V \subseteq U$. Since f^{-1} is continuous, $f[V]$ is open in B. Since $x \in V$, $f(x) \in f[V]$. Also, $f[V] \subseteq f[U] = f[f^{-1}[W]] = W$. Finally, since $(V, \mathcal{T}_V)$ is connected and connectedness is a topological property, $(f[V], \mathcal{T}_{f[V]})$ is connected. □

LEVEL 4

Prove each of the following:

46. Let $(S, \mathcal{T})$ be a topological space. An arbitrary intersection of closed sets in S is closed in S.

Proof: Let $\boldsymbol{X}$ be a nonempty set of closed sets in S. Then for each $A \in \boldsymbol{X}$, $S \setminus A$ is an open set in S. Since $(S, \mathcal{T})$ is a topological space, $\bigcup\{S \setminus A \mid A \in \boldsymbol{X}\}$ is open in S. Therefore, $S \setminus \bigcup\{S \setminus A \mid A \in \boldsymbol{X}\}$ is closed in S. By De Morgan' s Law, $\bigcap \boldsymbol{X} = S \setminus \bigcup\{S \setminus A \mid A \in \boldsymbol{X}\}$. Therefore, $\bigcap \boldsymbol{X}$ is closed in S. □

47. Let $(S, \mathcal{T})$ be a topological space. A finite union of closed sets in S is closed in S.

Proof: Let $\boldsymbol{X}$ be a finite set of closed subsets of S. Then for each $A \in \boldsymbol{X}$, $S \setminus A$ is an open set in S. Since $(S, \mathcal{T})$ is a topological space, $\bigcap\{S \setminus A \mid A \in \boldsymbol{X}\}$ is open in S. Therefore, $S \setminus \bigcap\{S \setminus A \mid A \in \boldsymbol{X}\}$ is closed in S. By De Morgan' s Law, $\bigcup \boldsymbol{X} = S \setminus \bigcap\{S \setminus A \mid A \in \boldsymbol{X}\}$. Therefore, $\bigcup \boldsymbol{X}$ is closed in S. □

48. $\mathcal{B}' = \{(a, b) \mid a, b \in \mathbb{Q} \wedge a < b\}$ is a countable basis for a topology on $\mathbb{R}$ that generates the standard topology on $\mathbb{R}$.

Proof: Define $g: \mathcal{B}' \to \mathbb{Q} \times \mathbb{Q}$ by $g((a, b)) = (a, b)$ (the open interval (a, b) is being sent to the ordered pair (a, b)—it is unfortunate that the notation for these two objects is identical). If $g((a, b)) = g((c, d))$, then $(a, b) = (c, d)$ (as ordered pairs). So, $a = c$ and $b = d$. Therefore, $(a, b) = (c, d)$ (as open intervals). This shows that g is injective. So, $\mathcal{B}' \preccurlyeq \mathbb{Q} \times \mathbb{Q} \sim \mathbb{Q}$. Since $\mathbb{Q}$ is countable, so is $\mathcal{B}'$.

If $x \in \mathbb{R}$, then $x - 1 < x < x + 1$. By the Density Theorem, we can choose $a, b \in \mathbb{Q}$ such that $x - 1 < a < x$ and $x < b < x + 1$. Then $x \in (a, b)$ and $(a, b) \in \mathcal{B}'$. So, $\mathcal{B}'$ covers $\mathbb{R}$. Now, let $x \in \mathbb{R}$ and $(a, b), (c, d) \in \mathcal{B}'$ with $x \in (a, b) \cap (c, d)$. By Problem 37 from Problem Set 6, we have $(a, b) \cap (c, d) = (e, f)$ for some $e, f \in \mathbb{R}$. By the Density Theorem, we can choose $g, h \in \mathbb{Q}$ such that $e < g < x$ and $x < h < f$. Then $x \in (g, h)$ and $(g, h) \subseteq (a, b) \cap (c, d)$. So, $\mathcal{B}'$ has the intersection containment property. It follows that $\mathcal{B}'$ is a basis for a topology on $\mathbb{R}$.

Since every open interval with rational endpoints is open in the standard topology on $\mathbb{R}$, the topology generated by $\mathcal{B}'$ is contained in the standard topology. Let $a, b \in \mathbb{R}$. For each $n \in \mathbb{Z}^+$, by the Density Theorem, we can choose $q_n, r_n \in \mathbb{Q}$ with $a < q_n < a + \frac{1}{n}$ and $b - \frac{1}{n} < r_n < b$. We will now show that $(a, b) = \bigcup\{(q_n, r_n) \mid n \in \mathbb{Z}^+\}$. If $x \in (a, b)$, then there is $n \in \mathbb{Z}^+$ with $x \in \left(a + \frac{1}{n}, b - \frac{1}{n}\right) \subseteq (q_n, r_n)$. So, $x \in \bigcup\{(q_n, r_n) \mid n \in \mathbb{Z}^+\}$. Therefore, $(a, b) \subseteq \bigcup\{(q_n, r_n) \mid n \in \mathbb{Z}^+\}$. If $x \in \bigcup\{(q_n, r_n) \mid n \in \mathbb{Z}^+\}$, then there is $n \in \mathbb{Z}^+$ such that $x \in (q_n, r_n) \subseteq (a, b)$. So, $\bigcup\{(q_n, r_n) \mid n \in \mathbb{Z}^+\} \subseteq (a, b)$. It follows that $(a, b) = \bigcup\{(q_n, r_n) \mid n \in \mathbb{Z}^+\}$. Since (a, b) is a union of sets in $\mathcal{B}'$, the standard topology is contained in the topology generated by $\mathcal{B}'$. So, $\mathcal{B}'$ generates the standard topology on $\mathbb{R}$. □

49. Let $(S, \mathcal{T})$ be a topological space, let A be a subspace of S, let B be a subset of A, and let $\overline{B}$ be the closure of B in S. Then the closure of B in A is $\overline{B} \cap A$.

Proof: Let Y be the closure of B in A. We want to show that $Y = \overline{B} \cap A$. Since $\overline{B}$ is closed in S, by Problem 27, $\overline{B} \cap A$ is closed in A. Since $B \subseteq \overline{B}$ and $B \subseteq A$, we have $B \subseteq \overline{B} \cap A$. By definition, Y is the intersection of all closed subsets of A containing B, we have $Y \subseteq \overline{B} \cap A$.

Also, Y is closed in A. So, by Problem 27, $Y = C \cap A$ for some set C that is closed in S. Then C is a closed set of S containing B. Since $\overline{B}$ is the intersection of all closed sets in S that contain B, we see that $\overline{B} \subseteq C$. So, $\overline{B} \cap A \subseteq C \cap A = Y$.

Since $Y \subseteq \overline{B} \cap A$ and $\overline{B} \cap A \subseteq Y$, we have $Y = \overline{B} \cap A$. □

50. Let $\mathcal{T}_L$ be the set generated by the half open intervals of the form $[a, b)$ with $a, b \in \mathbb{R}$. $\mathcal{T}_L$ is a topology on $\mathbb{R}$ that is strictly finer than the standard topology on $\mathbb{R}$ and incomparable with the topology $\mathcal{T}_K$ from part 4 of Example 14.29.

Proof: Let $\mathcal{B} = \{[a, b) \mid a, b \in \mathbb{R}\}$ and let $\mathcal{T}_L$ be the set generated by $\mathcal{B}$. Let $x \in \mathbb{R}$. Then $x \in [x, x + 1)$. This shows that $\mathcal{B}$ covers $\mathbb{R}$. If $[a, b), [c, d) \in \mathcal{B}$ with $[a, b) \cap [c, d) \neq \emptyset$, then $[a, b) \cap [c, d) = [e, f)$, where $e = \max\{a, c\}$ and $f = \min\{b, d\}$. To see this, let $x \in [a, b) \cap [c, d)$. Then we have $a \leq x < b$ and $c \leq x < d$. Since $a \leq x$ and $c \leq x$, $e \leq x$. Since $x < b$ and $x < d$, $x < f$. It follows that $x \in [e, f)$. Conversely, if $x \in [e, f)$, then $e \leq x < f$. Since $a \leq e$ and $c \leq e$, $a \leq x$ and $c \leq x$. Since $f \leq b$ and $f \leq d$, $x < b$ and $x < d$. So, $x \in [a, b)$ and $x \in [c, d)$. Thus, $x \in [a, b) \cap [c, d)$. It follows that $\mathcal{B}$ has the intersection containment property. Since $\mathcal{B}$ covers $\mathbb{R}$ and $\mathcal{B}$ has the intersection containment property, $\mathcal{B}$ is a basis for a topology on $\mathbb{R}$.

To see that $\mathcal{T}_L$ is finer than the standard topology on $\mathbb{R}$, note that each basic open set (a, b) in the standard topology is equal to the union $\bigcup\left\{\left[a + \frac{1}{n}, b\right) \,\middle|\, n \in \mathbb{Z}^+\right\}$. See Problem 48 from Problem Set 6 for a proof similar to what is needed to prove this result.

To see that $\mathcal{T}_L$ is **strictly** finer than the standard topology, just note that $[0, 1)$ cannot be written as a union of bounded open intervals, for 0 would need to be inside one of those open intervals, and it would then follow that there is an $x < 0$ with $x \in [0,1)$.

The set $(-1,1) \setminus K$ is open in $\mathcal{T}_K$. We show that $(-1,1) \setminus K$ is **not** open in $\mathcal{T}_L$. If $(-1,1) \setminus K$ is the union of sets of the form $[a, b)$, then 0 would need to be inside one of those half-open intervals, let's say that $0 \in [a, b)$. But then there is some $n > 0$ such that $\frac{1}{n} < b$ (use the Archimedean property). Therefore, $\frac{1}{n} \in [a, b)$, contradicting that $\frac{1}{n} \in K$. This shows that $\mathcal{T}_L$ is **not** finer than $\mathcal{T}_K$.

The set $[0,1)$ is open in $\mathcal{T}_L$. We've already seen that $[0,1)$ cannot be written as a union of bounded open intervals. If we throw additional sets of the form $(a, b) \setminus K$ into such a union, then we still run into the same issue with 0. If $0 \in (a, b)$ or $(a, b) \setminus K$, we would get an $x < 0$ with $x \in [0,1)$. □

51. Let $(S_1, \mathcal{T}_1)$ and $(S_2, \mathcal{T}_2)$ be T_2-spaces. $S_1 \times S_2$ with the product topology is also a T_2-space. Determine if the analogous statement is true for T_3-spaces.

Proof: Let $(x, y), (z, w) \in S_1 \times S_2$ with $(x, y) \neq (z, w)$. Then $x \neq z$ or $y \neq w$. Without loss of generality, assume that $x \neq z$. Since $(S_1, \mathcal{T}_1)$ is a T_2-space, there are $U, V \in \mathcal{T}_1$ with $x \in U$, $z \in V$, and $U \cap V = \emptyset$. Then $(x, y) \in U \times S_2$, $(z, w) \in V \times S_2$, and

$$(U \times S_2) \cap (V \times S_2) = (U \cap V) \times (S_2 \cap S_2) = \emptyset \times S_2 = \emptyset.$$

Since $(x, y), (z, w) \in S_1 \times S_2$ were arbitrary, $S_1 \times S_2$ with the product topology is a T_2-space.

Let $(S_1, \mathcal{T}_1)$ and $(S_2, \mathcal{T}_2)$ be T_3-spaces. We will show that $S_1 \times S_2$ with the product topology is also a T_3-space.

Let $(x, y) \in S_1 \times S_2$ and $B \subseteq (S_1 \times S_2) \setminus \{(x, y)\}$ with B closed in the product topology. Consider the open set $(S_1 \times S_2) \setminus B$. Since $(x, y) \in (S_1 \times S_2) \setminus B$, there are sets $U \in \mathcal{T}_1$ and $V \in \mathcal{T}_2$ with $x \in U$, $y \in V$, and $U \times V \subseteq (S_1 \times S_2) \setminus B$. Since U and V are open sets in $(S_1, \mathcal{T}_1)$ and $(S_2, \mathcal{T}_2)$, respectively, $S_1 \setminus U$ and $S_2 \setminus V$ are closed sets in $(S_1, \mathcal{T}_1)$ and $(S_2, \mathcal{T}_2)$, respectively. Also, $x \notin S_1 \setminus U$ and $y \notin S_2 \setminus V$. Since $(S_1, \mathcal{T}_1)$ and $(S_2, \mathcal{T}_2)$ are T_3-spaces, there are open sets W_1, Z_1, W_2, and Z_2 with $x \in W_1$, $y \in W_2$, $S_1 \setminus U \subseteq Z_1$, $S_2 \setminus V \subseteq Z_2$, $W_1 \cap Z_1 = \emptyset$, and $W_2 \cap Z_2 = \emptyset$.

Since $x \in W_1$ and $y \in W_2$, $(x, y) \in W_1 \times W_2$.

We now show that $B \subseteq (Z_1 \times S_2) \cup (S_1 \times Z_2)$. To see this, let $(a, b) \in B$. Then $(a, b) \notin U \times V$. So, $a \notin U$ or $b \notin V$. Without loss of generality, assume that $a \notin U$. Then $a \in Z_1$. So, $(a, b) \in Z_1 \times S_2$. It follows that $(a, b) \in (Z_1 \times S_2) \cup (S_1 \times Z_2)$.

Finally, we have

$$\begin{aligned}(W_1 \times W_2) \cap [(Z_1 \times S_2) \cup (S_1 \times Z_2)] &= [(W_1 \times W_2) \cap (Z_1 \times S_2)] \cup \big[(W_1 \times W_2) \cap \big((S_1 \times Z_2)\big)\big] \\ &= [(W_1 \cap Z_1) \times (W_2 \cap S_2)] \cup [(W_1 \cap S_1) \times (W_2 \cap Z_2)] \\ &= [\emptyset \times (W_2 \cap S_2)] \cup [(W_1 \cap S_1) \times \emptyset] = \emptyset \times \emptyset = \emptyset.\end{aligned}$$

Since $(x, y) \in S_1 \times S_2$ was arbitrary and $B \subseteq (S_1 \times S_2) \setminus \{(x, y)\}$ was an arbitrary closed set, $S_1 \times S_2$ with the product topology is a T_3-space. □

52. A second-countable T_3-space is a T_4-space.

Proof: Let $(S, \mathcal{T})$ be a T_3-space with a countable basis $\mathcal{B}$ and let A and B be disjoint closed sets in S. For each $x \in A$, $S \setminus B$ is an open set containing x. By Problem 15, there is $V_x \in \mathcal{B}$ containing x with $\overline{V}_x \subseteq S \setminus B$. The collection is countable, and so we can write $\{V_x \mid x \in S\} = \{V_k \mid k \in \mathbb{N}\}$. Note that we have $A \subseteq \bigcup\{V_k \mid k \in \mathbb{N}\}$ and for each $k \in \mathbb{N}$, $B \cap \overline{V}_k = \emptyset$.

In the same way, we can find $\{W_k \mid k \in \mathbb{N}\}$ with $B \subseteq \bigcup\{W_k \mid k \in \mathbb{N}\}$ and for each $k \in \mathbb{N}$, $A \cap \overline{W}_k = \emptyset$.

For each $n \in \mathbb{N}$, let $C_n = V_n \setminus \bigcup\{\overline{W_k} \mid k \leq n\}$ and $D_n = W_n \setminus \bigcup\{\overline{V_k} \mid k \leq n\}$. For each $n \in \mathbb{N}$, C_n and D_n are open in S. Let $V = \bigcup\{C_n \mid n \in \mathbb{N}\}$ and $W = \bigcup\{D_n \mid n \in \mathbb{N}\}$. Then V and W are open in S.

Since $A \subseteq \bigcup\{V_n \mid n \in \mathbb{N}\}$ and $A \cap \overline{W}_n = \emptyset$, we have $A \subseteq V$. Since $B \subseteq \bigcup\{W_n \mid n \in \mathbb{N}\}$ and $B \cap \overline{V}_n = \emptyset$, we have $B \subseteq W$.

If $x \in V \cap W$, then there are $m, n \in \mathbb{N}$ such that $x \in C_m \cap D_n$. Without loss of generality, assume that $m \geq n$. Then $x \in C_m = V_m \setminus \bigcup\{\overline{W_k} \mid k \leq m\} \subseteq V_m \setminus \overline{W_n}$ and $x \in D_n \subseteq W_n$, which is impossible. Therefore, $V \cap W = \emptyset$.

So, V and W are disjoint open sets with $A \subseteq V$ and $B \subseteq W$. □

53. Every metrizable space is a T_4-space.

Proof: Let $(S, \mathcal{T})$ be metrizable and let d be a metric on S that induces $\mathcal{T}$. Let A, B be disjoint closed subsets of S. Let $x \in A$. Since $A \cap B = \emptyset$, $x \notin B$. So, x is in the open set $S \setminus B$. Therefore, there is $r_x \in \mathbb{R}^+$ such that $B_{r_x}(x) \subseteq S \setminus B$. Let $U = \bigcup\left\{B_{\frac{r_x}{2}}(x) \,\middle|\, x \in A\right\}$. Then $U \in \mathcal{T}$, $A \subseteq U$, and $U \cap B = \emptyset$. Similarly, for each $x \in B$, let $V = \bigcup\left\{B_{\frac{r_x}{2}}(x) \,\middle|\, x \in B\right\}$, so that $V \in \mathcal{T}$, $B \subseteq V$, and $V \cap A = \emptyset$.

We now show that $U \cap V = \emptyset$. If $a \in U \cap V$, then there is $x \in A$ and $y \in B$ with $a \in B_{\frac{r_x}{2}}(x) \cap B_{\frac{r_y}{2}}(y)$. Then $d(x, y) \leq d(x, a) + d(a, y) < \frac{r_x}{2} + \frac{r_y}{2}$. Without loss of generality, assume that $r_y \leq r_x$. Then, we have $d(x, y) < \frac{r_x}{2} + \frac{r_x}{2} = r_x$. So, $y \in B_{r_x}(x)$. Since $B_{r_x}(x) \subseteq S \setminus B$, $y \in S \setminus B$. So, $y \notin B$, a contradiction. It follows that $U \cap V = \emptyset$. □

54. Let $(S_1, \mathcal{T}_1)$ and $(S_2, \mathcal{T}_2)$ be metrizable spaces. $S_1 \times S_2$ with the product topology is metrizable.

Proof: Let d_1 and d_2 be metrics that induce the topologies $\mathcal{T}_1$ and $\mathcal{T}_2$, respectively. Define $d: (S_1 \times S_2) \times (S_1 \times S_2) \to \mathbb{R}$ by $d\big((a, b), (c, d)\big) = \max\{d_1(a, c), d_2(b, d)\}$. We first show that d defines a metric on $S_1 \times S_2$. We have $d\big((a, b), (c, d)\big) = 0$ if and only if $\max\{d_1(a, c), d_2(b, d)\} = 0$ if and only if $d_1(a, c) = 0$ and $d_2(b, d) = 0$ if and only if $a = c$ and $b = d$ if and only if $(a, b) = (c, d)$. So, property 1 holds. Property 2 is clear. For property 3, Let $(a, b), (c, k), (e, f) \in S_1 \times S_2$, Then

$$d\big((a, b), (e, f)\big) = \max\{d_1(a, e), d_2(b, f)\} \leq \max\{d_1(a, c) + d_1(c, e), d_2(b, k) + d_2(k, f)\}$$
$$\leq \max\{d_1(a, c), d_2(b, k)\} + \max\{d_1(c, e), d_2(k, f)\} = d\big((a, b), (c, k)\big) + d\big((c, k), (e, f)\big).$$

We now show that d induces the product topology on $S_1 \times S_2$.

Let $B = B_r((x,y);d)$ be an arbitrary open ball in the topology induced by d. We show that $B = B_r(x;d_1) \times B_r(y;d_2)$. If $(a,b) \in B$, then $\max\{d_1(x,a), d_2(y,b)\} = d((x,y),(a,b)) < r$. So, $d_1(x,a) < r$ and $d_2(y,b) < r$. Thus, $a \in B_r(x;d_1)$, $b \in B_r(y;d_2)$. So, $(a,b) \in B_r(x;d_1) \times B_r(y;d_2)$. Therefore, $B \subseteq B_r(x;d_1) \times B_r(y,d_2)$.

Now, if $(a,b) \in B_r(x;d_1) \times B_r(y;d_2)$, then $a \in B_r(x;d_1)$ and $b \in B_r(y;d_2)$. So, $d_1(x,a) < r$ and $d_2(y,b) < r$. So, $d((x,y),(a,b)) = \max\{d_1(x,a), d_2(y,b)\} < r$. Therefore, $(a,b) \in B$. So, $B_r(x;d_1) \times B_r(y;d_2) \subseteq B$. Since $B \subseteq B_r(x;d_1) \times B_r(y,d_2)$ and $B_r(x;d_1) \times B_r(y;d_2) \subseteq B$, we have $B = B_r(x;d_1) \times B_r(y,d_2)$. This shows that B is open in the product topology. Therefore, the product topology is finer than the topology induced by d.

Conversely, a basic open set in the product topology has the form $B_r(x;d_1) \times B_r(y;d_2)$, and we saw that this equal to $B_r((x,y);d)$. It follows that each basic open set in the product topology is open in the topology induced by d. Therefore, the topology induced by d is finer than the product topology. So, d induces the product topology. □

55. Let $(S,\mathcal{T})$ be a compact T_2-space and let K, L be disjoint subsets of S that are each compact in the subspace topology. Then there are disjoint open sets U and V such that $K \subseteq U$ and $L \subseteq V$.

Proof: Let $(S,\mathcal{T})$ be a compact T_2-space and let K, L be subsets of S that are each compact in the subspace topology.

Let $x \in K$. Since $(S,\mathcal{T})$ is a T_2-space, for each $y \in L$, we can find open sets U_y and V_y such that $x \in U_y$, $y \in V_y$, and $U_y \cap V_y = \emptyset$. The collection $\mathcal{C} = \{V_y \mid y \in L\}$ is an open cover of L by open sets in S. By Lemma 14.54, there is a finite subcollection $\mathcal{D} \subseteq \mathcal{C}$ that still covers L. Let $V = \bigcup\mathcal{D}$ and let $W = \bigcap\{U_y \mid V_y \in \mathcal{D}\}$. Then V and W are disjoint open sets such that $x \in W$ and $L \subseteq V$.

Now, for each $x \in K$, by the previous paragraph, we can find disjoint open sets W_x and V_x such that $x \in W_x$ and $L \subseteq V_x$. The collection $\mathcal{E} = \{W_x \mid x \in K\}$ is an open cover of K by open sets in S. By Lemma 14.54, there is a finite subcollection $\mathcal{F} \subseteq \mathcal{E}$ that still covers K. Let $W = \bigcup\mathcal{F}$ and let $X = \bigcap\{V_x \mid W_x \in \mathcal{F}\}$. Then W and X are disjoint open sets such that $K \subseteq W$ and $L \subseteq V$. □

56. A compact T_2-space is a T_4-space.

Proof: Let $(S,\mathcal{T})$ be a compact T_2-space. Since every T_2-space is a T_1-space, it follows that $(S,\mathcal{T})$ is a T_1-space.

Let X and Y be disjoint closed subsets of S. By Theorem 14.56, X and Y are compact. By Problem 55, there are $U, V \in \mathcal{T}$ with $X \subseteq U$, $Y \subseteq V$, and $U \cap V = \emptyset$.

It follows that $(S,\mathcal{T})$ is a T_4-space. □

57. Let (A,d) be a subspace of $(\mathbb{R}^n, d)$, where d is the Euclidean metric. A is bounded if and only if A is totally bounded.

Proof: By Problem 37, if A is totally bounded, then A is bounded.

Now, let (A, d) be a bounded subspace of $(\mathbb{R}^n, d)$. Then there is $M > 0$ such that $A \subseteq [-M, M]^n$. Let $r \in \mathbb{R}^+$. Then $C = \left\{\frac{r}{2}k \,\middle|\, k \in \mathbb{Z}\right\}^n \cap [-M, M]^n$ is an r-net. □

58. Every compact metrizable space is separable. Is every compact topological space separable?

Proof: Let $(S, \mathcal{T})$ be a compact metrizable space and suppose that $\mathcal{T}$ is induced by the metric d. For each $n \in \mathbb{N}$, let $\mathcal{C}_n = \left\{B_{\frac{1}{n}}(x; d) \,\middle|\, x \in S\right\}$. $\mathcal{C}_n$ is an open covering of S, and so, by compactness, there is a finite subcollection $\mathcal{D}_n \subseteq \mathcal{C}_n$ that still covers S. Let $X = \left\{x \in S \,\middle|\, \text{there is } n \in \mathbb{N} \text{ with } B_{\frac{1}{n}}(x; d) \in D_n\right\}$. X is a countable union of countable sets, and so, X is countable. To see that X is dense in S, let $y \in S$ and let $U \in \mathcal{T}$ with $y \in U$. Then there is an open ball $B_r(y; d) \subseteq U$. By the Archimedean Property of $\mathbb{R}$, there is $n \in \mathbb{N}$ such that $n > \frac{1}{r}$, or equivalently, $\frac{1}{n} < r$. Since $\mathcal{D}_n$ is an open covering of S, there is $B_{\frac{1}{n}}(x; d) \in \mathcal{D}_n$ for some $x \in S$ with $y \in B_{\frac{1}{n}}(x; d)$. So, $x \in X$ and we have $y \in U \cap B_{\frac{1}{n}}(x; d)$. This proves that $y \in \overline{X}$. Since $y \in S$ was arbitrary, $\overline{X} = S$, and so, X is a countable dense subset of S. □

A compact topological space does **not** need to be separable. For example, if we let $S = \mathbb{R}$ and we let $\mathcal{T} = \{\mathbb{R}\} \cup \{A \subseteq \mathbb{R} \mid 0 \notin A\}$. It is easily checked that $(\mathbb{R}, \mathcal{T})$ is a topological space. To see that $(\mathbb{R}, \mathcal{T})$ is compact, just observe that any open covering of $\mathbb{R}$ must contain $\mathbb{R}$ because $\mathbb{R}$ is the only open set containing 0. We now show that $(\mathbb{R}, \mathcal{T})$ is **not** separable. To see this, let $D \subseteq \mathbb{R}$ be dense. If $x \neq 0$, then $\{x\}$ is open, and so, $\{x\} \cap D \neq \emptyset$. Therefore, $\mathbb{R} \setminus \{0\} \subseteq D$, showing that D is uncountable. Therefore, there is no countable dense subset of $\mathbb{R}$, and so, $(\mathbb{R}, \mathcal{T})$ is not separable.

59. The one-point compactification of a locally compact T_2-space is a compact T_2-space.

Proof: Let's first verify that $\overline{\mathcal{T}}$ defines a topology on $\overline{S}$.

Since $\mathcal{T} \subseteq \overline{\mathcal{T}}$ and $\emptyset \in \mathcal{T}$, we have $\emptyset \in \overline{\mathcal{T}}$. Since $\emptyset$ is compact, $\overline{S} = S \cup \{\infty\} = (S \setminus \emptyset) \cup \{\infty\} \in \overline{\mathcal{T}}$.

Let $\boldsymbol{X} \subseteq \overline{\mathcal{T}}$. If $\boldsymbol{X} \subseteq \mathcal{T}$, then since $\mathcal{T}$ is a topology on S, $\bigcup \boldsymbol{X} \in \mathcal{T} \subseteq \overline{\mathcal{T}}$. Otherwise, let $\boldsymbol{W} = \{A \mid A \in \boldsymbol{X} \cap \mathcal{T}\}$ and let $\boldsymbol{Z} = \{S \setminus K \mid (S \setminus K) \cup \{\infty\} \in \boldsymbol{X}\}$.

Let $\boldsymbol{C} = \bigcap\{S \setminus A \mid A \in \boldsymbol{X} \cap \mathcal{T}\}$, and $\boldsymbol{L} = \bigcap\{K \mid (S \setminus K) \cup \{\infty\} \in \boldsymbol{X}\}$. $\boldsymbol{C}$ is an intersection of closed sets in S. Since $\mathcal{T}$ is a topology on S, $\boldsymbol{C}$ is closed in S. $\boldsymbol{L}$ is an intersection of compact sets. Since $\boldsymbol{L} \subseteq S$ and $(S, \mathcal{T})$ is a T_2-space, by Theorem 14.55, $\boldsymbol{L}$ is closed in S. Therefore, $\boldsymbol{C} \cap \boldsymbol{L}$ is closed in S. Since $\boldsymbol{C} \cap \boldsymbol{L}$ is a subset of a compact set (any K being used in the intersection $\boldsymbol{L}$ will work here), by Theorem 14.56, $\boldsymbol{C} \cap \boldsymbol{L}$ is compact. Now, $\bigcup \boldsymbol{X} = [\bigcup \boldsymbol{W} \cup \bigcup \boldsymbol{Z}] \cup \{\infty\} = [(S \setminus \boldsymbol{C}) \cup (S \setminus \boldsymbol{L})] \cup \{\infty\} = [S \setminus (\boldsymbol{C} \cap \boldsymbol{L})] \cup \{\infty\}$. Since $\boldsymbol{C} \cap \boldsymbol{L}$ is compact, $\bigcup \boldsymbol{X} \in \overline{\mathcal{T}}$. Finally, let $\boldsymbol{Y} \subseteq \overline{\mathcal{T}}$ with $\boldsymbol{Y}$ finite. If $\infty \notin U$ for some $U \in \boldsymbol{Y}$, then $\bigcap \boldsymbol{Y}$ is a finite intersection of sets in $\mathcal{T}$, and since $\mathcal{T}$ is a topology on S, $\bigcap \boldsymbol{Y} \in \mathcal{T} \subseteq \overline{\mathcal{T}}$. So, assume that every set in $\boldsymbol{Y}$ has the form $(S \setminus K) \cup \{\infty\}$ for some compact $K \subseteq S$. Let's say the compact sets are $K_0, \dots, K_n$. Then we have $\bigcap \boldsymbol{Y} = [(S \setminus K_0) \cap \cdots \cap (S \setminus K_n)] \cup \{\infty\} = [S \setminus (K_0 \cup \cdots \cup K_n)] \cup \{\infty\}$. Since a finite union of compact sets is compact, we see that $\bigcap \boldsymbol{Y} \in \overline{\mathcal{T}}$.

Let's now show that $(\overline{S},\overline{\mathcal{T}})$ is a T_2-space. Let $x, y \in \overline{S}$. Since $(S,\mathcal{T})$ is a T_2-space, we can assume that one of these points is ∞, let's say $y = \infty$. Since S is locally compact, there is a compact set $K \subseteq S$ and an open set $U \subseteq K$ with $x \in U$. Then $(S \setminus K) \cup \{\infty\}$ is an open set containing ∞ and disjoint from U. Therefore, $(\overline{S},\overline{\mathcal{T}})$ is a T_2-space.

Finally, we show that $(\overline{S},\overline{\mathcal{T}})$ is compact. To see this, let $\mathcal{C}$ be an open covering of $\overline{S}$. If $\overline{S} \in \mathcal{C}$, then $\{\overline{S}\}$ is a finite subcover. So, we can assume that $\overline{S} \notin \mathcal{C}$. Since $\mathcal{C}$ is a covering of $\overline{S}$, there is $V \in \mathcal{C}$ with $\infty \in V$. Therefore, $\overline{S} \setminus V$ is a compact subspace of S. Now, $\overline{S} \setminus V$ is covered by the sets in $\mathcal{C}$, and so, there is a finite subset $\mathcal{D} \subseteq \mathcal{C}$ such that $\mathcal{D}$ still covers $\overline{S} \setminus V$. Then $\mathcal{D} \cup \{V\}$ is the desired finite subcover. □

60. The one-point compactification of a locally compact T_2-space is a Tychonoff space.

Proof: Let $(\overline{S},\overline{\mathcal{T}})$ be the one-point compactification of $(S,\mathcal{T})$. By Problem 59, $(\overline{S},\overline{\mathcal{T}})$ is a compact T_2-space. By Problem 56, $(\overline{S},\overline{\mathcal{T}})$ is a T_4-space. By Corollary 14.81, $(\overline{S},\overline{\mathcal{T}})$is a Tychonoff space. □

61. Let $(A,\mathcal{T})$ and $(B,\mathcal{U})$ be topological spaces and let $f: A \to B$. Then f is continuous if and only if for each $X \subseteq A$, $f[\overline{X}] \subseteq \overline{f[X]}$.

Proof: Let $(A,\mathcal{T})$ and $(B,\mathcal{U})$ be topological spaces and let $f: A \to B$. First assume that f is continuous. Let $f(x) \in f[\overline{X}]$ and let U be an open set containing $f(x)$. Then $f^{-1}[U]$ is an open set containing x. Since $x \in \overline{X}$, there is $z \in X \cap f^{-1}[U]$. Since $z \in X$, $f(z) \in f[X]$. Since $z \in f^{-1}[U]$, $f(z) \in f[f^{-1}[U]]$. By part 3 of Example 2.61, $f[f^{-1}[U]] \subseteq U$. So, $f(z) \in U$. Thus, $f(z) \in f[X] \cap U$. By part 5 of Theorem 14.4, $f(x) \in \overline{f[X]}$. Since $f(x) \in f[\overline{X}]$ was arbitrary, $f[\overline{X}] \subseteq \overline{f[X]}$.

Conversely, assume that for each $X \subseteq A$, $f[\overline{X}] \subseteq \overline{f[X]}$. Let C be a closed set in B and let $D = f^{-1}[C]$. By part 3 of Example 2.61, $f[D] = f[f^{-1}[C]] \subseteq C$. Now, if $x \in \overline{D}$, then we have $f(x) \in f[\overline{D}]$. By our assumption, $f[\overline{D}] \subseteq \overline{f[D]}$. So, $f(x) \in \overline{f[D]} \subseteq \overline{C} = C$ (because C is closed in B). Therefore, we have $x \in f^{-1}[C] = D$. Since $x \in \overline{D}$ was arbitrary, $\overline{D} \subseteq D$. Since $D \subseteq \overline{D}$ is always true (by part 1 of Theorem 14.4), we have $D = \overline{D}$. By part 4 of Theorem 14.4, $D = f^{-1}[C]$ is closed in A. By Problem 9 above, f is continuous. □

62. An injective continuous function from a compact space onto a T_2-space is a homeomorphism.

Proof: Let $(S_1,\mathcal{T}_1)$ be a compact space, let $(S_2,\mathcal{T}_2)$ be a T_2-space, and let $f: S_1 \to S_2$ be a bijective continuous function.

We first prove that if K is a compact subspace of S_1, then $f[K]$ is a compact subspace of S_2. So, let K be a compact subspace of S_1, and let $\mathcal{C}$ be an open covering of $f[K]$ by open sets in S_2. Let $\mathcal{D} = \{f^{-1}[U] \mid U \in \mathcal{C}\}$. Since f is continuous, every element of $\mathcal{D}$ is open in S_1. If $x \in K$, then $f(x) \in f[K]$. Therefore, there is $U \in \mathcal{C}$ with $f(x) \in U$. It follows that $x \in f^{-1}[U]$. Since $x \in K$ was arbitrary, we see that $\mathcal{D}$ is an open covering of K by open sets in S_1. By the compactness of K and Lemma 14.54, there is a finite subcollection $\mathcal{E}$ of $\mathcal{D}$ that still covers K. Let $\mathcal{F} = \{U \in \mathcal{C} \mid f^{-1}[U] \in \mathcal{E}\}$. Then U is a finite subcollection of $\mathcal{C}$. Let $f(x) \in f[K]$. Then $x \in K$. Since $\mathcal{E}$ covers K, there is $f^{-1}[U] \in \mathcal{E}$ such that $x \in f^{-1}[U]$. So, $f(x) \in U$ and $U \in \mathcal{F}$. It follows that $\mathcal{F}$ is a finite subcollection of $\mathcal{C}$ that still covers $f[K]$. So, $f[K]$ is a compact subspace of S_2.

Now, let C be a closed set in S_1. If $C = \emptyset$, then $f[C] = \emptyset$, which is closed. Assume that $C \neq \emptyset$. By Theorem 14.56, C is compact in S_1. By the preceding paragraph, $f[C]$ is compact in S_2. By Theorem 14.55, $f[C]$ is closed in S_2. By Problem 9 above, f^{-1} is continuous. □

63. The components of a totally disconnected space are its points.

Proof: Let $(S,\mathcal{T})$ be a totally disconnected topological space and let $(A,\mathcal{T}_A)$ be a subspace of $(S,\mathcal{T})$ such that $|A| \geq 2$. Let $x, y \in A$ with $x \neq y$. Since $(S,\mathcal{T})$ is totally disconnected, there is a disconnection (U,V) of S with $x \in U$ and $y \in V$. Then $A = (A \cap U, A \cap V)$ is a disconnection of A with $x \in A \cap U$ and $y \in A \cap V$. □

64. An arbitrary product of totally disconnected spaces is totally disconnected.

Proof: Let K be an index set, let $(S_k,\mathcal{T}_k)$ be totally disconnected for each $k \in K$, and assume that $X \subseteq \prod S_k$ is connected. Then for each $k \in K$, $X_k = \{x_k \in S_k \mid \mathbf{x} \in X\}$ is connected. Since S_k is totally disconnected, by Problem 63, $X_k = \{a_k\}$ for some $a_k \in S_k$. So, $X = \{\mathbf{a}\}$. Thus, $\prod S_k$ is totally disconnected. □

65. Let $(S,\mathcal{T})$ be a T_2-space and let $\mathcal{B}$ be a basis for $\mathcal{T}$ such that each set in $\mathcal{B}$ is both open and closed. Then $(S,\mathcal{T})$ is totally disconnected.

Proof: Let $(S,\mathcal{T})$ be a T_2-space, let $\mathcal{B}$ be a basis for $\mathcal{T}$ such that each set in $\mathcal{B}$ is both open and closed, and let $x, y \in S$ with $x \neq y$. Since $(S,\mathcal{T})$ is a T_2-space, there are open sets U, V in S such that $x \in U$, $y \in V$, and $U \cap V = \emptyset$. Let $W \in \mathcal{B}$ with $x \in W$ and $W \subseteq U$. Then $(W, S \setminus W)$ is a disconnection of S with $x \in W$ and $y \in S \setminus W$. Since $x, y \in S$ were arbitrary, $(S,\mathcal{T})$ is totally disconnected. □

66. $(C,\mathcal{T}_C)$ is totally disconnected, where the Cantor set C is being considered as a subspace of $\mathbb{R}$ with the standard topology.

Proof: Let $x, y \in C$ with $x \neq y$. Let $K \in \mathbb{N}$ be such that $|x-y| > \frac{1}{3^K}$. Then x and y belong to different intervals in C_K. There is at least one interval between x and y that is disjoint from C_K, and therefore, from C. Let z be an element of that interval. Then $\big((-\infty, z) \cap C, (z, \infty) \cap C\big)$ is a disconnection of C separating x and y. □

67. Let $(S, \mathcal{T})$ be a topological space, let $(A, \mathcal{T}_A)$ be a connected subspace of $(S, \mathcal{T})$, and let $(B, \mathcal{T}_B)$ be a subspace of $(S, \mathcal{T})$ such that $A \subseteq B \subseteq \overline{A}$. Then $(B, \mathcal{T}_B)$ is connected.

Proof: Let $(S, \mathcal{T})$ be a topological space, let $(A, \mathcal{T}_A)$ be a connected subspace of $(S, \mathcal{T})$, and let $(B, \mathcal{T}_B)$ be a subspace of $(S, \mathcal{T})$ such that $A \subseteq B \subseteq \overline{A}$. Assume toward contradiction that B is disconnected. Let $(U \cap B, V \cap B)$ be a disconnection of B. Since A is connected and $A \subseteq B = (U \cap B) \cup (V \cap B)$, we have $A \subseteq U \cap B$ or $A \subseteq V \cap B$. Without loss of generality, assume that $A \subseteq U \cap B$. Therefore, $A \cap V = A \cap (V \cap B) = \emptyset$. So, $\overline{A} \cap V = \emptyset$. Since $B \subseteq \overline{A}$, $V \cap B = \emptyset$, contradicting that $(U \cap B, V \cap B)$ is a disconnection of B. □

Let $(S, \mathcal{T})$ be a topological space. Prove each of the following:

68. Every $x \in S$ is contained in exactly one component of S.

Proof: Define the relation $\sim$ on S by $x \sim y$ if and only if $x, y \in A$ for some connected subspace $(A, \mathcal{T}_A)$ of $(S, \mathcal{T})$. By Problem 23, $\sim$ is an equivalence relation on S. Since the equivalence classes of $\sim$ form a partition of S, x is in exactly one component of S. □

69. If $(A, \mathcal{T}_A)$ is a connected subspace of $(S, \mathcal{T})$, then A is contained in a component of S.

Proof: Define the relation $\sim$ on S by $x \sim y$ if and only if $x, y \in A$ for some connected subspace $(A, \mathcal{T}_A)$ of $(S, \mathcal{T})$. By Problem 23, $\sim$ is an equivalence relation on S. Let $(A, \mathcal{T}_A)$ be a connected subspace of $(S, \mathcal{T})$. Let $B = \{x \mid \exists y \in A(x \sim y)\}$. If $x \in A$, then since $x \sim x$ (because $\sim$ is reflexive), $x \in B$. Therefore, $A \subseteq B$. Suppose toward contradiction that $(B, \mathcal{T}_B)$ is disconnected and let (U, V) be a disconnection of B. Let $a \in U$ and $b \in V$. Since $a, b \in B$, there are $y, z \in A$ with $a \sim y$ and $b \sim z$. Since y, z are both in A, which is connected, $y \sim z$. Since $\sim$ is symmetric and transitive, $a \sim b$. Therefore, there is a connected subspace $(C, \mathcal{T}_C)$ of $(S, \mathcal{T})$ with $a, b \in C$. But then $a \in U \cap C$, $b \in V \cap C$, and $(U \cap C) \cap (V \cap C) = \emptyset$, a contradiction. So, $(B, \mathcal{T}_B)$ is connected. Let $a \in A$. We will now show that $B = [a]$. First, suppose that $x \in B$. Then there is $y \in A$ with $x \sim y$. Since $a \in A$, $y \sim a$. Since $\sim$ is transitive, $x \sim a$. So, $x \in [a]$. Since $x \in B$ was arbitrary, $B \subseteq [a]$. Now, let $x \in [a]$. Then $x \sim a$. So, the statement $\exists y \in A(x \sim y)$ is true (let $y = a$). Therefore, $x \in B$. Since $x \in [a]$ was arbitrary, $[a] \subseteq B$. Since $B \subseteq [a]$ and $[a] \subseteq B$, we have $B = [a]$. □

70. If $(A, \mathcal{T}_A)$ is a connected subspace of $(S, \mathcal{T})$ and A is both open and closed in S, then A is a component of S.

Proof: Let $(A, \mathcal{T}_A)$ be a connected subspace of $(S, \mathcal{T})$ that is both open and closed in S and let $a \in A$. By Problem 69, A is contained in some component B. If $A \neq B$, then $\big((A \cap B), ((S \setminus A) \cap B)\big)$ would be a disconnection of B. Therefore, $A = B$. □

71. Every component of S is closed in S.

Proof: Let A be a component of S and assume toward contradiction that A is not closed in S. By Problem 67, $\overline{A}$ is a connected subspace of S, and since A is not closed, $A \neq \overline{A}$. Let $a \in A$ and $b \in \overline{A} \setminus A$. Then $a \sim b$, and so b is in the same component as a. Since $b \notin A$, this contradicts our assumption that A is a component. Therefore, A is closed in S. □

LEVEL 5

Prove each of the following:

72. Let K be an index set and let $(S_k, \mathcal{T}_k)$ be a topological space for each $k \in K$. The collection $\mathcal{B} = \{\prod U_k \mid \forall k \in K(U_k \in \mathcal{T}_k) \wedge (U_k = S_k \text{ for all but finitely many } k)\}$ is a basis for a topology $\mathcal{T}$ on $\prod S_k$. Furthermore, if $\mathcal{B}_k$ is a basis for $\mathcal{T}_k$ for each $k \in K$, then the collection $\mathcal{C} = \{\prod U_k \mid \forall k \in K(U_k \in \mathcal{B}_k) \wedge (U_k = S_k \text{ for all but finitely many } k)\}$ is a basis for the same topology (this topology is called the **product topology** on $\prod S_k$). In both cases, if we remove "$U_k = S_k$ for all but finitely many k," we get a topology (called the **box topology**) that is finer than the product topology.

Proof: Let $(s_k) \in \prod S_k$. Since $S_k \in \mathcal{T}_k$ for each $k \in K$, $\prod S_k \in \mathcal{B}$. So, $\mathcal{B}$ covers $\prod S_k$.

Now, let $(s_k) \in \prod U_k \cap \prod V_k$, where $\prod U_k, \prod V_k \in \mathcal{B}$. Since $U_k, V_k \in \mathcal{T}_k$ for each $k \in K$, $U_k \cap V_k \in \mathcal{T}_k$ for each $k \in K$. Also, since $U_k = S_k$ for all but finitely many k and $V_k = S_k$ for all but finitely many k, it follows that $U_k \cap V_k = S_k$ for all but finitely many k. Therefore, $\prod(U_k \cap V_k) \in \mathcal{B}$. Since $\prod U_k \cap \prod V_k = \prod(U_k \cap V_k)$ (Check this!), we have $\prod U_k \cap \prod V_k \in \mathcal{B}$. So, $\mathcal{B}$ has the intersection containment property.

Since $\mathcal{B}$ covers $\prod S_k$ and $\mathcal{B}$ has the intersection containment property, it follows that $\mathcal{B}$ is a basis for a topology on $\prod S_k$.

Since $\mathcal{B}_k \subseteq \mathcal{T}_k$ for each $k \in K$, $\mathcal{C} \subseteq \mathcal{B}$. Therefore, the set generated by $\mathcal{C}$ is contained in the set generated by $\mathcal{B}$, which is $\mathcal{T}$.

Now, let $\prod U_k \in \mathcal{B}$ and let $(x_k) \in \prod U_k$ Then $x_k \in U_k$ for each $k \in K$. Since for each $k \in K$, $U_k \in \mathcal{T}_k$, for each $k \in K$, there is $U_k^{x_k} \in \mathcal{B}_k$ with $U_k^{x_k} \subseteq U_k$. Let $W = \bigcup\{\prod U_k^{x_k} \mid x_k \in U_k \text{ for each } k \in K\}$. Then $\prod U_k = W$ (Check this!). Therefore, $\prod U_k$ is a union of elements from $\mathcal{C}$. Thus, $\mathcal{T}$ (the set generated by $\mathcal{B}$) is contained in the set generated by $\mathcal{C}$. It follows that $\mathcal{B}$ and $\mathcal{C}$ generate the same set. Since $\mathcal{B}$ generates $\mathcal{T}$, so does $\mathcal{C}$. Therefore, $\mathcal{C}$ is a basis for $\mathcal{T}$.

Let $\mathcal{D} = \{\prod U_k \mid \forall k \in K(U_k \in \mathcal{T}_k)\}$. An argument nearly identical to the one given in the first three paragraphs above shows that $\mathcal{D}$ is a basis for a topology on $\prod S_k$. Clearly $\mathcal{B} \subseteq \mathcal{D}$, and therefore, the topology generated by D is finer than $\mathcal{T}$. □

73. Let K be an index set, let $(S_k, \mathcal{T}_k)$ be a topological space for each $k \in K$, and let $A_k \subseteq S_k$ for each $k \in K$. Then $\overline{\prod A_k} = \prod \overline{A_k}$.

Proof: First note that for any set X in a topological space, $x \in \overline{X}$ if and only if for every open set U containing x, we have $U \cap X \neq \emptyset$.

To see this, first assume that $x \in \overline{X}$ and let U be an open set containing x. If $x \in X$, then $x \in U \cap X$. So, assume that $x \notin X$. Then by part 3 of Theorem 14.4, x is an accumulation point of X. So, there is $y \in U \cap X$ with $y \neq x$.

Conversely, assume that for every open set U containing x, $U \cap X \neq \emptyset$. If $x \in X$, then since $X \subseteq \overline{X}$, we have $x \in \overline{X}$. So, assume that $x \notin X$. Let U be an open set containing x. Since $U \cap X \neq \emptyset$, there is $y \in U \cap X$ with $y \neq x$. Then x is an accumulation point of X. By part 3 of Theorem 14.4, $x \in \overline{X}$.

Now, let $(x_k) \in \overline{\prod A_k}$. We will show that for each $n \in K$, $x_n \in \overline{A_n}$. To see this, let V_n be an open set contining x_n. Let $U = \prod U_k$, where $U_n = V_n$ and $U_k = S_k$ for all $k \neq n$. Then U is an open set in $\prod S_k$ containing (x_k). Since $(x_k) \in \overline{\prod A_k}$, there is $(y_k) \in U \cap \prod A_k$. Then $y_n \in V_n \cap A_n$. Therefore, $x_n \in \overline{A_n}$. Since $n \in K$ was arbitrary, we see that $(x_k) \in \prod \overline{A_k}$. It follows that $\overline{\prod A_k} \subseteq \prod \overline{A_k}$.

Now, let $(x_k) \in \prod \overline{A_k}$ and for each $k \in K$, let $U_k \in \mathcal{T}_k$ with $U_k = S_k$ for all but finitely many k. Then $U = \prod U_k$ is a basis element for the product topology on $\prod S_k$. Let $k \in K$. Since $x_k \in \overline{A_k}$, we can choose $y_k \in U_k \cap A_k$. Then $(y_k) \in U \cap \prod A_k$. Thus, $(x_k) \in \overline{\prod A_k}$. It follows that $\prod \overline{A_k} \subseteq \overline{\prod A_k}$.

Since $\overline{\prod A_k} \subseteq \prod \overline{A_k}$ and $\prod \overline{A_k} \subseteq \overline{\prod A_k}$, we have $\overline{\prod A_k} = \prod \overline{A_k}$. □

74. Let J and K be index sets. Then $\bigcap\{\prod_{k \in K} B_k^j \mid j \in J\} = \prod_{k \in K}(\bigcap\{B_k^j \mid j \in J\})$.

Proof: $\mathbf{x} \in \bigcap\{\prod_{k \in K} B_k^j \mid j \in J\}$ if and only if $\forall j \in J(\mathbf{x} \in \prod_{k \in K} B_k^j)$ if and only $\forall j \in J\ \forall k \in K(x_k \in B_k^j)$ if and only if $\forall k \in K\ \forall j \in J(x_k \in B_k^j)$ if and only if $\forall k \in K\ (x_k \in \bigcap\{B_k^j \mid j \in J\})$ if and only if $\mathbf{x} \in \prod_{k \in K}(\bigcap\{B_k^j \mid j \in J\})$ □

75. Consider the topological space $(\mathbb{R}, \mathcal{T}_L)$ (see Problem 50 above). $\mathbb{R}^2$ with the corresponding product topology is a T_3-space, but not a T_4-space.

Proof: We first show that $(\mathbb{R}, \mathcal{T}_L)$ is a T_4-space. Since $\mathcal{T}_L$ is finer than the standard topology on $\mathbb{R}$ (by Problem 50), and the standard topology is a T_1-space, $\mathcal{T}_L$ is also a T_1-space. Now, let A, B be disjoint closed subsets of $\mathbb{R}$. For each $a \in A$, we have $a \notin B$. So, $a \in \mathbb{R} \setminus B$. Since $\mathbb{R} \setminus B$ is open, there is a basic open set $[c, x_a)$ containing a such that $[c, x_a) \subseteq \mathbb{R} \setminus B$. Since $c \leq a$, we have $[a, x_a) \subseteq [c, x_a)$, and therefore, $[a, x_a) \subseteq \mathbb{R} \setminus B$. So, $[a, x_a) \cap B = \emptyset$. Similarly, for each $b \in B$, we can find x_b so that $[b, x_b) \cap A = \emptyset$. Let $U = \bigcup\{[a, x_a) \mid a \in A\}$ and let $V = \bigcup\{[b, x_b) \mid b \in B\}$. U and V are unions of basic open sets, thus open. Clearly, $A \subseteq U$ and $B \subseteq V$.

We show that $U \cap V = \emptyset$. Suppose toward contradiction that $z \in U \cap V$. Then there is $a \in A$ and $b \in B$ with $z \in [a, x_a)$ and $z \in [b, x_b)$. Without loss of generality, assume that $a < b$. Since $z \in [b, x_b)$, we have $b \leq z$. Since $z \in [a, x_a)$, we have $z < x_a$. So, $a < b \leq z < x_a$. It follows that $b \in [a, x_a)$, contradicting $[a, x_a) \cap B = \emptyset$. This contradiction shows that $U \cap V = \emptyset$. So, $(\mathbb{R}, \mathcal{T}_L)$ is a T_4-space.

Since $(\mathbb{R}, \mathcal{T}_L)$ is a T_4-space, it is also a T_3-space.

Let $\mathcal{T}$ be the product topology on $\mathbb{R}^2$ with respect to the topology $\mathcal{T}_L$. By the proof of Problem 51, $(\mathbb{R}^2, \mathcal{T})$ is a T_3-space.

We will now show that $(\mathbb{R}^2, \mathcal{T})$ is **not** a T_4-space.

Assume toward contradiction that $(\mathbb{R}^2, \mathcal{T})$ is a T_4-space. Let $D = \{(x, -x) \mid x \in \mathbb{R}\}$. D is a closed set in the standard product topology on $\mathbb{R}$ (as are all lines). Since $\mathcal{T}_L$ is finer than the standard topology on $\mathbb{R}$, D is also closed in $(\mathbb{R}^2, \mathcal{T})$.

Furthermore, $\mathcal{T}_D$ is the discrete topology on D. To see this, observe that the point $(x, -x)$ is equal to the intersection of D with the basic open set $[x, x+1) \times [-x, -x+1)$. Therefore, every singleton set $\{(x, -x)\}$ is open in $\mathcal{T}_D$. It follows that all subsets of D are both open and closed in $(D, \mathcal{T}_D)$.

If $A \subseteq D$, since A is closed in $(D, \mathcal{T}_D)$ and D is closed in $(\mathbb{R}^2, \mathcal{T})$, it follows that, A is closed in $(\mathbb{R}^2, \mathcal{T})$ (by Problem 28). So, for any $A \subseteq D$ with $A \neq \emptyset$ and $A \neq D$, both A and $D \setminus A$ are closed in $(\mathbb{R}^2, \mathcal{T})$. Since we are assuming that $(\mathbb{R}^2, \mathcal{T})$ is a T_4-space, we can find disjoint $U_A, V_A \in \mathcal{T}$ with $A \subseteq U_A$ and $D \setminus A \subseteq V_A$.

Define $f: \mathcal{P}(D) \to \mathcal{P}(\mathbb{Q} \times \mathbb{Q})$ by $f(\emptyset) = \emptyset$, $f(D) = \mathbb{Q} \times \mathbb{Q}$, and $f(A) = (\mathbb{Q} \times \mathbb{Q}) \cap U_A$ for $A \neq \emptyset$ and $A \neq D$. We show that f is injective.

Let $A, B \in \mathcal{P}(D)$, both nonempty, both not equal to D or each other. Without loss of generality, assume there is $(x, -x) \in A \setminus B$. Then $(x, -x) \in D \setminus B$. Therefore, $(x, -x) \in U_A \cap V_B$. Since $U_A \cap V_B$ is open and nonempty, by the Density Theorem, there is $(a, -a) \in (U_A \cap V_B) \cap (\mathbb{Q} \times \mathbb{Q})$. Therefore, we have $(a, -a) \in (\mathbb{Q} \times \mathbb{Q}) \cap U_A = f(A)$ and $(a, -a) \notin (\mathbb{Q} \times \mathbb{Q}) \cap U_B = f(B)$. So, $f(A) \neq f(B)$.

Also, note that if $A \in \mathcal{P}(D)$, with $A \neq \emptyset$ and $A \neq D$, then $f(A) = (\mathbb{Q} \times \mathbb{Q}) \cap U_A$ is not empty because $\mathbb{Q} \times \mathbb{Q}$ has nonempty intersection with any open set, and $f(A) = (\mathbb{Q} \times \mathbb{Q}) \cap U_A$ is not $\mathbb{Q} \times \mathbb{Q}$ because $(\mathbb{Q} \times \mathbb{Q}) \cap V_A \neq \emptyset$. It follows that f is injective.

So, $\mathcal{P}(D) \preccurlyeq \mathcal{P}(\mathbb{Q} \times \mathbb{Q}) \sim \mathcal{P}(\mathbb{Q}) \sim \mathcal{P}(\mathbb{N}) \sim \mathbb{R} \sim D$, contradicting Cantor's Theorem. □

76. $(\mathbb{R}, \mathcal{T}_L)$ is not metrizable (see Problem 50 above for the definition of $\mathcal{T}_L$).

Proof: Assume towards contradiction that $(\mathbb{R}, \mathcal{T}_L)$ is metrizable. By Problem 54, $\mathbb{R}^2$ with the corresponding product topology is metrizable. By Problem 53, $\mathbb{R}^2$ with the product topology is a T_4-space, contradicting Problem 75. So, $(\mathbb{R}, \mathcal{T}_L)$ is **not** metrizable. □

77. Let K be a nonempty set and let S_k be a T_2-space for each $k \in K$. Then $\prod S_k$ (with the product topology) is a T_2-space. Determine if the same result is true if we replace T_2 by each of T_3 and T_4.

Proof: Let K be a nonempty set, let S_k be a T_2-space for each $k \in K$, and let $\mathbf{x}, \mathbf{y} \in \prod S_k$ with $\mathbf{x} \neq \mathbf{y}$. Then there is $n \in K$ such that $x_n \neq y_n$. Since S_n is a T_2-space, there are open sets U_n and V_n in S_n such that $x_n \in U_n$, $y_n \in V_n$, and $U_n \cap V_n = \emptyset$. Then $\mathbf{x} \in \prod U_k$, where $U_k = S_k$ for each $k \neq n$, and $\mathbf{y} \in \prod V_k$, where $V_k = S_k$ for each $k \neq n$. We have

$$\prod U_k \cap \prod V_k = \prod (U_k \cap V_k) = \emptyset \text{ because } U_n \cap V_n = \emptyset.$$

Since $\mathbf{x}, \mathbf{y} \in \prod S_k$ were arbitrary, $\prod S_k$ with the product topology is a T_2-space.

Let K be a nonempty set, let S_k be a T_3-space for each $k \in K$. We will use Problem 15 to show that $\prod S_k$ with the product topology is also a T_3-space.

Let $\mathbf{x} \in \prod S_k$ and let $U = \prod U_k$ be a basic open set with $\mathbf{x} \in U$. If $U_k = S_k$, let $V_k = S_k$ as well. If $U_k \neq S_k$, by Problem 15, we can find an open set V_k with $x_k \in V_k$ and $\overline{V}_k \subseteq U_k$. Then $V = \prod V_k$ is a basic open set with $\mathbf{x} \in V$ and $\overline{V} \subseteq U$ (by Problem 73). Once again, by Problem 15, $\prod S_k$ is a T_3-space.

We can use the counterexample from Problem 75 to see that in general the result is **not** true for T_4-spaces. □

78. Let K be a nonempty set and let S_k be a nonempty topological space for each $k \in K$. If $\prod S_k$ (with the product topology) is a T_2-space, then so is S_k for each $k \in K$. Determine if the same result is true if we replace T_2 by each of T_3 and T_4.

Proof: Let K be a nonempty set, let S_k be a topological space for each $k \in K$, and suppose that $\prod S_k$ is a T_2-space. Let $k \in K$ and let $a, b \in S_n$ with $a \neq b$. For each $k \in K$ with $k \neq n$, choose $z_k \in S_k$. Define $\mathbf{x}, \mathbf{y} \in \prod S_k$ by $x_n = a$, $y_n = b$, and $x_k = z_k$, $y_k = z_k$ if $k \neq n$. Since $\prod S_k$ is a T_2-space, there are basic open sets $\prod U_k$ and $\prod V_k$ with $\mathbf{x} \in \prod U_k$, $\mathbf{y} \in \prod V_k$ and $\prod U_k \cap \prod V_k = \emptyset$. We then have $a = x_n \in U_n$, $b = y_n \in V_n$, and $U_n \cap V_n = \emptyset$. (If $c \in U_n \cap V_n$, then define $\mathbf{w}$ by $w_n = c$ and $w_k = z_k$ for $k \neq n$. Then we have $\mathbf{w} \in \prod U_k \cap \prod V_k$, a contradiction.)

Next, let K be a nonempty set, let S_k be a topological space for each $k \in K$, and suppose that $\prod S_k$ is a T_3-space. Let $k \in K$ and let $a \in S_n$ and let A be a closed set in S_n not containing a. For each $k \in K$ with $k \neq n$, choose $z_k \in S_k$. Define $\mathbf{x} \in \prod S_k$ by $x_n = a$ and $x_k = z_k$ if $k \neq n$. Define $C = \prod C_k$, where $C_n = A$ and $C_k = S_k$ for $k \neq n$. Then C is closed in $\prod S_k$ and C does not contain $\mathbf{x}$. Since $\prod S_k$ is a T_3-space, there are basic open sets $\prod U_k$ and $\prod V_k$ with $\mathbf{x} \in \prod U_k$, $C \subseteq \prod V_k$ and $\prod U_k \cap \prod V_k = \emptyset$. We then have $a = x_n \in U_n$, $A = C_n \subseteq V_n$, and $U_n \cap V_n = \emptyset$. (If $c \in U_n \cap V_n$, then define $\mathbf{w}$ by $w_n = c$ and $w_k = z_k$ for $k \neq n$. Then we have $\mathbf{w} \in \prod U_k \cap \prod V_k$, a contradiction.)

Finally, let K be a nonempty set, let S_k be a topological space for each $k \in K$, and suppose that $\prod S_k$ is a T_4-space. Let $k \in K$ and let A and B be a disjoint closed set in S_n. Define $C = \prod C_k$ and $D = \prod D_k$, where $C_n = A$, $D_n = B$, and $C_k = D_k = S_k$ for $k \neq n$. Then C and D are disjoint closed sets in $\prod S_k$. Since $\prod S_k$ is a T_4-space, there are basic open sets $\prod U_k$ and $\prod V_k$ with $C \subseteq \prod U_k$, $D \subseteq \prod V_k$ and $\prod U_k \cap \prod V_k = \emptyset$. We then have $A = C_n \subseteq U_n$, $B = D_n \subseteq V_n$, and $U_n \cap V_n = \emptyset$. (If $c \in U_n \cap V_n$, then define $\mathbf{w}$ by $w_n = c$ and for $k \neq n$, $w_k = z_k$ for some $z_k \in S_k$. Then we have $\mathbf{w} \in \prod U_k \cap \prod V_k$, a contradiction.) □

79. An uncountable product of first-countable topological spaces need not be first-countable.

Proof: Let K be an uncountable index set and for each $k \in K$, consider $S_k = \mathbb{Z}$ with the discrete topology. By part 1 of Example 14.34, each S_k is first-countable. We now show that $\prod S_k = {}^K\mathbb{Z}$ is not first-countable. To see this, let $\mathbf{x} \in {}^K\mathbb{Z}$ and let $\mathcal{D}$ be a countable family of basic open sets containing $\mathbf{x}$. Then for each $\prod B_k \in \mathcal{D}$, we have $B_k = S_k$ for all but finitely many $k \in \mathbb{Z}$. Therefore, there is $j \in K$ such that for every product $\prod B_k \in \mathcal{D}$, we have the jth coordinate set $B_j = S_j$. Let $C = \prod C_k$ be defined by $C_k = S_k$ for $k \neq j$ and $C_j = \{x_j\}$. Then C is an open set containing $\mathbf{x}$ that does not contain any open set in $\mathcal{D}$. It follows that $\mathcal{D}$ is not a countable basis at $\mathbf{x}$. Therefore, ${}^K\mathbb{Z}$ with the product topology is **not** first-countable. □

80. $(\mathbb{R}, d)$, where d is the Euclidean metric, has the Bolzano-Weierstrass Property.

You may use the following fact: if $C = \{[a_k, b_k] \mid k \in \mathbb{N}\}$ is a sequence of closed intervals in $\mathbb{R}$, such that $j < k \to [a_k, b_k] \subseteq [a_j, b_j]$, then $\bigcap C \neq \emptyset$ (This fact follows from Example 14.51 and the Heine-Borel Theorem (Theorem 6.28)).

Proof: Let (s_n) be a bounded sequence in $\mathbb{R}$. Since (s_n) is bounded, there are $a, b \in \mathbb{R}$ such that for all $n \in \mathbb{N}$, $a \leq s_n \leq b$, or equivalently, for all $n \in \mathbb{N}$, $s_n \in [a, b]$.

Let $a_0 = a$, $b_0 = b$, and $c_0 = \frac{a_0+b_0}{2}$. Either $[a_0, c_0]$ or $[c_0, b_0]$ contains infinitely many terms of the sequence. If $[a_0, c_0]$ contains infinitely many terms of the sequence, let $a_1 = a_0$, $b_1 = c_0$. Otherwise, let $a_1 = c_0$ and $b_1 = b_0$. Then let $c_1 = \frac{a_1+b_1}{2}$. Continuing in this way, we see that at stage n, we have that $[a_n, c_n]$ or $[c_n, b_n]$ contain infinitely many terms of the sequence. If $[a_n, c_n]$ contains infinitely many terms of the sequence, let $a_{n+1} = a_n$, $b_{n+1} = c_n$. Otherwise, let $a_{n+1} = c_n$ and $b_{n+1} = b_n$. Then let $c_{n+1} = \frac{a_{n+1}+b_{n+1}}{2}$. This gives us a sequence of nested intervals $[a_0, b_0] \supseteq [a_1, b_1] \supseteq \cdots \supseteq [a_n, b_n] \supseteq \cdots$, where each interval $[a_n, b_n]$ contains infinitely many terms of the given sequence and the length of $[a_n, b_n]$ is $\frac{b-a}{2^n}$. It follows that $\bigcap\{[a_n, b_n] \mid n \in \mathbb{N}\}$ contains at most one real number. By the fact given in the problem, we have $\bigcap\{[a_n, b_n] \mid n \in \mathbb{N}\} \neq \emptyset$. Therefore, $\bigcap\{[a_n, b_n] \mid n \in \mathbb{N}\}$ contains exactly one real number x.

We now construct a subsequence of (s_n) converging to x. Let $s_{n_0} = s_0$. Given $\left(s_{n_j}\right)_{j \leq k}$ with $n_0 < n_1 < \cdots$ and $s_{n_j} \in [a_j, b_j]$ for each $j \leq k$, we can choose $s_{n_{k+1}} \in [a_{k+1}, b_{k+1}]$ with $n_{k+1} > n_k$ (because $\left[a_{n_{k+1}}, b_{n_{k+1}}\right]$ contains infinitely many terms of the sequence).

We claim that $s_{n_k} \to x$. To see this, let $k \in \mathbb{N}^+$. Since $\frac{b-a}{2^n} \to 0$, there is $K \in \mathbb{N}$ such that $n > K$ implies $\frac{b-a}{2^n} = \left|\frac{b-a}{2^n}\right| < \frac{1}{k}$. Since $s_{n_k} \in [a_k, b_k]$ and $x \in [a_k, b_k]$, we have that $|s_{n_k} - x\,|$ is less than or equal to the length of $[a_k, b_k]$, and so, $|s_{n_k} - x\,| \leq \frac{b-a}{2^{n_k}} < \frac{1}{k}$. It follows that $s_{n_k} \to x$. □

81. Let (S, d) be a metric space. For any index set K, the uniform metric on ${}^K S$ is finer than the product topology on ${}^K S$.

Proof: Let $\mathbf{x} \in {}^K S$ and let $U = \prod U_k$ be a basic open set in the product topology of ${}^K S$. So, $U_k = S$ for all but finitely $k \in K$, say $U_k = S$ for all $k \in K \setminus L$, where L is a finite set.

For each $j \in L$, there is $r_j \in \mathbb{R}^+$ with $r_j < 1$ such that $B_{r_j}(x_j; d) \subseteq U_j$. Let $r = \min\{r_j \mid j \in L\}$. Suppose $\rho(\mathbf{x}, \mathbf{y}) < r$. Then for each $j \in L$, $d(x_j, y_j) = \overline{d}(x_j, y_j) \leq \sup\{\overline{d}(x_k, y_k) \mid k \in K\} = \rho(\mathbf{x}, \mathbf{y}) < r$. So, for all $j \in L$, $y_j \in B_r(x_j; d)$. Since $B_r(x_j; d) \subseteq U_j$, we have $y_j \in U_j$. If $j \in K \setminus L$, then $U_j = S$, and so, we trivially have that $y_j \in U_j$. So, if $\mathbf{y} \in B_r(\mathbf{x}; \rho)$, then $\mathbf{y} \in U$. Thus, $B_r(\mathbf{x}; \rho) \subseteq U$. Therefore, the uniform metric on ${}^K S$ is finer than the product topology on ${}^K S$. □

82. For any index set K, $({}^K\mathbb{R}, \rho)$ is a complete metric space, where ρ is the uniform metric on ${}^K\mathbb{R}$.

Proof: By Theorem 14.47, $(\mathbb{R}, d)$ is a complete metric space. By Problem 24 above, the Cauchy sequences in $(\mathbb{R}, d)$ and $(\mathbb{R}, \overline{d})$ are the same. It follows that $(\mathbb{R}, \overline{d})$ is also a complete metric space.

Now, let $(\mathbf{x}^n)$ be a Cauchy sequence in $({}^K\mathbb{R}, \rho)$, let $k \in K$ and let $r \in \mathbb{R}^+$. Then there is $L \in \mathbb{N}$ such that $m \geq n > L$ implies $\rho(\mathbf{x}^m, \mathbf{x}^n) < r$. Therefore, $m \geq n > L$ implies $\sup\{\overline{d}(x_j^m, x_j^n) \mid j \in \mathbb{N}\} < r$. So, for each $k \in K$, $\overline{d}(x_k^m, x_k^n) \leq \sup\{\overline{d}(x_j^m, x_j^n) \mid j \in \mathbb{N}\} = \rho(\mathbf{x}^m, \mathbf{x}^n) < r$. It follows that for each $k \in K$, (x_k^n) is a Cauchy sequence in $(\mathbb{R}, \overline{d})$. Since $(\mathbb{R}, \overline{d})$ is complete, for each $k \in K$, there is x_k such that $x_k^n \to x_k$ in $(\mathbb{R}, \overline{d})$. Let $\mathbf{x} = (x_k) \in {}^K\mathbb{R}$.

We now show that $\mathbf{x}^n \to \mathbf{x}$ in $({}^K\mathbb{R}, \rho)$. To that end, let $r \in \mathbb{R}^+$. Since $(\mathbf{x}^n)$ is a Cauchy sequence in $({}^K\mathbb{R}, \rho)$, there is $L \in \mathbb{N}$ such that $m \geq n > L$ implies $\rho(\mathbf{x}^n, \mathbf{x}^m) < \frac{r}{2}$.

Let $m \geq n > L$. Then for each $k \in K$, we have

$$\overline{d}(x_k^n, x_k) \leq \overline{d}(x_k^n, x_k^m) + \overline{d}(x_k^m, x_k) \leq \sup\{\overline{d}(x_j^n, x_j^m) \mid j \in \mathbb{N}\} + \overline{d}(x_k^m, x_k)$$
$$= \rho(\mathbf{x}^n, \mathbf{x}^m) + \overline{d}(x_k^m, x_k) < \frac{r}{2} + \overline{d}(x_k^m, x_k)$$

Let $k \in K$ and choose $m > L$ such that $\overline{d}(x_k^m, x_k) < \frac{r}{2}$ (we can do this because $x_k^n \to x_k$ in $(\mathbb{R}, \overline{d})$). Since the left -hand side does not depend on m, we get that for $n > L$ and $k \in K$,

$$\overline{d}(x_k^n, x_k) < \frac{r}{2} + \frac{r}{2} = r.$$

So, for $n > L$, we have $\rho(\mathbf{x}^n, \mathbf{x}) \leq r$. Therefore, $\mathbf{x}^n \to \mathbf{x}$ in the uniform metric on ${}^K\mathbb{R}$. It follows that $({}^K\mathbb{R}, \rho)$ is a complete metric space. □

83. A locally compact T_2-space is a T_3-space.

Proof: Let $(S, \mathcal{T})$ be a locally compact T_2-space, let $x \in S$, and let A be a closed set $A \subseteq S \setminus \{x\}$. Since $(S, \mathcal{T})$ is locally compact, there is a compact subspace $(K, \mathcal{T}_K)$ such that K contains a neighborhood W of x. By Theorem 14.55, K is closed in S. So, $A \cap K$ is closed in S. By Theorem 14.56, $(A \cap K, \mathcal{T}_{A \cap K})$ is a compact subspace of $(S, \mathcal{T})$. By Problem 55 above, there are disjoint open sets V_1 and V_2 with $x \in V_1$ and $A \cap K \subseteq V_2$. Let $U = V_1 \cap W$ and $V = V_2 \cup (S \setminus K)$. Then U and V are open and disjoint, $x \in U$ and $A \subseteq V$. So, $(S, \mathcal{T})$ is a T_3-space. □

84. Lebesgue's Covering Lemma (Theorem 14.60)

Proof: Let (S,d) be a sequentially compact metric space and let $\mathcal{C}$ be an open cover of S. If every subset of S is contained in a member of $\mathcal{C}$, then every positive real number is a Lebesgue number for $\mathcal{C}$. So, we may assume that there exist subsets of S that are not contained in any member of $\mathcal{C}$. Let $c = \inf\{\text{diam } A \mid A \subseteq S \wedge A \text{ is not contained in any member of } \mathcal{C}\}$. If $x = +\infty$, then once again every positive real number is a Lebesgue number for $\mathcal{C}$. So, we may assume that c is a nonnegative real number. Also, if $c > 0$, then c is a Lebesgue number for $\mathcal{C}$. Assume toward contradiction that $c = 0$. Then for each $n \in \mathbb{Z}^+$, there is a subset A_n of S such that A_n is not contained in any member of $\mathcal{C}$ and $0 < \text{diam } A_n < \frac{1}{n+1}$. Let $s_n \in A_n$ for each $n \in \mathbb{N}$. Since (S,d) is sequentially compact, the sequence (s_n) has a subsequence that converges to some $s \in S$. Since $\mathcal{C}$ is an open cover of S, there is $U \in \mathcal{C}$ such that $s \in U$. Since U is open, there is $r \in \mathbb{R}^+$ such that $B_r(s) \subseteq U$. Since $s_n \to s$, there are infinitely many natural number n for which $s_n \in B_{\frac{r}{2}}(s)$. Choose $K \in \mathbb{N}$ such that $s_K \in B_{\frac{r}{2}}(s)$ and $\frac{1}{K} < \frac{r}{2}$. Since $\text{diam } A_K < \frac{1}{K} < \frac{r}{2}$, we have $A_K \subseteq B_r(s)$ (check this!), and therefore, $A_K \subseteq U$, contradicting our assumption that A_K is not contained in any member of $\mathcal{C}$. □

85. Let $(S,\mathcal{T})$ be a topological space. A is nowhere dense in S if and only if for any nonempty open set $U \subseteq S$, there is a nonempty open set V with $V \subseteq U$ and $V \cap A = \emptyset$.

Proof: First suppose that A is nowhere dense in S and let $U \subseteq S$ be a nonempty open set. If $U \subseteq \overline{A}$, then since U is open, $U \subseteq (\overline{A})^\circ$, contradicting that $(\overline{A})^\circ = \emptyset$. So, $V = U \cap (S \setminus \overline{A}) \neq \emptyset$. Therefore, V is a nonempty open set with $V \subseteq U$. Since $A \subseteq \overline{A}$, we have $S \setminus \overline{A} \subseteq S \setminus A$. Since $V \subseteq S \setminus \overline{A}$, we have $V \subseteq S \setminus A$, and therefore, $V \cap A = \emptyset$.

Conversely, assume that A is **not** nowhere dense. Then there is $U \subseteq \overline{A}$ with $U \neq \emptyset$ and U open in S. Let $V \subseteq U$ be a nonempty open set. Since $V \subseteq \overline{A}$, $V \cap A \neq \emptyset$. So, U is an open set that does **not** contain an open set that does not intersect A. □

86. Let $(S,\mathcal{T})$ be a topological space. The union of finitely many sets that are nowhere dense in S is nowhere dense in S.

Proof: Let's first show that the union of two nowhere dense sets is nowhere dense. So, let A and B be nowhere dense in S. Let $U \subseteq S$ be a nonempty open set. Since A is nowhere dense, by Problem 85, there is a nonempty open set V such that $V \subseteq U$ and $V \cap A = \emptyset$. Similarly, since B is nowhere dense, by Problem 85, there is a nonempty open set W such that $W \subseteq V$ and $W \cap B = \emptyset$. It follows that $W \subseteq U$ and $W \cap (A \cup B) = (W \cap A) \cup (W \cap B) \subseteq (V \cap A) \cup (W \cap B) = \emptyset \cup \emptyset = \emptyset$. By Problem 85 again, $A \cup B$ is nowhere dense.

We proceed by induction. Assuming that the union of k sets that are nowhere dense in S is nowhere dense in S, let $A_0, \dots, A_k$ be sets that are nowhere dense in S. By the inductive assumption, $A_0 \cup \cdots \cup A_{k-1}$ is nowhere dense in S. Finally, by the base case, we have

$$A_0 \cup \cdots \cup A_k = (A_0 \cup \cdots \cup A_{k-1}) \cup A_k,$$

and so, $A_0 \cup \cdots \cup A_k$ is nowhere dense in S. □

87. Let $(S, \mathcal{T})$ be a topological space. A is nowhere dense in S if and only if $(S \setminus A)^\circ$ is dense in S.

Proof: Let $(S, \mathcal{T})$ be a topological space. First note that for any subset $X \subseteq S$, we have the following equalities: $S \setminus \overline{X} = (S \setminus X)^\circ$ and $S \setminus X^\circ = \overline{S \setminus X}$.

To see that $S \setminus \overline{X} = (S \setminus X)^\circ$, note that we have

$$x \in S \setminus \overline{X} \Leftrightarrow x \notin \overline{X} \Leftrightarrow \text{There is an open set } U \text{ with } x \in U \text{ and } U \cap X = \emptyset$$
$$\Leftrightarrow \text{There is an open set } U \text{ with } x \in U \text{ and } U \subseteq S \setminus X \Leftrightarrow x \in (S \setminus X)^\circ.$$

To see that $S \setminus X^\circ = \overline{S \setminus X}$, note that we have

$$x \in S \setminus X^\circ \Leftrightarrow x \notin X^\circ \Leftrightarrow \text{For every open set } U \text{ with } x \in U, U \cap (S \setminus X) \neq \emptyset \Leftrightarrow x \in \overline{S \setminus X}.$$

Now, let $A \subseteq S$. We have

$$A \text{ is nowhere dense in } S \Leftrightarrow \left(\overline{A}\right)^\circ = \emptyset \Leftrightarrow S \setminus \left(\overline{A}\right)^\circ = S \Leftrightarrow \overline{S \setminus \overline{A}} = S \Leftrightarrow \overline{(S \setminus A)^\circ} = S.$$

This proves the result. □

88. Let $(K, \mathcal{T})$ and $(B, \mathcal{U})$ be metrizable topological spaces with $(K, \mathcal{T})$ compact and let $f: K \to B$ be continuous. Then f is uniformly continuous.

Proof: Let $(K, \mathcal{T})$ and $(B, \mathcal{U})$ be metrizable topological spaces with $(K, \mathcal{T})$ compact induced by metrics d and ρ, respectively. Let $f: K \to B$ be continuous, let $\epsilon > 0$, and let $\mathcal{C} = \left\{B_{\frac{\epsilon}{2}}(y; \rho) \middle| y \in B\right\}$. Then $\mathcal{C}$ is an open covering of B. The set $\mathcal{D} = \left\{f^{-1}\left[B_{\frac{\epsilon}{2}}(y; \rho)\right] \middle| y \in B\right\}$ is an open covering of K. Let δ be a Lebesgue number for $\mathcal{D}$. If $a, b \in K$ with $d(a, b) < \delta$, then $\operatorname{diam}\{a, b\} < \delta$. So, there is $y \in B$ such that $\{a, b\} \subseteq f^{-1}\left[B_{\frac{\epsilon}{2}}(y; \rho)\right]$. Therefore, $\{f(a), f(b)\} \in f\left[f^{-1}\left[B_{\frac{\epsilon}{2}}(y; \rho)\right]\right] \subseteq B_{\frac{\epsilon}{2}}(y; \rho)$ (by part 3 of Example 2.61). It follows that $\rho\big(f(a), f(b)\big) \leq \rho(f(a), y) + \rho\big(y, f(b)\big) < \frac{\epsilon}{2} + \frac{\epsilon}{2} = \epsilon$. □

89. Let $(A, \mathcal{T})$ be a metrizable topological space and let $(B, \mathcal{U})$ be a completely metrizable topological space. Let $C \subseteq A$ be dense in A and suppose that $f: C \to B$ is uniformly continuous. Then f can be extended uniquely to a uniformly continuous function $g: A \to B$. If we replace "uniformly continuous" by "continuous," is the result still true?

Proof: Let $(A, \mathcal{T})$ be a metrizable topological space and let $(B, \mathcal{U})$ be a completely metrizable topological spaces induced by metrics d and ρ, respectively. Let $C \subseteq A$ be dense in A and let $f: C \to B$ be uniformly continuous. If $C = A$, we can let $g = f$. So, assume that $C \neq A$. If $x \in C$, let $g(x) = f(x)$. Now, let $x \in A \setminus C$. Since C is dense in A, $\overline{C} = A$. Therefore, there is a sequence (x_n) in C such that $x_n \to x$. By Theorem 14.42, (x_n) is a Cauchy sequence in C. By a proof similar to the solution to Problem 53 in Problem Set 6, $\big(f(x_n)\big)$ is a Cauchy sequence in B. Since (B, ρ) is a complete metric space, $\big(f(x_n)\big)$ converges to some point y in B. Let $g(x) = y$. If (z_n) is another Cauchy sequence in C such that $z_n \to x$, then $d(x_n, z_n) \to 0$. Since f is uniformly continuous, $\rho\big(f(x_n), f(z_n)\big) \to 0$. Therefore, $f(z_n) \to y = g(x)$.

We need to show that g is uniformly continuous. To this end, let $\epsilon > 0$. Since f is uniformly continuous, there is $\delta > 0$ such that for all $x, z \in C$, $d(x, z) < \delta \to \rho\big(f(x), f(z)\big) < \epsilon$. Suppose that $a, b \in A$ with $d(a, b) < \delta$. Let $(a_n), (b_n)$ be sequences in C such that $a_n \to a$ and $b_n \to b$. We then have that $d\{a_n, b_n) \leq d(a_n, a) + d(a, b) + d(b, b_n)$. Since $d(a_n, a) \to 0$, $d(b, b_n) \to 0$, and $d(a, b) < \delta$, we can find $K \in \mathbb{N}$ such that for all $n > K$, $d(a_n, b_n) < \delta$. So, $n > K \to \rho\big(f(a_n), f(b_n)\big) < \epsilon$. From this it is easy to show that $\rho\big(g(a), g(b)\big) \leq \epsilon$.

The proof that g is unique is straightforward.

If we replace "uniformly continuous" by "continuous," the result is still true. The proof is similar. □

90. Let $(S, \mathcal{T})$ be a compact T_2-space. $\mathcal{T}$ has a basis $\mathcal{B}$ such that each set in $\mathcal{B}$ is both open and closed if and only if $(S, \mathcal{T})$ is totally disconnected.

Proof: Let $(S, \mathcal{T})$ be a T_2-space. If $\mathcal{T}$ has a basis $\mathcal{B}$ such that each set in $\mathcal{B}$ is both open and closed, then by Problem 65 above, $(S, \mathcal{T})$ is totally disconnected.

Conversely, assume that $(S, \mathcal{T})$ is totally disconnected. We will show that the collection of clopen sets in S is a basis for S. Let $x \in S$ and let U be an open set in S containing x. If $U = S$, then S is a clopen set containing x with $S \subseteq U$. So, assume $U \neq S$. Since U is open in S, $S \setminus U$ is closed in S. Since S is compact, $S \setminus U$ is also compact. Since $(S, \mathcal{T})$ is totally disconnected, for each $y \in S \setminus U$, there is a clopen set V_y with $y \in V_y$ and $x \notin V_y$. The collection $\mathcal{C} = \{V_y \mid y \in S \setminus U\}$ is an open covering of $S \setminus U$ by open sets in S. By the compactness of $(S, \mathcal{T})$, there is a finite subcover $\mathcal{D}$ of $\mathcal{C}$. Let $V = \bigcup\mathcal{D}$. Since a union of open sets is open, V is open in S. Since a finite union of closed sets is closed, V is also closed in S. It follows that $S \setminus V$ satisfies $x \in S \setminus V \subseteq U$, as desired. □

91. An arbitrary product of connected spaces is connected.

Proof: Let K be an index set, let $(S_k, \mathcal{T}_k)$ be connected for each $k \in K$, and assume toward contradiction that $\prod S_k$ is disconnected. By Problem 22 above, there is a surjective continuous function $f: \prod S_k \to \{0, 1\}$ (here $\{0, 1\}$ is given the discrete topology).

Let $\mathbf{x} \in \prod S_k$ and for some $j \in K$, define $f_j: S_j \to \prod S_k$ by $f_j(s) = (z_k)$, where $z_k = x_k$ for $k \neq j$ and $z_j = s$. Then f_j is continuous, and so, the function $g_j: S_j \to \{0, 1\}$ defined by $g_j = f \circ f_j$ is continuous. So, by Problem 22 above, g_j is constant. Since $g_j(x_j) = f\left(f_j(x_j)\right) = f(\mathbf{x})$, it follows that $f\left(f_j(s)\right) = g_j(s) = f(\mathbf{x})$ for all $s \in S_j$. So, $f(\mathbf{s}) = f(\mathbf{x})$ whenever $\mathbf{s}$ agrees with $\mathbf{x}$ everywhere except possibly at s_j (f is constant on the jth "slice" of the product).

By repeating this process with another index, say $t \neq j$, we see that $f(\mathbf{s}) = f(\mathbf{x})$ whenever $\mathbf{s}$ agrees with $\mathbf{x}$ everywhere except possibly at s_t. It then follows that $f(\mathbf{s}) = f(\mathbf{x})$ whenever $\mathbf{s}$ agrees with $\mathbf{x}$ everywhere except possibly at s_j and s_t.

We can repeat this process finitely many times to see that $f(\mathbf{s}) = f(\mathbf{x})$ whenever $\mathbf{s}$ agrees with $\mathbf{x}$ everywhere except possibly at finitely many coordinates. The set of all $\mathbf{s}$ of this type is dense in $\prod S_k$ and so, f is constant, contradicting our assumption that f was surjective. □

Problem Set 15

LEVEL 1

Compute the derivative of each of the following complex-valued functions of the complex variable z:

1. $f(z) = 2z^3 + 5z^2 - 6z + 8$

$f'(z) = \mathbf{6z^2 + 10z - 6}$

2. $g(z) = (3 - z + z^2)^8$

$g'(z) = 8(3 - z + z^2)(-1 + 2z) = \mathbf{8(2z - 1)(3 - z + z^2)}$

3. $h(z) = \frac{5z-3}{11z+2}, z \neq -\frac{2}{11}$

$$h'(z) = \frac{(11z+2)(5) - (5z-3)(11)}{(11z+2)^2} = \frac{55z + 10 - 55z + 33}{(11z+2)^2} = \mathbf{\frac{43}{(11z+2)^2}}$$

4. $k(z) = \frac{z^2-5z+1}{(3z-1)^3}, z \neq \frac{1}{3}$

$$k'(z) = \frac{(3z-1)^3(2z-5) - (z^2 - 5z + 1) \cdot 3(3z-1)^2 \cdot 3}{(3z-1)^6}$$

$$= \frac{(3z-1)(2z-5) - (z^2 - 5z + 1) \cdot 9}{(3z-1)^4} = \frac{6z^2 - 17z + 5 - 9z^2 + 45z - 9}{(3z-1)^4} = \mathbf{\frac{-3z^2 + 28z - 4}{(3z-1)^4}}$$

5. $b(z) = \frac{e^z}{z(e^{2z}+e^{3z})}, z \neq 0$

$$b'(z) = \frac{z(e^{2z} + e^{3z})e^z - e^z[z(2e^{2z} + 3e^{3z}) + (e^{2z} + e^{3z})]}{z^2(e^{2z} + e^{3z})^2}$$

$$= \frac{ze^{3z} + ze^{4z} - 2ze^{3z} - 3ze^{4z} - e^{3z} - e^{4z}}{z^2(e^{2z} + e^{3z})^2} = \frac{-ze^{3z} - 2ze^{4z} - e^{3z} - e^{4z}}{z^2(e^{2z} + e^{3z})^2}$$

$$= \frac{-ze^{3z} - 2ze^{4z} - e^{3z} - e^{4z}}{z^2\left(e^{2z}(1 + e^z)\right)^2} = \frac{-ze^{3z} - 2ze^{4z} - e^{3z} - e^{4z}}{z^2e^{4z}(1 + e^z)^2} = -\mathbf{\frac{z + 2ze^z + 1 + e^z}{z^2e^z(1 + e^z)^2}}$$

Define the function $f: \mathbb{R}^2 \to \mathbb{R}$ by $f(x, y) = x^2 + 2xy + 3y^2$.

6. Compute f_x and f_y.

$f_x(x, y) = \mathbf{2x + 2y}$

$f_y(x, y) = \mathbf{2x + 6y}$

7. Evaluate f_x and f_y at the point $(-1, 2)$.

$f_x(-1,2) = 2(-1) + 2 \cdot 2 = \mathbf{2}$

$f_y(-1,2) = 2(-1) + 6 \cdot 2 = \mathbf{10}$

8. Compute f_{xx}, f_{xy}, f_{yx}, and f_{yy}.

$f_{xx}(x,y) = 2 \quad f_{xy}(x,y) = \mathbf{2} \quad f_{yx}(x,y) = \mathbf{2} \quad f_{yy}(x,y) = \mathbf{6}$

9. Compute all the third partials of f.

$f_{xxx}(x,y) = f_{xxy}(x,y) = f_{xyx}(x,y) = f_{yxx}(x,y) = f_{xyy}(x,y) = f_{yxy}(x,y) = f_{yyx}(x,y) = f_{yyy}(x,y) = \mathbf{0}$

10. Compute $D_{\mathbf{u}}f$ in the direction of the vector $\mathbf{u} = \left(\frac{\sqrt{3}}{2}, \frac{1}{2}\right)$.

$$D_{\mathbf{u}}f(x,y) = \nabla f \cdot \mathbf{u} = (2x+2y, 2x+6y) \cdot \left(\frac{\sqrt{3}}{2}, \frac{1}{2}\right) = \sqrt{3}x + \sqrt{3}y + x + 3y = (\mathbf{\sqrt{3}+1})\boldsymbol{x} + (\mathbf{\sqrt{3}+3})\boldsymbol{y}$$

11. Compute $D_{\mathbf{u}}f$ in the direction of the vector $\mathbf{v} = (3,4)$.

Let $\mathbf{u} = \frac{\mathbf{v}}{|\mathbf{v}|} = \frac{1}{\sqrt{3^2+4^2}}(3,4) = \frac{1}{5}(3,4) = \left(\frac{3}{5}, \frac{4}{5}\right)$.

$$D_{\mathbf{u}}f(x,y) = \nabla f \cdot \mathbf{u} = (2x+2y, 2x+6y) \cdot \left(\frac{3}{5}, \frac{4}{5}\right) = \frac{6}{5}x + \frac{6}{5}y + \frac{8}{5}x + \frac{24}{5}y = \frac{\mathbf{14}}{\mathbf{5}}\boldsymbol{x} + \mathbf{6}\boldsymbol{y}$$

Verify that each of the following functions $u: A \to \mathbb{R}$ is harmonic in A:

12. $u(x,y) = 3y(5-x)$; $A = \mathbb{R}^2$

$u(x,y) = 15y - 3xy$

$u_x = -3y$, $u_{xx} = 0$, $u_y = 15 - 3x$, and $u_{yy} = 0$. Therefore, $u_{xx} + u_{yy} = 0 + 0 = 0$.

13. $u(x,y) = \frac{x}{x^2+y^2}$; $A = \mathbb{R}^2 \setminus \{(0,0)\}$

$$u_x = \frac{(x^2+y^2)(1) - x(2x)}{(x^2+y^2)^2} = \frac{y^2 - x^2}{(x^2+y^2)^2}$$

$$u_{xx} = \frac{(x^2+y^2)^2(-2x) - (y^2-x^2) \cdot 2(x^2+y^2)(2x)}{(x^2+y^2)^4} = \frac{(x^2+y^2)(-2x) - (y^2-x^2) \cdot 2(2x)}{(x^2+y^2)^3}$$

$$= \frac{-2x^3 - 2xy^2 - 4xy^2 + 4x^3}{(x^2+y^2)^3} = \frac{2x^3 - 6xy^2}{(x^2+y^2)^3}$$

$$u_y = \frac{(x^2+y^2)(0) - x(2y)}{(x^2+y^2)^2} = \frac{-2xy}{(x^2+y^2)^2}$$

$$u_{yy} = \frac{(x^2+y^2)^2(-2x) - (-2xy)\cdot 2(x^2+y^2)(2y)}{(x^2+y^2)^4} = \frac{(x^2+y^2)(-2x) - (-2xy)\cdot 2(2y)}{(x^2+y^2)^3}$$

$$= \frac{-2x^3 - 2xy^2 + 8xy^2}{(x^2+y^2)^3} = \frac{6xy^2 - 2x^3}{(x^2+y^2)^3}$$

Therefore, we have

$$u_{xx} + u_{yy} = \frac{2x^3 - 6xy^2}{(x^2+y^2)^3} + \frac{6xy^2 - 2x^3}{(x^2+y^2)^3} = 0.$$

14. $u(x,y) = xe^x \cos y - ye^x \sin y;\ A = \mathbb{R}^2$

$$u_x = (xe^x + e^x)\cos y - ye^x \sin y$$

$$u_{xx} = (xe^x + e^x + e^x)\cos y - ye^x \sin y = (xe^x + 2e^x)\cos y - ye^x \sin y$$

$$u_y = -xe^x \sin y - e^x(y\cos y + \sin y)$$

$$u_{yy} = -xe^x \cos y - e^x(-y\sin y + \cos y + \cos y) = -xe^x \cos y - e^x(-y\sin y + 2\cos y)$$

Therefore, we have

$$u_{xx} + u_{yy} = (xe^x + 2e^x)\cos y - ye^x \sin y - xe^x \cos y - e^x(-y\sin y + 2\cos y) = 0.$$

Determine if each of the following regions of complex numbers is a domain. If so, determine if the domain is simply connected or multiply connected. Also determine if each region is bounded.

15. $\{z \in \mathbb{C} \mid |z| < 1\}$

Bounded simply connected domain

16. $\{z \in \mathbb{C} \mid |z| \leq 1\}$

Bounded region that is not a domain

17. $\{z \in \mathbb{C} \mid 1 < |z| < 3\}$

Bounded multiply connected domain

18. $\{z \in \mathbb{C} \mid 1 \leq |z| < 3\}$

Bounded region that is not a domain

19. $\{z \in \mathbb{C} \mid 0 < |z| \leq 3\}$

Bounded region that is not a domain

20. $\{z \in \mathbb{C} \mid \operatorname{Re} z \geq 0\}$

Unbounded region that is not a domain

21. $\mathbb{C}$

Unbounded simply connected domain

LEVEL 2

Use one of the limit definitions of the derivative to differentiate each of the following complex-valued functions of a complex variable or to show that the derivative does not exist:

22. $f(z) = iz^2 - 2z$

$$f'(a) = \lim_{h \to 0} \frac{f(a+h) - f(a)}{h} = \lim_{h \to 0} \frac{\left(i(a+h)^2 - 2(a+h)\right) - (ia^2 - 2a)}{h}$$

$$= \lim_{h \to 0} \frac{(ia^2 + 2iah + ih^2 - 2a - 2h) - (ia^2 - 2a)}{h}$$

$$= \lim_{h \to 0} \frac{(ia^2 - ia^2) + 2iah + ih^2 - (2a - 2a) - 2h}{h}$$

$$= \lim_{h \to 0} \frac{2iah + ih^2 - 2h}{h} = \lim_{h \to 0} \frac{h(2ia + ih - 2)}{h}$$

$$= \lim_{h \to 0} (2ia + ih - 2) = \mathbf{2ia - 2}.$$

23. $g(z) = \frac{1}{z}, z \neq 0$

$$g'(a) = \lim_{h \to 0} \frac{g(a+h) - g(a)}{h} = \lim_{h \to 0} \frac{\frac{1}{a+h} - \frac{1}{a}}{h}$$

$$= \lim_{h \to 0} \frac{\frac{1}{a+h} \cdot a(a+h) - \frac{1}{a} \cdot a(a+h)}{ha(a+h)}$$

$$= \lim_{h \to 0} \frac{a - (a+h)}{ha(a+h)} = \lim_{h \to 0} \frac{-h}{ha(a+h)} = \lim_{h \to 0} \frac{-1}{a(a+h)} = -\frac{1}{a(a+0)} = -\frac{\mathbf{1}}{\boldsymbol{a^2}}$$

24. $k(z) = \operatorname{Re} z$

$$k'(a) = \lim_{h \to 0} \frac{k(a+h) - k(a)}{h} = \lim_{h \to 0} \frac{\operatorname{Re}(a+h) - \operatorname{Re} a}{h}$$

$$= \lim_{h \to 0} \frac{\frac{a + h + \overline{a+h}}{2} - \frac{a + \overline{a}}{2}}{h} = \lim_{h \to 0} \frac{\frac{a + h + \overline{a} + \overline{h}}{2} - \frac{a + \overline{a}}{2}}{h} = \lim_{h \to 0} \frac{h + \overline{h}}{2h} = \frac{1}{2} \lim_{h \to 0} \left(1 + \frac{\overline{h}}{h}\right).$$

If we consider complex numbers of the form $x + 0i$, then $h = \overline{h}$, and therefore, $1 + \frac{\overline{h}}{h} = 1 + 1 = 2$. Thus, $\frac{1}{2} \lim_{h \to 0} \left(1 + \frac{\overline{h}}{h}\right) = \frac{1}{2} \cdot 2 = 1$.

If we consider complex numbers of the form $0 + yi$, then $\overline{h} = -h$, and so, $1 + \frac{\overline{h}}{h} = 1 - 1 = 0$. Thus, $\frac{1}{2}\lim_{h \to 0}\left(1 + \frac{\overline{h}}{h}\right) = \frac{1}{2} \cdot 0 = 0$.

Since limits are unique, we see that the limit does not exist for any complex number a. So, $k'(a)$ does not exist for any complex number a.

25. $b(z) = \text{Im } z$

$$b'(a) = \lim_{h \to 0} \frac{b(a+h) - b(a)}{h} = \lim_{h \to 0} \frac{\text{Im } (a+h) - \text{Im } a}{h}$$

$$= \lim_{h \to 0} \frac{\frac{a + h - \overline{a+h}}{2i} - \frac{a - \overline{a}}{2i}}{h} = \lim_{h \to 0} \frac{\frac{a + h - \overline{a} - \overline{h}}{2i} - \frac{a - \overline{a}}{2i}}{h} = \lim_{h \to 0} \frac{h - \overline{h}}{2ih} = \frac{1}{2i}\lim_{h \to 0}\left(1 - \frac{\overline{h}}{h}\right).$$

If we consider complex numbers of the form $x + 0i$, then $h = \overline{h}$, and therefore, $1 - \frac{\overline{h}}{h} = 1 - 1 = 0$. Thus, $\frac{1}{2i}\lim_{h \to 0}\left(1 + \frac{\overline{h}}{h}\right) = \frac{1}{2i} \cdot 0 = 0$.

If we consider complex numbers of the form $0 + yi$, then $\overline{h} = -h$, and so, $1 - \frac{\overline{h}}{h} = 1 + 1 = 2$. Thus, $\frac{1}{2i}\lim_{h \to 0}\left(1 + \frac{\overline{h}}{h}\right) = \frac{1}{2i} \cdot 2 = \frac{1}{i} = -i$.

Since limits are unique, we see that the limit does not exist for any complex number a. So, $b'(a)$ does not exist for any complex number a.

Define $f: \mathbb{R}^2 \to \mathbb{R}$ by $f(x, y) = x^2y^2$.

26. Use the limit definitions of the partial derivatives to compute f_x and f_y.

$$f_x(b, c) = \lim_{\Delta x \to 0} \frac{f(b + \Delta x, c) - f(b, c)}{\Delta x} = \lim_{\Delta x \to 0} \frac{(b + \Delta x)^2c^2 - b^2c^2}{\Delta x}$$

$$= \lim_{\Delta x \to 0} \frac{b^2c^2 + 2b\Delta xc^2 + \Delta x^2c^2 - b^2c^2}{\Delta x} = \lim_{\Delta x \to 0} \frac{2b\Delta xc^2 + \Delta x^2c^2}{\Delta x} = \lim_{\Delta x \to 0} \frac{\Delta x(2bc^2 + \Delta xc^2)}{\Delta x}$$

$$= \lim_{\Delta x \to 0}(2bc^2 + \Delta xc^2) = \mathbf{2bc^2}$$

$$f_y(b, c) = \lim_{\Delta y \to 0} \frac{f(b, c + \Delta y) - f(b, c)}{\Delta y} = \lim_{\Delta y \to 0} \frac{b^2(c + \Delta y)^2 - b^2c^2}{\Delta y}$$

$$= \lim_{\Delta y \to 0} \frac{b^2c^2 + 2b^2c\Delta y + b^2\Delta y^2 - b^2c^2}{\Delta y} = \lim_{\Delta y \to 0} \frac{2b^2c\Delta y + b^2\Delta y^2}{\Delta y} = \lim_{\Delta y \to 0} \frac{(2b^2c + b^2\Delta y)\Delta y}{\Delta y}$$

$$= \lim_{\Delta y \to 0}(2b^2c + b^2\Delta y) = \mathbf{2b^2c}$$

27. Use the limit definition of the directional derivative to compute $D_{\mathbf{u}}f$ in the direction of the vector $\mathbf{u} = \left(\frac{1}{\sqrt{2}}, \frac{1}{\sqrt{2}}\right)$.

$$D_{\mathbf{u}}f(b,c)=\lim_{h\to 0}\frac{f(b+u_1h,c+u_2h)-f(b,c)}{h}=\lim_{h\to 0}\frac{\left(b+\frac{1}{\sqrt{2}}h\right)^2\left(c+\frac{1}{\sqrt{2}}h\right)^2-b^2c^2}{h}$$

$$=\lim_{h\to 0}\frac{\left(b^2+\frac{2}{\sqrt{2}}bh+\frac{1}{2}h^2\right)\left(c^2+\frac{2}{\sqrt{2}}ch+\frac{1}{2}h^2\right)-b^2c^2}{h}$$

$$=\lim_{h\to 0}\frac{b^2c^2+\frac{2}{\sqrt{2}}b^2ch+\frac{1}{2}b^2h^2+\frac{2}{\sqrt{2}}bc^2h+2bch^2+\frac{1}{\sqrt{2}}bh^3+\frac{1}{2}c^2h^2+\frac{1}{\sqrt{2}}ch^3+\frac{1}{4}h^4-b^2c^2}{h}$$

$$=\lim_{h\to 0}\frac{\frac{2}{\sqrt{2}}b^2ch+\frac{1}{2}b^2h^2+\frac{2}{\sqrt{2}}bc^2h+2bch^2+\frac{1}{\sqrt{2}}bh^3+\frac{1}{2}c^2h^2+\frac{1}{\sqrt{2}}ch^3+\frac{1}{4}h^4}{h}$$

$$=\lim_{h\to 0}\frac{\left(\frac{2}{\sqrt{2}}b^2c+\frac{1}{2}b^2h+\frac{2}{\sqrt{2}}bc^2+2bch+\frac{1}{\sqrt{2}}bh^2+\frac{1}{2}c^2h+\frac{1}{\sqrt{2}}ch^2+\frac{1}{4}h^3\right)h}{h}$$

$$=\lim_{h\to 0}\left(\frac{2}{\sqrt{2}}b^2c+\frac{1}{2}b^2h+\frac{2}{\sqrt{2}}bc^2+2bch+\frac{1}{\sqrt{2}}bh^2+\frac{1}{2}c^2h+\frac{1}{\sqrt{2}}ch^2+\frac{1}{4}h^3\right)$$

$$=\boldsymbol{\frac{2}{\sqrt{2}}b^2c+\frac{2}{\sqrt{2}}bc^2}$$

Let e^z be the complex-valued exponential function. Prove each of the following:

28. For all $a,b\in\mathbb{C}$, $e^ae^b=e^{a+b}$.

Proof: Let $a=x_1+iy_1$ and $b=x_2+iy_2$. Then $a+b=(x_1+x_2)+i(y_1+y_2)$. So,

$$e^{a+b}=e^{x_1+x_2}e^{i(y_1+y_2)}.$$

Therefore,

$$e^{a}e^b=\left(e^{x_1}e^{iy_1}\right)\left(e^{x_2}e^{iy_2}\right)=\left(e^{x_1}e^{x_2}\right)\left(e^{iy_1}e^{iy_2}\right)=e^{x_1+x_2}e^{i(y_1+y_2)}=e^{a+b}.$$ □

29. For all $a,b\in\mathbb{C}$, $\frac{e^a}{e^b}=e^{a-b}$.

Proof: By Problem 28, $e^a=e^{(a-b)+b}=e^{a-b}e^b$. Therefore,

$$\frac{e^a}{e^b}=e^{a-b}.$$ □

30. For all $a\in\mathbb{C}$, $e^a\neq 0$.

Proof: From the equation $e^z=e^xe^{iy}$, we see that $|e^z|=e^x$ and $\arg e^z=y+2n\pi$, $n\in\mathbb{Z}$. Since $|e^z|=e^x$ is never 0, e^z is never 0 . Therefore, for all $a\in\mathbb{C}$, we have $e^a\neq 0$. □

31. e^z is periodic with pure imaginary period $2\pi i$.

Proof: $e^{z+2\pi i}=e^ze^{2\pi i}=e^z(\cos 2\pi+i\sin 2\pi)=e^z\cdot 1=e^z$. □

32. There are infinitely many complex numbers a such that e^a is equal to a negative real number.

Proof: $e^{\pi i} = \cos \pi + i \sin \pi = -1$.

More generally, $e^{(2n+1)\pi i} = e^{2n\pi i} e^{\pi i} = (1)(-1) = -1$. □

We define the complex-valued natural logarithm as follows: $w = \ln z$ if and only if $z = e^w$. In Problem 73 below, we will see that the complex-valued natural logarithm can be written $\ln z = \ln|z| + i \arg z$. In Problem 74 below, we will see that $\ln z$ is a multiple-valued function (in fact, for each nonzero complex number z, $\ln z$ takes on infinitely many values). We define the **principal value** of $\ln z$ to be the single-valued function $\text{Ln } z = \ln r + i \text{ Arg } z$, where $\text{Arg } z$ is the principal argument of z (in other words, $-\pi < \text{Arg } z \leq \pi$. Compute each of the following:

33. $\ln 1$

$\ln 1 = \ln 1 + i(0 + 2n\pi) = \boldsymbol{2n\pi i} \ (\boldsymbol{n} \in \mathbb{Z})$

34. $\text{Ln } 1$

$\text{Ln } 1 = \mathbf{0}$

35. $\ln(-1)$

$\ln(-1) = \ln 1 + i(\pi + 2n\pi) = \mathbf{(2n+1)\pi i} \ (\boldsymbol{n} \in \mathbb{Z})$

36. $\text{Ln}(-1)$

$\text{Ln}(-1) = \boldsymbol{\pi i}$

37. $\ln i$

$\ln i = \ln 1 + i\left(\frac{\pi}{2} + 2n\pi\right) = \frac{\mathbf{1}}{\mathbf{2}}\mathbf{(4n+1)\pi i} \ (\boldsymbol{n} \in \mathbb{Z})$

38. $\text{Ln } i$

$\text{Ln } i = \frac{\boldsymbol{\pi}}{\mathbf{2}}\boldsymbol{i}$

39. $\ln\left(-1 + \sqrt{3}i\right)$

$\ln\left(-1 + \sqrt{3}i\right) = \ln 2 + i\left(\frac{2\pi}{3} + 2n\pi\right) = \mathbf{\ln 2} + \frac{\mathbf{2}}{\mathbf{3}}\mathbf{(3n+1)\pi i} \ (\boldsymbol{n} \in \mathbb{Z})$

40. $\text{Ln}\left(-1 + \sqrt{3}i\right)$

$\text{Ln}\left(-1 + \sqrt{3}i\right) = \mathbf{\ln 2} + \frac{\mathbf{2\pi}}{\mathbf{3}}\boldsymbol{i}$

41. $\text{Ln}\, i^3$

$\text{Ln}\, i^3 = \text{Ln}(-i) = \ln 1 + i\left(-\frac{\pi}{2}\right) = -\boldsymbol{\frac{\pi}{2} i}$

42. $3\,\text{Ln}\, i$

$3\,\text{Ln}\, i = 3\left(\ln(1) + i\left(\frac{\pi}{2}\right)\right) = \boldsymbol{\frac{3\pi}{2} i}$

Note that $\text{Ln}\, i^3 \neq 3\,\text{Ln}\, i$.

Let c be a complex number. We define the complex-valued function z^c to be $e^{c \ln z}$ for all $z \neq 0$. Compute each of the following:

43. i^i

$i^i = e^{i \ln i} = e^{i\left(\ln 1 + i\left(\frac{\pi}{2} + 2n\pi\right)\right)} = e^{i^2\left(\frac{\pi}{2} + 2n\pi\right)} = \boldsymbol{e^{-\frac{1}{2}(1+4n)\pi}}$

44. The principal value of i^i

$\boldsymbol{e^{-\frac{\pi}{2}}}$

45. i^{-2i}

$i^{-2i} = e^{-2i \ln i} = e^{-2i\left(\ln 1 + i\left(\frac{\pi}{2} + 2n\pi\right)\right)} = e^{-2i^2\left(\frac{\pi}{2} + 2n\pi\right)} = \boldsymbol{e^{(1+4n)\pi}}$

46. The principal value of i^{-2i}

$\boldsymbol{e^{\pi}}$

For the given function $f\colon \mathbb{R} \to \mathbb{C}$, evaluate $f'(t)$ and $\int_a^b f(t)\, dt$.

47. $f(t) = e^{2it}$, $a = 0$, $b = \frac{\pi}{8}$

$$f'(t) = 2ie^{2it}$$

$$\int_0^{\frac{\pi}{8}} f(t)\, dt = \int_0^{\frac{\pi}{8}} e^{2it}\, dt = \left[\frac{1}{2i} e^{2it}\right]_0^{\frac{\pi}{8}} = \frac{1}{2i}\left(e^{\frac{\pi}{4}i} - e^0\right)$$

$$= \frac{i}{2(-1)}\left(\frac{\sqrt{2}}{2} + \frac{\sqrt{2}}{2} i - 1\right) = \boldsymbol{\frac{\sqrt{2}}{4} + i\left(\frac{1}{2} - \frac{\sqrt{2}}{4}\right)}$$

48. $f(t) = \left(\frac{3}{t^2} - 2t^2 i\right)^2, a = 1, b = 2$

$$f'(t) = 2\left(\frac{3}{t^2} - 2t^2 i\right)\left(-\frac{6}{t^3} - 4ti\right)$$

$$\int_1^2 f(t)\,dt = \int_1^2 (9t^{-4} - 12i - 4t^4)\,dt = \left[\frac{9t^{-3}}{-3} - 12it - \frac{4t^5}{5}\right]_1^2$$

$$= \left[-\frac{3}{t^3} - 12it - \frac{4t^5}{5}\right]_1^2 = \left(-\frac{3}{8} - 24i - \frac{128}{5}\right) - \left(-3 - 12i - \frac{4}{5}\right) = -\frac{\mathbf{887}}{\mathbf{40}} - \mathbf{12}\boldsymbol{i}$$

49. $f(t) = e^{imt}e^{-int}, a = 0, b = 2\pi, m, n \in \mathbb{Z}$

First note that $f(t) = e^{i(m-n)t}$. So, if $m = n$, then $f(t) = 1$, $f'(t) = 0$, and $\int_0^{2\pi} f(t)\,dt = t|_0^{2\pi} = 2\pi$. If $m \neq n$, then we have

$$f'(t) = i(m-n)e^{i(m-n)t},$$

$$\int_0^{2\pi} f(t)\,dt = \frac{1}{i(m-n)} e^{i(m-n)t}|_0^{2\pi} = \frac{1}{i(m-n)}\left(e^{i(m-n)2\pi} - e^0\right) = \frac{1}{i(m-n)}(1-1) = 0.$$

For the given function $f: \mathbb{C} \to \mathbb{C}$ and contour C, evaluate the contour integral $\int_C f(z)\,dz$.

50. $f(z) = \bar{z}$; C is the right half of the unit circle from $z = -i$ to $z = i$.

We can parametrize C as $z = e^{i\theta}$ $\left(-\frac{\pi}{2} \leq \theta \leq \frac{\pi}{2}\right)$. Therefore,

$$\int_C \bar{z}\,dz = \int_{-\frac{\pi}{2}}^{\frac{\pi}{2}} \overline{e^{i\theta}} ie^{i\theta}\,d\theta = i\int_{-\frac{\pi}{2}}^{\frac{\pi}{2}} e^{-i\theta}\,e^{i\theta}\,d\theta = i\int_{-\frac{\pi}{2}}^{\frac{\pi}{2}} d\theta = i\theta|_{-\frac{\pi}{2}}^{\frac{\pi}{2}} = i\left(\frac{\pi}{2} + \frac{\pi}{2}\right) = \boldsymbol{\pi i}.$$

51. $f(z) = \bar{z}$; C is the line segment along the imaginary axis from $z = -i$ to $z = i$.

We can parametrize C as $z = it$ $(-1 \leq t \leq 1)$. Therefore,

$$\int_C \bar{z}\,dz = \int_{-1}^{1} \overline{it} \cdot i\,d\theta = i\int_{-1}^{1} -it\,d\theta = \int_{-1}^{1} t\,d\theta = \left[\frac{t^2}{2}\right]_{-1}^{1} = \frac{1}{2} - \frac{(-1)^2}{2} = \frac{1}{2} - \frac{1}{2} = \mathbf{0}.$$

52. $f(z) = z$; C is the lower half of the circle with center $z = 2$ and radius $R = 2$ from $z = 0$ to $z = 4$.

We can parametrize C as $z = 2 + 2e^{i\theta}$ $(\pi \leq \theta \leq 2\pi)$. Therefore,

$$\int_C z\,dz = \int_{\pi}^{2\pi} \left(2 + 2e^{i\theta}\right) \cdot 2ie^{i\theta}\,d\theta = 4i\int_{\pi}^{2\pi} \left(e^{i\theta} + e^{2i\theta}\right) d\theta$$

$$= 4i\left[\left(\frac{1}{i}e^{i\theta} + \frac{1}{2i}e^{2i\theta}\right)\right]_{\pi}^{2\pi} = 4\left[e^{i\theta} + \frac{1}{2}e^{2i\theta}\right]_{\pi}^{2\pi} = 4\left[\left(1 + \frac{1}{2}\right) - \left(-1 + \frac{1}{2}\right)\right] = \mathbf{8}.$$

53. $f(z) = z$; C is the line segment along the real axis from $z = 0$ to $z = 4$.

We can parametrize C as $z = t$ $(0 \le t \le 4)$. Therefore,

$$\int_C z\,dz = \int_0^4 t\,dt = \left[\frac{t^2}{2}\right]_0^4 = \frac{16}{2} = \mathbf{8}.$$

Note: Since $f(z) = z$ is entire, we could also evaluate the integrals in Problems 52 and 53 simultaneously with the following computation:

$$\int_0^4 z\,dz = \left[\frac{1}{2}z^2\right]_0^4 = \frac{1}{2}\cdot 4^2 = \frac{16}{2} = \mathbf{8}$$

54. $f(z) = \frac{e^z}{(z^2-16)(z-2i)}$; C is the positively oriented circle $|z| = 3$.

By Cauchy's Integral Formula with $f(z) = \frac{e^z}{z^2-16}$ and $z_0 = 2i$, we have

$$\int_C \frac{e^z\,dz}{(z^2-16)(z-2i)} = \int_C \frac{\frac{e^z}{z^2-16}\,dz}{(z-2i)} = \int_C \frac{f(z)\,dz}{(z-z_0)} = 2\pi i f(z_0)$$
$$= 2\pi i\left(\frac{e^{2i}}{(2i)^2-16}\right) = 2\pi i\left(\frac{e^{2i}}{4i^2-16}\right) = \frac{2\pi i e^{2i}}{-20} = -\frac{\pi i e^{2i}}{10}$$
$$= -\frac{\pi i}{10}(\cos 2 + i\sin 2) = \boldsymbol{\frac{\pi}{10}\sin 2 - i\frac{\pi}{10}\cos 2}.$$

55. $f(z) = \frac{z^5}{(z-1)^2(z+3i)}$; C is the positively oriented circle $|z| = 2$.

By Cauchy's Extended Integral Formula with $f(z) = \frac{z^5}{z+3i}$, $z_0 = 1$, and $n = 1$, we have

$$\int_C \frac{z^5\,dz}{(z-1)^2(z+3i)} = \int_C \frac{\frac{z^5}{z+3i}\,dz}{(z-1)^2} = \int_C \frac{f(z)\,dz}{(z-z_0)^{1+1}} = \frac{2\pi i}{1!}f'(z_0)$$
$$= 2\pi i\left[\frac{(z+3i)\cdot 5z^4 - z^5\cdot 1}{(z+3i)^2}\right]_{z=1} = 2\pi i\frac{(1+3i)\cdot 5 - 1}{(1+3i)^2} = 2\pi i\frac{4+15i}{-8+6i}$$
$$= \pi i\frac{(4+15i)}{(-4+3i)}\frac{(-4-3i)}{(-4-3i)} = \frac{\pi i(29-72i)}{16+9} = \frac{\pi}{25}(72+29i) = \boldsymbol{\frac{72\pi}{25} + \frac{29\pi}{25}i}$$

56. $f(z) = \frac{e^{z^2}}{z-i}$; C is the square with vertices $-1, 1, -1+2i, 1+2i$.

By Cauchy's Integral Formula with $f(z) = e^{z^2}$ and $z_0 = i$, we have

$$\int_C \frac{e^{z^2}\,dz}{z-i} = \int_C \frac{f(z)\,dz}{(z-z_0)} = 2\pi i f(z_0) = 2\pi i e^{i^2} = 2\pi i e^{-1} = \boldsymbol{\frac{2\pi}{e}i}.$$

57. $f(z) = \frac{1}{(z^2+9)^2}$; C is the positively oriented circle $|z - i| = 3$.

By Cauchy's Extended Integral Formula with $f(z) = \frac{1}{(z+3i)^2}$, $z_0 = 3i$, and $n = 1$, we have

$$\int_C \frac{dz}{(z^2+9)^2} = \int_C \frac{dz}{(z-3i)^2(z+3i)^2} = \int_C \frac{\frac{1}{(z+3i)^2}\,dz}{(z-3i)^2} = \int_C \frac{f(z)\,dz}{(z-z_0)^{1+1}} = \frac{2\pi i}{1!} f'(z_0)$$

$$= 2\pi i \left[\frac{-2}{(z+3i)^3}\right]_{z=3i} = 2\pi i \left(\frac{-2}{(6i)^3}\right) = 2\pi i \left(\frac{-2}{-216i}\right) = \frac{\pi i}{54i} = \boldsymbol{\frac{\pi}{54}}$$

LEVEL 3

Solve each of the following equations for z.

58. $e^z = 1 + i$

$e^x e^{iy} = \sqrt{2}e^{\frac{\pi}{4}i} \Rightarrow e^x = \sqrt{2}$ and $y = \frac{\pi}{4} + 2n\pi\ (n \in \mathbb{Z}) \Rightarrow x = \frac{1}{2}\ln 2$ and $y = \frac{\pi}{4} + 2n\pi\ (n \in \mathbb{Z})$

So, $\boldsymbol{z = \frac{1}{2}\ln 2 + \left(\frac{\pi}{4} + 2n\pi\right) i\ (n \in \mathbb{Z})}$

59. $e^z = -5$

$e^x e^{iy} = \sqrt{5}e^{\pi i} \Rightarrow e^x = \sqrt{5}$ and $y = \pi + 2n\pi\ (n \in \mathbb{Z}) \Rightarrow x = \frac{1}{2}\ln 5$ and $y = (1 + 2n)\pi\ (n \in \mathbb{Z})$

So, $\boldsymbol{z = \frac{1}{2}\ln 5 + (1 + 2n)\pi i\ (n \in \mathbb{Z})}$

60. $e^z = 1 + \sqrt{3}i$

$e^x e^{iy} = 2e^{\frac{\pi}{3}i} \Rightarrow e^x = 2$ and $y = \frac{\pi}{3} + 2n\pi\ (n \in \mathbb{Z}) \Rightarrow x = \ln 2$ and $y = \frac{\pi}{3} + 2n\pi\ (n \in \mathbb{Z})$

So, $\boldsymbol{z = \ln 2 + \left(\frac{\pi}{3} + 2n\pi\right) i\ (n \in \mathbb{Z})}$

Use Theorem 15.11 or Theorem 15.13 to determine where each of the following functions is differentiable. Then evaluate the derivative of the function on its domain of definition.

61. $f(z) = iz^2$

$f(z) = iz^2 = i(x + iy)^2 = i\big((x^2 - y^2) + 2xyi\big) = -2xy + i(x^2 - y^2)$.

So, $u(x,y) = -2xy$, $v(x,y) = x^2 - y^2$, $u_x = -2y = v_y$, $u_y = -2x = -v_x$.

Since the partial derivatives are continuous everywhere, it follows from Theorem 15.11 that $f'(z)$ exists everywhere in the Complex Plane and $f'(z) = u_x + iv_x = -2y + 2ix$ for all $z \in \mathbb{C}$.

Note that $-2y + 2ix = 2i(x + iy) = 2iz$, and so, $f'(z) = 2iz$.

62. $g(z) = e^{-z}$

$g(z) = e^{-z} = e^{-(x+iy)} = e^{-x}e^{-iy} = e^{-x}\cos y - ie^{-x}\sin y$.

So, $u(x,y) = e^{-x}\cos y$, $v(x,y) = -e^{-x}\sin y, u_x = -e^{-x}\cos y = v_y$, $u_y = -e^{-x}\sin y = -v_x$.

Since the partial derivatives are continuous everywhere, it follows from Theorem 15.11 that $g'(z)$ exists everywhere in the Complex Plane and $f'(z) = u_x + iv_x = -e^{-x}\cos y + ie^{-x}\sin y$ for all $z \in \mathbb{C}$.

Note that $-e^{-x}\cos y + ie^{-x}\sin y = -e^{-x}(\cos y - i\sin y) = -e^{-x}e^{-iy} = -e^{-z}$, and so, $g'(z) = -e^{-z}$.

63. $h(z) = \frac{1}{z^2}$

Solution using Theorem 15.13:

$$f(z) = \frac{1}{z^2} = \frac{1}{(re^{i\theta})^2} = \frac{1}{r^2}e^{-2i\theta} = \frac{1}{r^2}(\cos 2\theta - i\sin 2\theta) = \frac{1}{r^2}\cos 2\theta - i\frac{1}{r}\sin 2\theta.$$

If we write $f(z) = u(r,\theta) + iv(r,\theta)$, then

$$u(r,\theta) = \frac{1}{r^2}\cos 2\theta\,, \qquad v(r,\theta) = -\frac{1}{r^2}\sin 2\theta.$$

For $r \neq 0$, we have

$$u_r = -\frac{2}{r^3}\cos 2\theta\,, u_\theta = -\frac{2}{r^2}\sin 2\theta\,, v_r = \frac{2}{r^3}\sin 2\theta\,, v_\theta = -\frac{2}{r^2}\cos 2\theta.$$

Therefore, for $r \neq 0$,

$$ru_r = r\left(-\frac{2}{r^3}\cos 2\theta\right) = -\frac{2}{r^2}\cos 2\theta = v_\theta, \qquad u_\theta = -\frac{2}{r^2}\sin 2\theta = -rv_r.$$

Since the partial derivatives are continuous for all nonzero points, it follows from Theorem 15.13 that $f'(z)$ exists for all complex numbers $z \neq 0$ and for all such z,

$$f'(z) = e^{-i\theta}(u_r + iv_r) = e^{-i\theta}\left(-\frac{2}{r^3}\cos 2\theta + i\frac{2}{r^3}\sin 2\theta\right)$$

$$= -e^{-i\theta}\left(\frac{2}{r^3}\cos 2\theta - i\frac{2}{r^3}\sin 2\theta\right) = -\frac{2}{r^3}e^{-i\theta}e^{-2i\theta} = -\frac{2}{r^3}e^{-3\theta i} = -\frac{2}{(re^{i\theta})^3} = -\frac{2}{z^3}.$$

Solution using Theorem 15.11:

$$h(z) = \frac{1}{z^2} = \frac{1}{(x^2 - y^2) + 2xyi} \cdot \frac{(x^2 - y^2) - 2xyi}{(x^2 - y^2) - 2xyi} = \frac{(x^2 - y^2) - 2xyi}{x^4 - 2x^2y^2 + y^4 + 4x^2y^2}$$

$$= \frac{(x^2 - y^2) - 2xyi}{x^4 + 2x^2y^2 + y^4} = \frac{x^2 - y^2}{(x^2 + y^2)^2} - i\frac{2xy}{(x^2 + y^2)^2}$$

So, we have

$$u(x,y) = \frac{x^2 - y^2}{(x^2+y^2)^2}, \quad v(x,y) = -\frac{2xy}{(x^2+y^2)^2};$$

$$u_x = \frac{(x^2+y^2)^2 \cdot 2x - (x^2-y^2)\cdot 2(x^2+y^2)(2x)}{(x^2+y^2)^4} = \frac{(x^2+y^2)\cdot 2x - (x^2-y^2)\cdot 2(2x)}{(x^2+y^2)^3}$$

$$= \frac{2x^3 + 2xy^2 - 4x^3 + 4xy^2}{(x^2+y^2)^3} = \frac{6xy^2 - 2x^3}{(x^2+y^2)^3}$$

$$v_y = \frac{(x^2+y^2)^2(-2x) - (-2xy)\cdot 2(x^2+y^2)(2y)}{(x^2+y^2)^4} = \frac{(x^2+y^2)(-2x) - (-2xy)\cdot 2(2y)}{(x^2+y^2)^3}$$

$$= \frac{-2x^3 - 2xy^2 + 8xy^2}{(x^2+y^2)^3} = \frac{6xy^2 - 2x^3}{(x^2+y^2)^3} = u_x$$

$$u_y = \frac{(x^2+y^2)^2(-2y) - (x^2-y^2)\cdot 2(x^2+y^2)(2y)}{(x^2+y^2)^4} = \frac{(x^2+y^2)(-2y) - (x^2-y^2)\cdot 2(2y)}{(x^2+y^2)^3}$$

$$= \frac{-2x^2y - 2y^3 - 4x^2y + 4y^3}{(x^2+y^2)^3} = \frac{2y^3 - 6x^2y}{(x^2+y^2)^3}$$

$$v_x = \frac{(x^2+y^2)^2(-2y) - (-2xy)\cdot 2(x^2+y^2)(2x)}{(x^2+y^2)^4} = \frac{(x^2+y^2)(-2y) - (-2xy)\cdot 2(2x)}{(x^2+y^2)^3}$$

$$= \frac{-2x^2y - 2y^3 + 8x^2y}{(x^2+y^2)^3} = \frac{6x^2y - 2y^3}{(x^2+y^2)^3} = -\frac{2y^3 - 6x^2y}{(x^2+y^2)^3} = -u_y$$

Since the partial derivatives are continuous for all $(x, y) \in \mathbb{R}^2 \setminus \{(0,0)\}$, it follows from Theorem 15.11 that $h'(z)$ exists for all $z \in \mathbb{C}^* = \mathbb{C} \setminus \{0\}$ and for all $z \in \mathbb{C}^*$,

$$h'(z) = u_x + iv_x = \frac{6xy^2 - 2x^3}{(x^2+y^2)^3} - i\frac{2y^3 - 6x^2y}{(x^2+y^2)^3}.$$

Note that

$$\frac{1}{z^3} = \frac{1}{z^2}\cdot\frac{1}{z} = \left[\frac{x^2-y^2}{(x^2+y^2)^2} - i\frac{2xy}{(x^2+y^2)^2}\right]\frac{1}{(x+iy)}$$

$$= \left[\frac{x^2-y^2}{(x^2+y^2)^2} - i\frac{2xy}{(x^2+y^2)^2}\right]\frac{1}{(x+iy)}\cdot\frac{x-iy}{(x-iy)} = \left[\frac{x^2-y^2}{(x^2+y^2)^2} - i\frac{2xy}{(x^2+y^2)^2}\right]\frac{x-iy}{x^2+y^2}$$

$$= \frac{(x^2-y^2)x - 2xy(y)}{(x^2+y^2)^3} - i\frac{(2xy)x + (x^2-y^2)y}{(x^2+y^2)^3} = \frac{x^3 - xy^2 - 2xy^2}{(x^2+y^2)^3} - i\frac{2x^2y + x^2y - y^3}{(x^2+y^2)^3}$$

$$= \frac{x^3 - 3xy^2}{(x^2+y^2)^3} - i\frac{3x^2y - y^3}{(x^2+y^2)^3}.$$

So,

$$h'(z) = \frac{6xy^2 - 2x^3}{(x^2 + y^2)^3} - i\frac{2y^3 - 6x^2y}{(x^2 + y^2)^3} = -2\left[\frac{x^3 - 3xy^2}{(x^2 + y^2)^3} - i\frac{3x^2y - y^3}{(x^2 + y^2)^3}\right] = -\frac{2}{z^3}.$$

64. $k(z) = z \cdot \text{Re } z$

$k(z) = z \cdot \text{Re } z = (x + iy) \cdot x = x^2 + xyi.$

So, $u(x,y) = x^2$, $v(x,y) = xy, u_x = 2x$, $u_y = 0$, $v_x = y$, $v_y = x$.

So, we see that $u_x = v_y$ if and only if $x = 0$ and $u_y = -v_x$ if and only if $y = 0$. Therefore, the Cauchy-Riemann equations are satisfied only at the point $(0, 0)$. Since the partial derivatives of u and v exist everywhere and are continuous at $(0, 0)$, by Theorem 15.11, k is differentiable at $z = 0$ and $k'(0) = u_x(0, 0) + iv_x(0, 0) = 0$.

Theorem 15.11 cannot be used to say anything about the differentiability of k. However, Theorem 15.9 tells us that $k'(z)$ does not exist for all nonzero complex numbers z.

65. $b(z) = (2 - x)^5 + i(y - 1)^8$

$$u(x,y) = (2 - x)^5,\ v(x,y) = (y - 1)^8,\ u_x = -5(2 - x)^4,\ u_y = 0,\ v_x = 0,\ v_y = 8(y - 1)^7.$$

$u_y = -v_x$ is always true (they are both 0).

However, $u_x = v_y$ only when $5(2 - x)^4 + 8(y - 1)^7 = 0$.

Therefore, the Cauchy-Riemann equations are only satisfied when $x = 2$ and $y = 1$. So, by Theorem 15.9, b is not differentiable for $z \neq 2 + i$.

Since the partial derivatives are continuous at $(2, 1)$, $b'(2 + i)$ exists and

$$b'(2 + i) = u_x(2, 1) + iv_x(2, 1) = 0.$$

For each of the following harmonic functions $u: A \to \mathbb{R}$, find a harmonic conjugate $v: A \to \mathbb{R}$ and an analytic function $f: B \to \mathbb{C}$ such that $f(z) = u(x,y) + iv(x,y)$ for all $z = x + iy \in \mathbb{C}$.

66. $u(x,y) = 3y(5 - x)$; $A = \mathbb{R}^2$, $B = \mathbb{C}$

Suppose that v is a harmonic conjugate of u. Then $v_y = u_x = -3y$.

It follows that $v(x,y) = -\frac{3}{2}y^2 + \phi(x)$ for some function $\phi: \mathbb{R} \to \mathbb{R}$.

Differentiating v with respect to x gives us $v_x = \phi'(x)$.

Since $v_x = -u_y$, we have $\phi'(x) = -3(5 - x)$.

So, $\phi(x) = -3\left(5x - \frac{x^2}{2}\right) + C$, where C is an arbitrary real constant. So, we see that there are infinitely many harmonic conjugates of u, one for each value of C. For example, setting C equal to 0, we see that $v(x,y) = -\frac{3}{2}y^2 + \phi(x) = -\frac{3}{2}y^2 - 3\left(5x - \frac{x^2}{2}\right)$ is a harmonic conjugate of u.

By Theorem 15.25, if we define $f\colon \mathbb{C} \to \mathbb{C}$ by $f(z) = 3y(5-x) + i\left(-\frac{3}{2}y^2 - 3\left(5x - \frac{x^2}{2}\right)\right)$, then f is entire. Let's use this definition to rewrite the function f in terms of z (instead of x and y).

We have the following:

$$f(z) = 3y(5-x) + i\left(-\frac{3}{2}y^2 - 3\left(5x - \frac{x^2}{2}\right)\right) = 15y - 3xy + i\left(\frac{3}{2}x^2 - \frac{3}{2}y^2 - 15x\right)$$
$$= i\left(\frac{3}{2}x^2 - \frac{3}{2}y^2\right) - 3xy + 15y - 15xi = \frac{3}{2}i[(x^2 - y^2 + i(2xy)] - 15i(x+iy)$$
$$= \frac{3}{2}iz^2 - 15iz$$

67. $u(x,y) = \frac{x}{x^2+y^2}$; $A = \mathbb{R}^2 \setminus \{(0,0)\}, B = \mathbb{C} \setminus \{0\}$

Suppose that v is a harmonic conjugate of u. Then $v_y = u_x = \frac{(x^2+y^2)\cdot 1 - x(2x)}{(x^2+y^2)^2} = \frac{y^2-x^2}{(x^2+y^2)^2}$.

It follows that $v(x,y) = \frac{-y}{x^2+y^2} + \phi(x)$ for some function $\phi\colon \mathbb{R} \to \mathbb{R}$.

Differentiating v with respect to x gives us $v_x = -y(-1)(x^2+y^2)^{-2}(2x) + \phi'(x) = \frac{2xy}{(x^2+y^2)^2} + \phi'(x)$.

Since $v_x = -u_y$, we have $\frac{2xy}{(x^2+y^2)^2} + \phi'(x) = -x(-1)(x^2+y^2)^{-2}(2y) = \frac{2xy}{(x^2+y^2)^2}$.

So, $\phi'(x) = 0$, and therefore, $\phi(x) = C$, where C is an arbitrary real constant. So, we see that there are infinitely many harmonic conjugates of u, one for each value of C. For example, setting C equal to 0, we see that $v(x,y) = \frac{-y}{x^2+y^2}$ is a harmonic conjugate of u.

By Theorem 15.25, if we define $f\colon \mathbb{C} \setminus \{0\} \to \mathbb{C}$ by $f(z) = \frac{x}{x^2+y^2} - i\frac{y}{x^2+y^2}$, then f is differentiable at all nonzero complex numbers. Let's use this definition to rewrite the function f in terms of z (instead of x and y).

We have

$$\frac{1}{z} = \frac{1}{x+iy} = \frac{1}{(x+iy)} \cdot \frac{x-iy}{(x-iy)} = \frac{x-iy}{x^2+y^2} = \frac{x}{x^2+y^2} - i\frac{y}{x^2+y^2}.$$

So, $f(z) = \frac{1}{z}$.

68. $u(x,y) = xe^x \cos y - ye^x \sin y$; $A = \mathbb{R}^2, B = \mathbb{C}$

Suppose that v is a harmonic conjugate of u. Then $v_y = u_x = xe^x \cos y + e^x \cos y - ye^x \sin y$.

It follows that

$$v(x,y) = xe^x \sin y + e^x \sin y + ye^x \cos y - e^x \sin y + \phi(x) = xe^x \sin y + ye^x \cos y + \phi(x)$$

for some function $\phi: \mathbb{R} \to \mathbb{R}$.

Differentiating v with respect to x gives us

$$v_x = xe^x \sin y + e^x \sin y + ye^x \cos y + \phi'(x).$$

Since $v_x = -u_y$, we have

$$xe^x \sin y + e^x \sin y + ye^x \cos y + \phi'(x) = xe^x \sin y + ye^x \cos y + e^x \sin y$$

So, $\phi'(x) = 0$, and therefore, $\phi(x) = C$, where C is an arbitrary real constant. So, we see that there are infinitely many harmonic conjugates of u, one for each value of C. For example, setting C equal to 0, we have that $v(x,y) = xe^x \sin y + ye^x \cos y$ is a harmonic conjugate of u.

By Theorem 15.25, if we define $f: \mathbb{C} \to \mathbb{C}$ by

$$f(z) = (xe^x \cos y - ye^x \sin y) + i(xe^x \sin y + ye^x \cos y),$$

then f is entire. Let's use this definition to rewrite the function f in terms of z (instead of x and y).

We have

$$ze^z = (x + iy)(e^x \cos y + ie^x \sin y) = (xe^x \cos y - ye^x \sin y) + i(xe^x \sin y + ye^x \cos y).$$

So, $f(z) = ze^z$.

Let $f, g: \mathbb{R} \to \mathbb{C}$ and let c be a complex constant. Prove each of the following:

69. $\frac{d}{dt}[cf(t)] = cf'(t)$

Let $f(t) = u(t) + iv(t)$ and $c = a + ib$.

$$\frac{d}{dt}[cf(t)] = \frac{d}{dt}\big[(a+ib)\big(u(t) + iv(t)\big)\big] = \frac{d}{dt}\big[\big(au(t) - bv(t)\big) + i\big(av(t) + bu(t)\big)\big]$$
$$= \frac{d}{dt}[au(t) - bv(t)] + i\frac{d}{dt}[av(t) + bu(t)] = [au'(t) - bv'(t)] + i[av'(t) + bu'(t)]$$
$$= (a+ib)\big(u'(t) + iv'(t)\big) = cf'(t)$$

70. $\frac{d}{dt}[e^{ct}] = ce^{ct}$

Let $c = a + ib$.

$$\frac{d}{dt}\left[e^{(a+ib)t}\right] = \frac{d}{dt}\left[e^{at}e^{ibt}\right] = \frac{d}{dt}[e^{at}(\cos bt + i\sin bt)] = \frac{d}{dt}[e^{at}\cos bt + ie^{at}\sin bt]$$
$$= [e^{at}(-b\sin bt) + ae^{at}\cos bt] + i[e^{at}(b\cos bt) + ae^{at}\sin bt]$$
$$= [ae^{at}\cos bt - be^{at}\sin bt] + i[ae^{at}\sin bt + be^{at}\cos bt]$$
$$= (a+ib)(e^{at}\cos bt + ie^{at}\sin bt) = (a+ib)e^{at}(\cos bt + i\sin bt)$$
$$= ce^{at}e^{ibt} = ce^{(a+ib)t} = ce^{ct}$$

71. $\frac{d}{dt}[f(t) + g(t)] = f'(t) + g'(t)$

Let $f(t) = u_1(t) + iv_1(t)$ and $g(t) = u_2(t) + iv_2(t)$.

$$\frac{d}{dt}[f(t) + g(t)] = \frac{d}{dt}\left[\big(u_1(t) + iv_1(t)\big) + \big(u_2(t) + iv_2(t)\big)\right]$$
$$= \frac{d}{dt}\left[\big(u_1(t) + u_2(t)\big) + i\big(v_1(t) + v_2(t)\big)\right] = \frac{d}{dt}[u_1(t) + u_2(t)] + i\frac{d}{dt}[v_1(t) + v_2(t)]$$
$$= \big(u_1'(t) + u_2'(t)\big) + i\big(v_1'(t) + v_2'(t)\big) = \big(u_1'(t) + iv_1'(t)\big) + \big(u_2'(t) + iv_2'(t)\big)$$
$$= f'(t) + g'(t)$$

72. $\frac{d}{dt}[f(t)]^2 = 2f(t)f'(t)$

$$[f(t)]^2 = (u + iv)^2 = u^2 - v^2 + 2uvi$$
$$\frac{d}{dt}[f(t)]^2 = (u^2 - v^2)' + (2uv)'i = 2uu' - 2vv' + (2uv' + 2u'v)i$$
$$= 2(u + iv)(u' + iv') = 2f(t)f'(t).$$

We define the complex-valued natural logarithm as follows: $w = \ln z$ if and only if $z = e^w$.

73. Show that $\ln z = \ln r + i\theta$, where $r > 0$ and $\theta = \arg z$ (note that $\ln r$ is the real-valued natural logarithm of r).

Let $z = re^{i\Theta}$ $(-\pi < \Theta \leq \pi)$ and $w = u + iv$. Then

$$w = \ln z \Rightarrow z = e^w \Rightarrow re^{i\Theta} = e^{u+iv} \Rightarrow re^{i\Theta} = e^u e^{iv}$$
$$\Rightarrow r = e^u \text{ and } v = \Theta + 2n\pi \ (n \in \mathbb{Z}) \Rightarrow u = \ln r \text{ and } v = \Theta + 2n\pi \ (n \in \mathbb{Z}).$$

So, $\ln z = w = u + iv = \ln r + i(\Theta + 2n\pi)$ $(n \in \mathbb{Z})$.

Therefore, $\ln z = \ln r + i\theta$, where $r = |z| > 0$ and $\theta = \arg z$.

74. Show that for each nonzero complex number z, $\ln z$ has infinitely many values. We say that $\ln z$ is a **multiple-valued function**.

If $\ln r + i\theta$ is a value of $\ln z$, then for each $n \in \mathbb{Z}$, $\ln r + i(\theta + 2n\pi)$ is another value of $\ln z$.

75. Show that the complex-valued natural logarithmic function is undefined at $z = 0$.

By Problem 30, the equation $e^w = 0$ has no solution. Therefore, there is no complex number w such that $\ln 0 = w$. So, $\ln z$ is undefined at $z = 0$.

76. Prove that $\ln(z_1 z_2) = \ln z_1 + \ln z_2$ in the sense that if two of the three terms in the equation are specified, then there is a value for the third term so that the equation holds.

Proof: Since $|z_1 z_2| = |z_1||z_2|$ (by Problem 61 in Problem Set 7), by laws of real-valued logarithms, we have $\ln|z_1 z_2| = \ln|z_1| + \ln|z_2|$.

We also have $\arg(z_1 z_2) = \arg z_1 + \arg z_2$ (by Problem 82 in Problem Set 7). So,

$$\ln(z_1 z_2) = \ln|z_1 z_2| + i\arg(z_1 z_2) = \ln|z_1| + \ln|z_2| + i(\arg z_1 + \arg z_2)$$
$$= (\ln|z_1| + i\arg z_1) + (\ln|z_2| + i\arg z_2) = \ln z_1 + \ln z_2.$$ □

77. Prove that $\ln z^{-1} = -\ln z$ in the sense that the set of all values on the left-hand side is the same as the set of all values on the right-hand side.

Proof:

$$\ln z^{-1} = \ln|z^{-1}| + i\arg z^{-1} = \ln\frac{1}{|z|} - i\arg z \text{ (by Problem 83 in Problem Set 7)}$$
$$= (\ln 1 - \ln|z|) - i\arg z = -\ln|z| - i\arg z = -\ln z.$$ □

78. Prove that $\ln\left(\frac{z_1}{z_2}\right) = \ln z_1 - \ln z_2$ in the sense that if two of the three terms in the equation are specified, then there is a value for the third term so that the equation holds.

Proof: By Problems 76 and 77, we have

$$\ln\left(\frac{z_1}{z_2}\right) = \ln(z_1 z_2^{-1}) = \ln z_1 + \ln z_2^{-1} = \ln z_1 + (-\ln z_2) = \ln z_1 - \ln z_2.$$ □

79. Prove that $e^{\ln z} = z$.

Proof: Recall that $w = \ln z$ if and only if $z = e^w$. So, if we let $w = \ln z$, we have $e^{\ln z} = e^w = z$. □

80. Prove that $\ln e^z = z + 2n\pi i$ $(n \in \mathbb{Z})$.

Proof: Let $z = x + iy$. Then $e^z = e^x e^{iy}$, and so, $|e^z| = e^x$ and $\arg e^z = y + 2n\pi$ $(n \in \mathbb{Z})$. So,

$$\ln e^z = \ln|e^z| + i\arg(e^z) = \ln e^x + i(y + 2n\pi) = x + iy + 2n\pi i = z + 2n\pi i \ (n \in \mathbb{Z}).$$ □

81. Prove that for each integer n, $z^n = e^{n\ln z}$.

Proof: Let $z = re^{i\theta}$ and let $n \in \mathbb{Z}$. Then $z^n = \left(re^{i\theta}\right)^n = r^n\left(e^{i\theta}\right)^n = r^n e^{n\theta i}$.

Also, $e^{n \ln z} = e^{n(\ln r + i(\theta + 2n\pi))} = e^{n \ln r} e^{in(\theta+2n\pi)} = \left(e^{\ln r}\right)^n e^{n\theta i} = r^n e^{n\theta i}$. So, $z^n = e^{n \ln z}$. □

82. Prove that $z^{\frac{1}{n}} = e^{\frac{1}{n}\ln z}$ $(n \in \mathbb{Z}^+)$ in the sense that the expression on the right has n distinct values, namely the nth roots of z.

Proof: Let $z = re^{i\theta}$, let $\Theta = \text{Arg } z$, and let $n \in \mathbb{Z}^+$. Then

$$e^{\frac{1}{n}\ln z} = e^{\frac{1}{n}(\ln r + i(\Theta + 2k\pi))} = e^{\frac{1}{n}\ln r} e^{\frac{i(\Theta+2k\pi)}{n}} = \left(e^{\ln r}\right)^{\frac{1}{n}} e^{\frac{i(\Theta+2k\pi)}{n}} = r^{\frac{1}{n}} e^{\frac{i(\Theta+2k\pi)}{n}} \quad (k \in \mathbb{Z}).$$

The last expression on the right has distinct values for $0 \leq k \leq n-1$ and is precisely the expression for the nth roots of z. □

Let $C: z = z(t)$ $(a \leq t \leq b)$ be a contour, let f and g be complex-valued functions, and let k be a complex constant. Prove each of the following:

83. $\int_C kf(z)\,dz = k\int_C f(z)\,dz$

Proof:

$$\int_C kf(z)\,dz = \int_a^b kf[z(t)]z'(t)\,dt = k\int_a^b f[z(t)]z'(t)\,dt = k\int_C f(z)\,dz$$ □

84. $\int_C [f(z) + g(z)]\,dz = \int_C f(z)\,dz + \int_C g(z)\,dz$

Proof:

$$\int_C [f(z) + g(z)]\,dz = \int_a^b \big[f[z(t)] + g[z(t)]\big]z'(t)\,dt$$

$$= \int_a^b f[z(t)]z'(t)\,dt + \int_a^b g[z(t)]z'(t)\,dt = \int_C f(z)\,dz + \int_C g(z)\,dz$$ □

85. If $-C$ is the contour that consists of the same points as C but with the order of the points reversed, then $\int_{-C} f(z)\,dz = -\int_C f(z)\,dz$.

Proof: Suppose that C is expressed parametrically as $z = z(t)(a \leq t \leq b)$.

A parametric representation for $-C$ is $z = z(-t)$ $(-b \leq t \leq -a)$. Therefore,

$$\int_{-C} f(z)\,dz = \int_{-b}^{-a} f[z(-t)]\frac{d}{dz}[z(-t)]\,dt = \int_{-b}^{-a} f[z(-t)]z'(-t)(-1)\,dt = \int_b^a f[z(u)]z'(u)\,du$$

$$= -\int_a^b f[z(u)]z'(u)\,du = -\int_C f(z)\,dz.$$ □

86. If C be consists of a contour C_1 from a point z_1 to a point z_2 followed by a contour C_2 from the point z_2 to a point z_3 (we will write $C = C_1 + C_2$), then $\int_C f(z)\,dz = \int_{C_1} f(z)\,dz + \int_{C_2} f(z)\,dz$.

Proof: Suppose that C is expressed parametrically as $z = z(t)$ $(a \leq t \leq b)$. Then $z(a) = z_1$ and $z(b) = z_3$. Let c be the real number such that $z(c) = z_2$. Then C_1 and C_2 can be expressed parametrically as follows:

$$C_1 \colon z = z(t) \ (a \leq t \leq c)$$
$$C_2 \colon z = z(t) \ (c \leq t \leq b)$$

Therefore,

$$\int_C f(z)\,dz = \int_a^b f[z(t)]z'(t)\,dt = \int_a^c f[z(t)]z'(t)\,dt + \int_c^b f[z(t)]z'(t)\,d$$
$$= \int_{C_1} f(z)\,dz + \int_{C_2} f(z)\,dz.$$ □

Prove each of the following:

87. Suppose that f is a complex-valued function that is analytic and real-valued everywhere in some domain D. Then f is constant throughout D.

Proof: Let f be a complex-valued function that is analytic and real-valued everywhere in some domain D. Since f is real-valued, the conjugate of f is equal to f. Therefore, f and its conjugate are both analytic everywhere in D. By Theorem 15.19, f is constant throughout D. □

88. An antiderivative of a complex-valued function $f(z)$ is unique up to an additive constant.

Proof: Suppose that F and G are both antiderivatives of f and define H by $H(x) = F(x) - G(x)$. Then we have $H'(x) = F'(x) - G'(x) = f(x) - f(x) = 0$. By Theorem 15.18, H is constant. So, there is a constant C such that $F(x) - G(x) = H(x) = C$, or equivalently, $F(x) = G(x) + C$. □

LEVEL 4

For each $\alpha \in \mathbb{R}$, define the complex-valued function $F_\alpha(z) = \ln r + i\theta$ $(r > 0, \alpha < \theta < \alpha + 2\pi)$.

89. Prove that for each $\alpha \in \mathbb{R}$, F_α is continuous.

Proof: $u(r,\theta) = \ln r$ is continuous for all $r > 0$ and $v(r,\theta) = \theta$ is continuous for $\alpha < \theta < \alpha + 2\pi$. By Problem 93 in Problem Set 7, F_α is continuous. □

90. Prove that if the function F_α were defined for $\theta = \alpha$, then it would **not** be continuous.

Proof: $\lim_{\theta \to \alpha^+} \theta = \alpha$ and $\lim_{\theta \to \alpha^-} \theta = \alpha + 2\pi$. So, the function $v(r,\theta) = \theta, \alpha \leq \theta < \alpha + 2\pi$ is not continuous at $\theta = \alpha$. □

91. A **branch** of a multiple-valued function f is any single-valued function F that is analytic in some domain at each point z of which the value $F(z)$ is one of the values of $f(z)$. Prove that for each $\alpha \in \mathbb{R}$, F_α is a branch of $\ln z$.

Proof: Let $\alpha \in \mathbb{R}$. $u(r,\theta) = \ln r$ and $v(r,\theta) = \theta$ have continuous partial derivatives for $r > 0$ and $\alpha < \theta < \alpha + 2\pi$. Furthermore, $ru_r = r\left(\frac{1}{r}\right) = 1 = v_\theta$ and $u_\theta = 0 = -rv_r$. So, the partial derivatives satisfy the polar form of the Cauchy-Riemann equations. By Theorem 15.13, $F'_\alpha(z)$ exists for all z in the domain $r > 0, \alpha < \theta < \alpha + 2\pi$. So, F_α is a branch of $\ln z$.. □

92. Compute $\frac{d}{dz}[\ln z]$ $(|z| > 0, \alpha < \arg z < \alpha + 2\pi)$.

$u(r,\theta) = \ln r$ and $v(r,\theta) = \theta$. So, $u_r = \frac{1}{r}$ and $v_r = 0$. By Theorem 15.13, we have

$$\frac{d}{dz}[\ln z] = e^{-i\theta}(u_r + iv_r) = e^{-i\theta}\left(\frac{1}{r} + 0i\right) = \frac{1}{re^{i\theta}} = \frac{1}{z} \ (|z| > 0, \alpha < \arg z < \alpha + 2\pi)$$

93. Compute $\frac{d}{dz}[\mathrm{Ln}\, z]$.

$$\frac{d}{dz}[\mathrm{Ln}\, z] = \frac{1}{z} \ (|z| > 0, -\pi < \mathrm{Arg}\, z < \pi)$$

Let c be a complex number. We define the complex-valued function z^c to be $e^{c \ln z}$ for all $z \neq 0$.

94. Prove that $\frac{1}{z^c} = z^{-c}$.

Proof: By Problem 29 above, for all $a, b \in \mathbb{C}$, we have $\frac{e^a}{e^b} = e^{a-b}$. Therefore, for all $z \in \mathbb{C}$, we have $\frac{1}{e^z} = \frac{e^0}{e^z} = e^{0-z} = e^{-z}$. So, $\frac{1}{z^c} = \frac{1}{e^{c \ln z}} = e^{(-c \ln z)} = z^{-c}$. □

95. Compute the derivative of $z^c = e^{c \ln z}$ $(|z| > 0, \alpha < \arg z < \alpha + 2\pi)$.

$$\frac{d}{dz}[z^c] = \frac{d}{dz}\left[e^{c \ln z}\right] = \frac{c}{z}e^{c \ln z} = \frac{c}{e^{\ln z}}e^{c \ln z}$$

$$= c\frac{e^{c \ln z}}{e^{\ln z}} = ce^{(c-1)\ln z} = cz^{c-1} \ (|z| > 0, \alpha < \arg z < \alpha + 2\pi)$$

96. Evaluate $\int_C z^{\frac{1}{2}}\, dz$ directly, where C is the semicircular path $z = 5e^{i\theta}$ $(0 \leq \theta \leq \pi)$ from the point $z = 5$ to the point $z = -5$ and we are considering the branch of $z^{\frac{1}{2}} = e^{\frac{1}{2}\ln z}$ such that $|z| > 0$ and $0 < \arg z < 2\pi$.

For $0 < \theta \leq \pi$, $f[z(\theta)] = f\left[5e^{i\theta}\right] = \left(5e^{i\theta}\right)^{\frac{1}{2}} = \sqrt{5}e^{\frac{\theta}{2}i}$ and $z'(t) = 5ie^{i\theta}$.

$$\lim_{\theta \to 0^+} f[z(\theta)]z'(\theta) = \lim_{\theta \to 0^+}\left[\sqrt{5}e^{\frac{\theta}{2}i} \cdot 5ie^{i\theta}\right] = \lim_{\theta \to 0^+}\left[5\sqrt{5}ie^{\frac{3\theta}{2}i}\right]$$

$$= 5\sqrt{5}i \lim_{\theta \to 0^+}\left[\cos\frac{3\theta}{2} + i\sin\frac{3\theta}{2}\right] = 5\sqrt{5}i$$

So, $f[z(\theta)]z'(\theta)$ is continuous on $[0, \pi]$ if we define it's value at $\theta = 0$ to be $5\sqrt{5}i$.

Therefore,

$$\int_C z^{\frac{1}{2}}\,dz = 5\sqrt{5}i\int_0^{\pi} e^{\frac{3\theta}{2}i}\,d\theta = 5\sqrt{5}i\cdot\frac{2}{3i}e^{\frac{3\theta}{2}i}|_0^{\pi} = \frac{10\sqrt{5}}{3}\left(e^{\frac{3\pi}{2}i} - e^0\right)$$

$$= \frac{10\sqrt{5}}{3}(-i-1) = -\frac{\mathbf{10\sqrt{5}}}{\mathbf{3}}(\mathbf{1}+\boldsymbol{i}).$$

97. Evaluate $\int_C z^{a-1}\,dz$ directly, where C is the positively oriented circle $z = e^{i\theta}$ $(-\pi \le \theta \le \pi)$, a is an arbitrary nonzero real number and we use the principal branch of z^{a-1} (the principal branch of z^c is $e^{c\,\mathrm{Ln}\,z}(z \neq 0)$).

Let $f(z) = z^{a-1} = e^{(a-1)\ln z}$ $(|z| > 0,\ -\pi < \arg z < \pi)$.

For $-\pi < \theta < \pi$, $f[z(\theta)]z'(t) = f\left[e^{i\theta}\right]z'(t) = e^{(a-1)(\ln 1 + i\theta)}ie^{i\theta} = e^{ai\theta}e^{-i\theta}ie^{i\theta} = ie^{a\theta i}$.

$$\lim_{\theta\to-\pi^+} f[z(\theta)]z'(\theta) = \lim_{\theta\to-\pi^+}\left[ie^{a\theta i}\right] = ie^{-a\pi i} \text{ and } \lim_{\theta\to\pi^-} f[z(\theta)]z'(\theta) = \lim_{\theta\to\pi^-}\left[ie^{a\theta i}\right] = ie^{a\pi i}.$$

So, $f[z(\theta)]z'(\theta)$ is continuous on $[-\pi,\pi]$ if we define it's values at $\theta = -\pi$ and $\theta = \pi$ to be $ie^{-a\pi i}$ and $ie^{a\pi i}$, respectively. Therefore,

$$\int_C z^{a-1}\,dz = i\int_{-\pi}^{\pi} e^{a\theta i}\,dt = \frac{1}{a}e^{a\theta i}|_{-\pi}^{\pi} = \frac{1}{a}\left(e^{a\pi i} - e^{-a\pi i}\right)$$

$$= \frac{1}{a}[(\cos a\pi + i\sin a\pi) - (\cos a\pi - i\sin a\pi)] = \frac{1}{a}[i\sin a\pi + i\sin a\pi] = \frac{\mathbf{2}}{\boldsymbol{a}}\boldsymbol{i}\,\mathbf{sin}\,\boldsymbol{a\pi}.$$

98. Let C be the semicircular path $z = Re^{i\theta}$ $(0 \le \theta \le \pi)$ and let $z^{\frac{1}{2}}$ denote the following branch of the square root function: $z^{\frac{1}{2}} = e^{\frac{1}{2}\ln z} = \sqrt{r}e^{\frac{i\theta}{2}}\left(r > 0, -\frac{\pi}{2} < \theta < \frac{3\pi}{2}\right)$. Show that

$$\lim_{R\to\infty}\int_C \frac{z^{\frac{1}{2}}}{z^2+1}\,dz = 0.$$

If z is on C, then $\left|z^{\frac{1}{2}}\right| = \left|\left(Re^{i\theta}\right)^{\frac{1}{2}}\right| = \left|R^{\frac{1}{2}}\left(e^{i\theta}\right)^{\frac{1}{2}}\right| = \left|R^{\frac{1}{2}}\right|\left|\left(e^{i\theta}\right)^{\frac{1}{2}}\right| = \sqrt{R}\cdot 1 = \sqrt{R}$ and by Problem 66 in Problem Set 7, $|z^2+1| \ge ||z|^2 - 1| = |R^2 - 1|$. So,

$$\left|\frac{z^{\frac{1}{2}}}{z^2+1}\right| \le \frac{\sqrt{R}}{|R^2-1|}.$$

Since the length of the semicircle is πR, by Theorem 15.31, we have

$$\left|\int_C \frac{z^{\frac{1}{2}}}{z^2+1}\,dz\right| \le \frac{R^{\frac{3}{2}}\pi}{|R^2-1|}.$$

The expression on the right approaches 0 as R approaches ∞.

99. Let C be the positively oriented circle $z = 2e^{i\theta}$ $(-\pi \le \theta \le \pi)$. Use Theorem 15.33 to evaluate $\int_C \frac{dz}{z}$.

The function $f(z) = \frac{1}{z}$ is continuous everywhere except at the origin and $F(z) = \ln z$ is an antiderivative of f in the domain $r > 0, \alpha < \theta < \alpha + 2\pi$ for any α. However, $F(z) = \ln z$ $(r > 0, \alpha < \theta < \alpha + 2\pi)$ is not differentiable (or even defined) along its branch cut $\theta = \alpha$. So, $F'(z)$ fails to exist at the point where the ray $\theta = \alpha$ intersects the circle C. Therefore, this F cannot be used in Theorem 15.33 for any α.

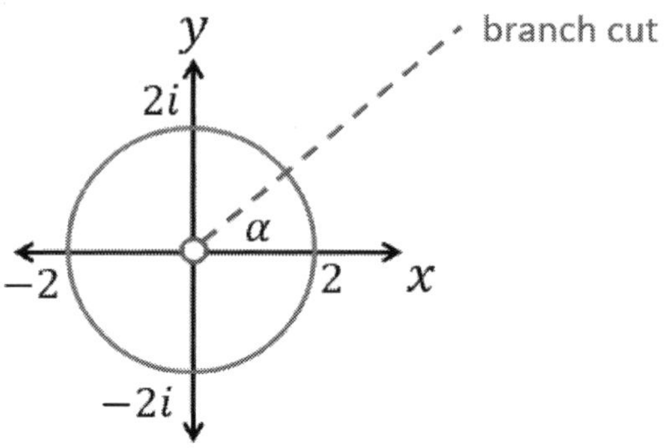

Let C_1 be the right half of C. That is C_1 is the semicircle $z = 2e^{i\theta}$ $\left(-\frac{\pi}{2} \le \theta \le \frac{\pi}{2}\right)$. Also, let F be the principal branch of the logarithm,

$$F(z) = \operatorname{Ln} z = \ln r + i\Theta \ (r > 0, \ -\pi < \Theta < \pi).$$

Then by Theorem 15.33, we have

$$\int_{C_1} \frac{dz}{z} = \int_{-2i}^{2i} \frac{dz}{z} = \operatorname{Ln} z|_{-2i}^{2i} = \operatorname{Ln} 2i - \operatorname{Ln}(-2i)$$
$$= \left(\ln 2 + \frac{\pi}{2}i\right) - \left(\ln 2 - \frac{\pi}{2}i\right) = \pi i.$$

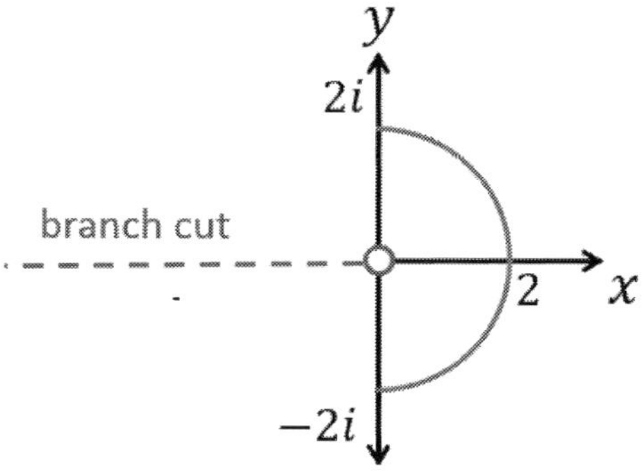

Next, let C_2 be the left half of C. That is C_2 is the semicircle $z = 2e^{i\theta}$ $\left(\frac{\pi}{2} \le \theta \le \frac{3\pi}{2}\right)$. Also, let G be the following branch of the logarithm: $G(z) = \operatorname{Ln} z = \ln r + i\Theta \ (r > 0, \ 0 < \theta < 2\pi)$. Then by Theorem 15.33, we have

$$\int_{C_2} \frac{dz}{z} = \int_{2i}^{-2i} \frac{dz}{z} = \operatorname{Ln} z|_{2i}^{-2i} = \operatorname{Ln}(-2i) - \operatorname{Ln} 2i$$
$$= \left(\ln 2 + \frac{3\pi}{2}i\right) - \left(\ln 2 + \frac{\pi}{2}i\right) = \pi i.$$

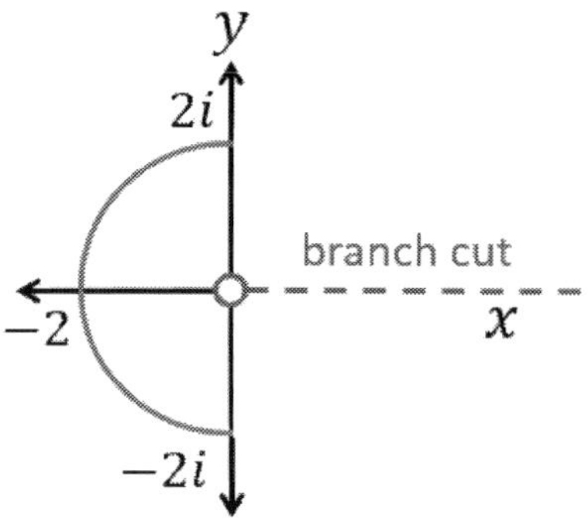

Thus,

$$\int_C \frac{dz}{z} = \int_{C_1} \frac{dz}{z} + \int_{C_2} \frac{dz}{z} = \pi i + \pi i = 2\pi i.$$

100. Let C be any contour from $z = -3$ to $z = 3$ that, except for its endpoints, lie entirely above the x-axis. Use Theorem 15.33 to evaluate $\int_C z^{\frac{1}{2}}\,dz$, where $z^{\frac{1}{2}}$ is the branch $z^{\frac{1}{2}} = e^{\frac{1}{2}\ln z} = \sqrt{r}e^{\frac{\theta}{2}i}$ $(r > 0,\ 0 < \theta < 2\pi)$.

$f(z) = z^{\frac{1}{2}} = e^{\frac{1}{2}\ln z} = \sqrt{r}e^{\frac{\theta}{2}i}$ $(r > 0,\ 0 < \theta < 2\pi)$ is piecewise continuous on C and therefore, $\int_C z^{\frac{1}{2}}\,dz$ exists. However, the given branch of $z^{\frac{1}{2}}$ is not defined on the ray $\theta = 0$ (at the point $z = 3$).

Let $g(z) = z^{\frac{1}{2}} = e^{\frac{1}{2}\ln z} = \sqrt{r}e^{\frac{\theta}{2}i}$ $\left(r > 0,\ -\frac{\pi}{2} < \theta < \frac{3\pi}{2}\right)$. This branch of $z^{\frac{1}{2}}$ is defined and continuous on C and $g(z) = f(z)$ for all z on C, except for $z = 3$. $G(z) = \frac{2}{3}z^{\frac{3}{2}} = \frac{2}{3}r^{\frac{3}{2}}e^{\frac{3\theta}{2}i}$ $\left(r > 0, -\frac{\pi}{2} < \theta < \frac{3\pi}{2}\right)$ is an antiderivative of g. By Theorem 15.33, we have

$$\int_C z^{\frac{1}{2}}\,dz = \int_C f(z)\,dz = \int_{-3}^{3} g(z)\,dz = G(z)|_{-3}^{3} = \frac{2}{3}\cdot 3^{\frac{3}{2}}\left(e^{0i} - e^{\frac{3\pi}{2}i}\right) = 2\sqrt{3}(1+i).$$

Prove each of the following:

101. Let $n \geq 1$, let $A \subseteq \mathbb{R}^n$ be a nonempty closed and bounded region, and let $f: A \to \mathbb{R}$ be continuous on A. Then $\sup f[A]$ and $\inf f[A]$ are both in $f[A]$.

Proof: By the Generalized Heine-Borel Theorem (Theorem 14.57), A is compact. Since f is continuous, by an argument similar to the proof of Theorem 6.39, $f[A]$ is compact. By the Heine-Borel Theorem, $f[A]$ is closed and bounded. By the proof of Corollary 6.42, $\sup f[A]$ and $\inf f[A]$ are both in $f[A]$. □

102. Let $A \subseteq \mathbb{C}$ be a nonempty closed and bounded region and let $f: A \to \mathbb{C}$ be continuous. Then there exists a nonzero real number M such that $|f(z)| \leq M$ for all $z \in A$, where equality holds for at least one value of $z \in A$.

Proof: Let $f: A \to \mathbb{C}$ be continuous and let $B = \{(x, y) \in \mathbb{R}^2 \mid x + iy \in A\}$. Then the function $g: B \to \mathbb{R}^2$ defined by $g(x, y) = \sqrt{[u(x, y)]^2 + [v(x, y)]^2}$ is a continuous real-valued function on A. By Problem 101 above, $\sup g[B]$ is in $g[B]$. Let $\sup g[B] = g(x, y)$. Then $x + iy \in A$ and we see that $M = \sup f[A] = f(x + iy) \in f[A]$. □

103. Let $R \subseteq \mathbb{C}$ be a nonempty closed and bounded region and let f be a complex-valued function that is continuous on R such that f is analytic and not constant on the interior of R. Then $|f(z)|$ has a maximum value in R and this maximum value occurs on the boundary of R and never in the interior of R. You may use the Maximum Modulus Principle described in Problem 123 below.

Proof: By Problem 102 above, $|f(z)|$ has a maximum value M in R. By the Maximum Modulus Principle (see Problem 123 below), $|f(z)| \neq M$ for any point in the interior of R. Therefore, the maximum value occurs on the boundary of R. □

104. Define $f: \mathbb{C} \to \mathbb{C}$ by $f(z) = (z + 1)^2$ and let R be the closed triangular region with vertices at $z = 0$, $z = 5$, and $z = 2i$. Then the maximum value of $|f(z)|$ occurs at $z = 5$ and the minimum value of $|f(z)|$ occurs at $z = 0$.

Proof: Let $f(z) = (z+1)^2$ defined on the closed triangular region R with vertices at $z = 0, z = 5$, and $z = 2i$. By Problem 103 above, a maximum value of $|f(z)|$ exists in R and this maximum value must occur on the boundary of R. Note that geometrically for $z \in R$, $|f(z)| = |z - (-1)|^2$ is the square of the distance between -1 and z. It follows that the maximum value of $|f(z)|$ occurs at $z = 5$. Similarly, the minimum value of $|f(z)|$ occurs at $z = 0$. □

Let $f(z) = u(r,\theta) + iv(r,\theta)$ be defined throughout some neighborhood of $a = se^{i\phi}$, let $x = r\cos\theta$, and let $y = r\sin\theta$. You may use Theorem 15.11 for the following problems.

105. Express the partials of u and v with respect to r and θ in terms of the partials with respect to x and y. You may use the Chain Rule described in Problem 119 below.

$$u_r = u_x x_r + u_y y_r = u_x \cos\theta + u_y \sin\theta, \qquad u_\theta = u_x x_\theta + u_y y_\theta = -u_x r\sin\theta + u_y r\cos\theta$$

$$v_r = v_x x_r + v_y y_r = v_x \cos\theta + v_y \sin\theta, \qquad v_\theta = v_x x_\theta + v_y y_\theta = -v_x r\sin\theta + v_y r\cos\theta$$

106. Prove that if the Cauchy-Riemann equations are satisfied, then the polar form of the Cauchy-Riemann equations are satisfied.

Proof: Assume that the Cauchy-Riemann equations are satisfied: $u_x = v_y$ and $u_y = -v_x$. Using Problem 105, we have

$$ru_r = r(u_x\cos\theta + u_y\sin\theta) = u_y r\sin\theta + u_x r\cos\theta = -v_x r\sin\theta + v_y r\cos\theta = v_\theta;$$

$$-rv_r = -r(v_x\cos\theta + v_y\sin\theta) = -v_y r\sin\theta - v_x r\cos\theta = -u_x r\sin\theta + u_y r\cos\theta = u_\theta.$$ □

107. Prove that if the polar form of the Cauchy-Riemann equations are satisfied, then the Cauchy-Riemann equations are satisfied.

Proof: Assume that the polar form of the Cauchy-Riemann equations are satisfied: $ru_r = v_\theta$ and $u_\theta = -rv_r$. From Problem 105, we have

$$u_r = u_x\cos\theta + u_y\sin\theta, \qquad u_\theta = -u_x r\sin\theta + u_y r\cos\theta,$$

$$v_r = v_x\cos\theta + v_y\sin\theta, \qquad v_\theta = -v_x r\sin\theta + v_y r\cos\theta.$$

Solving the system of equations consisting of the first two equations for u_x and u_y yields

$$u_x = u_r\cos\theta - u_\theta\frac{\sin\theta}{r};$$

$$u_y = u_r\sin\theta + u_\theta\frac{\cos\theta}{r}.$$

Solving the system of equations consisting of the second two equations for v_x and v_y yields

$$v_x = v_r\cos\theta - v_\theta\frac{\sin\theta}{r};$$

$$v_y = v_r\sin\theta + v_\theta\frac{\cos\theta}{r}.$$

So, we have

$$u_x = u_r \cos\theta - u_\theta \frac{\sin\theta}{r} = \left(\frac{v_\theta}{r}\right)\cos\theta - (-rv_r)\frac{\sin\theta}{r} = v_r \sin\theta + v_\theta \frac{\cos\theta}{r} = v_y;$$

$$u_y = u_r \sin\theta + u_\theta \frac{\cos\theta}{r} = \left(\frac{v_\theta}{r}\right)\sin\theta + (-rv_r)\frac{\cos\theta}{r} = -\left(v_r \cos\theta - v_\theta \frac{\sin\theta}{r}\right) = -v_x. \quad \square$$

108. Prove that if the polar form of the Cauchy-Riemann equations are satisfied and $b + ic = se^{i\phi}$, then $u_x(b,c) + iv_x(b,c) = e^{-i\theta}\big(u_r(s,\phi) + iv_r(s,\phi)\big)$.

Proof: By Problem 107, we have

$$u_x + iv_x = \left(u_r \cos\theta - u_\theta \frac{\sin\theta}{r}\right) + i\left(v_r \cos\theta - v_\theta \frac{\sin\theta}{r}\right)$$

$$= (u_r \cos\theta + v_r \sin\theta) + i(v_r \cos\theta - u_r \sin\theta) = (\cos\theta - i\sin\theta)(u_r + iv_r) = e^{-i\theta}(u_r + iv_r). \quad \square$$

109. Prove Theorem 15.13.

Proof: Suppose that $a = se^{i\phi} = b + ic$ and the following hold:

1. $f(z) = u(r,\theta) + iv(r,\theta)$ is defined throughout some neighborhood of $a = se^{i\phi}$,
2. the first order partial derivatives of u and v with respect to r and θ exist in the neighborhood,
3. the partial derivatives of u and v are continuous at (s,ϕ),
4. the partial derivatives of u and v satisfy the polar form of the Cauchy-Riemann equations $ru_r = v_\theta$, $u_\theta = -rv_r$ at (s,ϕ).

Then by Problem 107 (and its proof), we have the following:

1. $f(z) = u(x,y) + iv(x,y)$ is defined throughout some neighborhood of $a = b + ic$,
2. the first order partial derivatives of u and v with respect to x and y exist in the neighborhood,
3. the partial derivatives of u and v are continuous at (b,c),
4. the partial derivatives of u and v satisfy the Cauchy-Riemann equations $u_x = v_y$, $u_y = -v_x$ at (b,c).

By Theorem 15.11, $f'(a)$ exists and $f'(a) = u_x(b,c) + iv_x(b,c)$. By Problem 108, it follows that $f'(a) = e^{-i\theta}\big(u_r(s,\phi) + iv_r(s,\phi)\big)$. $\square$

LEVEL 5

Prove each of the following:

110. An open set of complex numbers is path connected if and only if it is connected.

Proof: Let A be an open set of complex numbers that is path connected and assume toward contradiction that A is not connected. Let (U, V) be a disconnection of A, let $x \in U$, and let $y \in V$. Since A is path connected, there is a continuous function $f: [0, 1] \to A$ such that $f(0) = x$ and $f(1) = y$ (a polygonal path is continuous). By the continuity of f, $f^{-1}[U]$ and $f^{-1}[V]$ are both open in $[0, 1]$. Also, $f^{-1}[U] \cap f^{-1}[V] = \emptyset$, $f^{-1}[U] \cup f^{-1}[V] = [0, 1]$, $0 \in f^{-1}[U]$ and $1 \in f^{-1}[V]$. Therefore, $(f^{-1}[U], f^{-1}[V])$ is a disconnection of $[0, 1]$, contradicting Theorem 6.34. It follows that A is connected.

Conversely, let A be a connected open set of complex numbers, let $x \in A$, and let

$$B = \{y \in A \mid \text{there is a polygonal path in } A \text{ from } x \text{ to } y\}.$$

Since $x \in B$ (the constant function $f: [0, 1] \to A$ defined by $f(t) = x$ is a polygonal path from x to itself), it follows that $B \neq \emptyset$.

Let $y \in B$ and let U be a neighborhood of y that is path connected (just take any open ball containing y and contained in A—see Example 15.16). If $z \in U$, then there is a polygonal path in A from y to z. Since there is also a polygonal path in A from x to y, we can combine paths to get a polygonal path in A from x to z. Therefore, $z \in B$. Since $z \in U$ was arbitrary, $U \subseteq B$. Since $y \in B$ was arbitrary, B is open in A.

Now, let $y \in A \setminus B$ and let V be a neighborhood of y that is pathwise connected (once again, we can take any open ball containing y and contained in A) . If $z \in V$, then there is a polygonal path in V from y to z. If there was a polygonal path in A from z to x, then we would get a polygonal path in A from y to x, contradicting the assumption that $y \notin B$. So, $z \in A \setminus B$. Since $z \in U$ was arbitrary, $U \subseteq A \setminus B$. Since $y \in A \setminus B$ was arbitrary, $A \setminus B$ is open, and therefore, B is closed in A.

Since A is connected and B is a nonempty clopen subset of A, we must have $B = A$. Since $x \in A$ was arbitrary, A is path connected. □

111. Let $D \subseteq \mathbb{R}^2$ be an open disk, let $f: D \to \mathbb{R}$, and assume that f_{xy} and f_{yx} exist and are continuous throughout D. Then $f_{xy} = f_{yx}$ in D. This is known as **Clairaut's Theorem**.

Proof: Let $(b, c) \in D$ and let U be an open interval containing 0 such that $f(b + h, c + k)$ is defined for all $h, k \in U$. Define $\Delta: U \to \mathbb{R}$ by

$$\Delta(h) = [f(b + h, c + h) - f(b + h, c)] - [f(b, c + h) - f(b, h)]$$

and for $h \in U$, define $g: [b, b + h] \to \mathbb{R}$ by $g(x) = f(x, c + h) - f(x, c)$. Then we have

$$\Delta(h) = g(b + h) - g(b).$$

Since g is continuous on $[b, b+h]$ and differentiable on $(b, b+h)$, by the Mean Value Theorem (Theorem 13.25), there is $d \in (b, b+h)$ such that

$$g(b+h) - g(b) = g'(d)h = [f_x(d, c+h) - f_x(d, c)]h.$$

Now, define $g^*: [c, c+h] \to \mathbb{R}$ by $g^*(y) = f_x(d, y)$

Since g^* is continuous on $[c, c+h]$ and differentiable on $(c, c+h)$, by the Mean Value Theorem (Theorem 13.25) again, there is $e \in (c, c+h)$ such that

$$f_x(d, c+h) - f_x(d, c) = g^*(c+h) - g^*(c) = (g^*)'(e)h = f_{xy}(d, e)h.$$

So, $\Delta(h) = g(b+h) - g(b) = [f_x(d, c+h) - f_x(d, c)]h = f_{xy}(d, e)h \cdot h = h^2 f_{xy}(d, e)$.

Now, as $h \to 0$, we have $(d, e) \to (b, c)$, and so, by the continuity of f_{xy} at (b, c), we have

$$\lim_{h \to 0} \frac{\Delta(h)}{h^2} = \lim_{(d,e) \to (b,c)} \frac{h^2 f_{xy}(d, e)}{h^2} = \lim_{(d,e) \to (b,c)} f_{xy}(d, e) = f_{xy}(b, c).$$

Similarly, if we let

$$\Delta(h) = [f(b+h, c+h) - f(b, c+h)] - [f(b+h, c) - f(b, h)],$$

an analogous procedure leads to

$$\lim_{h \to 0} \frac{\Delta(h)}{h^2} = \lim_{(d,e) \to (b,c)} \frac{h^2 f_{yx}(d, e)}{h^2} = \lim_{(d,e) \to (b,c)} f_{yx}(d, e) = f_{yx}(b, c).$$

Therefore, $f_{xy}(b, c) = f_{yx}(b, c)$. □

112. Let $A \subseteq \mathbb{R}^2$, let $f: A \to \mathbb{R}$, let (b, c) be an interior point of A, and let $\mathbf{u} = (u_1, u_2)$ be a unit vector in $\mathbb{R}^2$. If $f_x(b, c)$ and $f_y(b, c)$ exist, then $D_{\mathbf{u}} f(b, c) = f_x(b, c) \cdot u_1 + f_y(b, c) \cdot u_2$. You may use the Chain Rule described in Problem 119 below.

Proof: Let U be an open interval containing 0 such that $f(b+h, c+k)$ is defined for all $h, k \in U$. Define $\Delta: U \to \mathbb{R}$ by

$$\Delta(h) = f(b + hu_1, c + hu_2)$$

Then we have

$$\Delta'(0) = \lim_{h \to 0} \frac{\Delta(h) - \Delta(0)}{h} = \lim_{h \to 0} \frac{f(b + hu_1, c + hu_2) - f(b, c)}{h} = D_{\mathbf{u}} f(b, c).$$

If we write $\Delta(h) = f(x, y)$, where $x = b + hu_1$ and $y = c + hu_2$, then by the Chain Rule (Problem 119 below), we have

$$\Delta'(h) = \frac{\partial f}{dx} \cdot \frac{dx}{dh} + \frac{\partial f}{dy} \cdot \frac{dy}{dh} = f_x(x, y) \cdot u_1 + f_y(x, y) \cdot u_2.$$

If we let $h = 0$, then $x = b$, $y = c$, and $\Delta'(0) = f_x(b, c) \cdot u_1 + f_y(b, c) \cdot u_2$.

It follows that $D_{\mathbf{u}} f(b, c) = f_x(b, c) \cdot u_1 + f_y(b, c) \cdot u_2$. □

113. Theorem 15.11 (You may use Problem 118 below.)

Proof: Let $\Delta z = \Delta x + i\Delta y$, let $\Delta w = f(a + \Delta z) - f(a)$, let $\epsilon > 0$, and assume that $0 < |\Delta z| < \epsilon$. Then $\Delta w = \Delta u + i\Delta v$, where $\Delta u = u(b + \Delta x, c + \Delta y) - u(b, c)$ and $\Delta v = v(b + \Delta x, c + \Delta y) - v(b, c)$. Since the partial derivatives of u and v are continuous at (b, c), by Problem 118 below, we have

$$\Delta u = u_x(b,c)\Delta x + u_y(b,c)\Delta y + \epsilon_1\sqrt{(\Delta x)^2 + (\Delta y)^2}$$
$$\Delta v = v_x(b,c)\Delta x + v_y(b,c)\Delta y + \epsilon_2\sqrt{(\Delta x)^2 + (\Delta y)^2}$$

where $\epsilon_1, \epsilon_2 \to 0$ as $(\Delta x, \Delta y) \to (0,0)$.

By direct substitution, we have

$$\Delta w = \left[u_x(b,c)\Delta x + u_y(b,c)\Delta y + \epsilon_1\sqrt{(\Delta x)^2 + (\Delta y)^2}\right] + i\left[v_x(b,c)\Delta x + v_y(b,c)\Delta y + \epsilon_2\sqrt{(\Delta x)^2 + (\Delta y)^2}\right].$$

Since u and v satisfy the Cauchy-Riemann equations at (b, c), we can replace $u_y(b, c)$ by $-v_x(b, c)$ and $v_y(b, c)$ by $u_x(b, c)$ to get

$$\begin{aligned}\Delta w &= \left[u_x(b,c)\Delta x - v_x(b,c)\Delta y + \epsilon_1\sqrt{(\Delta x)^2 + (\Delta y)^2}\right] + i\left[v_x(b,c)\Delta x + u_x(b,c)\Delta y + \epsilon_2\sqrt{(\Delta x)^2 + (\Delta y)^2}\right] \\ &= [u_x(b,c)\Delta x + iu_x(b,c)\Delta y] + i[v_x(b,c)\Delta x + iv_x(b,c)\Delta y] + (\epsilon_1 + i\epsilon_2)\sqrt{(\Delta x)^2 + (\Delta y)^2}.\end{aligned}$$

Dividing each side of the equation by $\Delta z = \Delta x + i\Delta y$ and noting that $|\Delta z| = \sqrt{(\Delta x)^2 + (\Delta y)^2}$, we get

$$\frac{\Delta w}{\Delta z} = u_x(b,c) + iv_x(b,c) + (\epsilon_1 + i\epsilon_2)\frac{|\Delta z|}{\Delta z}.$$

If $\Delta z \to 0$, then $(\Delta x, \Delta y) \to (0,0)$, and so $\epsilon_1, \epsilon_2 \to 0$. Since $\left|\frac{|\Delta z|}{\Delta z}\right| = \frac{|\Delta z|}{|\Delta z|} = 1$, it follows that $\Delta z \to 0$ implies that $(\epsilon_1 + i\epsilon_2)\frac{|\Delta z|}{\Delta z} \to 0$. Therefore,

$$\lim_{\Delta z \to 0} \frac{\Delta w}{\Delta z} = u_x(b,c) + iv_x(b,c).$$

Therefore,

$$f'(a) = \lim_{\Delta z \to 0} \frac{f(a + \Delta z) - f(a)}{\Delta z} = \lim_{\Delta z \to 0} \frac{\Delta w}{\Delta z} = u_x(b,c) + iv_x(b,c),$$

as desired. □

114. Suppose that f is analytic everywhere on and inside a simple closed contour C with positive orientation. If a is any point in the interior of C, then

$$f'(a) = \frac{1}{2\pi i}\int_C \frac{f(z)dz}{(z-a)^2} \quad \text{and} \quad f''(a) = \frac{1}{\pi i}\int_C \frac{f(z)dz}{(z-a)^3}.$$

Proof: By Theorem 15.46 (Cauchy's Integral Formula), we have

$$f(a) = \frac{1}{2\pi i}\int_C \frac{f(z)\,dz}{z-a}.$$

Letting $d = \inf\{|w - a| \mid w \in C\}$, we have for $0 < |\Delta z| < d$,

$$\frac{f(a+\Delta z) - f(a)}{\Delta z} = \frac{1}{2\pi i}\int_C \left(\frac{1}{t - a - \Delta z} - \frac{1}{t - a}\right)\frac{f(t)}{\Delta z}dt = \frac{1}{2\pi i}\int_C \frac{f(t)dt}{(t - a - \Delta z)(t - a)}.$$

It follows that

$$\frac{f(a+\Delta z) - f(a)}{\Delta z} - \frac{1}{2\pi i}\int_C \frac{f(t)dt}{(t-a)^2} = \frac{1}{2\pi i}\int_C \frac{f(t)dt}{(t - a - \Delta z)(t - a)} - \frac{1}{2\pi i}\int_C \frac{f(t)dt}{(t-a)^2}$$
$$= \frac{1}{2\pi i}\int_C \frac{f(t)[(t-a) - (t - a - \Delta z)]dt}{(t - a - \Delta z)(t-a)^2} = \frac{1}{2\pi i}\int_C \frac{\Delta z f(t)dt}{(t - a - \Delta z)(t-a)^2}.$$

By Problem 66 from Problem Set 7, we have

$$|t - a - \Delta z| = |(t - a) - \Delta z| \geq \big||t - a| - |\Delta z|\big| \geq d - |\Delta z|.$$

Note that $|t - a| \geq d$ by the definition of d. Letting M be the maximum value of $|f(t)|$ on C (M exists by Problem 103) and letting L be the arc length of C, by Theorem 15.31, we have

$$\left|\int_C \frac{\Delta z f(t)dt}{(t - a - \Delta z)(t-a)^2}\right| \leq \frac{|\Delta z|M}{(d - |\Delta z|)d^2}L.$$

As $\Delta z \to 0$, we see that $\frac{|\Delta z|M}{(d - |\Delta z|)d^2}L \to \frac{0 \cdot M}{d^3}L = 0$. It follows that

$$\lim_{\Delta z \to 0}\left[\frac{f(a+\Delta z) - f(a)}{\Delta z} - \frac{1}{2\pi i}\int_C \frac{f(t)dt}{(t-a)^2}\right] = \frac{1}{2\pi i} \cdot 0 = 0.$$

Therefore,

$$f'(a) = \lim_{\Delta z \to 0}\frac{f(a+\Delta z) - f(a)}{\Delta z} = \frac{1}{2\pi i}\int_C \frac{f(t)dt}{(t-a)^2}.$$

Similarly, using the formula we just derived, we have for $0 < |\Delta z| < d$,

$$\frac{f'(a+\Delta z) - f'(a)}{\Delta z} = \frac{1}{2\pi i}\int_C \left(\frac{1}{(t - a - \Delta z)^2} - \frac{1}{(t-a)^2}\right)\frac{f(t)}{\Delta z}dt$$
$$= \frac{1}{2\pi i}\int_C \frac{(t-a)^2 - (t - a - \Delta z)^2}{(t - a - \Delta z)^2(t-a)^2} \cdot \frac{f(t)}{\Delta z}dt$$
$$= \frac{1}{2\pi i}\int_C \frac{[(t^2 - 2ta + a^2) - (t^2 - 2ta - 2t\Delta z + a^2 + 2a\Delta z + (\Delta z)^2)]f(t)dt}{(t - a - \Delta z)^2(t-a)^2\Delta z}.$$
$$= \frac{1}{2\pi i}\int_C \frac{(2t\Delta z - 2a\Delta z - (\Delta z)^2)f(t)dt}{(t - a - \Delta z)^2(t-a)^2\Delta z} = \frac{1}{2\pi i}\int_C \frac{(2t - 2a - \Delta z)f(t)dt}{(t - a - \Delta z)^2(t-a)^2}.$$

It follows that

$$\frac{f'(a+\Delta z)-f'(a)}{\Delta z}-\frac{1}{\pi i}\int_C\frac{f(t)dt}{(t-a)^3}=\frac{1}{2\pi i}\int_C\frac{(2t-2a-\Delta z)f(t)dt}{(t-a-\Delta z)^2(t-a)^2}-\frac{1}{\pi i}\int_C\frac{f(t)dt}{(t-a)^3}$$
$$=\frac{1}{2\pi i}\int_C\frac{[(2t-2a-\Delta z)(t-a)-2(t-a-\Delta z)^2]f(t)dt}{(t-a-\Delta z)^2(t-a)^3}$$
$$=\frac{1}{2\pi i}\int_C\frac{[2t^2-4ta+2a^2-t\Delta z+a\Delta z-2t^2+4ta+4t\Delta z-2a^2-4a\Delta z-2(\Delta z)^2]f(t)dt}{(t-a-\Delta z)^2(t-a)^3}$$
$$=\frac{1}{2\pi i}\int_C\frac{[3t\Delta z-3a\Delta z-2(\Delta z)^2]f(t)dt}{(t-a-\Delta z)^2(t-a)^3}=\frac{1}{2\pi i}\int_C\frac{[3(t-a)\Delta z-2(\Delta z)^2]f(t)dt}{(t-a-\Delta z)^2(t-a)^3}.$$

Once again, by Problem 66 from Problem Set 7, we have

$$|t-a-\Delta z|=|(t-a)-\Delta z|\geq\big||t-a|-|\Delta z|\big|\geq d-|\Delta z|.$$

Letting $D=\sup\{|w-a| \mid w\in C\}$, letting M be the maximum value of $|f(t)|$ on C (M exists by Problem 103) and letting L be the arc length of C, by Theorem 15.31 and the Triangle Inequality, we have

$$\left|\int_C\frac{[3(t-a)\Delta z-2(\Delta z)^2]f(t)dt}{(t-a-\Delta z)^2(t-a)^3}\right|\leq\frac{(3D|\Delta z|+2|\Delta z|^2)M}{(d-|\Delta z|)^2d^3}L.$$

As $\Delta z\to 0$, we see that $\frac{(3D|\Delta z|+2|\Delta z|^2)M}{(d-|\Delta z|)^2d^3}L\to\frac{(3D\cdot 0+2\cdot 0)M}{(d-0)^2d^3}L=0$. It follows that

$$\lim_{\Delta z\to 0}\left[\frac{f'(a+\Delta z)-f'(a)}{\Delta z}-\frac{1}{\pi i}\int_C\frac{f(t)dt}{(t-a)^3}\right]=\frac{1}{2\pi i}\cdot 0=0.$$

Therefore,

$$f''(a)=\frac{f'(a+\Delta z)-f'(a)}{\Delta z}=\frac{1}{\pi i}\int_C\frac{f(t)dt}{(t-a)^3}.$$ □

115. If f is analytic at a point, then its derivatives of all orders exist and are analytic at that point.

Proof: Assume that f is analytic at $a\in\mathbb{C}$. By definition, there is a neighborhood $V=N_r(a)$ such that f is analytic throughout V. Therefore, there is a positively oriented circle C with center a and radius $\frac{r}{2}$ such that f is analytic inside and on C. By Problem 114, for each b interior to C, we have

$$f''(b)=\frac{1}{\pi i}\int_C\frac{f(z)dz}{(z-b)^3}.$$

The existence of f'' throughout $N_{\frac{r}{2}}(a)$ implies that f' is analytic at a.

We can now finish the argument by induction. Assuming that $f^{(n)}$ is analytic at $a\in\mathbb{C}$, the same argument as above shows that $f^{(n+2)}$ exists throughout a neighborhood of a, proving that $f^{(n+1)}$ is analytic at a. □

116. If f is analytic at a point $a = b + ic$, then the component functions of f, namely u and v, have continuous partial derivatives of all orders at (b, c).

Proof: Assume that $f(z) = u(x, y) + iv(x, y)$ is analytic at $a = b + ic \in \mathbb{C}$. By Problem 115, $f'(a)$ exists. Therefore, by Note 4 following Example 15.1, f' is continuous at a. By Theorem 15.9, we have $f'(a) = u_x(b, c) + iv_x(b, c) = v_y(b, c) - iu_y(b, c)$. Since f' is continuous at a, it follows that the partial derivatives of u and v are continuous at (b, c).

We can now finish the argument by induction. Assuming that all nth partials of u and v are continuous at (b, c), assume that $f^{(n)}(z) = w(x, y) + q(x, y)$, where w and q are nth partials of u and v; specifically, either $w = \overline{w}_x, q = \overline{q}_x$ or $w = \overline{q}_y, q = -\overline{w}_y$ ($\overline{w}$ and $\overline{q}$ are $(n-1)^{\text{st}}$ partials of u and/or v). By the same argument as above, we have $f^{(n+1)}(a) = w_x(b, c) + q_x(b, c) = q_y(b, c) - w_y(b, c)$. It follows that the $(n+1)^{\text{st}}$ partials of u and v are continuous at (b, c). □

117. Theorem 15.33 (You may use the Chain Rule described in Problem 119 below.)

Proof: $(1 \to 2)$ Assume that F is an antiderivative of f throughout D and that C is a smooth arc from z_1 to z_2 lying entirely in D. If C is parametrized by $z = z(t)$ $(a \le t \le b)$, then for $a \le t \le b$, we have

$$\frac{d}{dt}\big[F[z(t)]\big] = F'[z(t)]z'(t) = f[z(t)]z'(t).$$

To see this, let $F(z) = U(x, y) + iV(x, y)$, let $z(t) = x(t) + iy(t)$, and let $w(t) = F[z(t)]$. Then we have $w(t) = U\big(x(t), y(t)\big) + iV\big(x(t), y(t)\big)$ and therefore, by the Chain Rule (see Problem 119 below), and Theorem 15.9, we have the following:

$$\begin{aligned} w'(t) &= \big[U_x\big(x(t), y(t)\big)x'(t) + U_y\big(x(t), y(t)\big)y'(t)\big] + i\big[V_x\big(x(t), y(t)\big)x'(t) + V_y\big(x(t), y(t)\big)y'(t)\big] \\ &= \big[U_x\big(x(t), y(t)\big)x'(t) - V_x\big(x(t), y(t)\big)y'(t)\big] + i\big[V_x\big(x(t), y(t)\big)x'(t) + U_x\big(x(t), y(t)\big)y'(t)\big] \\ &= \big[U_x\big(x(t), y(t)\big) + iV_x\big(x(t), y(t)\big)\big](x'(t) + iy'(t)] = F'[z(t)]z'(t). \end{aligned}$$

By the Fundamental Theorem of Calculus (see Note 3 before Example 15.27),

$$\int_C f(z)\,dz = \int_a^b f[z(t)]z'(t)\,dt = \big[F[z(t)]\big]_a^b = F[z(b)] - F[z(a)] = F(z_2) - F(z_1).$$

Therefore,

$$\int_C f(z)\,dz = \int_{z_1}^{z_2} f(z)\,dz = F(z)|_{z_1}^{z_2} = F(z_2) - F(z_1).$$

Now, if C is an arbitrary contour, say C consists of finite smooth arcs $C_1, C_2, .., C_n$, where for each $k = 1, 2, \dots, n$, C_k has endpoints w_k and w_{k+1} (note that $z_1 = w_1$ and $z_2 = w_{n+1}$), then we have

$$\begin{aligned} \int_C f(z)\,dz = \sum_{k=1}^{n}\int_{C_k} f(z)\,dz = \sum_{k=1}^{n}\int_{w_k}^{w_{k+1}} f(z)\,dz = \sum_{k=1}^{n}[F(w_{k+1}) - F(w_k)] \\ = F(w_{n+1}) - F(w_1) = F(z_2) - F(z_1). \end{aligned}$$

$(2 \to 3)$ Let C be any closed contour lying entirely in D and let z_1 and z_2 be any two distinct points lying on C. Let C_1 and C_2 be paths from z_1 to z_2 and from z_2 to z_1, respectively such that $C = C_1 - C_2$. By 2,

$$\int_{C_1} f(z)\,dz = \int_{C_2} f(z)\,dz.$$

Therefore, by Problems 85 and 86, we have

$$\int_C f(z)\,dz = \int_{C_1 - C_2} f(z)\,dz = \int_{C_1} f(z)\,dz + \int_{-C_2} f(z)\,dz = \int_{C_1} f(z)\,dz - \int_{C_2} f(z)\,dz = 0.$$

$(3 \to 1)$ Assume that the integral of f around any closed contour lying entirely in D is zero and let C_1 and C_2 be contours from z_1 to z_2 lying entirely in D. Let $C = C_1 - C_2$. Then C is a closed contour lying entirely in D and therefore, by Problems 85 and 86, we have

$$0 = \int_C f(z)\,dz = \int_{C_1 - C_2} f(z)\,dz = \int_{C_1} f(z)\,dz + \int_{-C_2} f(z)\,dz = \int_{C_1} f(z)\,dz - \int_{C_2} f(z)\,dz.$$

Thus,

$$\int_{C_1} f(z)\,dz = \int_{C_2} f(z)\,dz.$$

Since we have path independence, we can make the following definition:

$$F(z) = \int_a^z f(t)\,dt$$

where a is some complex number in D.

To see that $F'(z) = f(z)$ throughout D, we let $z \in D$ and we let Δz be small enough so that $z + \Delta z$ lies inside a neighborhood of z that is fully contained in D. Then we have

$$\frac{F(z+\Delta z) - F(z)}{\Delta z} = \frac{\int_a^{z+\Delta z} f(t)\,dt - \int_a^z f(t)\,dt}{\Delta z} = \frac{\int_z^{z+\Delta z} f(t)\,dt}{\Delta z} = \frac{1}{\Delta z}\int_z^{z+\Delta z} f(t)\,dt.$$

Therefore, we have

$$\frac{F(z+\Delta z) - F(z)}{\Delta z} - f(z) = \frac{1}{\Delta z}\int_z^{z+\Delta z} f(t)\,dt - \frac{1}{\Delta z}\int_z^{z+\Delta z} f(z)\,dt = \frac{1}{\Delta z}\int_z^{z+\Delta z} [f(t) - f(z)]\,dt.$$

Since f is continuous at z, given $\epsilon > 0$, there is $\delta > 0$ such that $|\Delta z| = |t - z| < \delta$ implies that $|f(t) - f(z)| < \epsilon$. So, if $|\Delta z| < \delta$, then we have

$$\left|\frac{F(z+\Delta z) - F(z)}{\Delta z} - f(z)\right| = \left|\frac{1}{\Delta z}\int_z^{z+\Delta z} [f(t) - f(z)]\,dt\right| \le \frac{1}{|\Delta z|}\left|\int_z^{z+\Delta z} |f(t) - f(z)|\,dt\right|$$

$$< \frac{\epsilon}{|\Delta z|}\left|\int_z^{z+\Delta z} dt\right| = \frac{\epsilon}{|\Delta z|}|z + \Delta z - z| = \frac{\epsilon}{|\Delta z|}|\Delta z| = \epsilon.$$

Therefore,

$$\lim_{\Delta z \to 0} \left[\frac{F(z+\Delta z) - F(z)}{\Delta z} - f(z) \right] = 0$$

It follows that

$$F'(z) = \lim_{\Delta z \to 0} \frac{F(z+\Delta z) - F(z)}{\Delta z} = f(z). \qquad \square$$

Problem Set 16

LEVEL 1

Compute the order of each of the following permutations:

1. (12345)

$(12345)^2 = (13524)$

$(12345)^3 = (13524)(12345) = (14253)$

$(12345)^4 = (14253)(12345) = (15432)$

$(12345)^5 = (15432)(12345) = (1)$

So, $|(12345)| = \mathbf{5}$.

Alternatively, the order of a cycle is simply the length of the cycle. So, $|(12345)| = \mathbf{5}$.

2. $(12)(345)(6789)$

$|(12)(345)(6789)| = \text{lcm}(2,3,4) = \mathbf{12}$.

Express each of the following permutations as a product of transpositions:

3. (123456)

$(123456) = (\mathbf{16})(\mathbf{15})(\mathbf{14})(\mathbf{13})(\mathbf{12})$

4. $(123)(4567)$

$(123)(4567) = (\mathbf{13})(\mathbf{12})(\mathbf{47})(\mathbf{46})(\mathbf{45})$

5. $(123)(345)(25)(1576)$

$(123)(345)(25)(1576) = (\mathbf{13})(\mathbf{12})(\mathbf{35})(\mathbf{34})(\mathbf{25})(\mathbf{16})(\mathbf{17})(\mathbf{15})$

Determine if each of the following permutations is even or odd:

6. (1837942)

Since the length of (1837942) is 7, which is odd, (1837942) is **even**.

7. $(123)(34)(4567)$

(123) is even, (34) is odd, and (4567) is odd. Therefore, (123) can be expressed as a product of an even number of transpositions and (34) and (4567) can each be expressed as a product of an odd number of transpositions. Since the sum of an even integer and two odd integers is even, $(123)(34)(4567)$ can be expressed as a product of an even number of transpositions. Therefore, $(123)(34)(4567)$ is **even**.

LEVEL 2

Prove each of the following:

8. The product of two even permutations is even.

Proof: Let π and σ be even permutations. Then $\pi = \pi_1\pi_2 \cdots \pi_{2j}$ and $\sigma = \sigma_1\sigma_2 \cdots \sigma_{2k}$, where $j, k \in \mathbb{Z}^+$, π_i is a transposition for each $i = 1, 2, \dots, 2j$ and σ_i is a transposition for each $i = 1, 2, \dots, 2k$. Then $\pi\sigma = \pi_1\pi_2 \cdots \pi_{2j}\sigma_1\sigma_2 \cdots \sigma_{2k}$, which is a product of $2j + 2k = 2(j + k)$ transpositions. Therefore, $\pi\sigma$ is even. □

9. The product of two odd permutations is even.

Proof: Let π and σ be odd permutations. Then $\pi = \pi_1\pi_2 \cdots \pi_{2j+1}$ and $\sigma = \sigma_1\sigma_2 \cdots \sigma_{2k+1}$, where $j, k \in \mathbb{Z}^+$, π_i is a transposition for each $i = 1, 2, \dots, 2j + 1$ and σ_i is a transposition for each $i = 1, 2, \dots, 2k + 1$. Then $\pi\sigma = \pi_1\pi_2 \cdots \pi_{2j+1}\sigma_1\sigma_2 \cdots \sigma_{2k+1}$, which is a product of $(2j + 1) + (2k + 1) = 2j + 2k + 2 = 2(j + k + 1)$ transpositions. So, $\pi\sigma$ is even. □

10. The product of an even permutation and an odd permutation is odd.

Proof: Let π be an even permutation and let σ be an odd permutation. Then $\pi = \pi_1\pi_2 \cdots \pi_{2j}$ and $\sigma = \sigma_1\sigma_2 \cdots \sigma_{2k+1}$, where $j, k \in \mathbb{Z}^+$, π_i is a transposition for each $i = 1, 2, \dots, 2j$ and σ_i is a transposition for each $i = 1, 2, \dots, 2k + 1$. Then $\pi\sigma = \pi_1\pi_2 \cdots \pi_{2j}\sigma_1\sigma_2 \cdots \sigma_{2k+1}$, which is a product of $2j + (2k + 1) = 2(j + k) + 1$ transpositions. So, $\pi\sigma$ is odd.

Similarly, $\sigma\pi$ is odd. □

11. For $n > 1$, the function $\Delta_n: S_n \to \{-1, 1\}$ is a surjective homomorphism, where Δ_n is defined by $\Delta_n(\pi) = \begin{cases} -1 & \text{if } \pi \text{ is odd.} \\ 1 & \text{if } \pi \text{ is even.} \end{cases}$

Proof: If π and σ are both even, then by Problem 8 above, $\pi\sigma$ is even. So,

$$\Delta_n(\pi\sigma) = 1 = 1 \cdot 1 = \Delta_n(\pi) \cdot \Delta_n(\sigma).$$

If π and σ are both odd, then by Problem 9 above, $\pi\sigma$ is even. So,

$$\Delta_n(\pi\sigma) = 1 = (-1) \cdot (-1) = \Delta_n(\pi) \cdot \Delta_n(\sigma).$$

If π is even and σ is odd, then by Problem 10 above, $\pi\sigma$ is odd. So,

$$\Delta_n(\pi\sigma) = -1 = 1 \cdot (-1) = \Delta_n(\pi) \cdot \Delta_n(\sigma).$$

If π is odd and σ is even, then by Problem 10 above, $\pi\sigma$ is odd. So,

$$\Delta_n(\pi\sigma) = -1 = (-1) \cdot 1 = \Delta_n(\pi) \cdot \Delta_n(\sigma).$$

To see that Δ_n is surjective, simply observe that (1) is even and (12) is odd, so that $\Delta_n\big((1)\big) = 1$ and $\Delta_n\big((12)\big) = -1$. □

12. $GL_2(\mathbb{Z}_2)$ is isomorphic to S_3.

Proof: $GL_2(\mathbb{Z}_2)$ consists of all 2×2 matrices with each entry equal to 0 or 1 and nonzero determinant. It is easy enough to write down this set explicitly:

$$GL_2(\mathbb{Z}_2) = \left\{\begin{bmatrix} 1 & 0 \\ 0 & 1 \end{bmatrix}, \begin{bmatrix} 1 & 0 \\ 1 & 1 \end{bmatrix}, \begin{bmatrix} 0 & 1 \\ 1 & 0 \end{bmatrix}, \begin{bmatrix} 0 & 1 \\ 1 & 1 \end{bmatrix}, \begin{bmatrix} 1 & 1 \\ 1 & 0 \end{bmatrix}, \begin{bmatrix} 1 & 1 \\ 0 & 1 \end{bmatrix}\right\}.$$

By Problem 32 in Problem Set 11 (and its proof), the only noncommutative group of order 6, up to isomorphism, is S_3. So, it suffices to prove that $GL_2(\mathbb{Z}_2)$ is not commutative.

$$\begin{bmatrix} 0 & 1 \\ 1 & 0 \end{bmatrix}\begin{bmatrix} 1 & 1 \\ 1 & 0 \end{bmatrix} = \begin{bmatrix} 1 & 0 \\ 1 & 1 \end{bmatrix} \qquad \begin{bmatrix} 1 & 1 \\ 1 & 0 \end{bmatrix}\begin{bmatrix} 0 & 1 \\ 1 & 0 \end{bmatrix} = \begin{bmatrix} 1 & 1 \\ 0 & 1 \end{bmatrix}$$

Since $\begin{bmatrix} 0 & 1 \\ 1 & 0 \end{bmatrix}$ and $\begin{bmatrix} 1 & 1 \\ 1 & 0 \end{bmatrix}$ do not commute, $GL_2(\mathbb{Z}_2)$ is a noncommutative group of order 6, and therefore, must be isomorphic to S_3. □

13. $U_2 = \{A \in GL_2(F) \mid A \text{ is upper triangular}\}$ is a matrix group.

Proof: Let $U_2 = \{A \in GL_2(F) \mid A \text{ is upper triangular}\}$. To see that U_2 is a matrix group, first observe that $U_2 \subseteq GL_2(F)$ and $I = \begin{bmatrix} 1 & 0 \\ 0 & 1 \end{bmatrix}$ is upper triangular. Let $A, B \in U_2$, say $A = \begin{bmatrix} a_{11} & a_{12} \\ 0 & a_{22} \end{bmatrix}$ and $B = \begin{bmatrix} b_{11} & b_{12} \\ 0 & b_{22} \end{bmatrix}$. By Theorem 16.23, B is invertible and by the proof of Theorem 16.23, $B^{-1} = \frac{1}{\det B}\begin{bmatrix} b_{22} & -b_{12} \\ 0 & b_{11} \end{bmatrix}$. Therefore,

$$AB^{-1} = \begin{bmatrix} a_{11} & a_{12} \\ 0 & a_{22} \end{bmatrix} \cdot \frac{1}{\det B}\begin{bmatrix} b_{22} & -b_{12} \\ 0 & b_{11} \end{bmatrix} = \frac{1}{\det B}\begin{bmatrix} a_{11}b_{22} & -a_{11}b_{12} + a_{12}b_{11} \\ 0 & a_{22}b_{11} \end{bmatrix},$$

which is upper triangular. By Problem 14 in Problem Set 9, we have $U_2 \leq GL_2(F)$. □

14. $L_2 = \{A \in GL_2(F) \mid A \text{ is lower triangular}\}$ is a matrix group.

Proof: Let $L_2 = \{A \in GL_2(F) \mid A \text{ is lower triangular}\}$. To see that L_2 is a matrix group, first observe that $L_2 \subseteq GL_2(F)$ and $I = \begin{bmatrix} 1 & 0 \\ 0 & 1 \end{bmatrix}$ is lower triangular. Let $A, B \in L_2$, say $A = \begin{bmatrix} a_{11} & 0 \\ a_{21} & a_{22} \end{bmatrix}$ and $B = \begin{bmatrix} b_{11} & 0 \\ b_{21} & b_{22} \end{bmatrix}$. By Theorem 16.23, B is invertible and by the proof of Theorem 16.23, $B^{-1} = \frac{1}{\det B}\begin{bmatrix} b_{22} & 0 \\ -b_{21} & b_{11} \end{bmatrix}$. Therefore,

$$AB^{-1} = \begin{bmatrix} a_{11} & 0 \\ a_{21} & a_{22} \end{bmatrix} \cdot \frac{1}{\det B}\begin{bmatrix} b_{22} & 0 \\ -b_{21} & b_{11} \end{bmatrix} = \frac{1}{\det B}\begin{bmatrix} a_{11}b_{22} & 0 \\ a_{21}b_{22} - a_{22}b_{21} & a_{22}b_{11} \end{bmatrix},$$

which is lower triangular. By Problem 14 in Problem Set 9, we have $L_2 \leq GL_2(F)$. □

15. $D_2 = \{A \in GL_2(F) \mid A \text{ is diagonal}\}$ is a matrix group.

Proof: Let $D_2 = \{A \in GL_2(F) \mid A \text{ is diagonal}\}$. To see that D_2 is a matrix group, first observe that $D_2 \subseteq GL_2(F)$ and $I = \begin{bmatrix} 1 & 0 \\ 0 & 1 \end{bmatrix}$ is diagonal. Let $A, B \in D_2$, say $A = \begin{bmatrix} a_{11} & 0 \\ 0 & a_{22} \end{bmatrix}$ and $B = \begin{bmatrix} b_{11} & 0 \\ 0 & b_{22} \end{bmatrix}$. Then B is invertible and $B^{-1} = \begin{bmatrix} \frac{1}{b_{11}} & 0 \\ 0 & \frac{1}{b_{22}} \end{bmatrix}$. Therefore, $AB^{-1} = \begin{bmatrix} a_{11} & 0 \\ 0 & a_{22} \end{bmatrix}\begin{bmatrix} \frac{1}{b_{11}} & 0 \\ 0 & \frac{1}{b_{22}} \end{bmatrix} = \begin{bmatrix} \frac{a_{11}}{b_{11}} & 0 \\ 0 & \frac{a_{22}}{b_{22}} \end{bmatrix}$, which is diagonal. By Problem 14 in Problem Set 9, we have $D_2 \leq GL_2(F)$. □

16. If X is a nonempty subset of a Banach space V, then X is bounded if and only if there is a real number M such that for all $x \in X$, $|x| \leq M$, where $|x|$ is the norm of x in V.

Proof: First assume that X is bounded. Then there is $L \in \mathbb{R}^+$ so that for all $x, y \in X$, $|x - y| \leq L$. Let $y \in X$ be any element of X. Then for all $x \in X$, we have

$$|x| = |(x - y) + y| \leq |x - y| + |y| \leq L + |y|.$$

So, if we let $M = L + |y|$. Then for all $x \in X$, $|x| \leq M$.

Conversely, suppose that there is $M \in \mathbb{R}$ such that $|x| \leq M$ for all $x \in X$. Then for all $x, y \in X$, we have

$$|x - y| \leq |x| + |y| \leq 2M.$$

So X is bounded. □

LEVEL 3

Write down a formula for the following in terms of the entries of the matrix:

17. The determinant of a 4×4 matrix

Permutation	Even or Odd	Term
(1)	Even	$a_{11}a_{22}a_{33}a_{44}$
(123)	Even	$a_{12}a_{23}a_{31}a_{44}$
(132)	Even	$a_{13}a_{21}a_{32}a_{44}$
(124)	Even	$a_{12}a_{24}a_{33}a_{41}$
(142)	Even	$a_{14}a_{21}a_{33}a_{42}$
(134)	Even	$a_{13}a_{22}a_{34}a_{41}$
(143)	Even	$a_{14}a_{22}a_{31}a_{43}$
(234)	Even	$a_{11}a_{23}a_{34}a_{42}$
(243)	Even	$a_{11}a_{24}a_{32}a_{43}$
$(12)(34)$	Even	$a_{12}a_{21}a_{34}a_{43}$
$(13)(24)$	Even	$a_{13}a_{24}a_{31}a_{42}$
$(23)(14)$	Even	$a_{14}a_{23}a_{32}a_{41}$
(12)	Odd	$-a_{12}a_{21}a_{33}a_{44}$
(13)	Odd	$-a_{13}a_{22}a_{31}a_{44}$
(14)	Odd	$-a_{14}a_{22}a_{33}a_{41}$
(23)	Odd	$-a_{11}a_{23}a_{32}a_{44}$
(24)	Odd	$-a_{11}a_{24}a_{33}a_{42}$
(34)	Odd	$-a_{11}a_{22}a_{34}a_{43}$
(1234)	Odd	$-a_{12}a_{23}a_{34}a_{41}$
(1243)	Odd	$-a_{12}a_{24}a_{31}a_{43}$
(1324)	Odd	$-a_{13}a_{24}a_{32}a_{41}$
(1342)	Odd	$-a_{13}a_{21}a_{34}a_{42}$
(1423)	Odd	$-a_{14}a_{23}a_{31}a_{42}$
(1432)	Odd	$-a_{14}a_{21}a_{32}a_{43}$

So, if A is a 4×4 matrix, then $\det A$ is the following:

$$\begin{aligned}
&a_{11}a_{22}a_{33}a_{44} + a_{12}a_{23}a_{31}a_{44} + a_{13}a_{21}a_{32}a_{44} + a_{12}a_{24}a_{33}a_{41} + a_{14}a_{21}a_{33}a_{42} + a_{13}a_{22}a_{34}a_{41}\\
&+a_{14}a_{22}a_{31}a_{43} + a_{11}a_{23}a_{34}a_{42} + a_{11}a_{24}a_{32}a_{43} + a_{12}a_{21}a_{34}a_{43} + a_{13}a_{24}a_{31}a_{42} + a_{14}a_{23}a_{32}a_{41}\\
&-a_{12}a_{21}a_{33}a_{44} - a_{13}a_{22}a_{31}a_{44} - a_{14}a_{22}a_{33}a_{41} - a_{11}a_{23}a_{32}a_{44} - a_{11}a_{24}a_{33}a_{42} - a_{11}a_{22}a_{34}a_{43}\\
&-a_{12}a_{23}a_{34}a_{41} - a_{12}a_{24}a_{31}a_{43} - a_{13}a_{24}a_{32}a_{41} - a_{13}a_{21}a_{34}a_{42} - a_{14}a_{23}a_{31}a_{42} - a_{14}a_{21}a_{32}a_{43}
\end{aligned}$$

Prove each of the following:

18. If π and σ are disjoint permutations, then for each $n \in \mathbb{Z}^+$, $(\pi\sigma)^n = \pi^n\sigma^n$.

Proof: We first prove that for each $n \in \mathbb{Z}^+$, $\pi\sigma^n = \sigma^n\pi$. We prove this by induction on $n \in \mathbb{Z}^+$.

Base case ($k = 1$): By Theorem 16.6, $\pi\sigma^1 = \pi\sigma = \sigma\pi = \sigma^1\pi$.

Inductive step: Assume that $k \in \mathbb{Z}^+$ and $\pi\sigma^k = \sigma^k\pi$. Then we have

$$\pi\sigma^{k+1} = \pi(\sigma^k\sigma) = (\pi\sigma^k)\sigma = (\sigma^k\pi)\sigma = \sigma^k(\pi\sigma) = \sigma^k(\sigma\pi) = (\sigma^k\sigma)\pi = \sigma^{k+1}\pi.$$

By the Principle of Mathematical Induction, for all $n \in \mathbb{Z}^+$, $\pi\sigma^n = \sigma^n\pi$.

Now, we prove that for each $n \in \mathbb{Z}^+$, $(\pi\sigma)^n = \pi^n\sigma^n$. We prove this by induction on $n \in \mathbb{Z}^+$.

Base case ($k = 1$): $(\pi\sigma)^1 = \pi\sigma = \pi^1\sigma^1$.

Inductive step: Assume that $k \in \mathbb{Z}^+$ and $(\pi\sigma)^k = \pi^k\sigma^k$. Then we have

$$(\pi\sigma)^{k+1} = (\pi\sigma)^k(\pi\sigma) = (\pi^k\sigma^k)(\pi\sigma) = \pi^k(\sigma^k\pi)\sigma = \pi^k(\pi\sigma^k)\sigma = (\pi^k\pi)(\sigma^k\sigma) = \pi^{k+1}\sigma^{k+1}.$$

By the Principle of Mathematical Induction, for all $n \in \mathbb{Z}^+$, $(\pi\sigma)^n = \pi^n\sigma^n$. □

19. If $n \geq 3$ and $s, t \in \{1, 2, \dots, n\}$ with $s \neq t$, then $A_n = \langle (stu) \mid 1 \leq u \leq n \wedge u \neq s, t \rangle$.

Proof: $A_3 = \{(1), (123), (132)\}$, and so, the result is clear for $n = 3$. Let $n \geq 3$. If $a, b, c, d \in \{1, \dots, n\}$ are distinct, then $(ab)(cd) = (acb)(acd)$. If $a, b, c \in \{1, \dots, n\}$ are distinct, then $(ab)(ac) = (acb)$. Since every element of A_n can be written as a product of elements of the form $(ab)(cd)$ and $(ab)(ac)$, we see that A_n is generated by all 3-cycles. Now, any 3-cycle has one of the forms (stu), (sut), (suv), (tuv), or (uvw), where u, v, w are distinct and $u, v, w \neq s, t$. We have $(sut) = (stu)^2$, $(suv) = (stv)(stu)^2$, $(tuv) = (stv)^2(stu)$, and $(uvw) = (stu)^2(stw)(stv)^2(stu)$. Therefore, $A_n = \langle (stu) | 1 \leq u \leq n \wedge u \neq s, t \rangle$. □

20. Let $\mathcal{A} = \{f_n : [0,1] \to \mathbb{R} \mid n \in \mathbb{Z}^+\}$, where for each $n \in \mathbb{Z}^+$, $f_n(x) = \begin{cases} nx & \text{if } 0 \leq x < \frac{1}{n}. \\ 1 & \text{if } \frac{1}{n} \leq x \leq 1. \end{cases}$

Then $\mathcal{A}$ is not equicontinuous.

Proof: Let $\epsilon = \frac{1}{2}$ and suppose toward contradiction that there is $\delta_1 > 0$ such that for all $x, y \in K$ and every $f \in \mathcal{A}$, $d(x, y) < \delta$ implies $|f(x) - f(y)| < \frac{1}{2}$. Let $\delta = \min\{\delta_1, 1\}$, let $x = 0$, and let $y = \frac{\delta}{2}$. Then $|x - y| = \left|0 - \frac{\delta}{2}\right| = \frac{\delta}{2} < \delta$. By the Archimedean Property of $\mathbb{R}$, we can choose $n \in \mathbb{N}$ so that $n > \frac{2}{\delta}$, or equivalently, $\frac{1}{n} < \frac{\delta}{2}$. Then $f_n\left(\frac{\delta}{2}\right) = 1$. So, $\left|f_n(0) - f_n\left(\frac{\delta}{2}\right)\right| = |0 - 1| = 1 \geq \frac{1}{2}$, a contradiction. So, $\mathcal{A}$ is not equicontinuous. □

LEVEL 4

Prove each of the following:

21. If π is a product of k disjoint cycles of lengths $n_1, n_2, \ldots, n_k$, respectively, then the order of π is $\operatorname{lcm}(n_1, n_2, \ldots, n_k)$.

Proof: First note that the order of an m-cycle is m. Let $\pi = \pi_1\pi_2 \cdots \pi_k$ be a product of k disjoint cycles of lengths $n_1, n_2, \ldots, n_k$, respectively. Let $t = \operatorname{lcm}(n_1, n_2, \ldots, n_k)$. By Problem 18 above, we have

$$\pi^t = \pi_1^t\pi_2^t \cdots \pi_k^t = (1)(1) \cdots (1) = (1).$$

So, $|\pi|$ divides t.

If $\pi^u = (1)$, then since the cycles are disjoint, we have $\pi_1^u = (1), \pi_2^u = (1), \ldots, \pi_k^u = (1)$. Therefore, for each $i = 1, 2, \ldots, k$, n_i divides u. Therefore, t divides u.

Thus, $|\pi| = t$. □

22. Two cycles in S_n have the same length if and only if they are conjugates of each other.

Proof: Let π be the k-cycle $(a_1a_2 \cdots a_k)$. Then $\sigma\pi\sigma^{-1}$ is the k-cycle $\big(\sigma(a_1)\sigma(a_2) \cdots \sigma(a_k)\big)$. To see this, observe that for each $i = 1, 2, \ldots, k-1$, $\sigma\pi\sigma^{-1}\big(\sigma(a_i)\big) = \sigma\pi(a_i) = \sigma(a_{i+1})$ and for $i = k$, $\sigma\pi\sigma^{-1}\big(\sigma(a_k)\big) = \sigma\pi(a_k) = \sigma(a_1)$. Since $\sigma: \{1, 2, \ldots, n\} \to \{1, 2, \ldots, n\}$ is a bijection, π and $\sigma\pi\sigma^{-1}$ both have length k.

Conversely, suppose that $\pi = (a_1a_2 \cdots a_k)$ and $\sigma = (b_1b_2 \cdots b_k)$. Define $\tau: \{1, 2, \ldots, n\} \to \{1, 2, \ldots, n\}$ by $\tau(a_i) = b_i$. Then $\tau\pi\tau^{-1}$ is the k-cycle $\big(\tau(a_1)\tau(a_2) \cdots \tau(a_k)\big) = (b_1b_2 \cdots b_k)$. So, $\tau\pi\tau^{-1} = \sigma$. □

23. $|S_n| = n!$

Proof: We prove this by induction on $n \in \mathbb{Z}^+$.

Base case ($k = 1$): Clearly there is just 1 bijection from $\{1\}$ to $\{1\}$.

Inductive step: Assume that $k \in \mathbb{Z}^+$ and $|S_k| = k!$.

Let $H \leq S_{k+1}$ consist of the permutations π such that $\pi(k+1) = k+1$. Then $|H| = |S_k|$. By Lagrange's theorem, $|S_{k+1}| = |H|[S_{k+1}: H] = (k!)[S_{k+1}: H]$. So, it suffices to show that there are $k+1$ cosets of H in S_{k+1}. For each $i = 1, 2, \ldots, k$, let π_i be the transposition $(i, k+1)$ (in other words, $\pi_i(i) = k+1, \pi_i(k+1) = i$, and $\pi_i(j) = j$ for $j \neq i, k+1$). Also, let $\pi_{k+1} = (1)$. We will show that $X = \{\pi_i H \mid i = 1, 2 \ldots, k+1\}$ is the set of cosets of H in S_{k+1}.

First, let $\sigma \in S_{k+1}$ and assume that $\sigma(k+1) = i$. Then $\pi_i^{-1}\sigma(k+1) = \pi_i^{-1}(i) = k+1$. Therefore, $\pi_i^{-1}\sigma \in H$. So, $\sigma \in \pi_i H$. This shows that every permutation in S_{k+1} is in a coset in X.

Next, suppose that $i, j \in \{0, 1, \ldots, k+1\}$ with $i \neq j$. Then $\sigma_j^{-1}\sigma_i(k+1) = \sigma_j^{-1}(i) \neq k+1$. Therefore, $\sigma_j^{-1}\sigma_i \notin H$. So, $\sigma_i H \neq \sigma_j H$. This shows that the cosets in X are distinct.

It follows that $[S_{k+1}: H] = |X| = k+1$, and so, $|S_{k+1}| = (k!)[S_{k+1}: H] = (k!)(k+1) = (k+1)!$. □

24. If $n \geq 3$ and N is a normal subgroup of A_n containing a 3-cycle, then $N = A_n$.

Proof: Let $(stu) \in N$ and let $a \in \{1, 2, \dots, n\}$ with $a \neq s, t, u$. Then we have

$$(sta) = (st)(ua)(stu)^2(ua)(st) = \big((st)(ua)\big)(stu)^2\big((st)(ua)\big)^{-1}.$$

Therefore, $(sta) \in N$. By Problem 19 above, $N = A_n$. □

25. The determinant of an $n \times n$ upper triangular matrix A is the product of the entries of A that lie on the main diagonal of A. Is the same result true for a lower triangular matrix A?

Proof: We prove this by induction on $n \in \mathbb{Z}^+$.

Base case ($k = 1$): The determinant of the 1×1 matrix $[a]$ is a (note that every 1×1 matrix is upper triangular).

Inductive step: Assume that for $k \in \mathbb{Z}^+$, the determinant of a $k \times k$ upper triangular matrix is the product of the entries on the main diagonal of the matrix.

Let A be a $(k+1) \times (k+1)$ matrix. We have

$$\det A = \sum_{\pi \in S_{k+1}} \Delta_k(\pi) a_{1,\pi(1)} a_{2,\pi(2)} \cdots a_{k+1,\pi(k+1)}$$

Now, if $\pi(k+1) \neq k+1$, then $\pi(k+1) < k+1$, and so, $a_{k+1,\pi(k+1)} = 0$. Therefore, if $\pi(k+1) \neq k+1$, then $\Delta_k(\pi) a_{1,\pi(1)} a_{2,\pi(2)} \cdots a_{k+1,\pi(k+1)} = 0$. So,

$$\det A = a_{k+1,k+1} \sum_{\pi \in S_k} \Delta_k(\pi) a_{1,\pi(1)} a_{2,\pi(2)} \cdots a_{k,\pi(k)} = a_{k+1,k+1} \det B,$$

where B is the $k \times k$ matrix that we get from removing the $(k+1)^{\text{st}}$ row and column of A. By the inductive hypothesis, $\det B = a_{11} a_{22} \cdots a_{kk}$. Therefore, $\det A = a_{1,1} a_{2,2} \cdots a_{k,k} a_{k+1,k+1}$, as desired.

The same result is true for a lower triangular matrix A. The proof is similar. □

LEVEL 5

Prove each of the following:

26. Let $(1) \in S_n$ be the identity and suppose that $(1) = \pi_1 \pi_2 \cdots \pi_k$, where for each $i = 1, 2, \dots, k$, π_i is a transposition. Then k is even. (This is Lemma 16.11.)

Proof: First note that if $a, b \in \{1, 2, \dots, n\}$ with $a \neq b$, then $(1) \neq (ab)$.

Suppose that $(1) = \pi_1 \pi_2 \cdots \pi_k$, where for each $i = 1, 2, \dots, k$, π_i is a transposition. By the note above, $k \geq 2$. We will now show that (1) can be written as a product of $k - 2$ transpositions.

Choose $a \in \{1, 2, \dots, n\}$ and the greatest $i \geq 2$ such that $\pi_i = (ab)$ moves a.

Case 1: If $\pi_{i-1} = (ab)$, then $\pi_{i-1}\pi_i = (ab)(ab) = (1)$, and so, $(1) = \pi_1\pi_2 \cdots \pi_{i-2}\pi_{i+1} \cdots \pi_k$. In this case, we have succeeded in writing (1) as a product of $k - 2$ transpositions.

Case 2: If $\pi_{i-1} = (ac)$, where $c \neq a, b$, then $\pi_{i-1}\pi_i = (ac)(ab) = (abc) = (ab)(bc)$. Replace π_{i-1} by (ab) and π_i by (bc). Now $i - 1$ is the greatest positive integer such that a is moved by π_{i-1}.

Case 3: If $\pi_{i-1} = (bc)$, where $c \neq a, b$, then $\pi_{i-1}\pi_i = (bc)(ab) = (acb) = (ac)(bc)$. Replace π_{i-1} by (ac) and π_i by (bc). Now $i - 1$ is the greatest positive integer such that a is moved by π_{i-1}.

Case 4: If $\pi_{i-1} = (cd)$, where $c, d \neq a, b$, then $\pi_{i-1}\pi_i = (cd)(ab) = (ab)(cd)$. Replace π_{i-1} by (ab) and π_i by (cd). Now $i - 1$ is the greatest positive integer such that a is moved by π_{i-1}.

If Case 2, 3, or 4 occurs, we continue to repeat the procedure with the same positive integer a until Case 1 occurs. Case 1 must eventually occur. Otherwise, we eventually wind up with $\pi_1 = (ab)$ with $a \neq b$, and π_j fixes a for all $j > i$. But then (1) moves a to b, which of course we know it does not.

So, we showed that if (1) can be written as a product of k transpositions, then it can be written as a product of $k - 2$ transpositions. If k were odd, then by repeating this process, we can eventually write (1) as a single transposition (xy). But then (1) moves x to y, which once again, we know it does not.

Therefore, k must be even. □

27. $S_n = \langle(12), (12 \cdots n)\rangle$.

Proof: Let $\pi = (12)$ and $\sigma = (12 \cdots n)$.

By the proof of Problem 22 above, $\sigma\pi\sigma^{-1} = \big(\sigma(1)\sigma(2)\big) = (23)$. We can then use induction to show that for each $k = 1, 2, \ldots, n - 2$, $\sigma^k\pi\sigma^{-k} = (k + 1\ k + 2)$. So, $\langle(12), (12 \cdots n)\rangle$ contains all the transpositions $(12), (23), \ldots, (n - 1\ n)$. By part 3 of Example 16.10, all that we need to show is that $(1n) \in \langle(12), (23), \ldots, (n - 1\ n)\rangle$. But $(1n) = (n - 1\ n) \cdots (34)(23)(12)(23)(34) \cdots (n - 1\ n)$. □

28. If H is a subgroup of S_n such that $[S_n : H] = 2$, then $H = A_n$.

Proof: Let H be a subgroup of S_n such that $[S_n : H] = 2$. Then there are exactly 2 cosets of H in S_n, namely H and $S_n \setminus H$. If $\pi \in S_n \setminus H$, then $\pi^2 \in H$. To see this, assume towards contradiction that $\pi^2 \notin H$. Then $\pi^2 \in \pi H$ (this is the only other coset of H in S_n), and so, $\pi^2 H = \pi H$. But then $\pi H = H$, and so, $\pi \in H$, contradicting our assumption that $\pi \in S_n \setminus H$. So, given $s, t, u \in \{1, 2, \ldots, n\}$, we have $(stu) = (sut)^2 \in H$. Therefore, H contains all 3-cycles, and so, by Problem 19 above, $H = A_n$. □

29. For $n \geq 5$, A_n has no proper normal subgroups.

Proof: Let N be a normal subgroup of A_n. There are five cases to consider.

Case 1: If N contains a 3-cycle, then by Problem 24 above, $N = A_n$.

Case 2: Suppose that N contains an element of the form $\pi = (a_1 a_2 \cdots a_k)\sigma$, where $k \geq 4$, σ and $(a_1 a_2 \cdots a_k)$ are disjoint, and σ is a product of disjoint cycles. Let $\tau = (a_1 a_2 a_3)$. Since N is normal in A_n, $\tau\pi\tau^{-1} \in N$. Since N is a subgroup of A_n, $\pi^{-1}\tau\pi\tau^{-1} \in N$. We have

$$\pi^{-1}\tau\pi\tau^{-1} = \sigma^{-1}(a_1 a_k a_{k-1} \cdots a_2)(a_1 a_2 a_3)(a_1 a_2 \cdots a_{k-1} a_k)\sigma(a_1 a_3 a_2) = (a_1 a_3 a_k).$$

By Problem 24 above, $N = A_n$.

Case 3: Suppose that N contains an element of the form $\pi = (a_1 a_2 a_3)(a_4 a_5 a_6)\sigma$, where σ and $(a_1 a_2 a_3)(a_4 a_5 a_6)$ are disjoint and σ is a product of disjoint cycles. Let $\tau = (a_1 a_2 a_4)$. Since N is normal in A_n, $\tau\pi\tau^{-1} \in N$. Since N is a subgroup of A_n, $\pi^{-1}\tau\pi\tau^{-1} \in N$. We have

$$\pi^{-1}\tau\pi\tau^{-1} = \sigma^{-1}(a_4 a_6 a_5)(a_1 a_3 a_2)(a_1 a_2 a_4)(a_1 a_2 a_3)(a_4 a_5 a_6)\sigma(a_1 a_4 a_2) = (a_1 a_4 a_2 a_6 a_3).$$

By Case 2, $N = A_n$.

Case 4: Suppose that N contains an element of the form $\pi = (a_1 a_2 a_3)\sigma$, where σ and $(a_1 a_2 a_3)$ are disjoint and σ is a product of disjoint transpositions. Then $\pi^2 \in N$ and we have

$$\pi^2 = (a_1 a_2 a_3)\sigma(a_1 a_2 a_3)\sigma = (a_1 a_2 a_3)^2\sigma^2 = (a_1 a_2 a_3)^2 = (a_1 a_3 a_2).$$

By Problem 24 above, $N = A_n$.

Case 5: Suppose that every element of N is a disjoint product of transpositions. Since $N \subseteq A_n$, each element of N must be a disjoint product of an even number of transpositions. Let $\pi \in N$ with $\pi = (a_1 a_2)(a_3 a_4)\sigma$, where σ and $(a_1 a_2)(a_3 a_4)$ are disjoint. Let $\tau = (a_1 a_2 a_3)$. Then N contains

$$\pi^{-1}\tau\pi\tau^{-1} = \sigma^{-1}(a_3 a_4)(a_1 a_2)(a_1 a_2 a_3)(a_1 a_2)(a_3 a_4)\sigma(a_1 a_3 a_2) = (a_1 a_3)(a_2 a_4).$$

Since $n \geq 5$, there is $c \in \{1, 2, \dots, n\}$ such that $c \neq a_1, a_2, a_3, a_4$. Let $\delta = (a_1 a_3)(a_2 a_4)$ and let $\rho = (a_1 a_3 c)$. We have $\delta \in N$ and $\rho \in A_n$. So, $\delta\rho\delta\rho^{-1} \in N$. Now,

$$\delta\rho\delta\rho^{-1} = (a_1 a_3)(a_2 a_4)(a_1 a_3 c)(a_1 a_3)(a_2 a_4)(a_1 c a_3) = (a_1 a_3 c).$$

By Problem 24 above, $N = A_n$. □

30. For each $n \in \mathbb{Z}^+$, the set of $n \times n$ permutation matrices forms a group under matrix multiplication that is isomorphic to S_n.

Proof: Let P_n be the set of permutation matrices. Define a function $f: S_n \to P_n$ by $f(\pi) = A_\pi$, where $(A_\pi)_{ij} = f(\pi)_{ij} = \begin{cases} 1 & \text{if } \pi(j) = i. \\ 0 & \text{if } \pi(j) \neq i. \end{cases}$

To see that f is surjective, let $A \in P_n$ and define $\pi: \{1, 2, \dots, n\} \to \{1, 2, \dots, n\}$ by $\pi(j) = i$ if and only if $a_{ij} = 1$. Since 1 appears exactly once in each row and column of A, $\pi \in S_n$. Clearly $f(\pi) = A$.

To see that f is injective, suppose that $f(\pi) = f(\sigma)$. Then $\pi(j) = i$ if and only if $f(\pi)_{ij} = 1$ if and only if $f(\sigma)_{ij} = 1$ if and only if $\sigma(j) = i$. So, $\pi = \sigma$.

To see that f is a homomorphism, let $\pi, \sigma \in S_n$. Then

$$\big(f(\pi)f(\sigma)\big)_{ij} = \sum_{k=1}^{n} f(\pi)_{ik} f(\sigma)_{kj} = \begin{cases} 1 & \text{if } \pi(t) = i \text{ and } \sigma(j) = t \text{ for some } t \\ 0 & \text{otherwise} \end{cases} = f(\pi\sigma)_{ij}$$

So, $f(\pi)f(\sigma) = f(\pi\sigma)$. □

31. $\mathcal{B}(S, \mathbb{R})$ together with the norm defined by $|f| = \sup|f(x)|$ is a Banach space.

Proof: By part 2 of Example 16.26, $\mathcal{B}(S, \mathbb{R})$ is a vector space. We now check that the 3 properties for a norm are satisfied.

If $f = \mathbf{0}$, then $f(x) = 0$ for all $x \in S$, and so, $|f| = \sup|f(x)| = 0$. On the other hand, if $|f| = 0$, then $f(x) = 0$ for all $x \in X$, and so, $f = \mathbf{0}$. So, Property 1 holds.

Let $f, g \in \mathcal{B}(S, \mathbb{R})$. For each $x \in X$, we have $|f(x) + g(x)| \leq |f(x)| + |g(x)|$, by the Triangle Inequality. By the definition of the norm, for each $x \in X$, $|f(x) + g(x)| \leq |f| + |g|$. Again, by the definition of the norm, $|f + g| \leq |f| + |g|$. So, Property 2 holds.

Let $f \in \mathcal{B}(S, \mathbb{R})$ and let $c \in \mathbb{R}$. For each $x \in X$, we have $|c \cdot f(x)| = |c| \cdot |f(x)|$. Therefore, we have $|cf| = \sup|(cf)(x)| = \sup(|c| \cdot |f(x)|) = |c| \cdot \sup|f(x)| = |c| \cdot |f|$. So, Property 3 holds.

We now show that $(\mathcal{B}(S, \mathbb{R}), d)$ is complete, where $d: \mathcal{B}(S, \mathbb{R}) \times \mathcal{B}(S, \mathbb{R}) \to \mathbb{R}$ is defined by $d(f, g) = |f - g|$. To see this, let (f_n) be a Cauchy sequence in $\mathcal{B}(S, \mathbb{R})$. Then for each $x \in S$, $\big(f_n(x)\big)$ is a Cauchy sequence in $\mathbb{R}$. Since $\mathbb{R}$ is complete, for each $x \in S$, there is $f(x) \in \mathbb{R}$ such that $f_n(x) \to f(x)$.

We first show that $f \in \mathcal{B}(S, \mathbb{R})$. So, we must show that f is bounded. Since (f_n) is a Cauchy sequence, there is $K \in \mathbb{N}$ such that $m, n > K$ implies $|f_m - f_n| < 1$. Then for all $x \in S$, we have

$$|f(x)| = \big|\big(f(x) - f_{K+1}(x)\big) + f_{K+1}(x)\big| \leq |f(x) - f_{K+1}(x)| + |f_{K+1}(x)|.$$

For $n > K$, we have $|f_n(x) - f_{K+1}(x)| < 1$, and so, $|f(x) - f_{K+1}(x)| \leq 1$ because $f_n(x) \to f(x)$. Therefore, $|f - f_{K+1}| = \sup|f(x) - f_{K+1}(x)| \leq 1$. It follows that for all $x \in S$,

$$|f(x)| \leq |f(x) - f_{K+1}(x)| + |f_{K+1}(x)| \leq 1 + |f_{K+1}|.$$

Thus, f is bounded.

Finally, we show that $f_n \to f$ in the metric d. Let $\epsilon > 0$. Since (f_n) is a Cauchy sequence, there is $K \in \mathbb{N}$ such that $m, n > K$ implies $|f_m - f_n| < \frac{\epsilon}{2}$. Then for $n > K$, we have $|f(x) - f_n(x)| \leq \frac{\epsilon}{2}$. Therefore, $|f_n - f| \leq \frac{\epsilon}{2} < \epsilon$. So, $f_n \to f$. □

32. $\mathcal{C}(S,\mathbb{R})$ is a closed subspace of $\mathcal{B}(S,\mathbb{R})$.

Proof: Let $f \in \overline{\mathcal{C}(S,\mathbb{R})}$. Then there is a sequence (f_n) in $\mathcal{C}(S,\mathbb{R})$ such that $f_n \to f$. Since $\mathcal{B}(S,\mathbb{R})$ is complete, $f \in \mathcal{B}(S,\mathbb{R})$. So, we just need to show that f is continuous. To this end, let $x \in S$ and let $\epsilon > 0$. Since $f_n \to f$, there is $K \in \mathbb{N}$ such that $n > K$ implies $|f_n - f| < \frac{\epsilon}{3}$. In particular, $|f_{K+1} - f| < \frac{\epsilon}{3}$. Since f_{K+1} is continuous at x, there is $\delta > 0$ such that $d(x,y) < \delta$ implies $|f_{K+1}(x) - f_{K+1}(y)| < \frac{\epsilon}{3}$. Therefore, if $d(x,y) < \delta$, then

$$\begin{aligned}
|f(x) - f(y)| &= \left|\big(f(x) - f_{K+1}(x)\big) + \big(f_{K+1}(x) - f_{K+1}(y)\big) + \big(f_{K+1}(y) - f(y)\big)\right| \\
&\leq |f(x) - f_{K+1}(x)| + |f_{K+1}(x) - f_{K+1}(y)| + |f_{K+1}(y) - f(y)| \\
&< 2\sup|f(y) - f_{K+1}(y)| + \frac{\epsilon}{3} \leq 2 \cdot \frac{\epsilon}{3} + \frac{\epsilon}{3} = \epsilon.
\end{aligned}$$

So, f is continuous at x. Since $x \in S$ was arbitrary, $f \in \mathcal{C}(S,\mathbb{R})$. Since $f \in \overline{\mathcal{C}(S,\mathbb{R})}$ was arbitarry, it follows that $\overline{\mathcal{C}(S,\mathbb{R})} \subseteq \mathcal{C}(S,\mathbb{R})$, and so $\mathcal{C}(S,\mathbb{R})$ is closed in $\mathcal{B}(S,\mathbb{R})$. □

33. There is a sequence (f_n) in $\mathcal{C}([0,1],\mathbb{R})$ such that f_n converges pointwise to $f \in \mathcal{C}([0,1],\mathbb{R})$, but (f_n) does **not** converge uniformly to f.

Proof: For each $n \geq 2$, define $f_n \in \mathcal{C}([0,1],\mathbb{R})$ by $f_n(x) = \begin{cases} nx & \text{if} \quad 0 \leq x < \frac{1}{n}. \\ 2 - nx & \text{if} \quad \frac{1}{n} \leq x < \frac{2}{n}. \\ 0 & \text{if} \quad \frac{2}{n} \leq x \leq 1. \end{cases}$

It is easy to see that this sequence converges to $\mathbf{0}$ pointwise.

However, (f_n) does not converge to $\mathbf{0}$ uniformly because for any $n \in \mathbb{N}$, we have

$$|f_n - f| = \sup|f_n(x) - f(x)| = \sup|f_n(x)| = 1$$ □

About the Author

Dr. Steve Warner, a New York native, earned his Ph.D. at Rutgers University in Pure Mathematics in May 2001. While a graduate student, Dr. Warner won the TA Teaching Excellence Award.

After Rutgers, Dr. Warner joined the Penn State Mathematics Department as an Assistant Professor and in September 2002, he returned to New York to accept an Assistant Professor position at Hofstra University. By September 2007, Dr. Warner had received tenure and was promoted to Associate Professor. He has taught undergraduate and graduate courses in Precalculus, Calculus, Linear Algebra, Differential Equations, Real Analysis, Complex Analysis, Mathematical Logic, Set Theory, and Abstract Algebra.

From 2003 – 2008, Dr. Warner participated in a five-year NSF grant, "The MSTP Project," to study and improve mathematics and science curriculum in poorly performing junior high schools. He also published several articles in scholarly journals, specifically on Mathematical Logic.

Dr. Warner has nearly two decades of experience in general math tutoring and tutoring for standardized tests such as the SAT, ACT, GRE, GMAT, and AP Calculus exams. He has tutored students both individually and in group settings.

In February 2010, Dr. Warner released his first SAT prep book "The 32 Most Effective SAT Math Strategies," and in 2012 founded Get 800 Test Prep. Since then Dr. Warner has written books for the SAT, ACT, SAT Math Subject Tests, AP Calculus exams, and GRE. In September 2018, Dr. Warner released his first advanced math book "Pure Mathematics for Beginners." Since then, he has released several more books in advanced mathematics.

Dr. Steve Warner can be reached at

steve@SATPrepGet800.com

BOOKS BY DR. STEVE WARNER

GET 800
Second Edition
320
SAT MATH PROBLEMS
arranged by Topic and Difficulty Level
By Dr. Steve Warner

GET 800
Beginner Course
28
2017 Edition
SAT MATH LESSONS
to Improve Your Score in One Month
By Dr. Steve Warner

GET 800
Intermediate Course
28
2017 Edition
SAT MATH LESSONS
to Improve Your Score in One Month
By Dr. Steve Warner

GET 800
Advanced Course
28
2017 Edition
SAT MATH LESSONS
to Improve Your Score in One Month
By Dr. Steve Warner

GET 800
320
SAT CHEMISTRY SUBJECT TEST PROBLEMS
arranged by Topic and Difficulty Level
Christopher Bozza
Dr. Steve Warner

GET 800
320
GRE MATH PROBLEMS
arranged by Topic and Difficulty Level
By Dr. Steve Warner

1000 Logic and Reasoning Questions for
GIFTED AND TALENTED
Elementary School Students

GET 800
PHYSICS MASTERY
for the Advanced High School Student
Complete Physics Review with 400 SAT and AP Physics Questions

Algebra Handbook
Dr. Steve Warner
Tayyip Oral
Serife Turan

555 Geometry Problems
for High School Students
Dr. Steve Warner
Tayyip Oral
Serife Turan

555 Math IQ Questions
for Middle School Students
Dr. Steve Warner
Tayyip Oral

555 Advanced Math Problems
for Middle School Students
Dr. Steve Warner
Tayyip Oral